Microstructural Science

Volume 19

Structure-Property Relationships and Correlations with the Environmental Degradation of Engineering Materials

Microstructural Science
Volume 19

Structure-Property Relationships and Correlations with the Environmental Degradation of Engineering Materials

Proceedings of the Twenty-Fourth
Annual Technical Meeting of the
International Metallographic Society

Edited by

D. A. Wheeler
Howmet Corporation

G. W. E. Johnson
Lockhead Missiles and Space Company

D. V. Miley
EG&G Idaho

M. R. Louthan, Jr.
Savannah River Laboratory

THE INTERNATIONAL
METALLOGRAPHIC SOCIETY
COLUMBUS, OHIO USA

AND

ASM INTERNATIONAL®
MATERIALS PARK, OHIO USA

The Materials
Information Society

Library of Congress Cataloging Card Number: 92-82801
ISBN: 0-87170-463-3

Production Manager
Linda Kacprzak

Production Assistant
Ann-Marie O'Loughlin

**The Materials
Information Society**

ASM International®
Materials Park, OH 44073-0002

Printed in the United States of America

CONTENTS

High Temperature Corrosion

CONTRIBUTORS

Alpas, A.T.
University of Windsor
Windsor, Ontario, Canada

Amouzouvi, K.F.
Whiteshell Labs
Whiteshell, Manitoba, Canada

Barbuto, A.T.
General Electric Corporate R & D
Schenectady, NY USA

Brooks, C.R.
The University of Tennessee
Knoxville, TN USA

Cao, S.
The University of Tennessee
Knoxville, TN USA

Chatteree, T.K.
US Army R & D & Engineering Center
Picatinny Arsenal, NJ USA

Chattopadhyay, R.
EWAC Alloys Limited
Bombay, India

Chen, P.S.
IIT Research Institute
Huntsville, AL USA

Choudhury, A.
Oak Ridge National Lab
Oak Ridge TN USA

Clausen, D.
Danfoss A/S
Nordborg, Denmark

Cole, S.J.
Imperial College
London UK

Cooke, C.M.
Metcut-Materials Research Group
Wright-Patterson AFB, OH USA

Damgaard, M.J.
Struers, Inc.
Westlake, OH USA

David, S.A.
Oak Ridge National Lab
Oak Ridge, TN USA

Diakow, D.A.
NOVA Corporation of Alberta
Calgary, Alberta, Canada

Ding, Y.
University of Windsor
Windsor, Ontario, Canada

Drexler, J.W.
University of Colorado
Boulder, CO USA

Dunn, K.A.
Westinghouse Savannah River Company
Aiken, SC USA

Eisenreich, N.
Fraunhofer-Institut ICT
Pfinztal, Germany

Ellis, F.V.
Tordonato Energy Consultants, Inc.
Chattanooga, TN USA

Engel, W.
Fraunhofer-Institut ICT
Pfinztal, Germany

Eylon, D.
Metcut-Materials Research Group
Wright-Patterson AFB, OH USA

Farrens, S.
University of California
Davis, CA USA

Fischer, K.S.
California Polytechnic State University
San Luis Obispo, CA USA

Forgeng, W.D., Jr.
California Polytechnic State University
San Luis Obispo, CA USA

Gammon, L.M.
Boeing Commercial Airplane Company
Seattle, WA USA

Gardner, M.J.
Oak Ridge National Lab
Oak Ridge, TN USA

Geary, A.R.
Pratt & Whitney
East Hartford, CT USA

Glancy, S.D.
Struers, Inc.
Westlake, OH USA

Grande, J.C.
General Electric Corporate R & D
Schenectady, NY USA

Groza, J.
University of California
Davis, CA USA

Hansen, H.
Danfoss A/S
Nordborg, Denmark

Ho, Y-T.
Oklahoma State University
Stillwater, OK USA

Hong, J-L.
Industrial Technology Research Institute
Chuttung, Hsinchu, Taiwan

Huang, H.
The University of Calgary
Calgary, Alberta, Canada

Hull, D.R.
NASA Lewis Research Center
Cleveland, OH USA

Juez-Lorenzo, M.
Fraunhofer-Institut ICT
Pfinztal, Germany

Kapitza, H-G.
Carl Zeiss
Oberkochen, Germany

Kestin, P.A.
Westinghouse Savannah River Lab
Aiken, SC USA

Kim, Y-W.
Metcut-Materials Research Group
Wright-Patterson AFB, OH USA

Kolarik, V.
Fraunhofer-Institut ICT
Pfinztal, Germany

Kothari, N.C.
James Cook University
Townsville, Queensland, Australia

Kumar, P.
Cabot Corporation
Boyertown, PA USA

Leonhardt, T.A.
Sverdrup Technology, Inc.
Brook Park, OH USA

Lin, C-A.
Industrial Technology Research Institute
Chuttung, Hsinchu, Taiwan

Ling, Y.
Hangzhou Institute of Electronic Engineering
Hangzhou, P.R. China

Lundin, C.D.
The University of Tennessee
Knoxville, TN USA

Mao, X.
The University of Calgary
Calgary, Alberta, Canada

Marra, J.C.
Westinghouse Savannah River Company
Aiken, SC USA

McCowan, C.N.
NIST
Boulder, CO USA

McKee, D.W.
General Electric Corporate R & D
Schenectady, NY USA

McNitt, R.P.
The Pennsylvania State University
University Park, PA USA

Mueller, H.J.
American Dental Association
Chicago, IL USA

Nisenholz, Z.
RAFAEL, Ministry of Defence
Haifa, Israel

Northwood, D.O.
University of Windsor
Windsor, Ontario, Canada

Pan, J.
The University of Calgary
Calgary, Alberta, Canada
Pangborn, R.N.
The Pennsylvania State University
University Park, PA USA

Pechersky, M.J.
Westinghouse Savannah River Company
Aiken, SC USA

Ploc, R.A.
Chalk River Labs
Chalk River, Ontario, Canada

Price, C.E.
Oklahoma State University
Stillwater, OK USA

Putatunda, S.K.
Wayne State University
Detroit, MI USA

Qiao, C.Y.P.
The University of Tennessee
Knoxville, TN USA

Queeney, R.A.
The Pennsylvania State University
University Park, PA USA

Rao, A.S.
Powertech Labs, Inc.
Surrey, BC Canada

Richman, R.H.
Daedalus Associates, Inc.
Mountain View, CA USA

Rubly, R.P.
Allied-Signal Aerospace Company
Torrance, CA USA

Ruffer, N.J.
Federal Mogul Research
Ann Arbor, MI USA

Sanders, W.A.
Analex Corporation
Brook Park, OH USA

Sayles, R.S.
Imperial College
London UK

Schaefer, J.
Detroit Edison
Detroit, MI USA

Schrems, K.K.
U.S. Bureau of Mines
Albany, OR USA

Sengupta, A.
Wayne State University
Detroit, MI USA

Shaw, W.J.D.
The University of Calgary
Calgary, Alberta, Canada

Sheasby, J.S.
The University of Western Ontario
London, Ontario, Canada

Smith, G.D.
Inco Alloys International
Huntington, WV USA

Soni, K.K.
The University of Chicago
Chicago, IL USA

Street, K.W.
NASA Lewis Research Center
Cleveland, OH USA

Swarr, T.E.
International Fuel Cells
South Windsor, CT USA

Swindeman, R.W.
Oak Ridge National Lab
Oak Ridge, TN USA

Tittmann, B.R.
The Pennsylvania State University
University Park, PA USA

Tomer, A.
Guest researcher at NIST
from the NRCN, Israel

Tosten, M.H.
Westinghouse Savannah River Lab
Aiken, SC USA

Turner, C.W.
Chalk River Labs
Chalk River, Ontario, Canada

Vitek, J.M.
Oak Ridge National Lab
Oak Ridge, TN USA

Wang, Y.
Hangzhou Institute of Electronic Engineering
Hangzhou, P.R. China

Whittaker, G.
Tennessee Eastman Corporation
Kingsport, TN USA

Wilcox, R.C.
Auburn University
Auburn, AL USA

Wilson, R.K.
Inco Alloys International
Huntington, WV USA

Yang, L.
Zhejiang University
Hangzhou, P.R. China

Yao, H.
Zhejiang University
Hangzhou, P.R. China

Yardy, J.W.
Danfoss A/S
Nordborg, Denmark

Zhao, W.
The University of Calgary
Calgary, Alberta, Canada

Preface

This volume of Microstructural Science contains selected manuscripts from the presentations made to the Technical Meeting of the 24th Annual Convention of the International Metallographic Society. The Technical Meeting meeting was on July 31-August 1, 1991, at the Monterey Convention Center, Monterey, California. These proceedings were prepared from camera ready maunscripts which were reviewed prior to inclusion in Volume 19 of the Microstructural Science series.

The Technical Meeting was chaired by Dean Wheeler and Mac Louthan and began with the ASM-IMS Distinguished Lecture in Metallography, "Role of Grain in High Temperature Deformation Processes" by Dr. Robert Gifkins, Institute of Metals and Materials, Australia. Dr. Gifkins spirited lecture provided an excellent keynote for the conference which was focussed on Structure-Property Relationships and Correlations with the Degradation of Engineering Materials. The conference included seventy eight presentations by authors representing nine countries and five continents.

The twelve Technical Sessions at the conference were: Characterization by Transmission Electron Microscopy, chaired by Mark Ransick; Fatigue Phenomena, chaired by Richard Ryan; Applied Metallography, chaired by Ann Kelly; Advanced Microscopy, chaired by Khalid Mansour; Wear and Errosion, chaired by Alan Place; Archaeometallurgy chaired by George VanderVoort; Environmentally Induced Cracking, chaired by Elliot Clark; Advanced Materials I, chaired by Bill Kanne; Coatings and Surface Modifications, chaired by Mary Doyle; High Temperature Corrosion, chaired by David Fitzgerald; Advanced Materials II, chaired by Mahmoud Shehata; and Novel Metallographic Techniques, chaired by Jim Bennett. Each chair reviewed abstracts of the papers prior to presentation and contributed significantly to the lively discussions that frequently followed a presentation. The arrangement of the papers in the proceedings does not correspond to the arrangement at the technical meeting, partially because every presentation is not included in the proceedings.

The success of the Technical Meeting was primarily due to the presenters and session chairs; however, many additional people made valuable contributions. Particular thanks are due to the Convention Co-Chairs, Garry Johnson and Del Miley and to the IMS Board of Directors.

This volume is the first camera ready proceedings prepared by the International Metallographic Society. Most of the real work associated with this publication was accomplished under the direction of Veronica Flint of ASM International and the editors are most greatful to her for that effort. The editors also give special thanks to those authors that followed all guidelines, meet all deadlines and provided an error free manuscript and gratefully acknowledge the commercial exhibitors and sustaining members for their support and donations to the Society.

D. A Wheeler
Howmet Corportation

G. W. E. Johnson
Lockhead Missiles and Space Co.

D. V. Miley
EG&G Idaho

M. R. Louthan, Jr.
Savannah River Laboratory

TEM Characterization

HELIUM BUBBLE DISTRIBUTIONS BENEATH GMA WELD OVERLAYS IN TYPE 304 STAINLESS STEEL

Michael H. Tosten[1] and Philip A. Kestin[1]

ABSTRACT

Transmission electron microscopy has been employed to characterize the microstructure and helium bubble distributions in Type 304 stainless steel used as substrate material for gas metal arc weld overlays. Helium-free and helium-bearing specimens, containing 48.5 and 147 appm helium were examined in both the welded and unwelded conditions. Helium bubble distributions were characterized according to bubble size and location within the material. Results showed that the initial bubble distributions were a result of tritium charging and aging (to produce helium-3) and not the overlay procedure. At distances greater than ~1.5 mm from the weld overlays the bubble distributions were generally unaffected by the temperature excursion and stresses produced during the welding process. However, within 0.5 mm of the overlay considerable bubble growth was observed, both at the grain boundaries and within the austenite matrix.

[1] Westinghouse Savannah River Company, Savannah River Laboratory, Aiken, SC 29808 USA.

INTRODUCTION

Nuclear reactors at the Savannah River Site (SRS) are used to generate radioisotopes for defense and space flight applications. Some primary components of the reactors, including the reactor tanks, are ultrasonically inspected for the presence of intergranular stress corrosion cracks. A research program was initiated to develop an in-tank repair process capable of repairing stress corrosion cracks that might be found within the reactor vessel walls.

A previous repair attempt in C-reactor tank at the SRS using a gas tungsten arc (GTA) weld technique was unsuccessful due to additional cracking in the heat-affected-zones of the repair welds [1]. It was determined that this additional cracking was a result of helium embrittlement caused by the combined effects of ^{4}He that exists within the reactor tank walls, the high heat input associated with the GTA process, and weld shrinkage stresses. (^{4}He is produced by thermal neutron reactions with nickel and boron.) These earlier studies suggested that the effects of helium-induced cracking could be minimized by a low heat input gas metal arc (GMA) overlay process [2,3].

In support of the in-tank repair program, the effects of the GMA overlay process on the microstructure and helium bubble distributions in Type 304 stainless steel were studied. Type 304 stainless steel was chosen as the material most representative of the actual SRS reactor tank walls. Helium was generated within the test specimens by tritium charging and subsequent aging. Helium build-in occurred from beta decay of tritium to ^{3}He (the tritium-trick). Due to the fine scale of the helium bubbles present after tritium decay, characterization of the helium bubble distributions was possible only through the use of transmission electron microcopy (TEM). The purpose of this paper is to describe the microstructure of the Type 304 stainless steel used as substrate material for the weld overlays and to analyze the bubble distributions in helium-containing specimens.

EXPERIMENTAL PROCEDURE

The composition of the Type 304 stainless steel plate used as a weld substrate is listed in Table 1. The as-received plate was approximately 12.7 mm thick and had been solution annealed at 1313K. Test specimens measuring ~119.4 mm x ~31.8 mm x ~6.4 mm were sectioned from this plate. These specimens were charged in high-pressure

TABLE 1. Chemical Analysis of Type 304 Substrate Material

Weight Percent

C	Mn	P	S	Si	Ni	Cr	N	Fe
0.073	1.390	0.023	0.016	0.052	8.280	18.22	0.026	Bal.

tritium for 30 days at 673K and subsequently aged at 233K for different lengths of time to generate several target helium concentrations. Following aging, the specimens were vacuum outgassed for 5 days at 723K to remove any tritium that remained.

TEM specimens were prepared from helium-free material and helium-bearing material that contained 48.5 and 147 appm helium. Specimens were sectioned from material both with and without GMA overlays. The heat input for the overlays was ~25.6 kJ/in^2. Disks for TEM examination from material with welds were punched from slices cut parallel to and at various distances from the weld fusion lines. Thin foils were electropolished at a potential of 12V with a solution of 7 vol% perchloric acid, 36 vol% butylcellosolve and 57 vol% methanol that had been cooled to 263K. All specimens were examined in a Philips EM400T operating at 120kV.

Helium bubbles were imaged with a technique that required viewing the specimens in an out-of-focused condition [4]. Using this technique, bubbles imaged in an under-focused condition are bright while those viewed in an over-focused condition are dark. The resolution of this imaging technique is ~0.8 nm. Bubbles smaller than this size are not visible due to their lack of contrast above background levels. Bubble size was determined by measuring from under-focused images where the diameter of the inner "bright" ring was taken as the actual bubble diameter [4].

RESULTS AND DISCUSSION

Helium-Free Material

The microstructure of the as-received material consisted predominantly of recrystallized grains ranging from 20 to 100 μm in diameter. Many of these grains contained annealing twins. In some areas, however, very high dislocation densities were observed. Selected area diffraction analysis of these regions revealed them to be either heavily cold-worked austenite or areas of α-martensite.

As-received plate containing a weld overlay was sectioned at 1 and 3 mm below the weld fusion line. The microstructure 3 mm below the weld overlay consisted of planar dislocation arrays and pile-ups at the grain boundaries. The dislocation density at this distance was representative of material that received little or no additional plastic strain as compared to the unwelded, as-received plate. The microstructure at 1 mm below the weld showed an increase in dislocation density and was characteristic of material that experienced an estimated 2-3% plastic strain. In general, the dislocations were restricted to planar arrays, but infrequently areas of dislocation tangles were observed. The higher temperatures experienced at this distance from the weld probably led to dislocation climb and the subsequent loss of the planar arrays in some areas.

Helium-Bearing Material

Specimen material that contained 147 appm helium was examined before welding to determine the helium bubble distribution present after tritium charging and aging. A well-developed bubble microstructure was observed. Bubbles measuring 2-4 nm in diameter were found on the high angle boundaries (Figure 1) and the incoherent twin boundaries. (Bubbles were not observed on the coherent twin boundaries.) Also, bubbles having diameters of 1-2 nm were observed within the grain interiors. Bubbles had nucleated on dislocations and were generally found in close association with dislocation loops. This latter observation is illustrated in Figure 2. Analysis of the dislocation loops indicated that they were Frank partial dislocation loops with $b=1/3<111>$ and that they were interstitial in nature. These results, in combination with the absence of dislocation loops in the helium-free plate, strongly suggest that the loops were "punched-out"

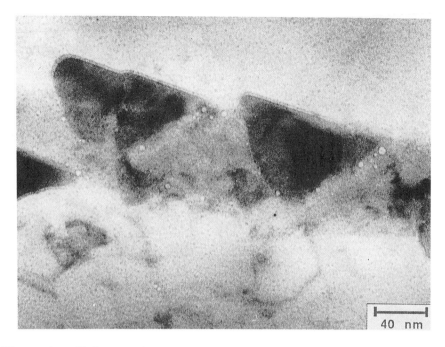

Figure 1. Helium bubbles on a grain boundary in the 147 appm He material (no weld overlay). The dark particles on this boundary are $M_{23}C_6$-type carbides.

during bubble growth.

A random distribution of loops existed throughout the grain interiors, except in the regions immediately adjacent to the grain boundaries, including incoherent and coherent twin boundaries. These regions were denuded in bubbles (and loops) except for those few bubbles associated with dislocations. The formation of these "bubble-free zones" may be related to the vacancy distribution in the near boundary regions. Since vacancies are known to act as strong traps for ^{3}He [e.g., 5], a reasonable assumption is that the presence of vacancy sinks, such as grain boundaries and dislocations, could result in a reduced number of vacancies and the concomitant reduction in the number of helium bubbles observed in these regions.

The effect of the overlay process on helium bubble distribution as a function of distance from the weld/substrate interface was also examined in specimens containing 147 appm helium. TEM specimens

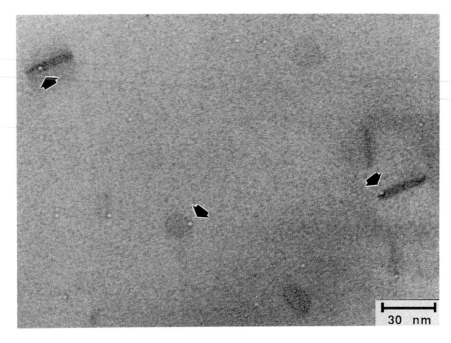

Figure 2. Helium bubbles and associated dislocation loops within a grain interior in the 147 appm He material (no weld overlay). Arrows indicate the position of several bubbles (and loops).

were prepared from slices taken from 3.5 and 2.3 mm below a weld overlay and compared to the unwelded material. Specimens from both distances below the overlay contained helium bubbles on all boundary types except the coherent twin boundaries. The distribution of bubbles on a typical grain boundary segment at 3.5 mm below the weld is shown in Figure 3. (The dark particles on this boundary are $M_{23}C_6$ carbides.) Bubble sizes ranged from ~2 to 4 nm on the grain boundaries. Additionally, helium bubbles were observed in the austenite matrix. Analogous to the unwelded material, these bubbles were 1-2 nm in diameter and closely associated with dislocations and dislocation loops. The bubble distribution in the welded material was essentially identical to that of the unwelded material. An attempt was made to examine 147 appm helium material at distances closer to the weld overlay; however, extensive cracking of the specimens occurred and foils could not be prepared.

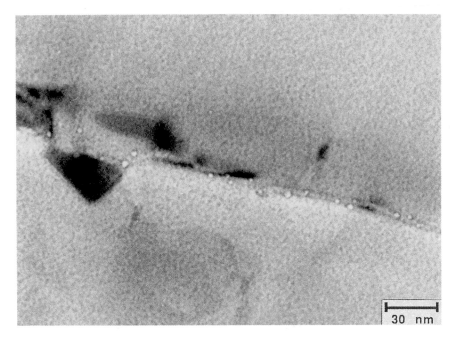

30 nm

Figure 3. Helium bubble distribution on a grain boundary in the 147 appm He material at 3.5 mm below a weld overlay. $M_{23}C_6$-type carbides are also located on this boundary.

In order to examine the helium bubble distribution in the near weld region, substrate material containing only 48.5 appm helium was sectioned at 1.5 and 0.5 mm below a weld overlay. (Lower helium-containing material was chosen to reduce the chance of cracking.) The bubble distribution at 1.5 mm was very similar to that observed in the 147 appm helium samples. Helium bubbles had nucleated on the high angle grain boundaries, incoherent twin boundaries and within the matrix. Matrix bubbles were found on dislocations and in the vicinity of dislocation loops.

The bubble microstructure at 0.5 mm from the overlay was strikingly different from the other material examined. Significant bubble growth had occurred in this region. This is evidenced in Figure 4, a micrograph which shows a high angle grain boundary. The average bubble size on this boundary measured ~10 nm, nearly 5 times the bubble diameter observed in specimens at 1.5 mm and below the

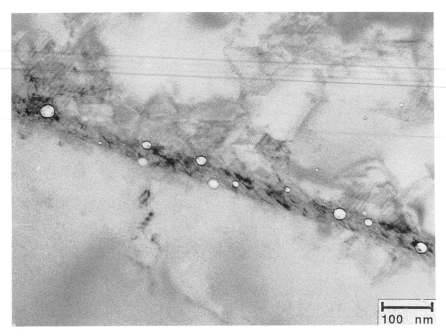

Figure 4. Helium bubble distribution on a grain boundary in the 48.5 appm He material at 0.5 mm below a weld overlay.

overlay. An increase in bubble diameter was also observed at the incoherent boundaries and within the matrix. At variance to the observations made from all other specimens, bubbles were observed at the coherent twin boundaries. These bubbles appeared to have nucleated on dislocations that had become incorporated into the twin boundary plane. Dislocations moved to the twin boundaries because of the "high" temperature deformation that had occurred at this distance from the overlay.

A summary of the average bubble diameters is presented in Table 2. As can be seen, there was very little change in the average bubble size in the material sectioned from distances greater than 1.5 mm from the overlay at either of the ^{3}He concentration. However, at 0.5 mm from the fusion line bubble size increased significantly over that measured at the other distances. The increased vacancy and ^{3}He fluxes as well as the weld shrinkage stresses in the near-weld region were most likely responsible for the increase in bubble size. Furthermore, these larger

TABLE 2. Bubble Diameter (nm)

Specimen	Matrix	HAGB*	ICTB†	CTB+
147/No Weld	2	4	2	-
147/3.5 mm	2	3	2	-
147/2.3 mm	2	3	-	-
48.5/1.5 mm	2	2	2	-
48.5/0.5 mm	9	10	8	5

*   High Angle Grain Boundary
†   Incoherent Twin Boundary
+   Coherent Twin Boundary

bubbles are believed to serve as nucleation sites for intergranular creep cavities, the growth and linkage of which can lead to grain boundary fracture.   Evidence for this latter phenomenon has been presented elsewhere [2].

CONCLUSIONS

The results of this study have shown that the low heat input GMA weld overlay technique produces limited changes in microstructure and helium bubble distribution as compared to unwelded material.   Plastic strain produced by the overlay process was estimated at no more than 2-3% at distances as close as 1 mm to an overlay and much less at 3 mm from an overlay.   A well-developed helium bubble microstructure was observed prior to welding, a result of the tritium charging and aging procedure.   After welding, significant helium bubble growth was limited to distances less than 1.5 mm from the overlay.   These observations support the use of the overlay technique as a possible reactor tank repair process.

ACKNOWLEDGEMENTS

This paper was prepared in connection with work done under Contract No. DE-AC09-89SR18035 with the U.S. Department of Energy. The authors are grateful to Drs. W. R. Kanne, Jr. and M. R. Louthan, Jr.

for their guidance throughout the course of this study. They also wish to express their thanks to F. E. Odom, A. W. Lacey and E. E. Kemper for their expert technical assistance.

REFERENCES

1) W. R. Kanne, Jr., "Remote Reactor Repair: GTA Weld cracking Caused by Entrapped Helium", **Welding Journal**, Vol. 67, p. 33, 1988.

2) W. R. Kanne, Jr., D. A. Lohmeier, K. A. Dunn, and M. H. Tosten, "Metallographic Analysis of Helium Embrittlement Cracking of repair Welds in Nuclear reactor Tanks", **Microstructural Science**, Vol. 19, ASM International, Metals Park, Ohio, in press.

3) E. A. Franco-Ferreira and W. R. Kanne Jr., "Remote reactor repair: Avoidance of Helium-Induced Cracking Using GMAW", WSRC-RP-89-312, Westinghouse Savannah River Company, Savannah River Laboratory, Aiken, SC.

4) M. F. Rhüle: "Transmission Electron Microscopy of Radiation-Induced Defects", Radiation-Induced Voids in Metals, Proc. Int. Conf. Radiation-Induced Voids in Metals, June 1971, U. S. Atomic Energy Commission, p. 255, 1972.

5) E. V. Kornelsen and A. A. van Gorkum, " A Study of Bubble Nucleation in Tungsten Using Thermal Desorption Spectrometry: Clusters or 2 to 100 Helium Atoms", **Journal of Nuclear Materials.**, Vol. 92, p. 79, 1980.

TEM CHARACTERIZATION OF THE OXIDES FORMED ON A

Zr-2.5WT%Nb ALLOY DURING AQUEOUS CORROSION

Yuquan Ding and Derek O. Northwood[1]

ABSTRACT

Zr-2.5wt%Nb alloy specimens were exposed to the pressurized lithiated water environment at 300°C for short (pre-transition) and long (post-transition) times, and the oxides at the oxide-metal interface were characterized by transmission electron microscopy (TEM) and selected area electron diffraction (SAED) techniques. The oxide close to the oxide-metal interface in the pre-transition specimens was tetragonal ZrO_2. Localized amorphous regions, which contained fine tetragonal ZrO_2, were also observed. This tetragonal ZrO_2 which formed as a thin continuous layer could act as a protective barrier to further oxidation. Long ridges of oxides are formed in the post-transition specimens as a result of the different oxidation rates of the different microstructural phases. The oxide at the oxide-metal interface in these post-transition specimens consist of both tetragonal and monoclinic ZrO_2. The destruction of the continuous tetragonal ZrO_2 layer results in a reduction of its protective nature and an increase in the oxidation rate.

---

[1]Engineering Materials Group, Dept. of Mechanical Engineering, University of Windsor, Windsor, Ontario, Canada N9B 3P4.

13

INTRODUCTION

A Zr-2.5wt%Nb alloy (hereafter referred to as Zr-2.5Nb) and other zirconium alloys such as the Zircaloys are used for the manufacture of nuclear reactor core components such as fuel cladding and pressure tubing because of their low neutron absorption cross-section, high strength and high corrosion resistance under reactor operating conditions [1]. The corrosion associated hydrogen ingress behaviour of zirconium alloys is often considered in terms of the effects of a "barrier layer" at the oxide-metal interface which can significantly reduce hydrogen ingress and oxidation rates with time, as long as it, i.e., the barrier layer, remains intact and protective. There is thus an increasing interest in developing specimen preparation techniques to examine the nature of the oxide-metal interface and the barrier oxide layer, if indeed it is present. There are relatively few reports of studies of the oxide near the oxide-metal interface in zirconium and its alloys, especially Zr-2.5Nb, as seen using TEM analysis techniques [2-7].

In the present work we have exposed Zr-2.5Nb pressure tubing specimens to a pressurized, lithiated water environment at 300°C for short (pre-transition) and long (post-transition) times and have studied the oxide-metal interface using TEM as well as SEM techniques.

EXPERIMENTAL DETAILS

1. Details of Corrosion Tests and Specimens

The starting material was commercial grade Zr-2.5Nb pressure tubing supplied by Ontario Hydro in the cold-worked and stress-relieved condition. Specimens were sections cut directly from pressure tubing with dimensions of approximately 20 mm x 15 mm x thickness of pressure tubing (~4mm). The corrosion/hydriding tests were carried out in high pressure (8.65 MPa), high temperature (300°C) static autoclaves of 50 cc capacity. A LiOH concentration of 4.8 g/ℓ deionized water was used in order to accelerate the corrosion reaction. The chemical analysis of the tubing and details of corrosion tests can be found in ref. [5]. The details of two groups A and B of the corrosion test specimens, which were exposed to the pressurized lithiated water environment for short (pre-transition) and long (post-transition) times are given in Table 1.

2. Preparation of Specimens for TEM and SEM Examination

Specimens were first cut from the corroded Zr-2.5Nb pressure tubing using a low speed diamond saw as shown in Figure 1.

Table 1 Details of Corrosion Specimens

Specimen Designation	Corrosion Environment	Exposure Time (hours)	Oxide Thickness (μm)
Pre-Transition (A)	300°C Lithiated Water	40	2.70
Post-Transition (B)	300°C Lithiated Water	480	43.0

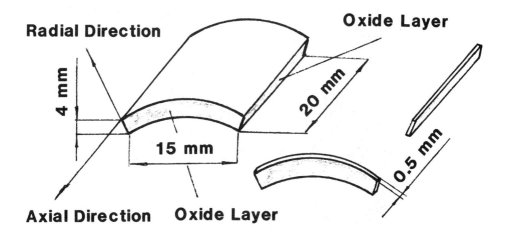

Figure 1. Sectioning of Corroded Specimen for Subsequent TEM and SEM Specimen Preparation.

Specimens for TEM and SEM examination include oxide layers both parallel to and perpendicular to the axial direction of pressure tubing. The TEM specimens containing the oxide-metal interface were then prepared using an ion milling technique that was modified from one originally developed by Moseley and Hudson [8]. The ion beam thinning was initially conducted from the metal side or the oxide side until the foil specimen was perforated at the edge [5]. The thin foil specimens were examined at 100 KV in a JEM-100CX microscope with a scanning attachment.

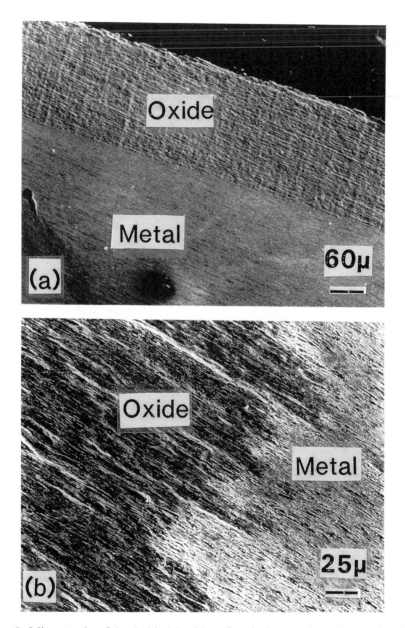

Figure 2. Micrographs of the Oxide-Metal Interface in the Specimen Exposed (a) For 40 Hours (Pre-Transition) Showing a Uniform Thickness of Oxide; (b) For 480 Hours (Post-Transition) Showing an Uneven Thickness of Oxide (Both Oxide Layers Parallel to the Axial Direction of Pressure Tubing).

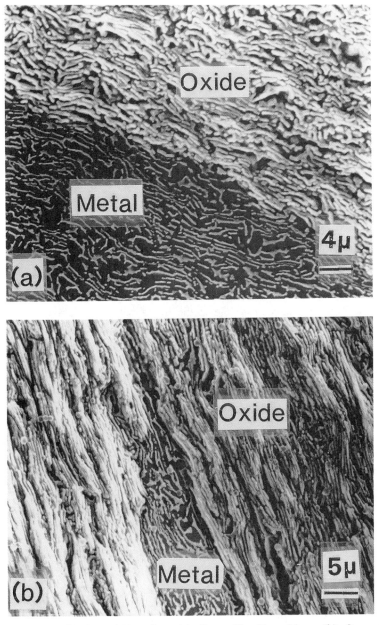

Figure 3. The Oxide-Metal Interface (a) On a Pre-Transition; (b) On a Post-Transition Specimens with Oxide Layers Perpendicular to the Axial Direction of Pressure Tubing.

The SEM tapered cross-section technique is particularly designed to observe the oxide-metal interface structure. Full details of the SEM specimen preparation can be found in ref. [9]. The SEM examination was performed on a Nanolab 7 Scanning Electron Microscope.

RESULTS

1. SEM Topography of the Oxide-Metal Interface Formed During an Aqueous Environment

Figures 2(a) and (b) show the typical SEM micrographs of tapered specimens for pre- and post-transition with oxide layers parallel to the axial direction of pressure tubing. From Figures 2(a) and (b) it can readily be seen that the oxide layer formed on the pre-transition specimens is characterized by a uniform thickness and a clear demarcation line (or surface) between oxide and metal, and that the oxide layer formed on the post-transition specimens is not uniform and the demarcation line between oxide and metal becomes uneven. Figures 3(a) and (b) show the oxide-metal interface on pre- and post-transition specimens where the oxide layers are perpendicular to the axial direction of pressure tubing. Figure 3(a) indicates that there is a definite demarcation line at the oxide-metal interface in pre-transition specimens. From Figure 3(b) it can be found that the demarcation line at the oxide-metal interface on post-transition specimens is not straight but is rather jagged. The characteristic with a straight demarcation line for pre-transition specimens may result from the same corrosion rates on an identical specimen surface because of short corrosion times. Therefore, the oxide front formed for pre-transition specimens is located at a constant level from the metal surface. The demarcation line between oxide and metal at the interface for post-transition specimens becomes uneven in both sections, i.e., oxide layers parallel to and perpendicular to the axial direction of pressure tubing, that were examined. This characteristic of the post-transition specimens is due to variations in corrosion rate at different locations for the long corrosion times.

2. TEM Topography of the Oxide-Metal Interface

Localized amorphous regions Figure 4 shows a localized amorphous region with fine equiaxed tetragonal crystallites (most are less than 10 nm in size) at the oxide-metal interface in a pre-transition specimen.

Crystalline oxide formed at the oxide-metal interface Figures 5 and 6 are the typical micrographs of pre- and post-transition specimens with oxide layers parallel to the axial direction of pressure tubing. The underlying metal structure of cold-worked Zr-2.5Nb alloy consists of elongated grains of α-Zr with filaments of ß-Zr at the grain boundaries [10]. Fine, relatively equiaxed grains of ZrO_2 are of a relatively uniform size ranging from about 30 to 50 nm in size (Figure 7). The microstructure of the oxide-metal interface where the oxide layer is perpendicular to the axial direction of the pressure tubing is shown in Figure 8. The oxide formed at the α-Zr grain boundaries has a variable grain size with grains as small as 10 nm. Some areas of the oxide formed at the grain boundaries, usually far from the oxide-metal interface, contained cracks. For the post-transition specimens, although the bulk of the ZrO_2 grains at the α-Zr grain boundaries are fine (30-50 nm), some of the grains are much larger than 50 nm (Figure 9). The SAED patterns of the oxide area shown in Figure 9 indicates that the oxide is a mixture of tetragonal and monoclinic ZrO_2. Figure 10 is another typical TEM micrograph of the oxide at the oxide-metal interface for a post-transition specimen. The SAED pattern shows that the oxide is tetragonal ZrO_2. As shown in Figures 9 and 10, there may be two typical structures for the oxide close to the interface in post-transition specimens.

Figure 4. Localized Amorphous Region with Small Tetragonal ZrO_2 Crystallites at the Oxide-Metal Interface (Pre-Transition Specimen).

19

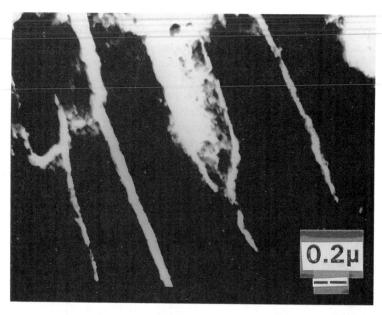

Figure 5. TEM Micrograph of the Oxide-Metal Interface on a Pre-transition Specimen with the Oxide Layer Parallel to the Axial Direction of Pressure Tubing.

Figure 6. TEM Micrograph of the Oxide-Metal Interface on a Post-Transition Specimen with Oxide Layer Parallel to the Axial Direction of Pressure Tubing.

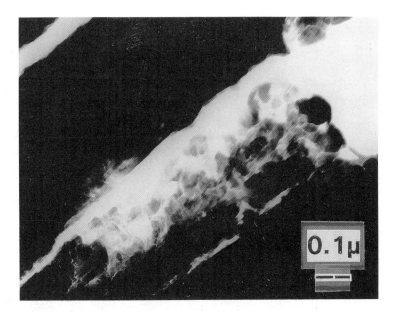

Figure 7. TEM Micrograph of Oxide Formed at the Oxide/α-Zr Interface on a Pre-transition Specimen with Oxide Layer Parallel to the Axial Direction of Pressure Tubing.

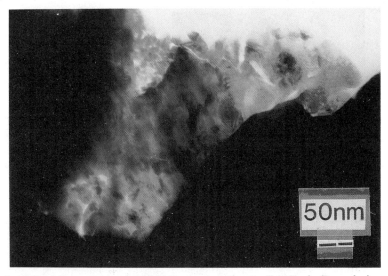

Figure 8. TEM Micrograph of Oxide Formed at the (α-Zr) Grain Boundaries on a Pre-transition Specimen with Oxide Perpendicular to the Axial Direction of Pressure Tubing.

21

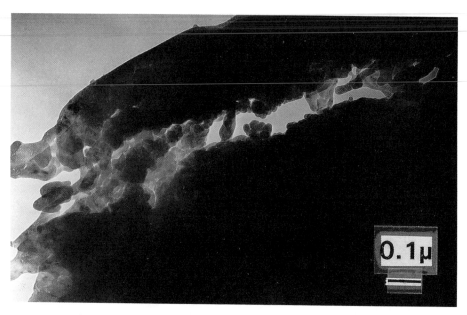

Figure 9. TEM Micrograph of Oxide (Tetragonal and Monoclinic ZrO_2) Formed at the (α-Zr) Grain Boundaries on a Post-Transition Specimen with Oxide Layer Parallel to the Axial Direction of Pressure Tubing.

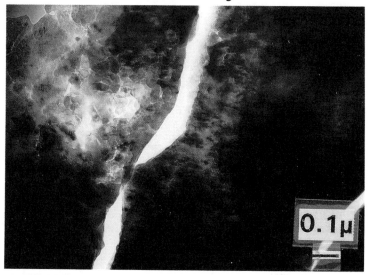

Figure 10. TEM Micrograph of Oxide (Tetragonal ZrO_2) Formed at the Oxide/α-Zr Interface on a Post-Transition Specimen with Oxide Layer parallel to the Axial Direction of Pressure Tubing.

DISCUSSION

1. Microstructure of the Oxides at the Oxide-Metal Interface on Post-Transition Zr-2.5Nb Specimens

Many investigators have put forward theories on the kinetics and mechanisms of corrosion in zirconium and its alloys. Some of the viewpoints are often contradictory, because of poor understanding of the microstructure of the barrier oxide layer at the oxide-metal interface.

Our early work [5] showed that the oxide at the oxide-metal interface on pre-transition Zr-2.5Nb specimens was tetragonal ZrO_2 whereas the oxide at the oxide-corrodent surface and away from the oxide-metal interface was monoclinic ZrO_2. This tetragonal ZrO_2 which formed as a thin continuous layer could act as a protective barrier to further oxidation. The retention of tetragonal ZrO_2 at the oxide-metal interface is attributed to the fine grain size of the ZrO_2 and the compressive stresses developed at the oxide-metal interface. Since the compressive stress gradient through the thickness is quite steep [11], the tetragonal ZrO_2 retained because of the compressive stresses, may only exist over a small distance from the oxide-metal interface. As shown in the Results section, the oxide at the oxide-metal interface on post-transition specimens consists of both tetragonal and monoclinic ZrO_2 (see Figure 9). The oxide front formed during corrosion of a pre-transition specimen is located at a constant level from the metal surface, whereas the oxide front formed during corrosion of post-transition specimens is uneven. Because of the unevenness, the oxide front for the post-transition specimen can deform, and hence compressive stresses formed at the oxide-metal interface may be relaxed. This results in transformation of some tetragonal ZrO_2 into monoclinic ZrO_2.

Garzarolli et al [6] have described the microstructure of the oxide-metal interface for the three different types of oxide structures: uniform oxides formed in (i) water and (ii) oxygen, and (iii) nodular oxide formed in steam for Zircaloy specimens. The main structure at the interface for the uniform oxide formed in water and for the nodular oxide formed in steam, is a columnar structure of monoclinic ZrO_2. This monoclinic columnar structure is considered to be formed because, as noted by Murase and Kato [12], moisture (water or steam) markedly accelerates crystallite growth for both monoclinic and tetragonal ZrO_2 and facilitates the tetragonal-to-monoclinic phase transformation. The normal structure formed by oxidation in oxygen primarily consists of monoclinic ZrO_2 as well as some tetragonal ZrO_2 and an amorphous structure. In an investigation of Zr-2.5Nb, Warr et al [4] also observed that amorphous regions were formed not only at the oxide-metal interface but also fairly generally at oxide grain boundaries. Recently Wadman and Andren [7] used atom probe analysis and TEM to study thin oxide layers grown in air and boiling water on Zircaloy-4 specimens and found that the 10 nm thick oxide layers were partially amorphous and contained small oxide crystallites. The oxygen content of the amorphous layers was low, less than 50 atomic percent.

In this study of pre-transition Zr-2.5Nb specimens, we also observed amorphous regions at the oxide-metal interface where very fine tetragonal ZrO_2 crystallites also formed (see Figure 4). A SEM examination of post-transition specimens [9] showed that the microstructure of metal close to the oxide-metal interface was different from that far away from the interface: the α-Zr grains close to the interface appeared "larger" than the α-Zr grains far from the interface. The solubility limit of oxygen in α-Zr, which is more than that of oxygen in ß-Zr, is 29at%O. Therefore, there may be an intermediate oxygen-rich layer in the metal (α-Zr) close to the oxide-metal interface for post-transition specimens. The formation of amorphous regions may be related to kinetics of the intermediate layer formation. The oxidation of cold-worked Zr-2.5Nb pressure tubing in pressurized lithiated water at 300°C proceeds first along grain boundaries at which there are ß-Zr and its decomposition products, and then continues on α-Zr grains. The oxide formed is mainly extended along grain boundaries [13]. Because the oxide formed during the aqueous corrosion of Zr-2.5Nb for long times is white and porous, lithiated water is probably able to penetrate much closer to the oxide-metal interface [14]. The corrosion resistance of Zr-2.5Nb depends mainly on structure and composition of the grain boundaries. Oxide films formed, especially at low temperatures, are initially continuous and amorphous, but the crystallization process is facilitated by water, heat, high electric fields and mechanical stress [14]. It is conjectured that the oxide formed at the α-Zr grain boundaries (ß-Zr and its decomposition products) consist mainly of crystallites rather than the amorphous phase. Further work is required to determine the extent and the location of the localized amorphous regions so as to better understand the role of these amorphous regions in the barrier oxide layer.

2. Proposed Mechanism of Breakaway Corrosion of Zr-2.5Nb in Aqueous Environments

Although a great amount of work on zirconium oxidation has been carried out in recent years, there is still a great deal of disagreement as to the exact cause of breakaway [16,17]. Different investigators hold strongly to their own theories. Comparing Figures 3(a) and (b) which show the oxide-metal interface in the pre- and post-transition regions, long ridges of oxides can be seen in the post-transition specimens. As shown in the Results section, oxides at or near the oxide-metal interface for post-transition specimens have two typical structures: either tetragonal ZrO_2, or a mixture of tetragonal and monoclinic ZrO_2. Figure 11 schematically illustrates the variation of the oxide front formed during corrosion in a pressurized lithiated water environment for pre- and post-transition specimens. The oxide front formed during corrosion of a pre-transition specimen is located at a constant level (from the metal surface) (see Figure 11(a)). From Figures 2(b) and 3(b) it can be seen that the demarcation line between oxide and metal at the interface on post-

(a) Pre-Transition

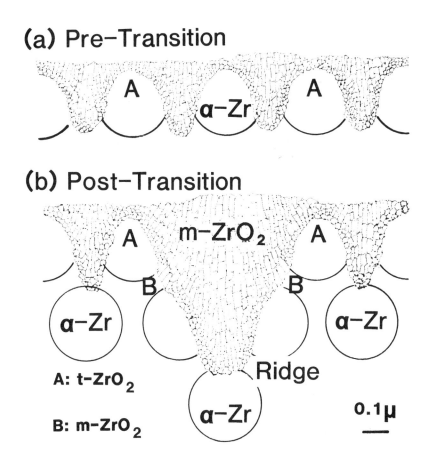

(b) Post-Transition

Figure 11. Comparison of Interface Oxide Layer Structure on (a) Pre- and (b) Post-Transition Specimens During the Aqueous Corrosion of Zr-2.5Nb.

transition specimens is not straight but is rather jagged. The oxide front formed during corrosion in post-transition specimen is uneven. The oxide front with regular corrosion rates may only consist of tetragonal ZrO_2 which can act as a protective layer. Movement of the oxide front (most likely in regions such as B in Figure 11(b) between the highest and the 'regular' corrosion rate regions can produce deformation of the oxide and compressive stresses in these regions are then relaxed. Hence, tetragonal ZrO_2 in these regions is transformed to monoclinic ZrO_2. The destruction of the continuous tetragonal ZrO_2 layer, which is formed in short (pre-transition) times, results in a reduction of its protective nature and an increase in the oxidation rates at the locations where tetragonal ZrO_2 transforms to monoclinic ZrO_2.

25

CONCLUSIONS

TEM and SEM studies show that the oxide close to the oxide-metal interface in the pre-transition corrosion specimens of Zr-2.5Nb pressure tubing exposed to a pressurized lithiated water environment is tetragonal ZrO_2. Localized amorphous regions, which contain fine tetragonal ZrO_2 crystallites, were also observed on these pre-transition specimens.

Long ridges of oxides are formed in the post-transition specimens as a result of corrosion of the different microstructural phases. The oxide at the oxide-metal interface in these post-transition specimens consist of both tetragonal and monoclinic ZrO_2. The destruction of the continuous tetragonal ZrO_2 layer results in a reduction of its protective nature and an increase in the oxidation rate.

ACKNOWLEDGEMENTS

This work was supported by the Natural Sciences and Engineering Research Council of Canada (Operating Grant A4391) and the CANDU Owner's Group (COG) through Working Party No. 35. Mr. John W. Robinson assisted with the corrosion tests and with the microscope examinations.

REFERENCES

1) C.E. Ells and Evans, "The Pressure Tubes in the CANDU Power Reactors," **The Canadian Mining and Metallurgical Bulletin,** Vol. 74, p. 105, 1981.

2) J.E. Bailey, "On the Oxidation of Thin Films of Zirconium," **Journal of Nuclear Materials,** Vol. 8, p. 259, 1963.

3) R.A. Ploc, "Transmission Electron Microscopy of Thin (<2000 Å) Thermally Formed ZrO_2 Films," **Journal of Nuclear Material,** Vol. 28, p. 48, 1968.

4) B.D. Warr, A.M. Brennenstuhl, M.B. Elmoselhi, E.M. Rasile, N.S. McIntyre, S.B. Newcomb and W.M. Stobbs, "Microstructure and Compositional Analysis of the Oxides on Zr-2.5wt%Nb in Relation to Their Role as Permeation Barriers to Deuterium Uptake in Reactor Application," Presented at "Microscopy of Oxidation" Conference, Cambridge, England, March 1990.

5) Y. Ding and D.O. Northwood, "TEM Study of the Oxide-Metal Interface Formed During the Aqueous Corrosion Zr-2.5wt%Nb Alloy," Presented at the 23rd Annual Meeting of the International Metallographic Society, Cincinnati, Ohio, U.S.A., July 1990. To be published in **Microstructural Science.**

6) F. Garzarolli, H. Seidel, R. Tricot and J.P. Gros, "Oxide Growth Mechanism on Zirconium Alloys," Presented at Ninth International Symposium on Zirconium in the Nuclear Industry, Kobe, Japan, November 1990.

7) Boel Wadman and Hans-Olof Andren, "Microanalysis of the Matrix and the Oxide-Metal Interface of Uniformly Corroded Zircaloy," ibid.

8) P.T. Moseley and B. Hudson, "Phases Involved in the Corrosion of Zircaloy by Hot Water (350°C)," **Journal of Nuclear Materials**, Vol. 99, p. 340, 1981.

9) Y. Ding and D.O. Northwood, "Method for Preparation of Cross-Sectional SEM Specimens and Their Application to Corroded Specimens of a Zirconium Alloy and TiN Coated Stainless Steel," **Materials Characterization**, Submitted for publication.

10) R.G. Fleck, E.G. Price and B.A. Cheadle, "Pressure Tube Development for CANDU Reactors," **Zirconium in the Nuclear Industry**: Sixth International Symposium, ASTM STP 824, D.G. Franklin and R.B. Adamson, Eds., American Society for Testing and Materials, p. 88, 1984.

11) Bang-Xin Zhou, "Electron Microscopy Study of Oxide Films Formed on Zircaloy-2 in Superheated Steam," Zirconium in the Nuclear Industry: Eighth International Symposium, ASTM STP 1023, L.F.P. Van Swam and C.W. Eucken, Eds., American Society for Testing and Materials, Philadelphia, p. 360, 1989.

12) Y. Murase and E. Kato, "Role of Water Vapour in Crystallite Growth and Tetragonal-Monoclinic Phase Transformation of ZrO_2," **Journal of the American Ceramic Society**, Vol. 66, p. 196, 1983.

13) Y. Ding and D.O. Northwood, "A SEM Examination of the Oxide-Metal Interface Formed During the Aqueous Corrosion of a Zr-2.5wt%Nb Alloy," Accepted for publication in **Journal of Materials Science**.

14) A.V. Mavnolescu, P. Mayer and C.J. Simpson, "Effect of Lithium Hydroxide on Corrosion Rate of Zirconium in 340°C Water," **Corrosion-NACE**, Vol. 38 [1], p. 23, 1982.

15) L.L. Shreir, In: **Corrosion Vol. 1 Metal/Environment Reaction**, Newnes-Butterworths, p. 1:23, 1976.

16) T. Ahmed and L.H. Keys, "The Breakaway Oxidation of Zirconium and Its Alloys: A Review, **Journal of the Less-Common Metals**, Vol. 39, p. 99, 1975.

17) B. Cox, "Mechanisms of Corrosion of Zirconium Alloys in High Temperature Water," Symposium Proceedings, Electrochemical Society Fall Meeting, 1981.

The Study of Degradation Phenomena of Cr-Mo Steels after Long-term High Temperature Service

Ching-An Lin[*] and Jain-Long Horng[*]

ABSTRACT

Cr-Mo steels have been used widely in electricity-generating and pertrochemical industry which require long-term, high-temperature service. The relationship between microstructure and mechanical properties has been investigated on 1Cr-0.5Mo, 2.25 Cr-1Mo, and 9 Cr-1Mo steels taken from the power plant after 25 to 30 years of service at 480 °C to 540 °C . The unexposed or reheat-treated steels were also studied for comparison.

From the X-ray diffraction and TEM experiments, steels with low contents of Cr and Mo after long-term service suffered carbide spheroidization, transformation, and coarsening. However, carbides in 9Cr -1Mo steel changed little, retaining the same forms of $M_{23}C_6$ and Cr_2 (C,N) throughout service. The partition of Cr/Mo in ferrite matrix shows that Cr and Mo contents in the matrix decrease after long term service. The extent of decrease in Cr-Mo steels is P12, T22 and T9 ordinal. The results of tensile properties were related to competitive strengthening mechanisms, specifically carbide dispersion, solid solution and dislocation strengthening. The creep acceleration test and the Cr/Mo partition were utilized to substantiate the degradation phenomena.

*

Materials Research Laboratories
Industrial Technology Research Institute
Chutung, Hsinchu 31015, Taiwan, R.O.C

INTRODUCTION

Cr-Mo steels are low carbon ferritic steels which have been successfully used in power and pertrochemical industry, requiring service at elevated temperature for many years. The life of equipment for elevated-temperature service is often limited by the degradation of these steels. Investigation of the degradation of Cr-Mo steels after long-term service would be useful for the prediction of the remaining life of all the structural components. Several papers have been published on microstructural studies of steels exposed to long-term service, including 1Cr-0.5Mo [1,2], 2.25 Cr-1Mo [3,4,5] and 9Cr-1Mo [6,7] steels. The purpose of this study is to study how the microstructure and mechanical properties of steels change during service and how such data relates to life-expectancy predictions. The present study deals with the advanced metallographic examination of structure with degradation ˌin mechanical properties. In addition, the unexposed and reheat-treated steels were also investigated to corroborate the comparison.

Experimetal

Materials

The materials examined in this study were 1 Cr-0.5 Mo(P12), 2.25Cr-1Mo(T22) and 9Cr-1Mo(T9) steel with compositions given in Table 1. Steels of T22-0 and T9-0 were taken from a reheater tube which had been operated at 540 °C for approximately 25 years. Steels designated by T22-N and T9-N were in the form of new tubes. P12-0 specimens were taken from a superheater outlet header which had been operated at 480 °C for approximately 30 years. Because of new P12 material unavailable, P12-R specimens were reheat-treated, duplication themanufacturer's initial heat treatment process of reproduce theoriginal microstructure of the P12 steel.

Metallographic Techniques

The metallographic techniques employed in this study consisted of examination of microstructures by optical and electron microscopy and identification of phase by X-ray and electron diffraction analysis. The specimens were mechanically polished and etched in a 2% nitricacid ethanol.

Carbon extraction replicas were prepared from the mounted metallographic specimens. The specimen polishing and etching techniques were the same as employed for optical microscopy. The specimen surface was coated with carbon by the evaporation method. The evaporated carbon film was then scored into a square grid of lines of $\sim$ 3mm separation, and the film was removed from the surface by dissolution of the matrix in a solution of 2% bromine in alcohol cooled to -20 ℃ . the replicas were then retrieved on copper grids and examined by a JOEL 2000 FX transmission electron microscope (TEM). Thephases present on the extraction replicas were identified by selected area electron diffraction and by energy dispersive spectroscopy (EDS).

Phase extracts were prepared by anodic dissolution of the matrix in a 10% HC1-methanol electrolyte. The extracts were separated from the electrolyte using a ceramic filter. Chromium and molybdenum in the supernate were analyzed through inductively coupled plasma atomic emission spectrometry (ICP-AES). The extracted precipitates (carbide) were then analyaed by a PHILTIPS X-ray difractometer.

Tensile and Creep Rupture Tests

Tesile testing was performed on three specimens each condition machined from test steels. Specimens were tested at room temperatureand at high temperature (540 ℃ or 565 ℃).

A creep rupture test was carried out in the air with a constant-load creep-testing machine that had a loading arm ratio of 1 to 3. The specimens from the P12 steel were 9 mm in diameter and 36 mm in gauge length.

RESULTS AND DISCUSSION

Characterization of Microstructure

The microstructure of service-exposed 1Cr-0.5 Mo steel (P12-0), illustrated from the optical and scnning electron micrographs as shown in Fig. 1, consisted of carbide-containing and spheroidized cementite on different ferrite grains. The larger carbide particles were also observed along grain boundaries. The microstructure of the reheat-treated P12-R steel (Fig. 1.) revealed isolated lamellar sturctures of perarlite distributed in the ferrite grains. The microstructure of 2.25 Cr-Mo steel after long-term service, consists of ferrite grains which contain relatively coarser precipitates dispersed throughout the matrix and along the grain boundaries.

The microsturcture of 9Cr-1Mo steel (Fig.2.) showed no clear difference in the size and distribution of carbide between T9-N and T9-0, but the grain boundaries of the T9-0 were more clearly revealed than T9-N's.

Carbide Analysis

The X-ray diffraction (XRD) results of extracted carbides in various Cr-Mo steels are summarized in Table 2. The carbide in the 1Cr-0.5Mo steel were mostly Fe_3C in the reheat-treated condition. The carbides extracted from the P12-0 specimen appeared to contain M_2C and M_7C_3 carbides, but the Fe_3C was still the major type present. No $M_{23}C_6$ and M_6C carbides were observed in the P12 specimens. Morphology of carbide in P12-0 TEM extraction replica is shown in Fig. 3. From an electron diffraction pattern, the needle-like carbide in Fig. 3(b), was identified as M_2C. The carbide shown in Fig. 3(c) is M_7C_3. Long-term service exposure resulted in the formation of alloy carbides, such as M_2C and M_7C_3. The present results were not completely in agreement with the results of V.A.Biss and T.Wada with 1Cr-0.5M steel service at 524 ℃ for 20 years, which reported that the Fe_3C carbide was not remained the main type. This discrepancy can be easily explained. It was because the temperature the P12-0 steel had to sustain was 480 ℃ (compared to 524 ℃) 。

The micrographs of the TEM replica for the 2.25Cr-1Mo steels are shown in Fig 4. Very fine and uniformly distributed carbides could be seen through the ferrite matrix in the T22-N steel-Fig.4(a). In the T22-0 specimen the carbides coarsened significantly after exposure service-Fig. 4(b). The predominate carbide in T22 steel was $M_{23}C_6$. After long-term service, the Mo_2C carbide was found in the T22-0 specimen. Concentractions of Mo and C in solid solution decreased as Mo-containing carbide formed. Consequently, the solid-solution strengthening effect diminshed for the T22-O specimen, a factor will be discussed further 。

The carbide transformation in the Cr-Mo steels has been extensively studied. The results are sumarized below:

(1) For P12 steel [10]

Pearlite Fe_3C → spheroidization of Fe_3C → $M_7C_3 + M_2C$

→ M_6C

Ferrite → $M_7C_3 + M_2C$ → M_6C

(2) For T22 Steel [3]

Bainite

$$M_7C_3 \longrightarrow M_6C$$

$$\varepsilon \text{ carbide} \rightarrow M_3C \rightarrow M_3C + M_2C$$

$$\longrightarrow M_{23}C_6$$

Ferrite
$M_2C \rightarrow M_6C$

Interestingly, the major carbides were Fe_3C in P12 steel and $M_{23}C_6$ in T22 steel. In both steels, ratios of Cr to Mo were nearly 2,and the equilibrium carbide was M_6C. Our experimental results show that for P12 and T22 both steels sustained at medium high temperatures, 480 ℃ and 540 ℃ , respectively, the carbide can be regarded as quasi-equilibrium carbide, half way to the equilibrium carbide.

For the T9 steel, the TEM extracted replica in Fig.5. shows the carbide morphogy. Form X-ray results only two types of carbides, $M_{23}C_6$ and $Cr_2(C,N)$ occurred in both unused (T9-N) and exposed (T9-0) steels.There is no noticeable difference in carbide size bewteen T9-N and T9-0 steels. However, in the T9-0 steel large carbides were clearly observed along the grain boundaries-Fig.5(b). The cuboid-like carbidesin the T9-N steel were identified from TEM as $(Cr, Fe, Mo)_{23}C_6$ by EDS interfaced with a STEM mode; an example of one particle is shown inFig. 6. Similar results were found in Sanderson's paper [7]. The contents of Mo and Fe varied from one particle to another. This explains where Mo goes other than remaining in the solution.

Carbide Content and Cr/Mo Partition in Matrix

The carbide content data are shown in Table 3. In the P12 and T22 steels, it increased appreciably after service exposure. In contrast, T9's carbide content decreased a little after service exposure. These results were consistent with TEM observations.

The results of Cr/Mo partition in the ferrite solid solution are also shown in Table 3. In P12 steel, the Cr concentration decreased from 96% partitioning to 83%, and Mo from 98% to 75% when P12-R and P12-0 steels were compared. Similarly, in the T22 steel, the Crconcentration fell from 83% to 75%, and Mo from 82% to 74% when T22-Nand T22-0 steels were compared. The decrease of Cr/Mo partition to a solid solution was mainly attributed to the increase in carbide content after long-term service.

In T9 Steel, it is interesting to find that service exposure caused Cr/Mo partitioning behavior in different directions. The Cr content in the matrix increased after long-term service, but the Mo content decreased. It was found that the $M_{23}C_6$ particle consisted of Mo and Fe (Fig. 6). The variation of Mo in $M_{23}C_6$, especially for the carbides on grain boundaries, may be an important parameter for controlling the strengthening mechanism; further study is required.

Mechanical Properties and Creep Rupture Test

The tensile test results conducted at room and high temperature are shown in Table 4. In P12 and T22 steels, the yield strength and tensile strength of unexposed specimens were always higher than high temperature exposed specimens, especially the difference in tensile strength. In the T9-N and T9-0 specimens, the mechanical properties were quite similar, due to the microstructureal stability even after exposure. There was still a little difference between the T9-N and T9-0 steels where at room temperature T9-0 was stronger, but at 540 °C T9-N was stronger.

Several papers discussed the strengthening mechanism in this series of steels [4,5,6,9]. Three mechanisms have been identified as the dominant sources: dispersion hardening due to carbide, solid solution strengthening due to the soluble species of Mo, C and N, and dislocation strengthening from lath martensite and transformation dislocation [6]. In P12 and T22 specimens, all three sources mentioned above were more preferrable for the unexposed specimens than for theexposed specimens. Thus, the mechanical strength in unexposed specimens was always higher than in exposed ones. In the T9 specimenthe three sources changed relatively less and cancelled each other out. since solid solution strengthening was a more dominate factor atelevated temperatures, this fact can explain the significant decreasein tensile strength of 29.0 kg/mm^2 in the T9-0 steel due to temperature change compared to the decrease of 24.1 kg/mm^2 in the T9-N steel. The partitioning of Mo to ferrite matrix would then be an important factor.

The results of the P12-O steel creep rupture tests are shown in Fig. 7. the time-temperature parameter (Larson-Miller parameter), P, has been plotted against stress for different test conditions. Our experimetal results are plotted and compared with the ASTM Data [8]. The steel exposed to service possessed lower rupture strength, especially in the short rupture time range. As compared with the ASTM reference data. The low rupture strength and low remaining life strongly indicated material degradiation.

The creep data of P12-0 and P12-R steel are show in Fig 8. The creep stain rate of exposed speimen were much higher than unexposed specimen, this result also indicated the material degradation phenomena.

CONCLUSION

(1)In the 1Cr-0.5Mo steel, Mo_2C and M_7C_3 were formed after long-term service at 480 ℃ . In the 2.25Cr-1Mo steel, Mo_2C was observed in the second predominat phase after long-term service at 540 ℃ . Carbide coarsening was clearly observed in both steels.

(2)The service exposure caused changes in the local chemistry of these two steels, transferring significant amounts of Cr and Mo from the matrix to the carbides.

(3)The tensile test results of the three steels can be explained from the combined strengthening effects among dispersion hardening, solid solution strengthening and dislocation strengthening.

(4)The microstructure of 9Cr-1Mo steel showed higher stability then those of 1Cr-0.5 Mo and 2.25Cr-1Mo steels. Although carbide coarsening strongly occurred along the grain bondaries, there was no noticeable change in carbide type during exposure.

(5)Creep rupture tests indicated that deterioration had already taken place in the exposed 1Cr-0.5Mo steel.

REFERENCES

(1)V. A. BISS and T. WADA, "Microstructural change in 1Cr-0.5Mo steel after 20 Years of Service", Metall. Trans., 16A, P109 (1985)

(2)R. A. Varin and J. Haftek, "Structural Changes in a Ferritic Heat-resistant steel", Mat. Sci. Engng., Vol 62, P129 (1984)
(3)R. G. Baker and J. Nutting, J. Iron steel Inst., 192, P257(1959)

(4)T. Wada and V. A. Biss, "Restoration of Elevated Temperature Tensile Strength in 2.25 Cr-1 Mo Steel", Metall. Trans., 16A, P845 (1983)

(5)N.S. Cheruru, "Degradation of Mechanical properties of Cr-Mo-V and 2.25 Cr-1 Mo Steel Components After Long-term Service at Elevated Temperatures", Metall. Trans., 20A, P87 (1989)

(6)W.B. Jones, Ferritic Steels for High Temperature Applications, ASM, 1983.

(7)S.J. Sanderson, ibid.

(8)Metal Handbook, 8th Edition, Vol.1, ASM, 1964.

(9)W.B. Jones and J.A. Van Den Avyle, "Substructure and strengthening Mechanims in 2.25 Cr-1M o Steel at Elevated Temperatures", Metall. Trans., 11A, P1275 (1980)

(10)L.H. Toft, and R.A. Marsden, Iron steel Inst. Specical report No.70, P276 (1961)

Table 1. Compositions of Test Steels (in wt%)

Element	P12-0	T9-N	T9-0	T22-N	T22-0
C	0.1	0.154	0.096	0.063	0.120
Si	0.19	0.54	0.42	0.4	0.31
Mn	0.45	0.45	0.46	0.46	0.45
P	0.007	0.03	0.03	0.025	0.015
S	0.11	<0.01	<0.01	<0.01	<0.01
Cr	1.02	9.28	9.28	2.1	2.1
Mo	0.53	0.9	0.96	0.9	0.96
Fe	bal.	bal.	bal.	bal.	bal.

Table 2. X-ray Diffraction Analysis of Extracted Carbide

specimens	P12-R	P12-0	T9-N	T9-0	T22-N	T22-0	Amount
carbide	Fe_3C	Fe_3C	$M_{23}C_6$	$M_{23}C_6$	$M_{23}C_6$	$M_{23}C_6$	
type	—	Mo_2C	—	—	—	—	
	—	M_7C_3	$Cr_2(C,N)$	$Cr_2(C,N)$	—	Mo_2C	
	—	—	—	—	Fe_3C	Fe_3C	decrease

Table 3. The Carbide Content and Concentration of Cr and Mo in Ferrite Matrix.

Specimens		P12-R	P12-O	T22-N	T22-O	T9-N	T9-O
Amount of carbide wt%		0.28	0.62	1.10	2.12	2.20	2.04
Cr	Total amount in steel wt%	1.02	1.02	2.1	2.1	9.28	9.28
	Amount in solid solution wt %	0.98	0.85	1.74	1.58	8.06	8.62
	Partition-ing to solid solution	96 %	83%	83%	75%	87%	93%
Mo	Total amount in steel wt%	0.53	0.53	0.9	0.96	0.9	0.96
	Amount in solid solution wt %	0.52	0.40	0.74	0.71	0.81	0.81
	Partition-ing to solid solution	98 %	75 %	82 %	74 %	90%	84%

Table 4. Mechanical Properties of Test Steels

| | Testing at room temperature | | | |
	YS(kg/mm^2)	UTS(kg/mm^2)	Elong(%)	hardness(Hv)
P12-R	28.3	42.3	35.8	124
P12-0	21.5	41.0	36.8	117
T9 -N	25.7	57.2	32.9	
T9 -0	29.2	59.0	28.8	
T22-N	30.0	52.0	32.5	
T22-0	24.8	51.7	33.6	

| | Testing at 540°C | | |
	YS(kg/mm^2)	UTS(kg/mm^2)	Elong(%)
P12-R[*]	16.2	33.5	33.1
P12-0[*]	15.3	22.1	38.0
T9-N	18.2	33 1	34.5
T9-0	18.2	30.0	33.0
T22-N	20.5	35.2	30.7
T22-0	17.1	29.7	30.3

* testing at 565 °C

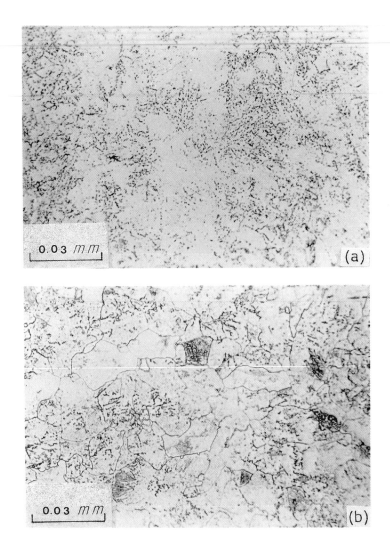

Fig.2. Microstructure of T9 steels (a) unexposed steel (T9-N) (b) exposed steel (T9-0)

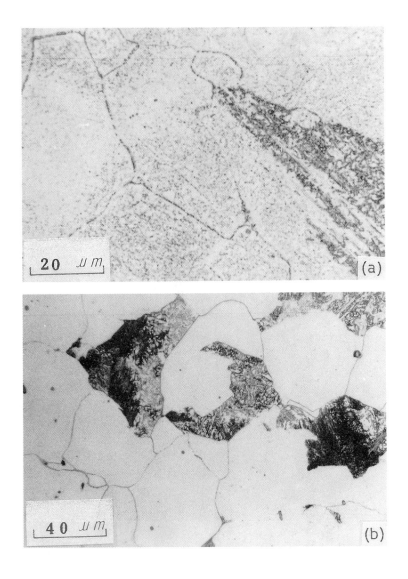

Fig.1. Optical microstructure of P12 steel (a) service-exposed (b) reheat -treated

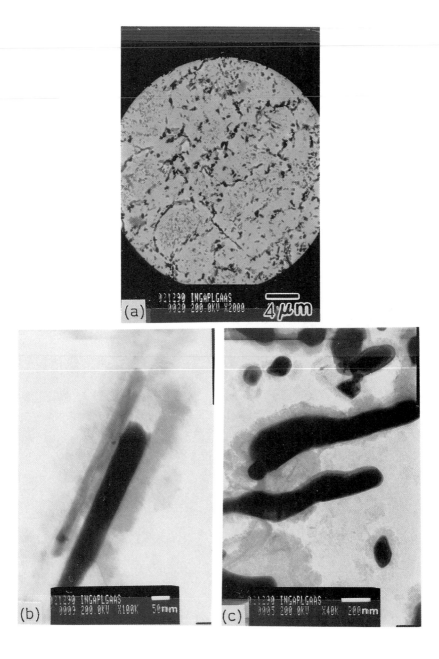

Fig.3. Morphology of extracted carbides in P12-0 steel (a) at low magnifation (b) needle-like M_2C (c) M_7C_3 carbide.

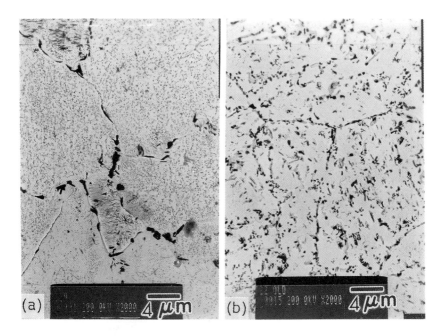

Fig.4. Morphology of carbide in T22 steel. Carbon extractionreplica.
(a) T22-N (b) T22-0

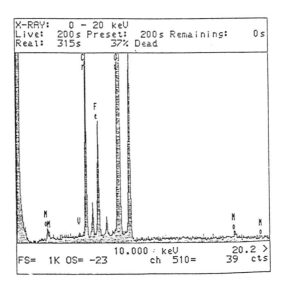

Fig.6. EDS spectrum of the cuboid-like carbides in the Fig 6(c).

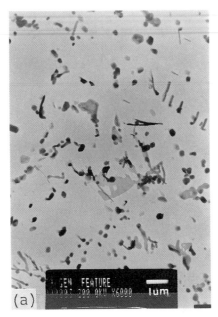

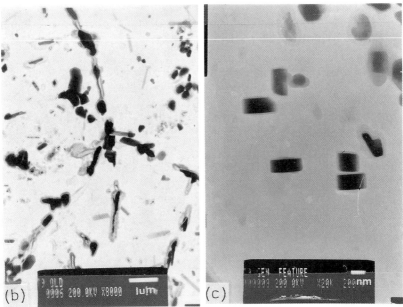

Fig.5. Morphology of carbide in T9 steel. Carbon extraction replica. (a) T9-N (b) T9-0 (c) cuboid-like carbides of T9-N are $M_{23}C_6$.

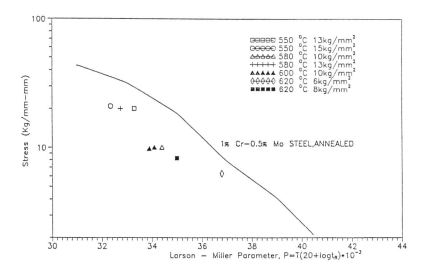

Fig.7. Variation of Larson-Miller parameter with various temperatures and stresses for 1Cr-0.5Mo steel.

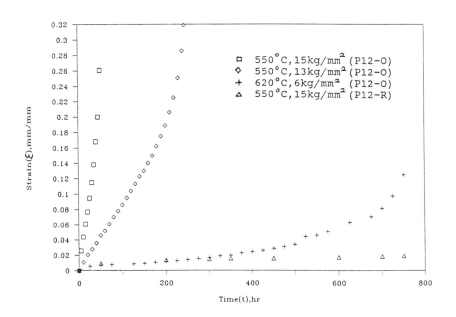

Fig.8. Creep data for P12-0 and P12-R steels.

EFFECT OF Mg ADDITION ON THE PRECIPITATION FEATURES

AND PROPERTIES OF Cu—1. 94wt%Be—0. 35wt%Ni ALLOY

Yong LING[1], Liu YANG[2], Yiyun WANG[1]

and Hongnian YAO[2]

ABSTRACT

This work has investigated the effect of small amount of Mg addition on the precipitation features and properties of Cu—1. 94wt%Be—0. 35wt% Ni alloy. The results show that after solution and aging treatments under 320℃ for 2 hours, the mechanical properties of the alloy can be greatly improved by means of 0. 1wt% Mg addition into the alloy. By use of XRD and TEM analysis, it has been shown that after the aging treatment, only dispersed particles of metastable phase γ'' and γ' occur in the matrix of the alloy with 0. 1wt% Mg addition; while for the alloy without Mg addition, there are some discontinuous equilibrium precipitates of γ phase at grain boundaries besides γ'' and γ' precipitates in grains. It has been concluded that small amount of Mg addition can effectively suppress the discontinuous precipitation at grain boundaries and therefore greatly improve the mechanical properties of the alloy.

1 Deparment of Mechanicl Engineering, Hangzhou Institute of Electronic Engineering, Hangzhou 310012, P. R. China.

2 Central Laboratory, Zhejiang University, Hangzhou 310027, P, R. China.

INTRODUCTION

The copper — beryllium alloys are typical age — hardening materials and have been used extensively for their combined high strength, excellent conductivity and corrosion resistance. Since the discovery of age — hardening in copper — beryllium alloys[1], much work has been done for determination of the precipitation of the alloys [2—4], from the results, the overall precipitation sequence may be described by:

$$\alpha_{ss} \rightarrow \text{G. P zones} \rightarrow \gamma'' \rightarrow \gamma' \rightarrow \gamma$$

however, various interpretation of the structure and orietation of precipitates have been proposed [3—7].

During the aging, the binary copper — beryllium alloy has considerable tendency to form the equilibrium phase γ from the discontinuous precipitation at grain boundaries while the metastable phase γ'' and γ' precipitate in grains. It is generally suggested [2,8] that the age — hardening copper — beryllium alloy is relative to the precipitation of semi — coherent phases γ' in grains, if the γ phases are formed at grain boundaries, the degratation of mechanical properties occurs. Several investigators [9—11] have found that by means of small amount of Mg addition into copper — beryllium alloy, the mechanical properties can be improved significantly due to the suppression of formation of equilibrium phase γ at grain boundaries.

However, the influence of magnesium on the continuous precipitation in grains is not clear, the main propose of present work is to investigate the effect of small amount of Mg addition on the precipitation features and properties of Cu — 1. 94wt% Be — 0. 35wt% Ni alloy, and then the relationship of microstructure and mechanical properties and the mechanism of small amount of Mg addition are discussed in this paper.

EXPERIMENTAL PROCEDURES

The alloys investigated are polycrystalline commercial copper — beryllium alloy strips with thickness of 0. 8mm, the chemical composition is listed in Table 1. The alloys are solution treated at 780℃ for 30 min in nitrogen atmosphere followed by a cold water quench, and then age—hardening treated for 2 hr at 280, 300, 320, 340, 360, 380℃ under dynamic vacuum of 10^{-2} Torr.

Table 1, Chemical composition (wt%)

Alloy	Be	Ni	Mg
Cu—Be—Ni	1. 95	0. 34	0. 0002
Cu—Be—Ni—Mg	1. 94	0. 35	0. 10

Specimens for tensile text are cut parallel to the rolling direction from the strips and are prepared according to GB 6397—86 standard, the tensile properties are determined on Shimadzu Autograph DCS—25T tensile machine according to the procedure of GB 228—87 standard.

Thin foils for electron microcopy are prepared by mechanical thining, chemical thining and ion milling. The foils are examined in a JEM 200cx electron microscope operated at 200 KV with a double tilt stage capable of $\pm 30^{\cdot}$.

The X—ray diffraction measurements are carried out on a 12 KW D/max—γ_A X—ray diffractometer with a graphite monochrometer mounted on the diffracted beam path, opreating at 40 KV and 30 mA, Cu radiation and Ni filter are used. The body samples with 10×15 mm used for XRD analysis are prepared by carefully mechanical and chemical polishing.

RESULTS

TEM analysis

The typical diffuse striations along with the traces of $(01\bar{1})_a$ plane and dislocation loops contrast in grains are imaged in Cu—Be—Ni alloy aged at 320℃ for 2 hr (Fig 1a) ,in corresponding selected area diffraction patterns, the orientation of foil plane is $(111)_a$ with $\mathbf{g}(\bar{2}20)$ active reflection.

The metastable phase γ″ formed in the earlier stage of precipitation present the fine dislocation loops contrast in matrix and make the $(\bar{2}20)^*$ reflection enlongate along the $[\bar{1}10]$ direction as the γ″ phase is coherent on (100) matrix plane and the distortion of matrix crystallic lattice is relatively small. While the precipitation con-

49

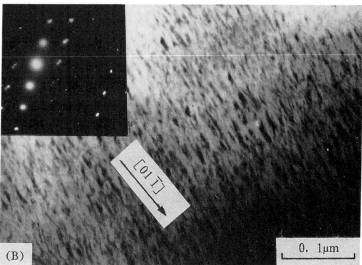

Fig. 1 Electron micrograph and SADP of precipitation microstructure in grain,
after solution treated at 780℃ for 30 min, and then aged at 320℃ for 2hr.
(111)$_a$ foil plane g($\overline{2}$20) active reflection.
a) Cu—Be—Ni   b) Cu—Be—Ni—Mg

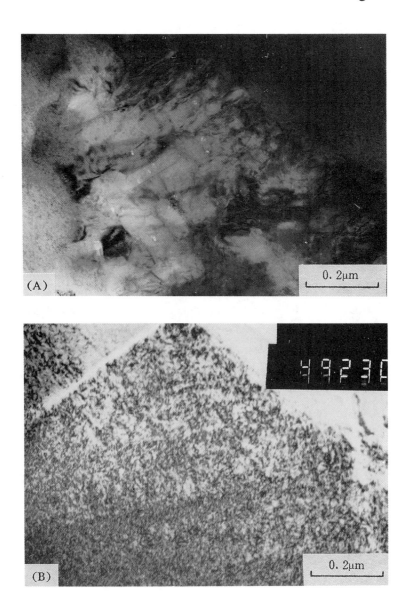

Fig. 2　Electron micrograph of grain boundary structure, aged at 320℃ for 2 hr

a)Cu—Be—Ni　　　　　b) Cu—Be—Ni—Mg

tinues, however, the semi—coherent precipitates γ' are transformed from the coherent precipitates γ''. Because of intense lattice distortion, the relots of $[20\,\overline{2}]^*$ and $[02\,\overline{2}]^*$ seperate and leave two extra faint satellite spots, from the bright—field microscopy, the plate—like γ' phases reveal dark diffuse striations aligment with the $[01\,\overline{1}]$ direction, which is the traces of $(01\,\overline{1})$ planes. The size of γ' phase at present heat treatment has a plate—like shape of $\sim$ 500 Å in diameter and $\sim$ 100 Å in thickness.

Fig. 1b shows the microstructure and SADP of Cu—Be—Ni—Mg alloy after aged at 320℃ for 2 hr, which is similar to that of Cu—Be—Ni alloy, except that due to the addition of 0. 1wt% magnesium, the γ' phases change into much finer and more dispersed striations with average diameter of $\sim$250 Å and thickness of $\sim$ 50 Å.

On aging at 320℃ for 2 hr, the Cu—Be—Ni alloy precipitates the equilibrium phases γ at grain boundaries (Fig. 2a), while the metastable phases γ'' and γ' precipitate in grains. The lamellar shaped γ phases nucleate at grain boundaries and grow into adjoining grains. Similar observation is made for the Cu—Be—Ni—Mg alloy, the characteristic feature is absence of γ phases at grain boundaries (Fig. 2b), that is to say that the discontinuous precipitation at grain boundaries of Cu—Be—Ni alloy is effectively suppressed by small amount of Mg addition.

To examine the position of magnesium, the micro—area chemical composition analysis is employed by means of the STEM—EDX, which has the sentivity of 0. 05wt% alloy element concentration in an area of $\sim$100 Å diameter, the results show that no magnesium segregation at grain bourdaries or in grains is detected.

X—ray diffraction

Fig. 3 show the X—ray diffraction profiles of two alloys samples after solution treatment at 780℃ for 30 min, the $(200)_a$ diffraction peak of α—Cu matrix is presented. The profiles of two peaks are similar to each other, and the peak position of Cu—Be—Ni alloy is deviated to the high degree by 0. 1wt% Mg addition, this is relative to the smaller lattice distance of $(200)_a$ planes. The reason of this phenomenon may be the lattice contraction caused by formation of high vaccum concentration after solution treatment.

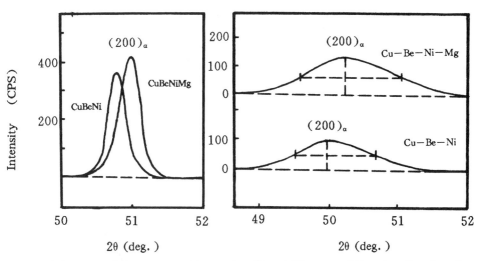

Fig. 3 X — ray diffraction profiles, solu-tion treated at 780℃ for 30 min.

Fig. 4 X — ray diffraction line broaden-ing, aged at 320℃ for 2 hr.

After aging at 320℃ for 2hr, the X — ray diffraction profiles of two alloys change obviously. Firstly. the $(200)_a$ peak lines are broadened as shown in Fig. 4, according to the theory of Enzo [12] and Wilkens [13], the X — ray line broaden-ing is caused by the mosaic of small size and dislocation of high density in samples. As the average grain diameter of α — Cu matrix remained the same before and after age — hardening traetment, it can be believed that the line broadening is a result of high density of dislcation which are formed by the distortion of matrix lattice in [100] direction. Because the width of peak is propertional to the density of disloca-tion, and the $W_{1/2}$ of $(200)_a$ peak of Cu — Be — Ni alloy is increased from 22 to 27 mm by 0. 1wt% Mg addition, the density of dislocation in matrix of Cu — Be — Ni — Mg alloy is greater than that of Cu — Be — Ni alloy . In addition, the new two over-lapping diffraction peaks occur at 33 — 35 degree of 2θ (Fig. 5) they are (100) an (010) peaks of γ″ and γ′ precipitates by theoretical calculation [14]. The X — ray diffraction peaks of γ″ and γ′ phases are also broadened due to the very fine parti-cles of γ″ and γ′ phases.

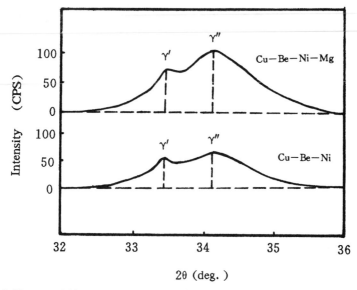

Fig. 5 X—ray diffraction profiles of γ'' and γ' precipitates of two alloys after age—hardening treatment at 320℃ for 2hr.

Mechanical properties

The mechanical properties of two alloys changed as a function of aging temperature are given in Fig. 6, the results of tensile and hardness texts indicate that the tensile strength (σ_b), 0. 2 percent proof stress $(\sigma_{0.2})$ and hardness (V. H. N.) of Cu—Be—Ni—Mg alloy are greater that of Cu—Be—Ni alloy after age—hardening treatment.

The maximum strength of Cu—Be—Ni alloy is attained at 320℃ for 2 hr as a result of precipitating the metastable phases in grains and relatively small amount of equilibrium phases γ at grain boundaries (about 2 percet in volumn). When the alloy is aged at higher temperature, its strength degrates greatly because the large amount of γ phases are produced by the discontinuous precipitation at grain boundaries.

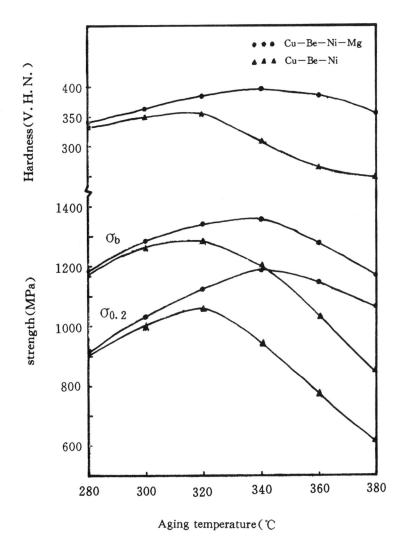

Fig. 6 Mechanical properties versus aging temperature (for 2 hr)

σ_b: tensile strength, $\sigma_{0.2}$: 0.2 pct proof stress

Compared with Cu—Be—Ni alloy, the peak hardening temperature of Cu—Be —Ni—Mg alloy extends to a platform of 320 to 340℃, the maximum strength increases significantly as a result of adequate precipitation of fine and dispersed γ' particles and suppression of precipitation of equilibrium phases γ. Even at the aging temperature higher than 340℃, its strength reduces slowly because the discontinuous precipitation is retardated by the 0.1wt% magnesium addition.

DISCUSSION

We will first discuss the mechanism of precipitation hardening in copper—beryllium alloy, and then the effect of 0.1wt% Mg addition on the precipitation features and properties of Cu—Be—Ni alloy.

During isothermal aging of copper — beryllium alloy, the metastable precipitates, G. P. zones, γ'' and γ' phases, are sequentially formed by the continuous precipitation in grains. The nucleation of γ' precipitate is predicted to take place at the interphase boundaries of γ'' phases [15]. Present work investigates that the γ'' and γ' phases both present in the peak age — hardening samples, but the γ' precipitates make major contribution to overall strength [2,14]. As the γ' precipitates are partially coherent with the matrix, the misfitting of lattice results in elastic internal stresses which interact with glide dislocation.

Lloyd[16] suggested that complex commercial alloys contained a combination of strengthening mechanisms. As considering copper — beryllium alloys, besides the coherency strengthening which results from the partially coherency of γ' precipitates, it also contains the Orowan strengthening in which the γ' precipitate can be considered as imperetrable obstacles on glide dislocation, therefore, the size and distribution of γ' phases control the strength.

Present study convinces that the addition of a small amount of Mg into Cu—Be —Ni alloy not only effectively suppresses the formation of equilibrium phases γ at grain boundaries, and also accelerates the continuous precipitation in grains.

From the results of STEM—EDX analysis, there is no magnesium segregation in grain boundaries, which is contradictory to the suggestion of Jitsu et al[9]. We

56

propose that a small amount of Mg addition into Cu—Be—Ni alloy can increase the concentration of vaccum in as—quench solution, the magnesium atoms occurpy the sizes of vaccum and dislocation, and retardate the diffusion of Be atoms, besides this, as one of most active interface elements in copper—beryllium alloy, magnesium can probably reduce the ΔG^*, the free energy of activation for formation of the critical nucleus of G. P. zones. For this reason, the addition of Mg can accelerate the nucleation of G. P. zones, γ'', γ' phases and make eventually the size and distribution of strengthening phases γ' finer and more dispersed.

In addition, because of suppression of discontinuous precipitation at grain boundaries by a small amount of Mg addition, the peak age—hardening temperature extends to a platform of 320 to 340℃. As the aging temperature increases to 340℃, the γ' phases can precipitate adequately. The 0.1wt% Mg addition decreases the sensitivity of overage in Cu—Be—Ni alloy, even at aging temperature higher than 340℃, its mechanical properties decreases slowly.

Above all, by addition of 0.1wt% magnesium, the microstructure of Cu—1.94wt%Be—0.35wt%Ni alloy after peak age—harening treatment is improved obviously, consequently, the mechanical properties such as tensile strength, 0.2 pct proof stress and hardness are improved significantly.

CONCLUSION

1. By addition of 0.1wt% magnesium into Cu—1.94wt%Be—0.35wt%Ni alloy, the microatructure of the alloy after age—hardening treatment is improved obviously, not only the eguilibrium phase γ at grain boundaries is effectively suppressed, and the continuous preipitation in grains is also accelerated.
2. Both methstable precipitates γ'' and γ' present after age treatment at 320℃ for 2 hr, as a result of 0.1wt% magnesium addtion, the size and distribution of γ' particles in alloy become finer and more dispersed.
3. The sensitivity of discontinuous precipitation of γ phases at grain boundaries is abated by addition of a small amount of Mg, the peak age—hardening temperature expends to a platform of 320 to 340℃.
4. As a result of 0.1wt% magnesium addition, the mechanical properties of Cu—

1. 94wt％Be — 0. 35wt％Ni alloy after peak age — hardening treatment are improved greatly.

REFERENCE

1. A. Guiner and P. Jacquet, *Rev. de Metall.* , **41**, 1, 1944
2. B. C. Edwards and W. Bonfield, *J. of Met. Sci.* **19**, 398, 1974
3. R. J. Rioja and D. E. Laughlin, *Acta Metall.* , **28**, 1301, 1980
4. Y. M. Koo and J. B. Cohen, *ibid*, **37**, 1295, 1989
5. A. H. Geisler, J. H. Mallery and F. E. Steigert, *Trans. Met. Sci. AIME*, **194**, 307, 1952
6. S. Yamamoto, M. Matsui and Y. Murakami, *Trans. Jap. Inst. Metals*, **12**, 159, 1971
7. K. Shimizu, Y. Mikami, H. Mitani and K. Otsuka, *ibid*, **12**, 206, 1971
8. A. R. Entwisle and J. K. Wyrn, *J. of Inst. Met.* , **89**, 24, 1960
9. H. Jitsu, T. Agatsuma and K. Hashizume, *J. of Mitsubithi Electr. Tech.* , **40**, 1075, 1966
10. M. Sygiyama and T. Furukawa, *J. of Jap. Inst. of Met.* , **28**, 530, 1964
11. M. Miki, S. Hori and Y. Amano, *ibid*, **44**, 160, 1980
12. S. Enzo, G. Fagherazzi, A. Benedetti and S. Polizzi, *J. of Appli. Cryst.* , **21**, 536, 1988
13. M. Wilkens, *ibid*, **12**, 119, 1979
14. L. Yang, Y, Ling, Y. Wang and H. Yao, The influence of prior — deformation on precipitation of Cu — 1. 9wt％Be — 0. 1wt％mg alloy, *J. of Mat. Sci. Letter*, accepted for publication, 1991
15. K. C. Russell and H. I. Aaronson, *J. of Mat. Sci.* **10**, 1991, 1975
16. D. J. Lloyd, Strength of materials and alloys, Vol. 3, p. 1745, ed. by H. J. Mc Queen et al, pergamon Press, Oxford, England, 1985

Advanced Microscopy

MICROSTRUCTURAL 3D ANALYSIS OF

MATERIALS SURFACES

USING THE

CONFOCAL LASER SCAN MICROSCOPE LSM

Hans-G. Kapitza
Department of Microscopy
Carl Zeiss
Oberkochen, Germany

ABSTRACT

Confocal laser scanning microscopy has been recently applied to characterize surface structures of materials. This type of scanning imaging system can discriminate small height differences down to about 50 nm and provides a lateral optical resolution between 0.3 and 0.2 μm depending on optics and light wavelength used. Different types of surface structures on materials have been analyzed: In addition to the improved imaging and the computer-based 3D reconstruction of surfaces quantitative analysis methods have been established, including measurement of local micro-roughness.

INTRODUCTION

The advantages of a "flying spot microscope" have been understood already in the early 50's (1). With the availability of small and reliable gas cw-lasers and since appropriate scanning systems have been developed (2) the technology has been utilized for light microscopy in the late 70's (3). At this time the advantages of confocal microscopy - used up to this time only in quantitative microscope photometry - have been demonstrated also for imaging in the microscope. Finally in 1982, the first commercial LSM has been brought to the market by Carl Zeiss (4). The confocal instrument described in this paper is available since 1988 (5). It is unique in so far, as many functions are motorized and routines for investigations on materials can be pre-programmed by the operator using macros. The applications are developed first for semiconductor research, development and testing, then for biomedical use and very recently for an increasing demand in materials microscopy.

EXPERIMENTAL PROCEDURES

The confocal LSM represents a highly integrated compact microscope system. The basic optical beam path diagram is shown in Figure 1.

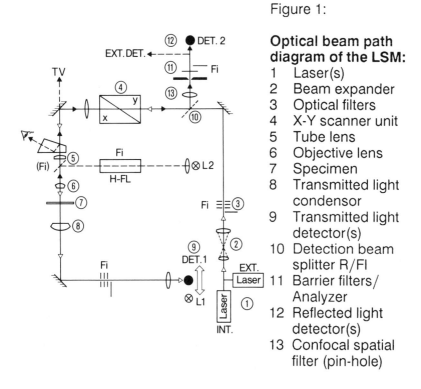

Figure 1:

Optical beam path diagram of the LSM:
1 Laser(s)
2 Beam expander
3 Optical filters
4 X-Y scanner unit
5 Tube lens
6 Objective lens
7 Specimen
8 Transmitted light condensor
9 Transmitted light detector(s)
10 Detection beam splitter R/Fl
11 Barrier filters/ Analyzer
12 Reflected light detector(s)
13 Confocal spatial filter (pin-hole)

The components indicated as numbers 11 through 13 in this figure are responsible for the confocal imaging effect: Only light irradiated into and emanating from the objectives lens' focal plane is able to pass the "pin hole" (position 11) and therefore to arrive at the detector (position 12). Out-of-focus contributions are blocked by the pinhole. Only a small fraction of this light can reach the detector. The effect is strongly dependent on the numerical aperture (N.A.) of the objective lens used. The basic laws ruling the depth selection capability of the confocal LSM are shown below:

Figure 2:

Theoretical limits of depth resolution.

$$dz = \frac{n \cdot 1.27 \cdot \lambda}{(N.A.)^2} \quad (1)$$

$$dz = \frac{n \cdot 0.89 \cdot \lambda}{(N.A.)^2} \quad (2)$$

λ = wavelength
n = refractive index

The terms describe limiting cases for ideal optics and zero detector pin-hole diameter.

Upper term (1): Fluorescence / Stray light situation.

Lower term (2): Reflected light situation.

It has been demonstrated (6), that there is a difference between the stray light/ fluorescence case (term 1) and reflected light situation, leading to different numerical factors in the formula. Since these terms have been derived for the limiting case of zero pin-hole diameter, practical values of the z-resolution (in μm) are slightly higher. With the LSM values of 0.5 m can be achieved with high N.A. lenses. This value is defined as the axial distance for which the intensity reflected from a surface is above 50% of the maximum value found exactly in the focal plane. This distance is a reproducible measure of confocal z-resolution, meaning the capability to resolve two different objects in vertical direction. The capability in locating a surface is at least by a factor of 10 better, because small changes in intensity can already be used by the computer to find the surface within the vertical intensity distribution.

Practical values have been found to be in the range of 30nm or less for the detection of coating layer edge on flat substrates. For the recording of "stacks" of confocal images, the z-motor drive of the LSM stage (DC type) is equipped with an optical encoder reporting steps of 50 nm within a range of at least 20 mm to the computer. If in addition a scanning stage is used, the 4"x4" travel range version offers a step size of 0.25 μm an a typical revisit accuracy of +/-1 μm for the full range.

The computer contains two independent processors, one for system control under DOS and a second, fast digital image processor for imaging including some image analysis. This design gives the possibility for extensive macro programming used for industrial testing jobs. For storage floppy, hard and WORM optical disks are available. Image output is standard RGB signal for easy connection to commercial video printers and monitors.

RESULTS

Typical results from the evaluation of the confocal image stacks created by the LSM are:

* 3D representations like cumulated images showing extended depth of focus resembling SEM images or stereo images.

* Quantitative data about distances, angles, areas and voluming, all given directly in μm, μm^2 or μm^3, degrees or percentage.

* Statistical curves for height (z-value) distribution, slope distribution and volume distribution.

* Numerical evaluation of parameters describing micro-roughness.

The images shown in Figure 3 give a typical set of results found after confocal analysis of the texture of an Al sample. The final result hasbeen created in a single run, where the operator only called a macro programmed previously: Then the LSM did a fully automated analysis using its 3D motor controls and the autofocus function without any interaction of the operator:

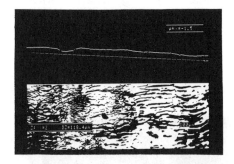
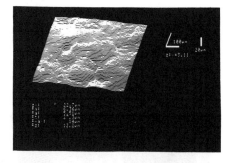
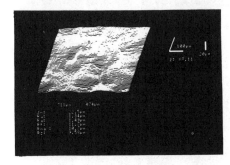
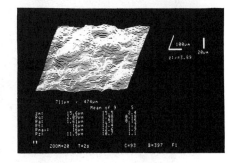

Figure 3: **Micro-roughness of an Al surface measured with the confocal LSM.**
(Sample: Aluminium Ranshofen / Austria)
Upper left: Profile scan shows single surface profile and allows determination of tilt angle.
Upper right: 3D profile plot of the tilted Al surface before digital correction. The micro-roughness parameters calculated are increased
Lower left: As above, but digitally corrected for tilt. The micro-roughness parameters are now minimized and therefore appropriate.
Lower right: After a fully automated measurement cycle performed at nine different locations of one sample mean values and standard deviations of micro-roughness are calculated and displayed.
The total time for a 256x256 (64k) point 3D analysis from start to final result takes less than 60 seconds

On (semitransparent) layers or foil materials in addition to surface testing local thickness measurements and the analysis of inner structures of multi-layered materials have been performed. Here especially single- and multilayer polymer foils have been tested for surface and volume analysis.

Special applications include semiconductor testing, where confocal 3D analysis is directly combined with the Optical Beam Induced Current (OBIC) method or even emission microscopy for failure analysis purposes.

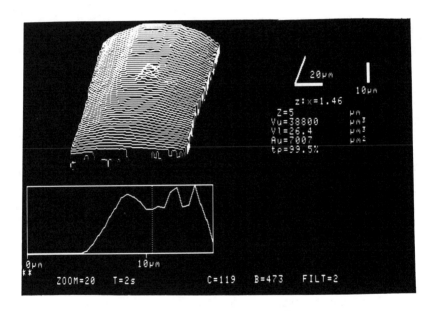

Figure 4:   **Non-contact, non-destructive optical measurement of solder tin volume found on a contact (Vu = 38 800 μm^3).**
Sample: Siemens Augsburg / Germany

A very late example is the fully automated surface analysis on MYLAR foil material, where a small number of surface bumps is found and analyzed according to their height distribution finally listed in classes of 1,2,3 or more μm.

DISCUSSION

The confocal microscopy on materials has been around for a couple of years and is now enhanced very much by the use of the highly automated LSM. We observe, that especially during the last three years powerful solutions for -sometimes old- problems have been developed.

On the other hand still many - even the most experienced - microscopist working in materials research are not aware of the new possibilities offered by the confocal methods. It is very important to realize, that confocal microscopy is not a way in metallography to get much better images from decent polished samples: Here the samples are two-dimensional and the resolution of digital imaging today doesn't reach the level of large format or even 35 mm photographs, not to mention color imaging at all. However at very high magnification and after analog and digital image processing one can see details not seen before. Also the LSM is a very powerful confocal laser scanning fluorescence microscope, which is giving new ways in crack analysis or computer board inspection.

Also, the LSM is not going to replace the SEM in its specific applications since resolution and depth of field are inherently far superior in electron beam systems, whereas laser scanning remains within the limits of light microscopy.

On the other hand the LSM is unique in 3D analysis of surfaces and layer systems. In addition the LSM contains a conventional research grade light microscope for best results, if traditional methods are used.

CONCLUSION

Confocal laser scanning microscopy as represented by the LSM now offers new ways for microscopy with the possibilities of quick and accurate quantitative 3D analysis. In addition the traditional methods are included. The high degree of motorization and the possibility of running automated routines by using macro's have opened the way to routine applications in industry.

REFERENCES

1) Young J.Z., Roberts F., "A flying spot microscope." **Nature**, Vol.167, p.231, 1951

2) Beiser L., "Laser scanning systems", **Laser Applications**, Vol.2, Academic Press, p.53, 1974

3) Brakenhoff G.J., Blom P., Barends P., "Confocal scanning microscopy with high aperture immersion lenses", **Journal of Microscopy**, Vol.117, p.219, 1979

4) Wilke V., "Laser scanning in microscopy", **Proceedings of SPIE**, Vol.396, p.164, 1983

5) Kapitza H.G., Wilke V., "Applications of the microscope system LSM", **Proceedings of SPIE**, Vol.1028-Scanning imaging, p.173, 1988

6) Hellmuth T., Siegel A., Seidel P., "Generation of 3D images via laser scanning microscopy", **European Journal of Cell Biology** Vol.48 (Suppl.25), p.35, 1988

FLUORESCENCE MICROSCOPY FOR THE CHARACTERIZATION

OF STRUCTURAL INTEGRITY

Kenneth W. Street[1] and Todd A. Leonhardt[2]

ABSTRACT

The absorption characteristics of light and the optical technique of fluorescence microscopy to enhance metallographic interpretation are presented. Characterization of thermally sprayed coatings by optical microscopy suffers because of the tendency for misidentification of the microstructure produced by metallographic preparation. Gray scale, in bright-field microscopy, is frequently the only means of differentiating the actual structural details of porosity, cracking, and debonding of coatings. Fluorescence microscopy provides a technique to help distinguish the artifacts of metallographic preparation (pullout, cracking, debonding) and the microstructure of the specimen by color contrasting structural differences. Alternative instrumentation and the use of other dye systems are also discussed. The combination of epoxy vacuum infiltration with fluorescence microscopy to verify microstructural defects is a powerful means for characterization of advanced materials and for assessment of structural integrity.

INTRODUCTION

The underlying principles and necessary equipment for the performance of the fluorescence microscopic technique have been presented [1] and numerous manufacturers supply fluorescence attachments for microscopes and the accompanying dyes. The original paper does not discuss the theory of how molecules absorb light or fluoresce and how these processes govern the appropriate selection of equipment necessary to

---

[1]   National Aeronautics and Space Administration, Lewis Research Center, MS 6-4, 21000 Brookpark Road, Cleveland, Ohio 44135.
[2]   Sverdrup Technology, Inc., Lewis Research Center Group, 2001 Aerospace Parkway, Brook Park, Ohio 44142.

effectively perform the technique. In this paper some of the fundamentals of spectroscopy, the interaction of light with molecules, and how to select appropriate dyes and associated optical components for optimization of the viewing of a fluorescence microscopy experiment are presented.

THEORY

The process by which molecules absorb energy in the form of light and emit energy in the form of fluoresced light are illustrated in Figure 1, a Jablonski diagram [2]. The

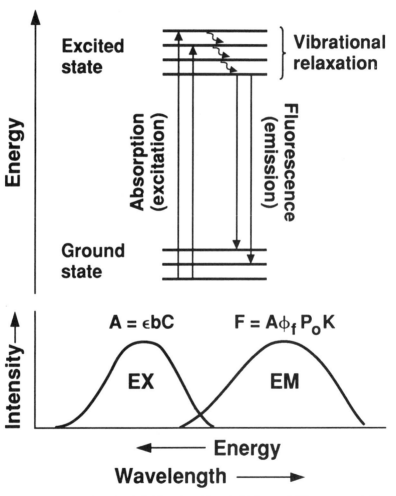

Figure 1. The Process by which Molecules Absorb Light and Fluoresce. Jablonski Diagram of Electronic Transitions (Top). Absorption or Excitation and Fluorescence Emission Spectra (Bottom).

energy diagram is divided into various electronic energy levels, the ground and various excited states, and vibrational levels within the various electronic states. Under normal conditions the electrons capable of absorption of ultraviolet or visible light energy reside in the lowest vibrational energy level of the ground state or, to a small extent by thermal processes, in the second lowest vibrational level of the ground state. Upon absorption of the correct amount of energy, the electron will be promoted to the appropriate vibrational level of an excited state. The molecule will now possess too much energy and immediately begin to dissipate the energy by any number of processes. The first process occurs rapidly and is called vibrational relaxation which is a nonradiative or non-light-emitting process leaving the molecule in the lowest vibrational level of the first excited state. The molecule may then internally convert, another nonradiative process, to the various vibrational levels of the ground state and, subsequently, more vibrational relaxation will return the electron to the lowest vibrational level of the ground state. This process is the primary competing process to fluorescence, which is the return of the electron from the lowest vibrational level of the excited state to the various vibrational levels of the ground state with the accompanying release of light energy or fluorescence. The electron will then return to the lowest vibrational level of the ground state as discussed previously. Because the vibrational levels are so closely spaced in most molecules (with respect to the energy required for the electron transitions between electronic states), the resulting spectrum, a plot of absorbed or fluoresced light intensity versus the energy of the light, resembles broad Gaussian distributions. It is obvious from this diagram that the fluoresced light will generally be of lower energy than the absorbed or excitation light. The absorption (or excitation) spectrum will overlap to a small degree with the fluorescence spectrum due to the possibility of excitation from the second lowest vibrational level of the ground state to the lowest vibrational level of the excited state. Fluorescence then occurs by transition from the lowest vibrational level of the excited state to the lowest vibrational level of the ground state. The relationship of the energy of the light involved in the transitions and the wavelength of the light wave are given according to:

$$E = h\nu \qquad\qquad (1)$$

where E is the energy of the light wave, h is Planck's constant and ν is the frequency of the light wave. The frequency and wavelength of light, λ, are related to the speed of light, c, according to:

$$c = \lambda\nu \qquad\qquad (2)$$

The intensity of fluorescence, F, is given by:

$$F = A\Phi_f P_o K \qquad\qquad (3)$$

71

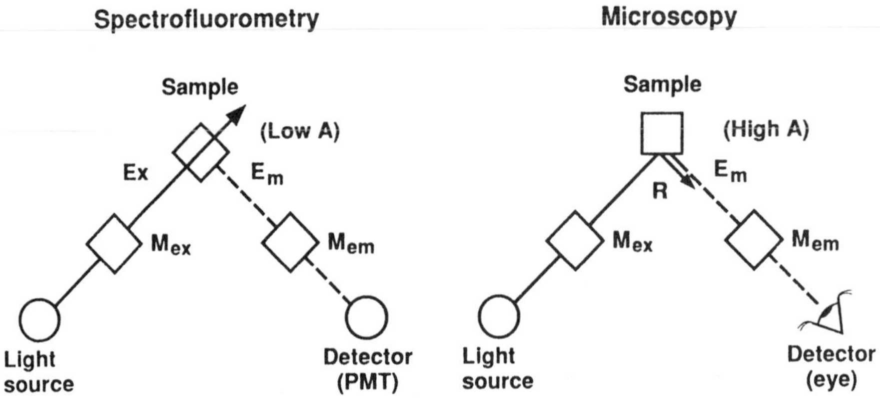

Spectrofluorometry   **Microscopy**

Figure 2. Simplified Schematic Optical Diagram for Observation of Fluorescence. Spectrofluorometer (Left) and Fluorescence Microscope (Right). See text for other details.

where Φ_f is the efficiency of the molecule for conversion of absorbed light into fluoresced light; P_o is the intensity of the exciting light; K is the efficiency of the instrument for collection of the fluorescence; and absorbance, A, is defined by:

$$A = \varepsilon b C \qquad (4)$$

where ε relates to the probability of a molecule absorbing a light wave, b is the depth of the light penetration into the sample, and C is the concentration of absorbing molecules.

The application of fluorescence is done through an instrument constructed such that two specific regions of the spectrum are isolated -- one to probe the sample (absorption) and the other to interrogate it (fluorescence). Two variations of this type of instrument are illustrated in Figure 2. The left frame depicts a fluorometer, an instrument frequently employed by chemists for quantifying trace levels of constituents in samples. The unit consists of a light source and monochromator (M_{ex}) to isolate light capable of being absorbed by molecules of interest within the sample. The light is then allowed to excite the sample. Because of the low concentrations determined, much of the light is not absorbed by the sample according to equation 4. Excited molecules in the sample emit fluorescence in all directions, some of which is isolated by the emission monochromator (M_{em}) and is then measured by a sensitive photodetector, usually a high-sensitivity photomultiplier tube (PMT). The fluorescence microscope, right frame, is remarkably similar with two exceptions. First, relatively high concentrations of the fluorescing dye are employed causing all of the exciting light to be absorbed at the surface of the sample; hence, the b term of equation 4 is small. As the fluorescence is

72

Wavelength	Response	Color
400 NM	0.0004	Violet
420	0.004	Indigo
440	0.023	Blue
460	0.060	Blue
480	0.14	Blue
500	0.32	Green
520	0.71	Green
540	0.95	Green
560 **	1.0 **	** Green
580	0.87	Yellow
600	0.63	Orange
620	0.38	Red
640	0.18	Red
660	0.060	Red
680	0.017	Red
700	0.004	Red

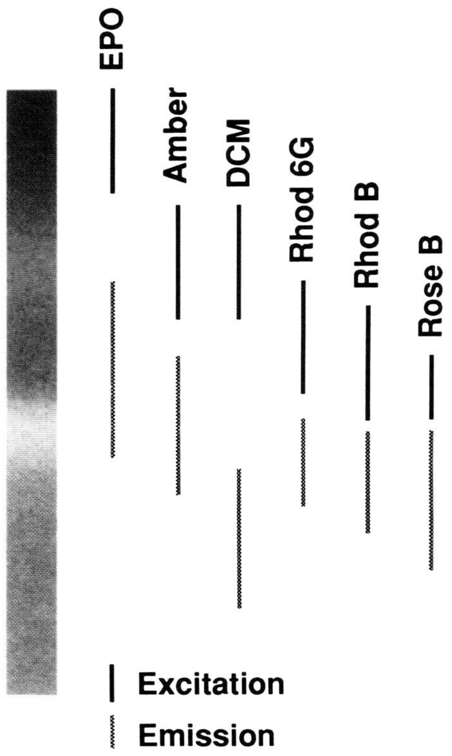

EPO
Amber
DCM
Rhod 6G
Rhod B
Rose B

— Excitation
~~~ Emission

Figure 3. Relative Response of the Cones of the Eye to Light of Various Wavelength (Left) and Spectral Characteristics of Fluorescent Dyes (Right). Note: The response of the eye at 550nm is taken as unity and the response at all other wavelengths is calculated relative to that at 550nm.

proportional to P_o, according to equation 3, then the higher the intensity light source employed, the more fluorescence will be observed for a given concentration of dye. A second major difference is in the method of light detection where the photodetector is replaced by the eye. In this respect, the eye is a superior detector as it will respond not only to light intensity but also to different colors and different hues of the same color. This will provide two-dimensional information and, in some cases, even three-dimensional information regarding the structure of the material under investigation. Drawbacks associated with the use of the eye include rapid fatigue associated with correct light intensity determinations and differing response to different colors as indicated in Figure 3. The eye is considerably more sensitive to green, yellow, and orange than other colors in the visible spectrum; and, hence, the fluorescence experiment is optimized by viewing in this region of the spectrum. The same effect is noted with photographic film, much of which is optimized for response in the green region of the spectrum.

The function of the monochromators is to isolate a single color or, in practice, a single region of the spectrum. Figure 4 shows the two types of filters employed for

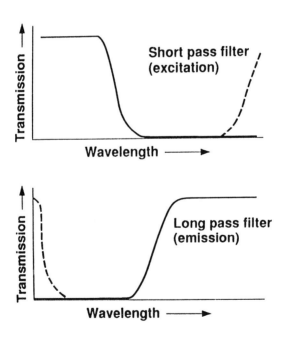

Figure 4. Filter Monochromators used in Fluorescence Instrumentation. Short Pass or Excitation Filter (Top) and Long Pass or Emission Filter (Bottom). Solid lines indicate ideal spectral characteristics and solid plus dashed indicates actual behavior.

isolation of excitation light (short pass filters) and fluorescence light (long pass filters) where the solid lines represent the ideal filter and the combined solid plus dashed curves represent what are actually available. The function of the short pass filter is to allow transmission of those wavelengths which excite sample molecules from the light source to the sample and block transmission of all light having wavelength similar to the fluoresced light. The long pass filter blocks excitation light reflected from the sample and allows transmission of the fluoresced light through to the detector. ie. These two filters are respectively wavelength matched to the absorption or excitation and fluorescence emission characteristics of the fluorescent molecules in the sample. In true fluorescence mode, top frame of Figure 5, the combination of short and long pass filters is such than none of the excitation light is transmitted through to the detector.

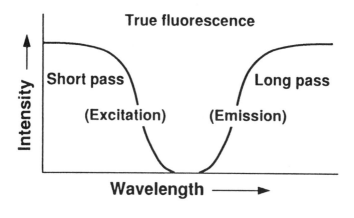

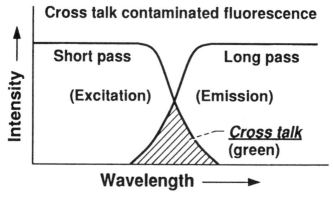

Figure 5. Combinations of Excitation and Emission Filters Employed for Viewing Fluorescence. Short pass transmits blue light and long pass transmits yellow, orange, and red. True fluorescence (Top) will not transmit green whereas "Crosstalk" mode (Lower) transmits substantial green intensity.

In the lower frame, the filter combination is such that a small portion of the excitation light is allowed to pass through to the detector, an artifact known as "crosstalk". The problem with crosstalk in a fluorescence instrument is that the detector is unable to distinguish between the crosstalk and the fluorescence wavelengths. Low concentrations of fluorescent materials in combination with the facts that Φ_f is typically much less than 1 and the collection efficiency of the unit, K, is very much less than 100%, produces fluorescence which can be orders of magnitude smaller than the directly reflected crosstalk. This effect is beneficial in fluorescence microscopy as it allows direct viewing of the specimen at the crosstalk color and simultaneous observation of the fluorescence experiment occurring at a different wavelength; i.e., color, providing that the intensity of light from the two experiments are approximately comparable. One of the commonly overlooked aspects of using filter monochromators is the undesirable characteristic of allowing transmission of light in undesirable regions of the spectrum as indicated by transmission associated with the dashed lines in Figure 4. This effect is frequently minimized or eliminated by virtue of the fact that the light source may not produce light in the region where undesirable transmission would occur; and, in the event that it did get through both the excitation and emission filters, the detector might not be sensitive to that wavelength anyway. For example, the short pass excitation filter given in Figure 4 will also transmit violet which will also transmit through the long pass emission filter. Even though the intensity of the violet may be comparable to the intensity of orange fluorescence transmitted by the emission filter, the eye response (given in Figure 3) would be swamped by the orange intensity. This effect can also be employed to generate useful crosstalk similar to that which was previously discussed in association with Figure 5, lower frame. A combination of the red transmission of the excitation filter in Figure 4 with the solid-line profile of the emission filter would produce a system similar to that illustrated in the top frame of Figure 6. This degree of crosstalk would cause the crosstalk intensity to swamp the fluorescence intensity even though the eye response is poorer to the crosstalk than the fluorescence. First, the fluorescence is typically much lower in intensity due to the nondirectional fluorescence emission versus the direct reflectance of the crosstalk; and, second, a tungsten bulb, for example, will produce higher intensity red output than the blue which is used to excite the sample as given in equation 3. The amount of crosstalk generated by this effect can also be adjusted by using two excitation filters having different transmission and blocking ranges as in Figure 6, lower frame. This modified system consists of the same system as in the upper frame with the addition of a second short pass filter (short pass #2) that begins to block in the region where the undesirable red light begins to be transmitted by the first short pass filter.

In order to efficiently correlate the structural information obtained from the bright-field observation of the specimen to the porosity information obtained from fluorescent viewing, it is necessary to view both modes of observation simultaneously. The use of crosstalk permits simultaneous viewing with both modes of observation. The Bright-field mode appears at the crosstalk color and the fluorescence mode appears at the same color as the fluorescence of the dye incorporated into the infiltrated epoxy.

PROCEDURE

Specimen Preparation. The thermally sprayed tungsten carbide coating (88-12) was first mounted in an epoxy resin by the vacuum infiltration method as in the literature [1]. After curing, the specimens were metallographically prepared by the procedure outlined in Table 1. The specimens were provided by General Electric Aircraft Engine Division. This procedure produced a uniform microstructure which correlated to the

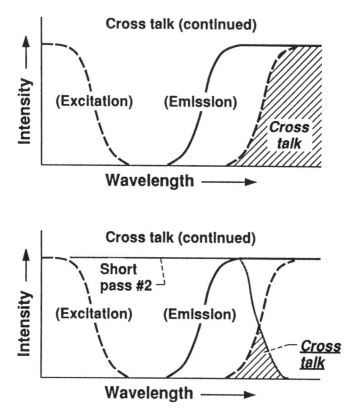

Figure 6. Combinations of Excitation and Emission Filters Which Produce Crosstalk at Longer Wavelengths than Fluorescence Wavelengths. Two filter combination (Top) including actual transmission characteristics of the excitation filter. Three filter combination (Bottom) where the excitation monochromator consists of the actual characteristics of two short pass filters.

processing parameters employed in the manufacturing of the spray coat specimens. Some of the specimens exhibited a large amount of pullout indicating a cold plasma jet and other specimens showed a fully dense microstructure. The specimens were examined by optical microscopy in bright-field and fluorescence methods to determine the microstructure. In the bright-field mode, it was difficult to determine if porosity or pullout was present in the microstructure. Pullout will occur if the metallographic procedure is too aggressive for the given material or, in the case of thermally sprayed coatings, the individual particles will be pulled out because they are weakly bonded together.

The fluorescent method helps distinguish between pullout versus filled porosity but cannot distinguish between isolated, nonconnecting porosity and pullout. The fluorescent method uses color differentiation of the specimen and the fluorescent dye, by the use of filters, to enhance the differences in the microstructure. A systematic study of fluorescent dyes and filtration systems (both commercial systems and systems constructed from commercially available filters) was undertaken in order to enhance viewing conditions such that the bright-field information and fluorescence information could be acquired simultaneously and improve conditions for either color or black and white photography.

Reagents. The dyes and solvents were of reagent quality and used as received. Fluorescent dyes were mixed with a dilute resin to a saturated solution of dye in resin and let stand for 48 hours at which time the resin was decanted into a clean container. The epoxy was blended in a ratio of 6:1 Struers, Inc., Epofix epoxy resin to hardener prior to vacuum infiltration of epoxy into the specimens. The fluorescent dyes examined were Rhodamine B and 6G, Rose Bengal, DCM, Amber Dye, and epodye. (See Figure 3 for spectral characteristics of dyes. Solid vertical lines indicate the wavelength range where dye molecules absorb light with high efficiency and hatched lines indicate spectral ranges where they fluoresce with high efficiency.) For spectrophotometric determinations, the appropriate amount of dye-containing resin was diluted to read on scale with ethanol and the spectrum was obtained in 1 cm quartz cells in the normal manner. The spectrophotometer was blanked with an appropriate ethanolic resin mixture. The absorption samples were further diluted 1:100 with ethanol for reading in the spectrofluorometer in 1 cm rectangular quartz cells. The

---

Table 1. Metallographic Preparation of the Thermally Sprayed Tungsten Carbide Cobalt Coatings.

---

1   Mount in fluorescent epoxy (epofix-two part epoxy) and vacuum infiltrated

2   Metallographic preparation (automated):
      150 grit Al_2O_3 until planar
      6 micron diamond on a petro disc M for 9 minutes
      6 and 3 micron diamond on a synthetic hard cloth for 1 minute each
      0.05 micron colloidal silica plus 10% V:V of 30% hydrogen peroxide on a
      chemically resistant cloth for 1 minute
      (All polishing at 150 rpm with 200 N force)

78

fluorescence for an appropriate blank was used to correct spectral data for the dyes examined.

Instrumentation. The Struers abraplan was used for initial grinding with 150 grit alumina wheel, and all other preparation step were done with Struers abrapol. The microscope used for all of the specimen examination and photographic work (except the true fluorescence figure) was the Reighert-Jung MeF$_3$ equipped with both tungsten and 450 W high-intensity xenon sources. All figures presented herein were taken under

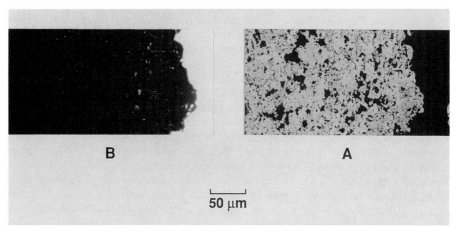

B   A

50 μm

Figure 7. Bright-Field Viewing (Right) and True Fluorescence Viewing (Left) of Specimen Mounted in Epodye/Epoxy Using LECO Filter System.

xenon lighting with either the 4 x 5 Polaroid 545 Land film holder or FT-1 Konica 35 mm camera using Polaroid 53 black and white or 59 color and 35 mm Kodak Ektar 125 film respectively. This unit was used with either the B5 fluorescence module (red filter) system or a similar module block was fitted with excitation filters of 520, 540, or 560nm and emission filters of 520, 540, or 560nm. The microscope used to obtain the true fluorescence photographs was the Olympus PMG3 equipped with a 150 W high-intensity xenon arc light source and PMG3 DMB blue excitation filter.

All absorption spectra were recorded on a Shimadzu model UV-160 spectrophotometer, operated in the normal fashion. All fluorescence spectra were recorded on a Perkin Elmer, Model MPF-44B, spectrofluorometer which was operated in the normal manner.

RESULTS AND DISCUSSION

The approach taken in this work was to systematically examine various combinations of excitation and emission filter systems (including the degree of direct crosstalk allowed between the filters) while also varying the region of spectrum where the fluorescent dyes excite and fluoresce in order to find the optimum system for simultaneous direct and fluorescent viewing of the sample. It is fortunate that the response characteristics of the eye parallel those of the color and black and white film employed such that optimization of direct viewing would also optimize conditions for photographically recording the work. Figure 3 also includes the approximate wavelength ranges for excitation and fluorescence observation of the dyes surveyed. Epodye is commercially available, moderately inexpensive, excites and fluoresces in a good spectral region but, more importantly, is relative soluble in the epoxy resin and is chemically stable in that medium. Figure 7 is the direct view (right) and true fluorescence viewing in black and white (left) of a specimen mounted in epodye-epoxy With a filtration system similar to the one depicted in Figure 5 (top). It is evident from the bright-field viewing that the sample contains a great deal of porosity throughout from left to right; however, the fluorescence viewing indicates that a good deal of the left porosity is pullout resulting from metallographic preparation. Furthermore this system indicates the need for simultaneous viewing of both experiments in order to easily correlate pullout regions of the specimen versus true porosity regions. Figure 8 shows the direct view and fluorescence views of a similar specimen preparation using the two crosstalk modes discussed in association with Figure 6 (lower), representing viewing with a B5 module (red filter system), and Figure 5 (lower). With epodye, the combination of the blue excitation/yellow emission (green crosstalk and later referred to as the green filter system) is a superior fluorescence viewing in that the details of the bright-field experiment are considerably more apparent under both visual inspection and color photography. Unfortunately, the details of the direct viewing of the specimen are not as clear in the fluorescence view as they appear in the bright-field view.

In terms of the less expensive, and easier to use black and white photographic recording, the blue/yellow filter system is also superior to the B5 module as demonstrated in Figure 9. The color view of fluorescence provides an indication of the degree of true porosity versus pullout and is unambiguous as compared to bright-field viewing. The black and white of the fluorescence also demonstrates the effect in that the bright-field information appears in gray, the pullout is in black, and the porosity is in white eg. lower right corner of the frames. The same effect is noted when employing different dyes as in Figure 10 where the epodye is replaced with Rhodamine B. The use of Rhodamine 6G produced unusual results in that the color photographs come out with more contrast using the B5 or red module but the black and white photographs of fluorescence come out better using the green filter system. This effect may be due to subtle differences in the sensitivity of the different films to different regions of the spectrum.

We have examined specimens prepared in all of the dyes listed in Figure 3, and it appears that all dyes are suitable with the filtration systems employed but with varying degrees of acceptability. This is not totally unexpected in that the dyes have broad excitation and emission ranges with respect to the ranges covered by the excitation and emission filters investigated. Other chemistry problems hamper the use of the various dyes. The dye DCM is moderately soluble in the resin and retains spectral quality with respect to excitation and fluorescence; but when hardener, a chemical base, is added, the dye molecule is deprotonated and unacceptable changes occur in both the excitation and emission spectra. Rose Bengal and the rhodamines are not as soluble as some of

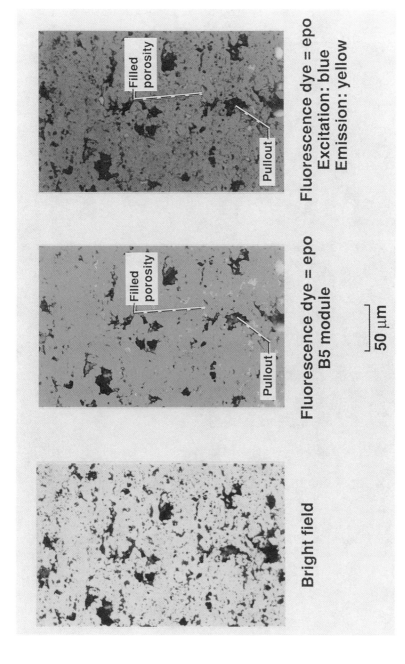

Figure 8. Bright-Field (Left) and Fluorescence (Center and Right) Viewing of Specimen.

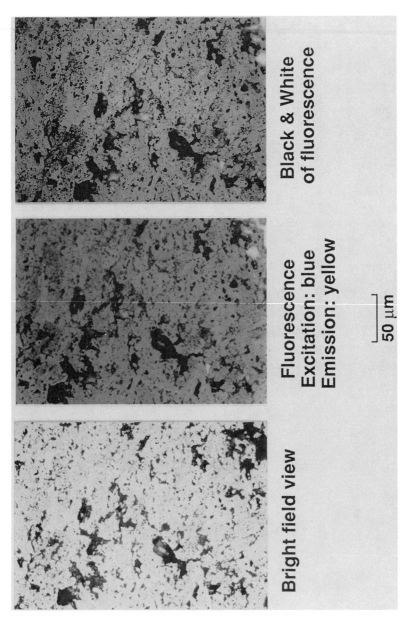

Figure 9. Bright-Field (Left) and Fluorescence (Center and Right) Viewing of Specimen mounted in Epodye/Epoxy.

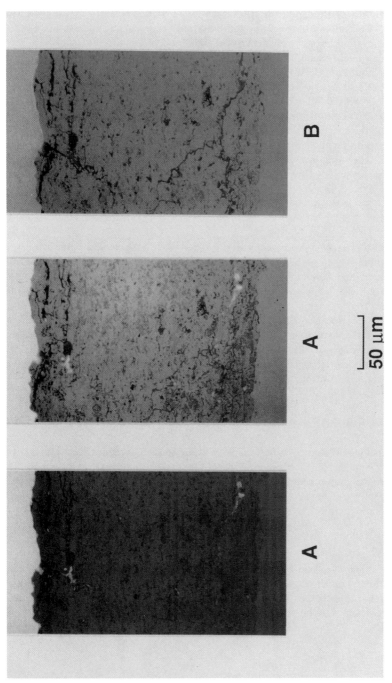

50 μm

Figure 10. Fluorescence View using Green Filter System (Left and Center) and B5 Module (Right) with Specimen Mounted in Rhodamine B/Epoxy. Center is Black and White Photograph of Fluorescence.

the other dyes resulting in lower florescence levels than anticipated. In an attempt to improve solubility of rhodamine B, the acetate salt was synthesized to replace the chloride salt. This effort improved the solubility only slightly, and the acetate altered the pH of the system once the hardener was added.

The final problem encountered in this study was the effect of prolonged exposure of the dyes to high intensity light. The tungsten source was of lower intensity and the effect of photobleaching, the loss of fluorescence, was unnoticed using this light source. The high intensity (450 W) xenon caused rapid, irreversible bleaching of all dyes. For example, after three minutes exposure to the high intensity xenon, the amber dye lost half of its fluorescence intensity, and no fluorescence was observed after six minutes. Of all dyes tested, the epodye appeared to be least susceptible to this process. On the other hand, the tungsten source was not intense enough to take high-quality photomicrographs with either filter system, especially the green system when using color film.

CONCLUSION

The use of fluorescence microscopy in the crosstalk mode provides a powerful tool for the identification of porosity, cracking, and debonding versus artifacts, such as pullout, cracking, and debonding occurring during mounting and polishing. Two modes of crosstalk are identified, and the crosstalk appearing at shorter wavelength than the fluorescence of the mounting dye appears to be the most promising for simultaneous direct viewing, fluorescence viewing, and/or photographic recording (either color or black and white). Because these detectors have similar response curves, the optimal viewing conditions consist of mounting in an epoxy with a dye that excites in the blue and fluoresces in the yellow combined with a filtration system constructed of a blue short pass filter, a yellow long pass filter, and such that crosstalk appears in the green region of the spectrum.

ACKNOWLEDGEMENT

The authors would like to express their gratitude to the Cleveland State University Chemistry Department for their generous use of the spectrofluorometer and their NASA colleagues for their support and collaboration on this research. The thermally sprayed coating samples were courtesy of Walter Riggs of General Electric Aircraft Engines Division, Evendale, Ohio.

REFERENCES

1) J.H. Richardson, **Optical Microscopy for the Materials Sciences**, Marcel Dekker, Inc., NY, p 165 and p 261, 1971
2) G.H. Schenk, **Absorption of Light and Ultraviolet Radiation: fluorescence and phosphorescence emission**, Allyn and Bacon, Inc., Boston, p 154, 1973

SCANNING ACOUSTIC MICROSCOPY OF ENGINEERED MATERIALS

B. R.Tittmann, R. N. Pangborn, and R. P. McNitt
Engineering Science and Mechanics
The Pennsylvania State University
University Park, Pa. 16802

ABSTRACT

Use of Scanning acoustic microscopy (SAM) in the characterization of several types of engineered materials is described. The materials include polymer, metal, and ceramic matrix composites, vapor deposited thin films, and adhesive bonds. The ability of SAM to provide high resolution (1μm) images of surface and subsurface regions has been exploited to give insight into the processing quality of the engineered materials. Selected examples of SAM images are presented and discussed.

INTRODUCTION

Engineered materials such as composites, coating, films, and bonded structures are emerging as one of the most important groups of materials in contemporary technology. Applications for jet engines, hypersonic aircraft structures and space components require significantly higher strength-to-weight ratios and higher temperature exposures and conventional monolithic materials are frequently unable to meet all the requirements. The use of fiber reinforced composites, particle reinforced composites, and coated materials has made some of these applications possible. Composites are also finding wide acceptance in the automotive industry for structural and engine component applications.

Scanning Acoustic Microscopy is emerging as an important new diagnostic tool, allowing high resolution imaging of subsurface regions, otherwise not accessible by conventional techniques such as optical and scanning electron (SEM) microscopy [1-7]. SAM is accomplished by the use of an acoustic lens which transmits an ultrasonic tome burst, focussed into the material. The beam is mechanically scanned on a plane parallel to the surface of the specimen. The acoustic energy back reflected by the material in the focus zone is recaptured by the lens and sent to a receiver for signal processing and storage. As the beam is scanned the signal strength of the reflected signal varies with the mechanical properties (such as density, elastic moduli, acoustic attenuation) of the material. A significant interrogation parameter is the frequency of the acoustic energy, typically chosen from two ranges, depending on the application: low frequencies (15-100MHz), high frequencies (100MHz-5GHz). Typical applications for the low frequency range include characterization of deep layers such as diffusion bonds, electronic packaging, solder bands, plastic encapsulation, cracks, delaminations, etc. Typical applications for high frequency operation include examination of grain structure, film adhesion, interfaces between the reinforcement material and the matrix, porosity, microfractures, short fiber distribution, anisotropy, etc. An important consideration is the trade-off between resolution and depth penetration. The higher the frequency, the higher the resolution...but the smaller is the depth to which the acoustic energy can penetrate. Depth penetration is in part limited by the ultrasonic attenuation which for many materials increases with the square of the frequency and therefore reduces the signal-to-noise ratio. Another consideration is the choice of acoustic propagation mode. Different lens designs emphasize different types of ultrasonic waves, such as Rayleigh, longitudinal, shear, pseudo-longitudinal. Depending on the objectives of the diagnostic examination, specific modes may be selected to emphasize a particular feature.

Samples are typically polished to optical flatness and smoothness. Whenever sample size is less than 1cm x 1cm, the sample is mounted in an epoxy mold prior to polishing. An important procedure is to adjust the stage tilt so that the scanning plane is parallel to the sample surface. With the Olympus UH-3, which was used for the work described here, tilt adjustment was accomplished iteratively by manual adjustment and the UH-3 interactive software package. The acoustic coupling between the lens and the sample was achieved via a droplet of deionized water administered with a syringe. The lens was first focussed on the surface and then adjusted from there to the desired depth. A special feature of the UH-3 is the ability to superimpose three images, while assigning to each a different color. Thus by choosing a different depth for each of the images, a "composite" image is obtained which presents a quasi-3D view of the material under examination.

RESULTS and DISCUSSION

1) *Diamond Thin Film.* Fig 1 is an image of a diamond thin fllm (7μm thick) on a substrate of Si_3N_4-Al_2O_3 (Sialon) in the vicinity of a diamond indentation. This image was obtained with a 400MHz lens, having a 60 degree

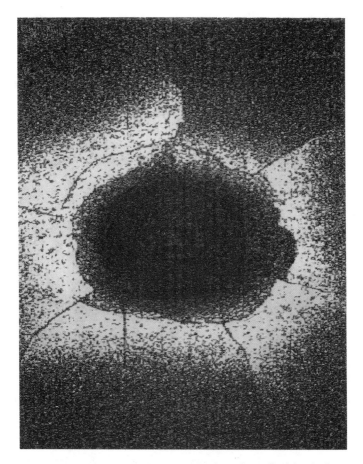

FIG 1   Diamond thin film on Sialon, cracked and spalled by indentation.
400MHz, 500μm scan width

half aperature angle.   With most materials this numerical aperature is
sufficient to produce acoustic rays at the Rayleigh critical angle; that is
acoustic surface waves are produced.   However, because of the high velocity
of sound in diamond (VL= 13.5km/sec) this criterion is not met and thus the
primary mode of wave propagation is that of longitudinal waves.   The figure
shows a color scale at the top (grey scale here), which relates the relative
reflected acoustic amplitude to the tint, with white representing the highest
relative amplitude (240 in arbitrary units) and black (0) representing the
lowest relative amplitude.   The width of the color image is equal to the scan
width (500μm for this figure).   The blue elliptical area, in the center of the
image, represents the area of the indentation where the diamond thin film

spalled away and the Sialon substrate lies exposed. Circumferential and radial cracks resulting from the indentation are seen. Beyond the crack tips one sees the unperturbed diamond film, which is mottled because of its fine gained ($\approx$ 5 μm) polycrystalline structure. The most unique feature is the white area or halo surrounding the indentation, representing a zone in which the film/substrate bond is broken or weakened, so that the ultrasonic energy cannot penetrate into the substrate where for the most part it would be absorbed. Instead, the energy is reflected from the upper and lower surfaces of the film, giving rise to an interference phenomenon producing sharp nulls and high peaks. To obtain this bright halo-image, the transducer was slightly defocussed; i.e. focussed above the film surface to take advantage of the interference peak. The appearance of the halo is important to the materials engineer as the total damage area can now be calculated and related to key process parameters in the film growth process.

2) *Carbon-Carbon Composite.* Fig 2 shows a cross-section of a carbon-carbon composite (CCC) after first carbonization. This material consists of a 2-D woven fiber fabric infiltrated with phenolic resin that is cured, repeatedly carbonized, and finally graphitized to convert the component into pure carbon. Between carbonizations the structure is re-infiltrated with resin in an attempt to fill cracks and voids created by the removal of all the volatiles during the high temperature disintegration of the resin. In the end, the structure is a light-weight, fully dense, and strong material for high temperature applications such as space shuttle leading edges, rocket inlet nozzles and exicones, aircraft brakes, etc. A critical step in the processing of CCC is the first carbonization during which microcracking can lead to delamination (which is the leading cause of CCC component failure). The SAM is an ideal instrument for trying to understand the process by which intra-laminar or cross ply cracks lead to inter-laminar cracks or delaminations. Fig 2 shows a 400MHz image with a scan width of 500μm. Visible are the dark strands of fibers ($\approx$3,000 fibers per strand) parallel to the image plane and the lighter cross sections of fibers perpendicular to the plane. Quite noticeable are the cross-ply cracks which form dark sinuous lines between strands. In optical microscopy, the contrast between the parallel and perpendicular strands is so great that the inter-laminar cracks are not visable. It is interesting to study the relationship between the corss-ply and inter-ply cracks. The image actually shows such an example in the middle of the left hand portion where a cross-ply crack appears to bifurcate and then reaches the strand, whereupon it appears to lead into an interlaminar crack going in both directions for a total length of about two to three lamina widths. Modelling for stress distribution associated with crack propagation during carbonization is well underway. Sensors, such as acoustic emission transducers are being developed to monitor the microcracking. Considerable progress [x,y] has been made in

FIG 2   Carbon-Carbon composite after first carbonizations
400MHz, 500μm scan width

recognizing delaminations and even precursors to delaminations.

    3) *Ceramic Particle/Metal Matrix Composite.* Fig 3 shows a polished surface of a sample consisting of an aluminum matrix and alumina particles acting as reinforcement. The automotive industry is interested in such materials for lighter, cooler, engine materials. The Al alloy matrix provides rapid and efficient heat conduction, whereas reinforcements consisting of

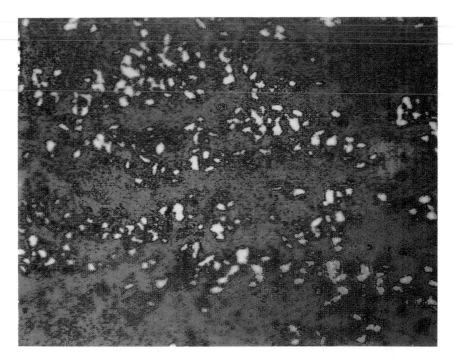

FIG 3   Ceramic Particle/Metal Matrix (Alumina/Aluminum) composite
showing particles, voids, and interfaces
400MHz, 500μm scan width

Al_2O_3 ceramic particles or fibers provide the strength necessary to absorb the thermal-mechanical shocks due to the explosions in the cylinders. Because of the similarity in optical reflectivity, the reinforcing particles are not easy to distinguish from the matrix and studies of the particle distribution, the presence of voids, the nature of the interface, are not readily feasible.   In contrast, the acoustic impedances (i.e. product of density and acoustic velocity) are significantly different for these entities and acoustic microscopy can provide vivid images with good contrast.   Fig. 3 readily shows the Al alloy matrix (red in color) the alumina particles (light, white and yellow), voids(dark, blue and black) and the interfaces in green. The images have sufficient contrast to permit statistical processing with a computerized image analyzer to provide average particle size and distribution for example.

For use in automobiles, low cost is crucial both in terms of ceramic preform fabrication and metal infiltration.   For the inexpensive discontinuous alumina fiber preforms that typically exhibit rather poorly defined fiber architectures, quality control through inspection is particularly important. Nondestructive inspection techniques which could detect flaws such as inclusions, large scale pores or internal cavities and substantial deviation from the nominal volume fraction of fibers, would allow defective preforms to be screened out of production before further investment. Alumina particles (shot) with sizes two to three orders of magnitude larger than the average fiber diameter serve as stress raisers that lead to thermal fatigue driven crack initiation. Cavities can lead to preform crushing during infiltration or unreinforced regions in the final composite. The SAM can be used to detect the presence of such flaws: Fig 4 shows a shot particle (an unfiberized inclusion of aluminum) embeddded in a near-surface layer, while Fig 5 illustrates a sizable cavity in the preform; both images taken with a 400 MHz lens. Flaws at deeper sites in the preform may be detected with lower frequency lenses, albeit with somewhat reduced resolution in the highly attenuative material.

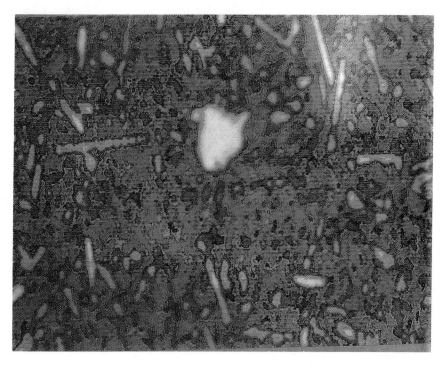

FIG 4   "Shot" of alumina embedded in near surface area
400MHz, 500μm scan width

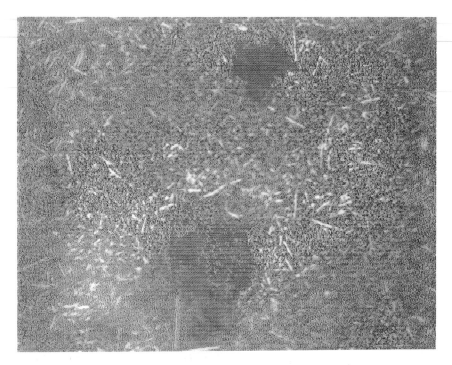

FIG 5   Sizable Cavity in preform
        2,000 MHz, 500µm scan width

4) *Epoxy matrix Composites.* The development of short glass fiber reinforced thermoplastics for injection molding of large parts presents a number of distinct advantages such as relatively low cost and ease in fabricating complex shapes [10]. Fiber incorporation enhances properties such as stiffness (increases) and thermal expansion (decreases) when compared to the unreinforced polymer, however some loss of ductility and resistance to fatigue damage usually occurs. In order to better understand the role of the fibers in fatigue crack initiation and propagation, it is most helpful to have a technique which will image the bulk material, not just the surface. An non destructive technique would also be desirable for studies of progressive or cumulative fatigue damage. The SAM is well suited for this task. Figures 6a and 6b show scanning acoustic micrographs of an injection molded composite prior to, and after fatigue cycling, respectively. Fig 6b discloses evidence of debonding around the fiber, a precursor to fiber pullout typical of the failure of short-fiber composites.

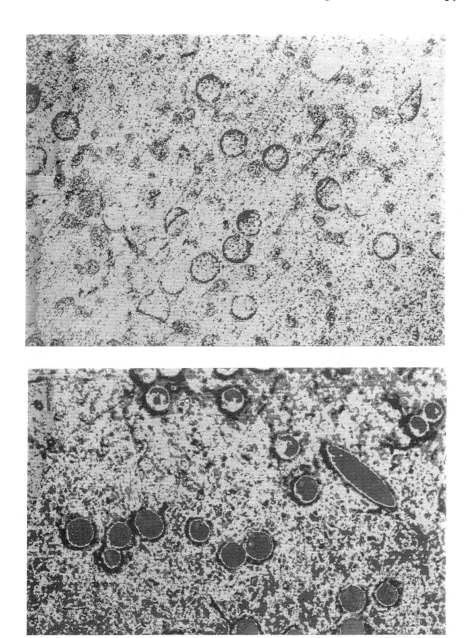

FIGURES 6a, 6b "Before and After" micrograpshs of an injection molded
     composite prior to and after subjecting specimen to fatigue cycling.
     Fig 6b shows debonding.
     400MHz, 500μm scan width

CLOSURE

A brief review of applications of SAM on engineered materials at an academic institution is given. SAM appears to be useful for the study of many aspects of current material processing with emphasis on thin films, adhesive bonds, and interfaces in composite materials. Although all SAM images are not necessarily quantitative, through the experienced eye of the materials engineer, they provide a view of the microstructure which is complementary to the SEM and the optical microscope. The key features that make this technique unique are that the acoustic waves can interrogate the subsurface regions and that contrast is derived from the differences in the acoustic impedances of the constituents. As might be expected, there is a trade-off between resolution and depth of penetration, i.e. higher acoustic frequencies provide finer detail but suffer from higher attenuation (thus shallower penetration).

REFERENCES

[1]   S. Sokolov, "The Ultrasonic Microscope", Adademia Nauk SSR, Doklady, 64, 333 (1940)

[2]   F. Dunn and W. J. Fry, "Ultrasonic Absorption Microscope", J. Acoust. Soc. Am, 31, 632 (1959)

[3]   R. A. Lemon and C. F. Quate, "Acoustic Microscope-Scanning Version", Appl. Phys. Letters 24, 163 (1974)

[4]   A. Atalar, C. F. Quate, and H. K. Wickramasinghe, "Phase Imaging in Reflection with the Acoustic Microscope",Appl. Phys. Lttrs., 31,791, (1977)

[5]   A. Briggs, "An Introduction to Scanning Acoustic Microscopy", New York:Oxford 1985

[6]   M. G. Somekh, H. L. Bertoni, G. A. D. Briggs and N. J. Burton, Proc. R. Soc London, A401, 29 (1985)

[7]   K. Yamanaka, Y. Nagata, and T. Koda, "Low Temperature Acoustic Microscopy with continuous temperature control", Ultrasonics Inter. Proc., 744 (1989)

[8]   B. R. Tittmann, "Acoustical Studies of Damage Mechanisms in Carbon-Carbon During First Carbonization", IEEE Ultrasonics Sump. Proc. (B. R. McAvoy Ed.) Cat. No. 89CH2791-2, 627 (1989)

[9]   B. R. Tittmann, "Ultrasonic Sensors for Process Monitoring and Control of Carbon/Carbon Pyrolysis", Rev. of Progress in Quantitative Nondestructive Evaluation (edited by D. O. Thompson and D. E. Chimenti) Plenum Press, New York, Vol 10A, 1119, (1991)

[10]   B. L. Petereson, R. N. Pangborn and C. G. Pantano, "Static and High Strain Rate Response of a Glass Reinforced Thermoplastic", J. Comp. Materials (in press (1991)

A SECONDARY ION MASS SPECTROMETRY (SIMS)

ANALYSIS OF THE MICROSTRUCTURE OF A PLAIN

CARBON STEEL AND A 12% Cr STEEL

Ashok Choudhury [1] and Charlie R. Brooks [2]

ABSTRACT

A 0.47% C, annealed steel, with a microstructure of primary ferrite and pearlite, has been analyzed using SIMS. The purpose was to determine which sputtered mass best represents carbon, which was accomplished by imaging the surface with $^{12}C^-$ and $^{24}C^-$ ions. The latter proved to give the best correlation with the image and the known location of carbon (i.e., in the pearlite). However, the cratering induced by the continued sputtering during analysis masked a clear correlation with the microstructure.

The microstructure of a 12% Cr (about 0.2% C, 1% Ni, 1% Mo, 1.5% W) steel bolt removed from a fossil-fired power plant after about 30 years service, and which was seriously

[1] Metals and Ceramics Division, Oak Ridge National Laboratory, Oak Ridge, Tennessee 37831.

[2] Materials Science and Engineering Department and Materials Processing Center, The University of Tennessee, Knoxville, Tennessee 37996.

embrittled, was examined using SIMS. The microstructure was known to consist mainly of Cr-rich $M_{23}C_6$ carbides in ferrite, with a high concentration of particles at the prior austenite grain boundaries. However, particles of Laves phase and a Ni-rich phase had also been found, and SIMS was used to try to locate these particles in the microstructure. The $^{24}C^-$ and $^{52}Cr^+$ images were consistent with a high concentration of Cr-rich carbides at the grain boundaries. The signal corresponding to Ni, Mo and W was too weak to discriminate their location in the microstructure. There was a relatively intense signal of mass 51, corresponding to V, even though this element was not reported in the chemical analysis. Imaging with this mass showed that V was concentrated with Cr in particles inside of the grains, but which had no carbon; these were probably oxides.

INTRODUCTION

Secondary ion mass spectrometry (SIMS) has become a recognized tool for microstructural analysis (1, 2). In this technique, a source of primary ions is focused onto the surface of a sample, causing sputtering of atoms from the surface and their ionization. These atoms form a plethora of ionized complexes, both negatively and positively charged, and these secondary ions are passed through an energy analyzer and a mass spectrometer to obtain an energy distribution spectrum based on mass and charge. In principle, all elements are detectable, and detection limits are in the ppmm range.

Carbon is an element the location of which is frequently sought in microstructural analysis. However, in SIMS the sputtered carbon may form many complexes, such as the dimmer $^{24}C^-$ and hydrogen complexes (e.g., CH), and other complexes may have identical mass to these. Thus in attempting to detect and locate carbon in a microstructure, it is necessary to know which mass is suitable to use. One purpose of the present study was to conduct a SIMS analysis of an annealed plain carbon steel of known microstructure, and hence of carbon location, and determine which mass to use to image the carbon.

A related problem is the effect of the surface topology on the carbon analysis. The analysis process intrinsically involves a continued sputtering of atoms from the surface, so that a crater is formed. We have determined the carbon image as a function of sputtering time, and have examined the crater morphology after the SIMS analysis.

We also report a SIMS analysis of the microstructure of an 12% Cr steel which was embrittled after 30 years service life in a fossil-fired power plant.

EXPERIMENTAL PROCEDURE

Materials:

A 0.47% C plain carbon steel was annealed to produce a microstructure of primary ferrite and pearlite, and hence of known carbon location. Samples were given standard metallographic preparation, then lightly etched. A secondary electron detector on the SIMS system allowed imaging the surface structure.

Samples from a 12% Cr steel bolt which had been in service for 30 years in a fossil-fired power plant were examined. The steel contained 0.24% C, 12% Cr, 1% Ni, 1% Mo and 1.5% W. The heat treatment of the bolt prior to installation involved austenitizing followed by quenching toform martensite, then tempering at a temperature above the subsequent service temperature. The original microstructure probably consisted of a uniform distribution of $M_{23}C_6$ Cr-rich carbides in ferrite. During service, the steel became embrittled, and fracture occurred along the prior austenite grain boundaries. The microstructure consisted mainly of a distribution of

$M_{23}C_6$ Cr-rich carbides in ferrite, but particles of Laves phase and of a Ni-rich phase of undetermined structure were also found. A microstructural analysis has been given by Brooks and Zhou. (3).

SIMS analysis:

A V.G. quadrupole based SIMSLAB was used in this study. A rastererd 25 KV, 0.5 nA Ga primary beam was used. Positive and negative secondary ions were monitored as necessary.

RESULTS AND DISCUSSION

Plain carbon steel:

The microstructure of the plain carbon steel is shown in Fig. 1. The location of carbon is easily distinguished to be restricted to the Fe_3C containing pearlite; the primary ferrite contains little carbon (e.g., 0.02 wt.%).

Typical spectra from an area of about 0.0075 x 0.0075 cm (encompassing the primary ferrite and the pearlite) are shown in Fig. 2. In the negative mass spectrum, the strong peaks are at mass 16, due to O, and 24. Note that the intensity at mass 12 is relatively low. In the positive mass spectrum, the strong $^{23}Na^+$ and $^{30}K^+$ peaks are due to the ubiquitous presence of these elements and to their high secondary ion yields. It is suspected that the $^{27}Al^+$ peak comes from residual Al_2O_3 from polishing the sample. Iron is clearly identified by $^{56}Fe^+$.

Imaging the surface using mass 12 gave no indication of carbon segregation. However, using mass 24 the micrograph in Fig. 3a was obtained. The corresponding electron image is shown Fig. 3b. There is clear segregation of carbon. Fig. 4a shows the same area imaged with mass $^{56}Fe^+$, and Fig. 4b shows the corresponding electron image. Note the inverse intensity of C and Fe in Figs. 3a and 4a.

The electron image resolution of the SIMS system was not adequate to clearly reveal the pearlite and primary ferrite. Upon completion of the analysis, the sample was examined with a scanning electron microscope (SEM), and the same area was found from which the images in Figs. 3 and 4 were obtained. The SEM image is shown in Fig.

5. The sputtering produced a marked crater with a very uneven surface topology. The primary ferrite and pearlite are readily discernable on the adjacent surface which has undergone little sputtering, but in the crater the specific details of the structure are masked by faceting. This is the surface topology from which the images in Figs. 3 and 4 were obtained. It is seen that due to this it is difficult to correlate the location of the primary ferrite and pearlite with the carbon distribution (Figs. 3a).

   12% Cr Steel:
   The microstructure of the as-received 12% Cr steel is shown in Fig. 6. There is a high density of particles in the grains, and the prior austenite grain boundaries have a higher density of particles. Spectra typical of those from a large area of the sample of the 12% Cr steel are shown in Fig. 7. Based on the results of the study of the plain carbon steel, the surface was imaged using the mass 24 peak (C_2^-). The image is shown in Fig. 8a, and the electron image of the same area is in Fig. 8b. The bright lines clearly correspond to the location of the prior austenite grain boundaries, and hence it is assumed that these boundaries are continuously lined with high carbon particles. Fig. 9 shows the image using mass $^{52}Cr^+$. The grain boundaries are clearly delineated, and hence the particles are assumed to be Cr-rich carbides. Note in Fig. 9a that there are a few particles inside the grains which are rich in Cr, but not in carbon (Fig. 8a). These may be Cr oxide.

   The location of the Laves phase (rich in W and Mo) and the Nickel-rich phase particles was to be sought by imaging the appropriate mass for these elements. However, the intensity of Ni, W and Mo was too weak to detect any segregation. In the spectrum, though, V showed a strong peak, even though it was not reported to be present by chemical analysis (3). Fig. 10 shows that V is concentrated in the relatively large particles in the matrix which contained Cr (Fig. 9a).

CONCLUSIONS

   The results show that in steels the best ion to use to image carbon is negative 24. However, attempting to build up an increasingly intense signal by continued sputtering will not necessarily prove useful because the sputtering produces an uneven surface topology which masks the true structure. Such an effect has been reported and discussed by McPhail et al. (4).

   The SIMS examination proved useful in locating Cr and C in the 12% Cr steel. However, the spatial resolution and sensitivity (intensity) for Ni, W and Mo were not sufficient to locate particles rich in these elements.

intense signal by continued sputtering will not necessarily prove useful because the sputtering produces an uneven surface topology which masks the true structure. Such an effect has been reported and discussed by McPhail et al. (4).

The SIMS examination proved useful in locating Cr and C in the 12% Cr steel. However, the spatial resolution and sensitivity (intensity) for Ni, W and Mo were not sufficient to locate particles rich in these elements.

ACKNOWLEDGEMENT

Dr. Y. C. Lin prepared the samples. The research was sponsored in part by the U. S. Department of Energy, Assistant Secretary for Conservation and Renewable Energy, Office of Transportation Technolgies, as part of the High Temperature Materials Laboratory User Program, under Contract DE-AC05-840R21400, managed by Martin Marietta Energy Systems, Inc.

REFERENCES

1) R.G. Wilson, F.A. Stevie and C.W. Magee, **Secondary Ion Mass Spectrometry, A Handbook for Depth Profiling and Bulk Impurity Analysis**, John Wiley & Sons, New York, 1989.

2) A. Benninghoven, F.G. Rudenauer and H.W. Werner, **Secondary Ion Mass Spectrometry, Basic Concepts, Instrumental Aspects, Applications and Trends**, John wiley & Sons, New York, 1987.

3) C. R. Brooks and J.-P. Zhou, "Microstructural Analysis of an Embrittled 422 Stainless Steel Stud Bolt after Approximately 30 years Service in a Fossil Power Plant", **Metallography**, Vol. 23, p. 27 (1989).

4) D. S. McPhail, E. A. Clark, J. B. Clegg, M. G. Dowsett, J. P. Gold, G. D. T. Spiller and and D. Sykes, "The Depth Resolution of Secondary Ion Mass Spectrometers: A Critical Review", **Scanning Microscopy**, Vol. 2, p. 639 (1988).

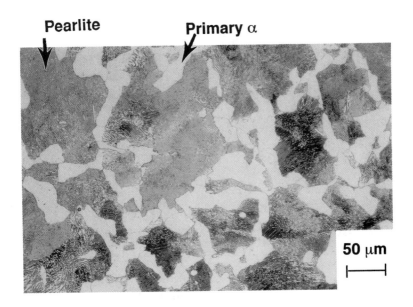

Figure 1.   Optical Micrograph of the Plain Carbon Steel.

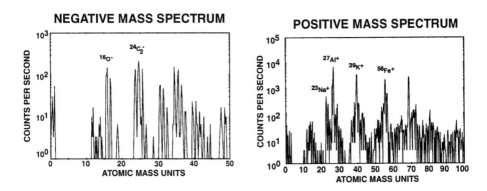

Figure 2.   A Positive and a Negative Mass Spectrum from a Large
             Area of the Surface of the Plain Carbon Steel.

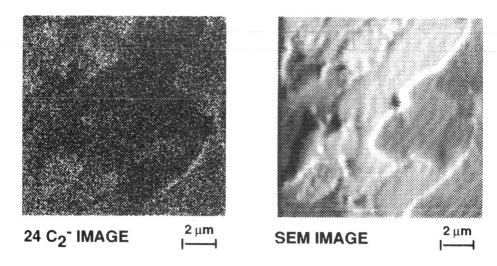

24 C$_2^-$ IMAGE   2 μm   **SEM IMAGE**   2 μm

Figure 3.   A ^{24}C$_2^-$ Image and the Corresponding Secondary
Electron Image from the Plain Carbon Steel.

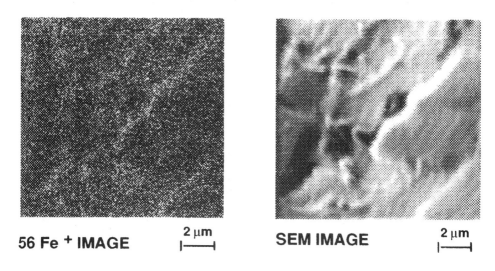

56 Fe$^+$ IMAGE   2 μm   **SEM IMAGE**   2 μm

Figure 4.   A ^{56}Fe$^+$ Image and the Corresponding Secondary
Electron Image from the Plain Carbon Steel. This is the
Same Location as that in Fig. 3.

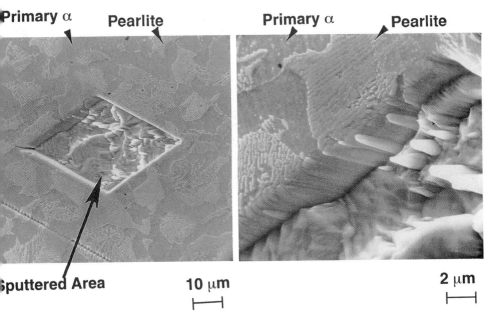

Figure 5.   Scanning Electron Micrograph of the Region of the
Images in Figs. 3 and 4.

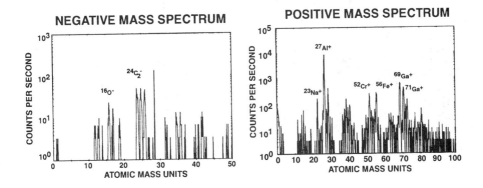

Figure 6.   A Positive and a Negative Mass Spectrum from a Large
Area of the Surface of the 12% Cr Steel.

25 μm   **Grain Boundary**   **2 μm**
                                        Particles

Figure 7.   Micrographs Showing the Structure of the 12% Cr Steel.

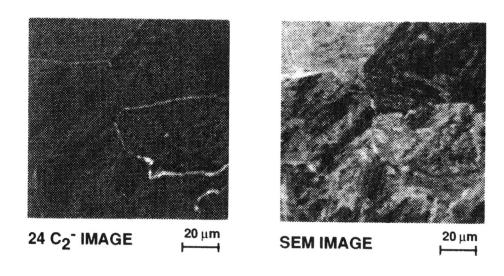

24 C₂⁻ IMAGE   **20 μm**   **SEM IMAGE**   **20 μm**

Figure 8.   A $^{24}C_2^-$ Image and the Corresponding Secondary
                      Electron Image from the 12% Cr Steel.

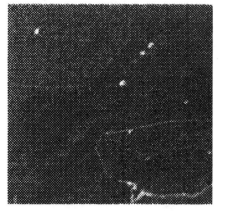

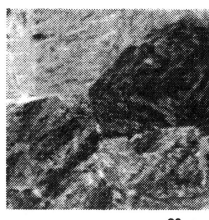

52 Cr + IMAGE   20 μm

SEM IMAGE   20 μm

Figure 9.   A $^{52}Cr^+$ Image and the Corresponding Secondary Electron Image from the 12% Cr Steel. This is the Same Area as that in Fig. 8.

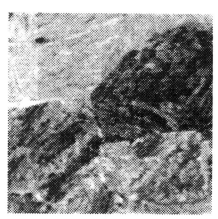

51 V+ IMAGE   20 μm

SEM IMAGE   20 μm

Figure 10.   A $^{51}V^+$ Image and the Corresponding Secondary Electron Image from the 12% Cr Steel. This is the Same Area as in Figs. 8 and 9.

COHERENT OPTICAL METHODS FOR METALLOGRAPHY

Martin J. Pechersky[1]

ABSTRACT

Numerous methods based on coherent optical techniques have been developed over the past two decades for nondestructive evaluation, vibration analysis and experimental mechanics. These methods have a great deal of potential for the enhancement of metallographic evaluations and for materials characterization in general. One such technique described in this paper is the determination of the material damping factors in metals. Damping loss factors as low as 10-5 were measured on bronze and aluminum specimens using a technique based on laser vibrometry. Differences between cast and wrought bronze were easily distinguishable as well as the difference between the bronze and aluminum. Other coherent optical techniques may be used to evaluate residual stresses and to locate and identify microcracking, subsurface voids and other imperfections. These techniques and others can serve as a bridge between microstructural investigations and the macroscopic behavior of materials.

INTRODUCTION

Optical techniques such as microscopy and other types of visual inspection and analysis have been in common usage in metallography and materials characterization for a very long time. Other optical

[1] Westinghouse Savannah River Company,
Savannah River Laboratory, Aiken SC, 29802

techniques such as infrared, visible and ultraviolet spectroscopy determine related information with regard to the material under investigation. Radiology for imaging, residual stress measurement and compositional analysis are additional techniques which are also fundamentally optical in nature. The use of laser light to enhance metallographic and materials analysis is therefore a natural extension of more traditional techniques.

The use of coherent optical techniques which has become possible with the advent of the laser now make it possible not only to perform morphological analysis [1] but also to obtain additional complementary information with relative ease. Holographic interferometry [2],[3] is an example of a widely used and well understood technique in this regard. This technique may be used to measure strain and therefore the stress state by inference. It has also been used to detect microcracking in metals and debonding in composite materials. Less widely known but related techniques known as speckle interferometry and speckle photography [4] are also in use to characterize the stress state of materials and for nondestructive evaluation. Speckle techniques can yield measurement sensitivities similar to holographic interferometry and they are more suitable for the measurement of in-plane deformations. It is also often easier to use; especially electronic speckle pattern interferometry (ESPI). Speckle does not however give as detailed spatial resolution as holographic interferometry.

Holography and speckle are full field techniques. That is, they are used to gather an image of a portion of a specimen or structure which can then be analyzed to determine the state of stress; the deformation; or defects over a relatively large area. Other coherent optical techniques may be used to analyze materials at a single point. Laser ultrasonics [5],[6] is a relatively new technique in which ultrasonic waves are generated in a material by the absorption of laser light. Laser interferometry is then used to sense the waves after they have passed through the material. Laser-extensometers [7] are commercially available devices which are used to measure strain in a local region. Other examples of the more established techniques include laser velocimetry and laser vibrometry [8]. These methods can measure the velocity at a point on the surface of an object which has been mechanically or thermally excited. These techniques are based on the measurement of the Doppler shift of laser light scattered from the test object. By measuring the response of a test object to small excitations information with regard to the elastic and anelastic properties of metals and other materials can be determined. A description of this measurement technique for some common metals is discussed below. An overview of many of the other optical methods mentioned here and their relationship to more conventional techniques is given by Cielo [9].

LASER MEASUREMENTS OF DAMPING IN METALS

This section will give a fairly detailed example of a method in which a remote and non-contacting laser measurement can determine a material property of a specimen which is intimately related to the microstructure of the material. Since the method is remote and noncontacting it may be applied in situations in which it may not be practical to perform normal metallographic analysis. A complete description of the measurement technique with extensive results on metal matrix composites can be found in references [10],[11] and [12]. This technique has also been used to measure the damping of polymer composites and viscoelastic materials.

The linear response of metals to vibrational excitation consists of both a deformation and the generation of heat due to the dissipation of energy within the bulk of the material. Several mechanisms for damping in metals are discussed by Zener [13]. Since many of these mechanisms are dependent on the microstructure of the material it is possible to deduce something about the microstructure of the material by a nondestructive measurement. Because the damping of metals is usually very small the method to do this type of measurement must not remove an appreciable amount of mechanical energy from the specimen. Therefore it is natural to chose a noncontact method for both excitation and measurement. Such a method based on laser vibrometry is described in the following paragraphs. This method is based on traditional methods of measuring modal parameters of vibrating structures [14]; the main differences being the noncontacting nature of the measurement and the sample mounting methods.

The technique is based on vibrating a specimen in a small frequency range which is centered at one of the specimens natural frequencies. This is sometimes referred to as the resonance dwell method. The shape of the specimen is in the form of a slender beam which usually has a rectangular cross section. While the specimen is vibrating a measurement of the drive point mobility is performed. The drive point mobility, M is simply the ratio of velocity (V) at the point at which the specimen is being excited to the excitation force (F).

111

Thus $$M(f) = V(f)/F(f) \quad (1)$$

Where the mobility is a function of the excitation frequency f. A complicated time varying signal is obtained from this measurement which is Fourier transformed so that a plot of M versus frequency is obtained. This plot, which has a single peak at the resonance frequency, is usually referred to as the Frequency Response Function, FRF. Once this plot is obtained, the damping coefficient is obtained by dividing the full width half maximum (FWHM) frequency bandwidth of the curve by the center (resonant) frequency. This ratio is called the loss factor and is a measure of the damping in the material. If one accurately measures the geometry of the specimen and its weight the modulus of elasticity can also be determined. Since the specimen is a continuous system several resonant frequencies can be excited so that the loss factor and elastic modulus can be determined as a function of frequency. One may also perform this measurement by setting the frequency at a fixed value and measuring the velocity then increasing the frequency and repeating the measurement until a frequency range which contains a resonant frequency has been covered. This second, is often referred to as the frequency sweep method. While conceptually simpler it is not as fast or precise.

From a metallographic point of view the peaks in the damping versus frequency curve can result from different size features in the microstructure of the material. A schematic of the experimental set-up is shown in figure 1.

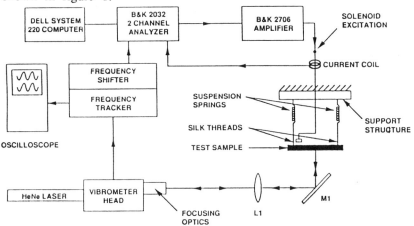

Figure 1 - Schematic of experimental set-up for material damping measurements.

There are no specific size limitations on the specimens however if the length to width or length to thickness ratio is to small a more complicated analysis is required. For the measured results presented in this paper the specimen length was about 25 cm; its width was 1.25 cm and it was one half of one centimeter in thickness.

The experimental set-up consists of: a specimen suspension system; an electromagnetic drive coil; a laser vibrometer, and signal analysis electronics. The suspension system supports the specimen from fine silk threads which in turn are attached to springs. The spring rates are chosen to minimize the flow of energy into or out of the specimen. The suspension points along the length of the specimen are chosen so that they correspond to nodal lines for the vibrational mode being excited thereby further reducing extraneous energy losses. The mechanical excitation system is also designed to minimize extraneous sources of energy flow. The excitation is accomplished by gluing a small rare earth magnet onto the specimen and the driving it with an electric coil. The driving force is determined from the current in the coil. The only extraneous sources or sinks of energy would come from hysteresis losses in the magnet, ohmic losses in the coil or electromotive forces generated by the magnet/coil combination. These losses were found by experience to be negligible.

Having taken so much care in both the suspension and excitation of the specimen it is natural to select a measurement technique which will also add negligible energy loss or gain into the system. Laser vibrometry satisfied this requirement in every way. Since the measurement is not only noncontacting but also remote this measurement can be performed at high temperature or under other hostile environmental conditions. There are other non-contact methods but none of these are remote. The beam from the vibrometer is focused onto the specimen and some of the light scattered from the focal spot on the specimen is collected through the same optics that are used to focus the beam. The collected light is passed through a small aperture and sensed a by a set of photodiodes. At the same time the photodiodes are illuminated by a reference beam. The fluctuations in the optical interference between the reference beam and the object beam from the specimen provides a measure of the velocity at the illuminated point on the specimen. It should be noted that the surface of the specimen need not have a mirror finish. That is, a flat or brushed finish provides sufficient back scattering of light with the proper coherence properties so that efficient interference between the object beam and the reference beam is possible. The output of the vibrometer electronics is simply a voltage which is proportional to the velocity being measured. The velocity and force signals are fed into an FFT analyzer which automatically determines the frequency response function over the desired frequency range. A typical plot of the FRF for wrought bronze is shown in figure 2.

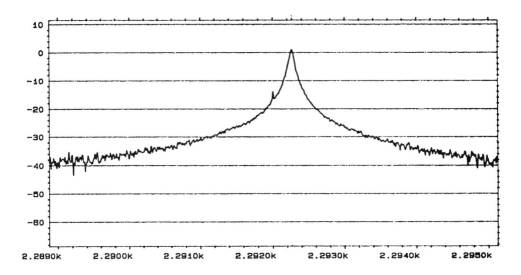

Figure - 2 Plot of drive point mobility as a function of drive frequency for a wrought bronze specimen vibrating at its second normal mode.

The ordinate in figure 2 is the Mobility in dB and the abscissa is the frequency in kHz. Notice that the total frequency range covered is about 6 Hz centered around the resonance frequency of 2292.5 Hz. The frequency resolution in this plot is about 7.8 mHz. The exact frequency of the resonance point is 2292.257 Hz and the FWHM bandwidth (-3 dB on either side of the peak) is 59 mHz which yields a loss factor of 2.57*10-5. Based on this type of measurement a plot of loss factor versus frequency can be obtained for any material so long as it can be adequately excited. The results of this type of measurements are shown in figure 3 for wrought bronze, cast bronze and T6 aluminum.

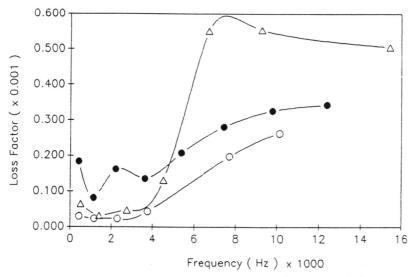

Figure 3 - Flexural Damping loss factor as a function of frequency for three metallic specimens.

In this figure the loss factor for wrought bronze, cast bronze and T6 Aluminum are plotted as a function of frequency. The open circle plot is for the wrought bronze, the closed circle plot is for the cast bronze and the open triangle plot is for the T6 aluminum. It can be clearly seen from the plot that the loss factor for the cast bronze is significantly higher than the rolled bronze by about the same amount over the measured frequency range. The T6 aluminum has about the same loss factor as the wrought bronze at the low end of the spectrum but increases more rapidly at the higher end. While metallographic analysis was not performed on these samples it well known that cast metals have higher damping coefficients than formed metals.

Notice that for all three samples there is a sudden rise in the loss coefficient around 4 kHz. This sudden rise is not due to a material property but results from the onset of appreciable acoustic radiation from the specimens. This sudden increase in acoustic radiation is a well known phenomena and the onset frequency is called the coincidence frequency. The frequency at which this occurs is proportional to the square root of the ratio of the mass density to Young's Modulus and inversely proportional to the thickness of the material. Therefore if measurements at high frequencies are required one may wish to do them in vacuo. This would not present a problem for this technique since the laser beam and return signal can be passed through any transparent material. As an aside; this measurement technique can provide a new method to measure the acoustic radiation properties of structures by doing the measurement in both air and in vacuo.

CONCLUDING REMARKS

The technique described above is just one of many coherent optical measurement techniques that has significant potential for materials characterization and metallographic analysis. The purpose of concentrating on this particular technique was because it is novel and because the type of results obtained are not as obvious as some of the more well known techniques that were mentioned in the introduction. The fact that most coherent optical methods can perform precise and accurate measurements remotely is a major feature which can be taken advantage of in many ways. Evaluations of materials at their operating temperatures and in harsh environments allows for information to be gathered which before these techniques emerged could not be accomplished. Since the measurement techniques often supply different but complimentary information new ways of interpreting data for materials characterization are possible. There is little doubt that many further developments along these lines will occur in the future.

REFERENCES

1) Glen M. Robinson, David M. Perry and Richard W. Peterson, "Optical Interferometry of Surfaces," **Scientific American**, Vol. 265, p. 66, July, 1991

2) Charles M. Vest, **Holographic Interferometry**, John Wiley & Sons, New York, NY, (1979)

3) **Holographic Nondestructive Testing**, Robert K. Erf Ed., Academic Press, Inc., Orlando, FA (1974)

4) **Speckle Metrology**, Robert K. Erf, Ed., Academic Press, New York, NY (1978)

5) L.E. Drain, **The Laser Doppler Technique**, John Wiley & Sons Ltd., London, England (1980)

6) G.V. Garcia, N.M. Carlson, K.L. Telschow, and J.A. Johnson, "Noncontacting Laser Ultrasonic Generation and Detection At The Surface Of Molten Metals," in **Review of Progress in Quantitative Nondestructive Evaluation**, Vol. 9, p. 1981, Plenum Press, New York, NY, (1990)

7) K.L. Telschow, J.B. Walter, and G.V. Garcia, "Laser Ultrasonic Monitoring Of Ceramic Sintering, " **J. Applied Physics**, Vol. 68 p. 6077, (1990)

8) Albert M. Creighton, "Zeeman-split-laser system measures material strains," **Laser Focus World**, Vol. 27, p. 87, June 1991

9) P. Cielo, **Optical Techniques for Industrial Inspection**, Academic Press, San Diego, CA, (1988)

10) M.J. Pechersky, R.B. Bhagat, C.A. Updike, and M.F. Amateau, "Control of Damping Characteristics of Graphite Fiber Reinforced Aluminum Composites," in **Metal & Ceramic Matrix Composites: Processing, Modeling & Mechanical Behavior**, p 641 The Minerals, Metals & Materials Society, Warrendale, PA (1990)

11) M.J. Pechersky, J.H. Ostar and S.J. Wroblewski, "Vibration Of Metal Matrix Composite Beams," in **Proceedings of NOISE-CON 8 8**, p 323, Noise Control Foundation, New York, NY (1988)

12) Clark A. Updike, Ram B. Bhagat, Martin J. Pechersky and Maurice F. Amateau, "The Damping Performance of Aluminum-Based Composites," **Journal of The Minerals, Metals & Materials Society**, Vol. 42, p 42, March 1990

13) Clarence Zener, **Elasticity and Anelasticity of Metals**, The University of Chicago Press, Chicago, IL (1948)

14) D.J. Ewins, **Modal Testing: Theory and Practice**, Research Studies Press, LTD., Letchworth, England (1986)

117

CONFOCAL LASER SCANNING MICROSCOPY IN THE

CHARACTERIZATION AND FAILURE OF ADVANCED MATERIALS

R. N. Pangborn, R.A.Queeney, R.P.McNitt
Engineering Science & Mechanics Dept.
The Pennsylvania State University, University Park, Pa 16801

ABSTRACT
The use of the Scanning Conformal Laser Microscope in typical failure analyses and material surface characterizations by faculty and students in the ESM department at Penn State is described. Examples given are:

"Rolling Contact Fatigue of Ausrolled Powder Metals",
"Etching Study of Polycrystalline Silicon Wafers",
"Interface Studies of Metal Matrix Composites", and
"Dissolution Studies of Calcite".

INTRODUCTION

The Scanning Confocal Laser Microscope (SCLM) has special features desirable to those interested in failure analysis and characterization of surface features. Advantages of this instrument include;

High resolution (to 0.25μm)
Good depth of focus (at least 1mm depth)
Real time imaging (15kHz horizontal, 60Hz vertical scan rates)
Heatless illumination (laser light lacks infrared rays)
As compared to the SEM, nonconductive samples need not be coated;
    no drying (biological samples) or degassing necessary,
Non contact surface profile measurements
Large specimens can be easily accommodated by the metallograph
    stage
Curved surfaces present no optical difficulties
Rough surfaces are also easily examined.

119

A focus of the Penn State ESM dept. is "Engineered Materials", including the modelling, fabrication, testing, evaluating, modification and characterization of special materials and materials systems. Thus the SCLM provides an additional tool for surface examination. The instrument used is the Lasertek 1L11M (operating through an Olympus BMJ optical microscope) utilizing a He-Ne 1.5 milliwatt laser. The sketch of the system shows the laser attached to the camera port of the optical microscope, the viewing monitor, and the control panel and computer. Essentially the system takes 277 "slices", forming an image at

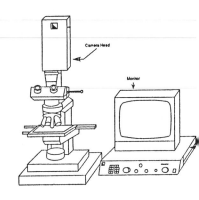

the first level, storing it, forming an image at the next level, comparing it to the first and replacing any less bright sections with a brighter pixel. The conformal imaging system is designed such that image intensity is a strong function of spatial distance, thus brightness (i.e. focus) is associated with depth and only that portion of the image that is "most in focus" is stored. By the time all 277 vertical scans (a motor drives the fine adjustment of the optical microscope stage) are accomplished, one has an in-focus image, as well as information on the depth at which the image was formed! This depth information, when coupled with information from the horizontal scan can clearly allow the instrument to be used as a profile measuring device .

ROLLING CONTACT FATIGUE OF AUSROLLED POWDER METALS

Using powdered alloy steels as the starting point in fabricating gears is a well-established technology. A low grade of gearing can be produced simply by cold pressing of the powders followed by sintering: the residual porosity, up to 15% by volume, is the performance degradation factor that may be offset by low manufacturing costs. Better gearing, including gears that are superior to those fashioned from wrought stock, may be produced by compacting the powders to full or near full, density by employing one of the viable hot pressing schemes, or through hot forging of sintered preforms. The additional thermomechanical processing is relatively costly.

An alternative approach to near full density gears may be to ausform sintered preforms in the same manner as gears produced from wrought alloy stock by machining [1]. In ausforming, the alloy is heated above the A3 temperature and quenched into heated oil (230 C) held above the Ms temperature; in this latter condition, it is still austenitic in structure, albeit unstable. If the steel is plastically deformed in this state, a refined martensite, nearly free of retained anstenite, will be induced upon subsequent quenching to below Ms. In addition, fine precipitated carbides and inherited dislocation substructures produce an end product that is harder and tougher than conventionally marquenched steel [2]. Finally, ausrolling produces surface densification to depths of 50 mils, or more.

Rolling contact fatigue (RCF) is a prevalent failure mode for the surfaces of gear teeth and bearing elements. The laser light scanning microscope possesses several features that lend it to a facile study of RCF damage development. Most RCF testing machines demand large cylindrical test samples that are run immersed in oil. The *curved surfaces of the RCF specimens present no difficulties* to the confocal microscope optics, nor is the roughened surface developed during testing problematical. *Specimen cleaning can be relatively easy,* as there is no vacuum jacketing to be contaminated, and the *large specimen geometry can be accommodated easily* by the metallograph stage. Finally, t*he noncontact profilometer yields a quantitative assessment of developing surface roughness prior to the appearance of gross pitting and spalling.*

Figure 1 is the as-ground surface of a 46100 steel powder specimen blended with 1.0% carbon by weight (in the form of graphite), pressed at 100 ksi and sintered for one hour at 1,200 C in flowing hydrogen gas. The specimen was ausrolled as noted above, and the surface regions were found to contain residual porosity of one to two volume percent. After 3,000,000 cycles in rolling contact, with a maximum Hertzian contact stress of 300 ksi, the surface has roughened slightly, as seen in Figure 2a. The surface roughness is indicated in Figure 2b for the same specimen. Contrasting sharply with the durability of ausrolled surface is that of the conventionally quenched surface depicted in Figure 3a, 3b. Tested under identical conditions, this sample evidences marked microdeterioration at one-tenth the testing previously shown, or only 300,000 cycles. Ongoing studies seek to characterize the effects of ausrolling parameters on endurance.

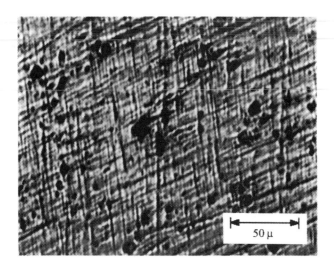

Fig. 1   As ground surface 46100 steel powder specimen

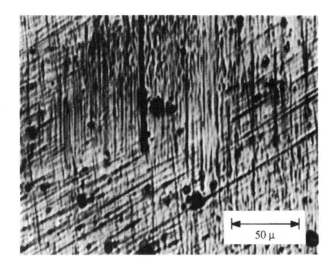

Fig. 2a Ausrolled powder specimen after 3 million cycles rolling contact

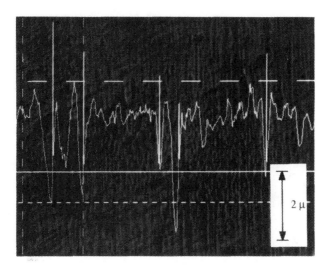

Fig. 2b Surface profile as obtained from SCLM of specimen in 2a.

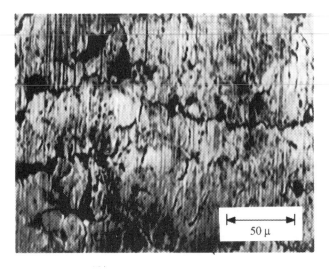

Fig. 3a Specimen not ausrolled, after 300,000 cycles of rolling contact

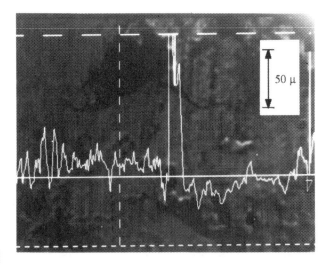

Fig. 3b Surface profile as obtained from SCLM of specimen in 3a

ELECTRONIC DEVICES

The purpose of this study was to relate changes in surface structure and topography associated with wet chemical etching of polycrystalline silicon in terms of effects on the photoresponse of devices prepared from the samples [3]. Waker cast polycrystalline silicon is a low cost alternative for photovoltaic applications, which require very large area solar arrays. The samples were cut from bulk polycrystalline material into wafers of about 0.5 mm thickness. In order to remove the damage (manifested as defects which serve as recombination sites for minority charge carriers at the surface) due to cutting, the samples were etched in a 10:4:1 solution of $HNO3:HAc:HF$. The perfection of the crystal structure of the surface layer was monitored by X-ray double crystal diffractometry, while the electrical characterization was performed by preparing MIS devices on samples etched for various times.

RESULTS: Figure 4 shows the wafer surface after exposure to the etchant for 25 minutes (removal of about 75 μm). A textured surface is produced by the anisotropic etch. The surface profile in Figure 5 shows the oblong depressions to vary between about 0.2 and 0.4 μm in depth. An increase in the short circuit current (I_{sc}) and open circuit voltage (V_{oc}) for samples etched up to 5 minutes may be attributed to the better coupling of light into the silicon afforded by the textured surface and the reduction of recombination sites, as evidenced by the sharper X-ray diffraction patterns. However, if the etching is allowed to continue for longer than 5 minutes, the Isc declines again. This may be accounted for by the introduction of grooves up to 3 μm in depth at grain and twin boundaries due to preferential etching. These degrade the photoresponse by causing an enhancement in the leakage current, thus suggesting that there is an optimum etching time to achieve the best response.

INTERFACE STUDIES OF METAL MATRIX COMPOSITES

It is well known that the interface between the metal matrix and reinforcing fibers in advanced, high-temperature composites can greatly influence the resultant component properties and performance in service. Although the fiber is an efficient form of reinforcement, its cylindrical geometry when incorporated into a composite is not well suited to many characterization techniques useful in interfacial studies. Thus, laminate structures can be made to "simulate" the interface, but in a planar conformation that is conducive to study with such techniques as optical microscopy, Rutherford backscattering spectroscopy and X-ray diffraction.

These techniques provide vital information regarding the morphology, interfacial diffusion and reaction, presence of interphases, degree of crystallographic accommodation and other features unique to the interfaces in metal matrix composites (MMCs).

RESULTS: Figure 7 shows the edge of an RF sputtered aluminum coating applied to a pyrolytic graphite substrate. The basal plane orientation of the substrate is representative of the onion-skin orientation of basal planes for pitch-type graphite fibers. The sample may then be given heat treatments typical of the processing of MMCs, or compared to similar samples in which a diffusion barrier has been deposited between the graphite and the aluminum "matrix" material. In this case, SCLM is used to provide a convenient measure of the thickness of the aluminum layer (approximately 50 μm) and a view of the morphology of both the substrate and deposited materials.

DISSOLUTION STUDIES OF CALCITE

An important field of study in the geosciences is the dissolution of minerals commonly found in soil and rock, their transport in aquifers and other groundwater and their eventual redeposition at new sites. To understand the rate by which the first of these processes takes place, the relative importance of both diffusion and surface reaction controlled dissolution mechanisms must be determined.

RESULTS: This study involved the dissolution of calcite (calcium carbonate) in a 0.7 molar solution of KCl in deionized water. The ion concentration level was chosen to simulate sea water, while the particular species were dictated by the electrochemical analyses to be performed. During the dissolution experiment, the sample was spun to reduce the concentration gradient and emphasize the surface reaction. As can be seen in Figure 8, the result is a process of successive etch pitting, with small pits forming at the base of larger ones. Reduced agitation at the bottom of the deepest craters may lead to a more diffusion-controlled reaction. Preferential dissolution at these sites may be due to the presence of accumulations of point defects or dislocation lines intersecting the surface. SCLM aids in the examination of surfaces exhibiting extreme topographical relief, so that the distribution, configuration and "coalescence" of pits with progressive dissolution can be accurately characterized. In addition, the shape and depth of the pits can be quantitatively evaluated, as depicted by the 1.5 μm deep pit in Figure 9, which lies at the bottom of the larger, dual crater of Figure 8.

CLOSURE

The Scanning Confocal Laser Microscope has proved to be a valuable tool in the failure analysis and surface characterization of a wide range of materials and applications. As it is a "user-friendly" instument, it has proved to be quite valuable in the education of both undergraduates and graduate students interested in Engineered Materials. The examples given above for powdered materials, microelectronic materials, composites, and geoscience materials demonstrate the wide applicability of this instrument.

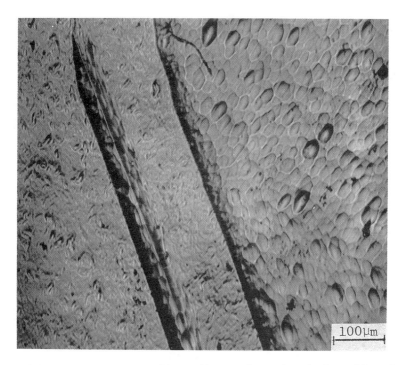

Fig. 4. Waker cast polycrystalline silicon after chemical etching for 25 minutes. Grain boundary at center separates regions with different textures due to anisotropic etch. Double boundary at left corresponds to lenticular twin.

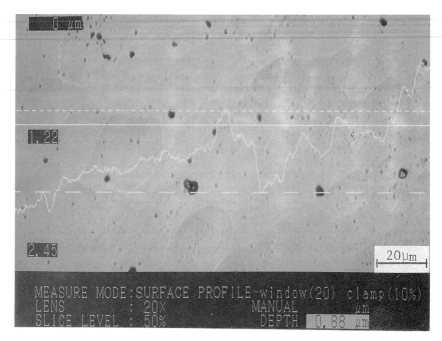

Fig. 5. Ten times higher magnification of oblong depressions with depths ranging from 0.2 to 0.4 μm. Solid line gives location of traverse corresponding to depth profile. Vertical distance between two broken lines represents 0.88 μm in depth.

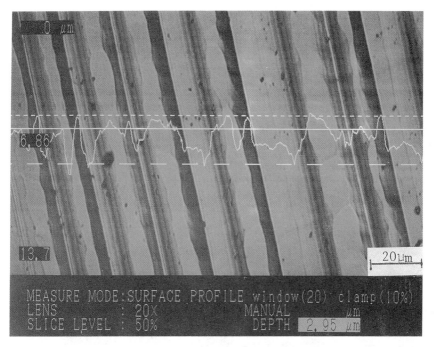

Fig. 6   Parallel grooves caused by preferential etching of twin boundaries.
Typical depths are 2 to 3 μm.

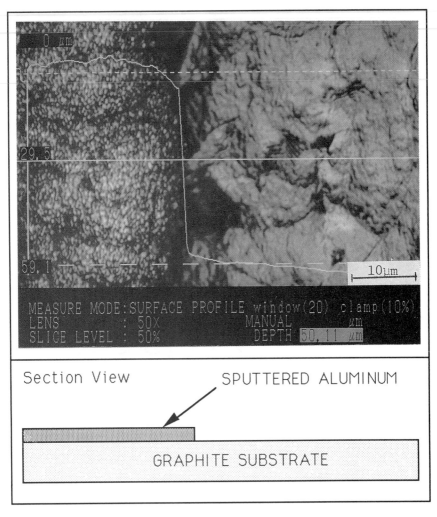

Fig. 7   Above: Top view of RF sputtered aluminum (left) on pyrolytic
                  graphite substrate. Deposited layer is about 50 µm thick.
         Below: Schematic drawing (side view) of substrate and aluminum
                  coating.

Fig. 8. Etch pits in calcite due to dissolution in 0.7 molar solution of KCl.

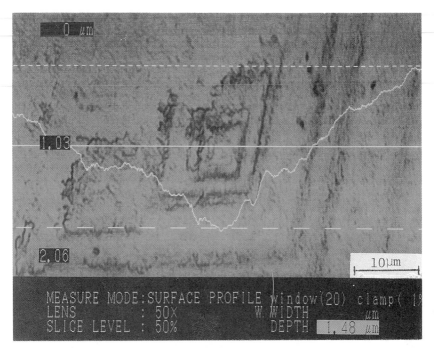

Fig.9. Higher magnification image of etch pit at bottom of large, dual crater in Fig. 8(arrow) and corresponding depth profile.

REFERENCES

1.   M. F. Amateau and R. A. Cellitti, "Finishing of Gears by Ausforming," SME Paper #MF86-922, Society of Manufacturing Engineers, Dearborn, Michigan 48121 (1986).

2.   G. Thomas, D. Schmatz, and W. Gerberich, "Structure and Strength of Some Ausformed Steels," High Strength Materials, V. F. Zackey, Ed., John Wiley and Sons, New York, NY,   251-326 (1965).

3.   R. N. Pangborn, B. L. Peterson, "Precision Measurement of Microplasticity by Computer Aided Rocking Curve Analysis", Crystal Prop. & Prep. Vol 16, 185 (1988)

Applied Metallography

PLASMA ETCHING OF POLYMERIC MATERIALS

H.J. Mueller[1]

ABSTRACT

Four types of polymeric dental materials, including
polymethylmethacrylate (PMMA) and PMMA with butadiene denture resins, a quartz particle polymer
matrix composite filling material, and a carboxylate adhesive cement were submitted to a radio
frequency argon plasma for various times. Results
revealed detailed microstructures after plasma
etching. For PMMA, the unreacted polymer beads, the
polymer matrix, and the reaction zone between the
two were clearly defined. At short times, the
polymeric matrix was more sensitive to the plasma,
while at longer times the polymer beads showed
increased reaction. By controlling the etching
time, enhancement of specifics became possible. The
impact resin required very short times to delineate
the rubbery component. Colorimetry with time
revealed PMMA to increase in brightness, decrease
in reddness, and decrease and then increase in
yellowness. A decarboxylation reduction has been
proposed to explain PMMA interactions with Ar
plasma. An O_2 plasma proved less effective in
generating overall microstructure but was effective
for revealing porosity.

American Dental Association, 211 E. Chicago, Ave.,
Chicago, IL 60611.

INTRODUCTION

Materials for microstructural characterization require special preparations which are dependent upon the particular techniques to be used. Transmission microscopy requires thin sections of materials while reflection microscopy requires features on the surfaces or within the outermost surface layers to reveal contrast. Enhancement of contrast can be achieved by chemical/electrochemical etching, relief polishing, generating fractured surfaces, deposition of interference films, polarization of the radiation being used, and back scattered electrons.

Plasma etching of materials to reveal microstructural features is a newer technique with much potential that has only been used limitedly [1,2]. Plasma etching to strip away photoresist masks in the lithography industry and to etch selectively silicon and other semiconductor components in the integrated circuits industry has gained widespread usage [3,4]. Plasma etching has also gained in popularity for removal of organics from surfaces.

Metallography and ceramography methods have been used with the above techniques for characterizing microstructural details with a variety of alloys and ceramics [5]. No such data base is available for most engineering polymeric materials. In order to characterize microstructural details with polymeric materials, a trial and error method often must be used. That is, a variety of methods are used until acceptable microstructural features are obtained.

Resinography has been used to enhance details about macromolecular domains, phases, interfaces, fibers and films [6]. Additional systems also need characterization. With the introduction of newer polymeric systems, a continual need exists to be able to define their microstructures accurately and efficiently.

Optical instead of spectroscopic methods are often desirable to use for characterizing polymers. Applications involving the distribution of phases, reinforcing fibers and fillers and others find much need for optical methods. Plasma etching provides the means to obtain contrast. With the development of digital image analysis techniques, an even greater need is placed upon the ability to reveal contrast within polymeric materials.

Polymers are used in numerous dental applications. For example, dentures are fabricated from poly(methylmethacrylate). High impact resins also contain a rubbery phase, like butadiene. Composite filling materials are composed of bis GMA polymeric matrix and glass particles. Materials for bonding and adhesion are polymeric in nature, such as carboxylate. With all of these materials the ability to use microscopic methods is a valuable aid in defining their property-structure relationships.

The purpose of this project was to determine the feasibility in using plasma glow electrical discharge as an etching technique for revealing microstructural features with a number of polymeric dental materials.

EXPERIMENTAL PROCEDURES

Two poly(methylmethacrylate) denture resins, Polytone cold-cure (General Dental) and Special 99 heat-cure high impact containing a butadiene additive (General Dental), one quartz filled polymeric-matrix, 2,2-BIS[p-(γ-methacryloxy-β-hydroxypropoxy)phenyl]propane (BIS-GMA) composite filling material, Adaptic (Johnson and Johnson), and one poly(carboxylate) cement, Durelon (Premier) were used. The denture resins were supplied as a powder and liquid. That is, prepolymerized methacrylate beads were mixed with methacrylate monomer. The composite was supplied as two pastes while the carboxylate material contained a zinc oxide powder and polyacrylic acid liquid. Samples of all four materials measuring 15 x 15 x 3 mm thick were prepared by following manufacturers' instructions.

One face surface of each sample was used for microscopic examination and plasma etching by first grinding on silicon carbide papers of 240, 400, 600, 2400, and 4000 grit sizes, followed by polishing with 1.0, 0.3, and 0.05 x 10^{-6} m alumina particles on micro cloth (Buehler). Surfaces were sonicated in water prior to cold air drying.

A Leitz orthoplan optical microscope was used to initially assess the quality of the polished surfaces.

Plasma etching was conducted with a Harrick model PDC-23G pyrex barrel-type (76 mm diameter and 0.028 m³ volume) radio frequency machine. Up to five samples positioned on a microscopic slide were placed within the center of the glass chamber and etched for various times up to several hours in duration. The power selector switch was either set on high (100 watts) or medium. Both argon and oxygen gases were used at pressures of 0.1-0.2 torr which permitted a reddish-bluish electrical glow to occur. At 100 watt power, temperatures of 50-55 °C were reached and maintained after about 10 min. At medium power, temperatures of 40-45 °C were generated.

At regular intervals during the etching process, the samples were taken out of the chamber to check the surface condition with optical microscopy and to also measure the CIE 1976 L*a*b* color vectors with a Minolta model CR-100 chroma meter. The face of a sample was large enough in area to completely cover the apeture of the meter.

At completion of plasma etching, the samples were sputtered coated with a thin gold film prior to scanning electron microscopy with a Cambridge Mk II machine operating at 10 Kv.

RESULTS

Figures 1-7 present SEM micrographs for Polytone resin after various times of argon plasma etching at 100 watts. The as-polished surface is revealed in Figure 1, while etching for times of 40, 80, 100, 120 and 240 min are shown in Figures 2, 3, 4, 5-6, and 7 respectively. Figures 8a and 8b present plots of the color vectors L*, a*, and b* after various times of etching. Figure 9 reveals an argon plasma etched Polytone resin surface reinforced with carbon fibers, while Figure 10 reveals a surface etched with an oxygen plasma for 20 min.

Figures 11-13 present the as-polished surface and a 10 min argon plasma etched (100 watts) surface for Special 99. Figures 14 and 15 present the as-polished surface and an argon plasma etched surface for Adaptic, while Figures 16 and 17 present the as-polished surface and an argon plasma etched surface for Durelon.

Figure 1.
Polytone
as-polished

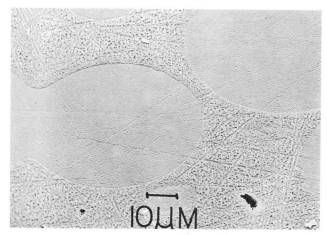

Figure 2.
Polytone
40 min
Ar-Plasma

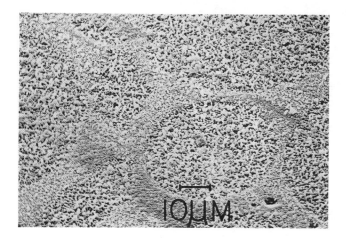

Figure 3
Polytone
80 min
Ar-Plasma

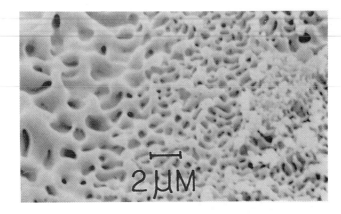

Figure 4.
Polytone
100 min
Ar-Plasma

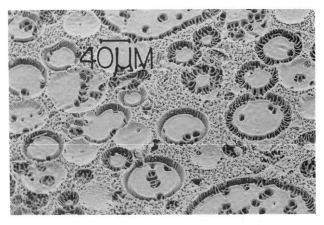

Figure 5.
Polytone
120 min
Ar-Plasma

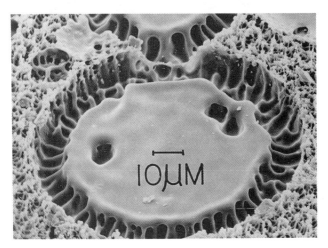

Figure 6
Polytone
120 min
Ar-Plasma

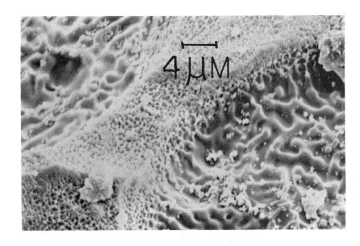

Figure 7.
Polytone
240 min
Ar-Plasma

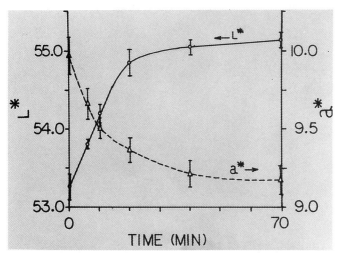

Figure 8a.
Polytone
L* & a*
Colorimetry

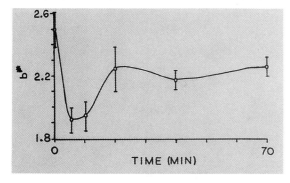

Figure 8b.
Polytone
   b*
Colorimetry

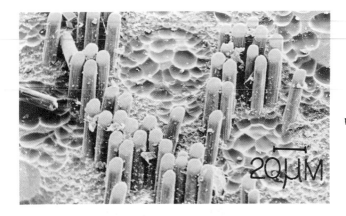

Figure 9.
Polytone
with carbon
Ar-Plasma

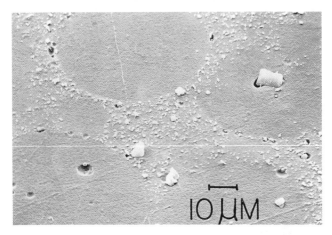

Figure 10.
Polytone
O_2-Plasma

Figure 11.
Special 99
as-polished

Figure 12.
Special 99
 10 min
Ar-Plasma

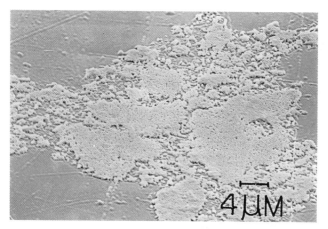

Figure 13.
Special 99
 10 min
Ar-Plasma

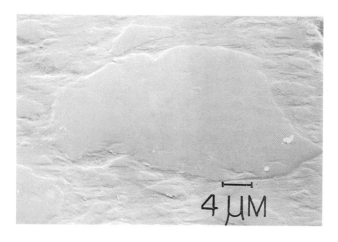

Figure 14.
 Adaptic
as-polished

Figure 15.
Adaptic
Ar-Plasma

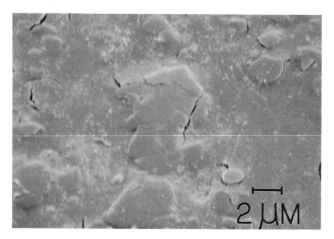

Figure 16.
Durelon
As-polished

Figure 17.
Durelon
Ar-Plasma

DISCUSSION

An oxygen plasma retained debris over the matrix as shown in Figure 10 which was likely due to oxidizations with the polymeric surface. The oxygen plasma did, however, delineate the polymer beads and the pores in the structure. The inability for the argon plasma to define pores was likely the result form the bombardment of the surface with the heavier argon ions. The features of interest will dictate the type of plasma to be used.

As shown in the series of micrographs in Figures 1-7, the ability for an argon plasma to enhance microscopic features in poly(methylmethacrylate) is exceptional. With no microstructural detail on the as-polished surface, unreacted polymer beads and matrix were progressively revealed as the etching time was increased. Even a reaction zone between polymer beads and matrix was clearly detected (Figure 4). Even though the polymer matrix is the first to become etched (Figure 2), longer times reveal an increased etching rate with polymer beads. (Figures 5-7). This is first shown by the reaction rings around the polymer beads and latter by the complete etching of the polymer beads from the structure.

Optical microscopy generally revealed microstructural details at shorter etching times. For example with PMMA, optical microscopy showed good contrast already after 20 min of etching, while SEM of the same surface showed poor definition of the features.

All three color vectors L^*, a^*, and b^* were affected by the argon plasma. The $L^*a^*b^*$ color space is defined [7] as white ($+L^*$), black ($-L^*$), red ($+a^*$), green ($-a^*$), yellow ($+b^*$), and blue ($-b^*$). This means that the brightness of the surface increased linearly with time (t) for about the first 20 min with a slope of 0.09 min^{-1} and thereafter decreased to another linear function with a much smaller slope of 0.007 min^{-1}. The reddness decreased in a nonlinear manner which was fitted with the following polynominal expression.

$$a^* = 9.8 - 3.8t + 7.6t^2 - 5.1t^3 \qquad (1)$$

The yellowness decreased within the first 6 min, there-

after increasing to a constant value reached after 20 min.
From a multispectroscopic analysis [8], the initial
mechanism for the surface modification of PMMA to an
argon plasma was reduction via decarboxylation by argon
bombardment. This could involve crosslinking or the
formation of reactive intermediates involving oxygen-rich
functionality like O-H. The topmost modified surface
layer was hydrophilic as compared to the unmodified
hydrophobic bulk material. Isotatic PMMA revealed a
higher reactivity than the atatic form which was related
to the higher chain mobility for the former isomer.

For the butadiene-containing high impact resin,
Special 99, only short times of the order of 10 min were
required to delineate the butadiene additive with the
argon plasma operating at 100 watts (Figure 12-13).
Longer etching times obscured the additive and at much
longer times of the order of 90 min these particles were
degraded with no detail remaining. Like with the Polytone
resin, longer etching times did reveal the polymer beads
in much detail.

For the two remaining materials, Adaptic and Durelon,
exceptional definition of the microstructural features
were obtained with the argon plasma. The as-polished
surfaces did reveal some evidence of the filler
particles, but lacked complete detail within the matrix.

CONCLUSIONS

1. A radio frequency argon plasma proved to generate
exceptional microstructural details with a number of
polymeric dental materials, including PMMA and PMMA with
butadiene denture resins, a BIS-GMA polymeric matrix
quartz filled material, and a zinc polycarboxylate
adhesive cement.
2. Due to the argon plasma interacting with the
denture resins, colorimetry with time revealed the resins
increased in brightness, decreased in reddness, and
decreased (short times) and then increased (longer times)
in yellowness.
3. An oxygen plasma proved not to be as effective as
an argon plasma in generating overall microstructural
detail for microstructural analysis with PMMA. The O_2

plasma retained oxidation debris overlying the polymeric matrix. The O_2 plasma was, however, more effective in delineating the pores within the structure.

ACKNOWLEDGEMENTS

A portion of this project was supported by a grant from the National Institutes of Health, DE 05761. John Lenke from the American Dental Association provided microscopy support.

REFERENCES

1) A. Joshi and R. Nimmagadda, "Erosion of Diamond Films and Graphite in Oxygen Plasma", **J. Mater. Res.**, Vol. 6, p 1484, 1991.
2) C.J. Zimmermann, N. Ryde, N. Kallay, R.E. Partch, and E. Matijevic, "Plasma Modification of Polyvinyltoluene and Polystyrene Latices", **J. Mater. Res.**, Vol 6, p 855, 1991.
3) H.F. Winters and J.W. Coburn, "Etching Reactions at Solid Surfaces", In: **Plasma Synthesis and Etching of Electronic Materials**, R.P.H. Chang and B. Abeles (Ed.), Vol. 38, Materials Research Society, Pittsburgh, PA, 1984, p 189.
4) T. Yogi, K. Saenger, S. Purushothaman, and C.P. Sun, "Polyimide Etching in O_2/CF_4 RF Plasmas", In: **Plasma Processing**, G.S. Mathead and G.C. Schwartz (Ed.), Proc. Vol. 85-1, The Electrochemical Society, Pennington, NJ, 1985, p 216.
5) American Society for Metals, "Materials and Processing Databook", **Metal Progress**, Vol 122 (1), p. 122, 1982.
6) T. Rochow, "Resinography", In: **"Encyclopedia of Polymer Science and Engineering"**, **J Wiley, NY, 1988, p 88.**
7) Minolta, **Chroma Meter CR-100/CR-110 Operation Manual E**, Minolta, Japan, 1984, p. 6.
8) T.J. Hook, J.A. Gardella, Jr. and L. Salvati, Jr. "Multi-technique Surface Spectroscopy Studies of Plasma-Modified Polymers I: H_2O/Ar Plasma-Modified Polymethylmeth-acrylates, **J. Mater. Res.**, Vol 2, p 117, 1987.

FAILURE OF HIGH TENSILE BOLTS - A CASE STUDY
by
N.C. Kothari [1]

ABSTRACT

Six galvanised high tensile steel bolts were used to hold the wheels of a Four-Wheel Drive vehicle. The right hand rear wheel of this vehicle detached causing the vehicle to roll and resulting in considerable damage to the body. The wheel was detached by shearing of four of the bolts and stripping the nuts from the other two bolts, which remained unbroken. SEM fractography of the fracture surfaces of the four broken bolts indicated that the failure was due to reversed bending fatigue. Optical microscopy indicated that the bolts were heat treated to a tempered martensite structure and that the nuts were manufactured from low carbon steel. The paper discusses the influence of the microstructure on the failure process the events surrounding the nature of incident and the analysis of in-service failure of the failed components utilizing conventional metallurgical techniques.

INTRODUCTION
Analysis of service failures is a complicated business because several mechanisms can operate simultaneously or sequentially. For example a structural component, bolt (fastener) holding the wheel of a vehicle can be simultaneously subjected to stress, an environment and fatigue loading with varying frequencies while a vehicle is moving.

---

[1] Department of Civil & Systems Engineering,
     James Cook University of North Queensland,
     Townsville Qld 4811, Australia.

A series of bolts holding the rear right hand wheel of a Four-Wheel Drive vehicle failed causing the wheel to be detached and as a result the vehicle rolled over sustaining extensive damage to the vehicle and the occupants (Figs. 1 and 2).

Figure 1.   Damaged Four- Wheel Drive Station Wagon.

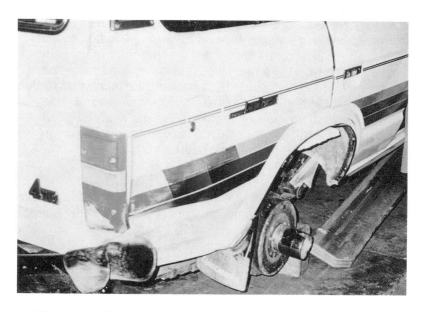

Figure 2.   Showing right hand rear-wheel hub after detachment of the wheel.

The vast majority of fasteners (bolts, screws, studs, etc) used in vehicle industry are made from one of several grades of carbon or alloy steels. Fatigue is the most common mechanism of failure in fasteners but sometimes more than one cause is involved in fracture, whether it is ductile or brittle. The most common locations for threaded fasteners to fracture are either in the head-to-shank-fillet, through the first thread inside the nut, or at the transition point, from the thread to the shank.

I receive many requests for failure analysis either from the insurance assessors, solicitors or metal fabricators every year and get specimens which already have undergone examination by other people or organisations. Thus, the specimens received are seldom in virgin condition. Nevertheless, it is possible to obtain considerable information from the fracture surface of severely damaged specimens. Furthermore the submitter would like to know the cause/s of failure in hopes that the events leading to failure can be determined. Sometime it is difficult to distinguish between stress corrosion cracking and corrosion fatigue. Furthermore a knowledge of circumstances surrounding failure and the operating history of the machine or component can generally be of great help in diagnosing a failure mechanism.

This paper discusses in-service failure of the galvanised bolts from a Four-Wheel drive vehicle.

The vehicle was purchased as new in August 1988 and was kept in excellent condition. The vehicle was serviced regularly and the wheels were rotated at the time of the service. The accident took place on the 23rd October, 1990 and the vehicle was serviced a day before (i.e. 22nd October, 1990). During this service, the wheels were rotated and the bolts were tightened using the manufacturer's wrench. All tyres were in new condition. A speedometer reading was of 76376 kilometer at the time of accident. The right hand rear wheel was detached by shearing of four bolts and stripping the nuts from the other two remaining unbroken bolts. The original nuts and the broken halves of the failed bolts were not found.

EXPERIMENTAL FAILURE ANALYSIS AND RESULTS

The right hand rear wheel hub with four failed bolts and two unbroken bolts, brakedrum and the wheel with tyre were submitted for examination to determine the possible cause/s of failure. Submitted specimens were photographed as completely as possible without initially cleaning or other treatment, and examined visually, macro-and microscopically. Chemical analysis and hardness determination were also made on the failed bolts.

Specifications for the threaded bolts called for medium carbon low alloy steel quenched and tempered to a hardness of Rockwell C - 38-40 with a maximum tensile strength of 1150 MN/m^2.

VISUAL AND MACROSCOPIC EXAMINATION

To identify the failure origin and type, it is imperative that a careful visual and macroexamination of the fracture surface to be done.

The visual examination of the submitted tyre, wheel rim and the hub showed no wear (Figs. 3 and 4). The wheel rim showed some pounding damage at the holes D and F where the bolts remained unbroken. The pounding damage at the hole F seemed to have occurred as the nut was becoming unscrewed (Figure 3). The wheel nut from the bolt D seemed to have completely unscrewed before the failure as there was no apparent damage on the bolt thread or on the rim

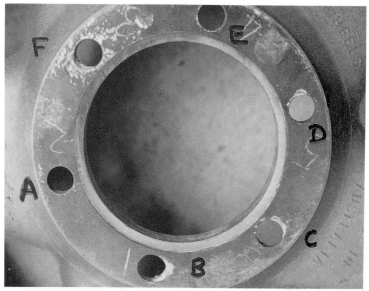

Figure 3   Showing pounding damage on the wheel rim plate hole from the loosened nut of the bolt F.

plate. Furthermore as this was happening (i.e. unscrewing) the play between the rim plate and the hub had developed, causing the adjacent bolt E to snap at the transition point, the first tread to the shank. When the bolt E snapped the nut from the bolt F started to unscrew while pounding on the wheel - rim surface depositing a large amount of material from the rim plate onto the bolt threads (Figs. 3 and 4).

Figure 4.   Unbroken bolt F showing excessive chewing and smearing damage with a large deposit of material from the rim plate.

Figure 5.   The bolt threads showing deposited material from the rim plate as the wheel nut unscrewed.

The unbroken bolts (D and F) showed considerable chewing and smearing damage with some clogging from the wheel-rim alloy (Figs. 4 and 5). The chewing and clogging damage was on the portion of threads that was in the wheel-rim. This strongly indicated the movement of the rim on the bolt threads as the wheel nut being loosened slowly and unscrewed. This caused the loss of intactness between the wheel rim and the brakedrum - hub assembly establishing fatigue condition in the remaining bolts at the interface of the outerface of the brake drum and the wheel rim. Furthermore, the wheel nuts that were on the bolts F and D appeared to be unscrewed first before the detachment of the wheel as the threads on these bolts apparently showed no sign of being stripped. There was also no damage observed on the other areas of these bolts. Furthermore with a loss of intactness of the hub to the wheel, a severe strain at the transition point, thread to shank on the remaining bolts A, B, and C had occurred.

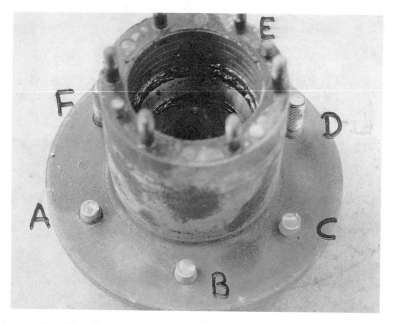

Figure 6.   The hub with four broken bolt heads (A, B, C, and E) and two unbroken bolts (D and F).

As the vehicle still moving, the intactness between the hub and the wheel was further lost resulting a considerable play between the hub and the wheel initiating larger stresses at the interface. This situation established a fast fatigue condition on the remaining three bolts A, B and C. As the vehicle was still moving, many small fatigue cracks at the transition point of thread to shank occurred, and seemed to have propagated rapidly causing failure of the remaining bolts A, B and C and the detachment of the wheel.

The hub with the broken and unbroken bolt-heads was examined for the damage (Fig. 6). The hub showed no apparent damage. The bolt heads that remained in the hub were tight showing no evidence of any apparent movement at the hub brake interface. The bolts A, B, C and E broke off level with the outside edge of the brake drum (Fig. 7). The failure also corresponded with the end of threads, the transition point thread to shank. The broken bolts A, B and C caused the abrasion marks on the brake drum surface as they broke off.

Figure 7. Assembly of brake drum and hub. The failure occurred at the transition point, from the thread to shank just outside edge of the brake drum as shown in the photograph

Visual, macroscopic and scanning electron microscopy of the fracture surfaces of the failed bolts A, B and C disclosed beach marks, indicative of fatigue cracking. The fatigue loading appeared to be mostly reversed bending in nature. However the area of the fatigue zones varied on the fracture surface of each bolt suggesting that the loading on each bolt was different. The fracture had occurred transversley across the bolt shank at the first thread. The fracture surface of the bolt B exhibited beach marks characteristics of high stress reversed - bending fatigue failure, the width of the final rupture was only 25 percent of the shank diameter (Fig. 8).

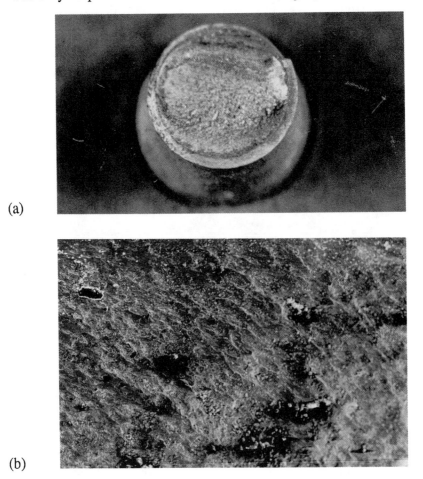

(a)

(b)

Figure 8 (a) Showing beach marks characteristics of fatigue failure.
(b) Coarse fatigue striation "tyre track" markings (1000X).

Failure of the bolt B caused severe stress on the adjacent bolts A and C and as a result these bolts (bolts A and C) were overloaded and failed.

Fatigue markings (tyre tracks) shown in Fig. 8b seemed to be caused by small hard particles in the matrix interupting fatigue crack propagation. The spacing between the track is an indication of the relative shear displacement with each successive cycle. Furthermore, if shear stress components are high, tyre track will be absent or very widely spaced. However their existence indicated that the crack opened and closed with very little permanent displacement per cycle.

A single crack origin can be seen in Fig. 8,with no evidence of fretting at the crack origin. The very initial cracking seemed to have propagated straight across the bolt diameter for approximately 2mm, as if a bending moment were applied only at the crack original location. Cracking than changed abruptly to a manner more typical of rotating bending (Fig 8). The final fracture region was somewhat larger thus indicating sudden high service stresses.

Careful examination of the bolt C rim plate location revealed some brinneling. This seemed to be caused by a sudden load high enough to fracture the bolt C at the transition point, the shank first thread. The cracking mechanism was intergranular at the origin (Fig. 9a), became mixed mode cleavage with dimples (Fig. 9b), and ultimately changed to shear. Thus the fracture surface of the bolt C had a mixture of cleavage, macrovoid coalescence and intergranular cracking mode of failure, a typical fast fatigue fracture (Fig. 9). Although all the bolts had been galvanized, there was no sign of corrosion fatigue in these bolts.

CHEMICAL ANALYSIS AND HARDNESS
Chemical analysis and hardness determinations were made on the failed bolts. The bolts were made from a high tensile steel equivalent to AISI 4140. While the wheel nuts were made from mild steel (AISI 1025). The bolts were heat treated and tempered to Rockwell hardness Rc 30-32. The hardness adjacent to the fracture surface was in the range of RC 25-28 which seemed to be somewhat lower for the hardened and tempered AISI 4140 steel and may have contributed to overload failure.

MICROSCOPIC EXAMINATION
Microscopic examination of the transverse and longitudinal section revealed that the bolts were in hardened and tempered condition with a

microstructure of tempered martensite while the microstructure of the wheel-nut consisted of ferrite and pearlite grains, a low carbon steel (Fig. 10).

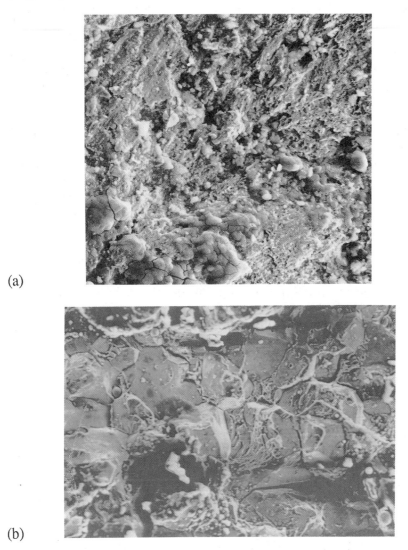

(a)

(b)

Figure 9. Showing mixed fracture (a) mostly intergranular (b) mixed intergranular with transgranular cleavage and dimples in the bolt C. (1200X).

(a)

(b)

Figure 10   (a) Micrograph of the failed bolt B adjacent to the fracture surface showing structure consisted of tempered martensite. (650X). (b) Micrograph of the wheel-nut consisted of ferrite and pearlite (250X).

The SEM micrographs of the bolt A polished section from the fracture surface indicated that the surface consisted of a multiple faceted blocky texture with some extremely fine river markings which were variable in orientation relative to the failure direction. The morphology of these flat blocky regions seemed to be controlled by the martensitic lath structure as shown in Fig. 11. The fracture and microstructural characteristics suggest that crack propagation occurred by the formation and linking of very small cleavage cracks ahead of the main front.

Furthermore, the SEM micrograph shows the blocky feature which appears to be caused by substructure within the tempered martensitic packets (Fig. 11).

Figure 11   SEM Micrograph showing martensite lath structure (8000X).

CONCLUSIONS

-       The bolts failed by reverse bending fatigue cracking initiated at the transition point (the shank and the first thread).

-       The wheel nuts that were on the bolts D and F were not tightened adequately and uniformly. They appeared to be unscrewed setting up fatigue cracking in other bolts (A, B and C).

-       The bolts were hardened and tempered. The microstructure consisted of tempered martensite. The microstructure showed no deformation and the zinc coating was intact indicating no sharp notch had occurred during service.

-       The detachment of the right side rear wheel was a result of unscrewing the wheel nuts on the bolts D and F which in turn promoted the failure of bolts A, B and C by fatigue.

METALLOGRAPHIC EVALUATION OF TEMPER EMBRITTLEMENT

R.H. Richman[1]

ABSTRACT

Temper embrittlement, which is caused by the segregation of impurity elements to prior-austenite grain boundaries, can substantially reduce the toughness of low-alloy steel components in high-temperature service. A method was developed for nondestructive measurement of phosphorus segregation at grain boundaries in alloy steels. This method consists of preferential etching at prior-austenite boundaries, replication of the etched grooves with acetate, and measurement of protrusion heights on the replicas by scanning electron-microscopy. Feasibility was demonstrated with specially-prepared, NiCrMoV steels. Excellent correlations were obtained between etch depth and phosphorus segregation measured by Auger-electron spectroscopy. Subsequent experiments with aged core samples from NiCrMoV rotor forgings confirmed that etch depths are strongly correlated with changes in fracture-appearance transition temperature. Etching is even deeper in aged 1CrMoV than in NiCrMoV steels, and the results are harder to interpret. This uncertainty can be attributed to the greater variability of microstructures in 1CrMoV rotor steels and to the tendency for corrosion products to occlude the developing grain-boundary grooves.

INTRODUCTION

Long-term exposures of alloy steels to temperatures in the range 300° - 600°C can result in a progressive reduction in toughness known as temper embrittlement. It is caused by segregation of impurities, principally phosphorus, to prior-austenite grain boundaries. The main concern is that as fracture toughness is lowered, the critical flaw size for unstable fracture becomes progressively smaller.

---

[1] Daedalus Associates, Inc., 1674 N. Shoreline Blvd., Mountain View, CA 94043.

Traditionally, the degree of embrittlement is determined by Charpy impact tests with V-notched specimens. A common criterion is the temperature at which the fracture surface appears to be 50-percent brittle, i.e., the 50% Fracture Appearance Transition Temperature (FATT). FATT has been shown to be directly related to metalloid concentration at prior-austenite boundaries [1].

Since the steels that are used for steam-turbine rotors are known to be susceptible to in-service temper embrittlement [2,3], powerplant operators need a way to assess the current status of their machinery. Excision of Charpy bars for impact tests would be destructive. In this paper we describe a metallographic method as one candidate route to nondestructive evaluation of temper embrittlement.

BACKGROUND

It has been known for some years that phosphorus segregation in steels accelerates intergranular attack by certain picric-acid-based solutions [4,5]. An extensive investigation by Bruemmer, *et al.* [6] established that the depth of penetration by the picric-acid etchant consistently reflected both the concentration of phosphorus at prior-austenite boundaries and the change of FATT of 3.5NiCrMoV steels. Penetration depths were determined by progressively polishing with 1 μm diamond paste until the intergranular grooves were removed.

Picric-acid etching (PAE) followed by iterative polishing is relatively simple, rapid, and inexpensive. However, it cannot readily be accomplished on large components in place, since the 2.5-hour etching exposures, microhardness indentations to keep track of surface recession during iterative polishing, and microscopic magnifications of at least 200 diameters to observe progressive disappearance of grain-boundary grooves, are difficult to implement in power plants. In order to improve the field usability of the PAE method, the approach taken here was to replicate the etched surfaces and then measure the groove depths by scanning-electron microscopy (SEM).

EXPERIMENTAL PROCEDURES

Materials. Samples from three laboratory heats of NiCrMoV steel, eight commercial heats of NiCrMoV steel, and six rotors made of 1CrMoV steel were included in the study. Bulk compositions of the three laboratory heats are listed in Table 1. Phosphorus concentrations in heats 10 and 13 represent levels near the maximum expected for rotors still in service, while a more typical level is reflected in heat 14. Samples from the three heats were austenitized at 985°C for two hours, air cooled, and tempered at 670°C for three hours to produce an upper bainitic microstructure with a hardness of 22 Rc. They were then embrittled at 480°C for 100 hours and 5000 hours.

Commercial heats of the NiCrMoV steels had essentially the same base composition as the laboratory heats and they are described in detail in [6]. Bore corings from actual rotor forgings were aged at temperatures in the range 343°-454°C for times up to 39000 hours. Samples were halves of Charpy bars machined from the bored corings. Specimens from CrMoV rotors (typically 1% Cr, 1% Mo, and 0.25% V) represent service embrittlement; the number of years in

Table 1. Chemical Composition of Laboratory Heats of NiCrMoV Steel.										
Heat	C	Ni	Cr	Mo	V	Mn	Si	Sn	P	S
10	0.242	3.28	1.60	0.61	0.10	0.35	0.02	0.021	0.031	0.006
13	0.232	3.31	1.59	0.60	0.10	0.35	0.01	0.004	0.029	0.005
14	0.240	3.25	1.58	0.58	0.10	0.35	0.01	0.018	0.011	0.005

service, service temperature at the sampled location, and bulk concentrations of metalloid constituents are given in Table 2.

Auger-Electron Spectroscopy. Detailed characterization of grain-boundary segregation by Auger-electron spectroscopy (AES) was performed only for the laboratory heats of NiCrMoV steel. All measurements were taken with Physical Electronics Models 545 and 560 at a primary beam voltage of 5kV. Details of the experimental parameters and calculation methods are given in [7].

Picric-Acid Etching. PAE tests were conducted with aqueous, saturated, picric-acid solution containing 10 grams per liter of sodium tridecylbenzene sulfonate at 22°C for an exposure time of 2.5 hours. Long etch exposures at closely controlled temperature are readily performed in the laboratory with small specimens immersed in a stirred etchant that is surrounded by a thermostat. For PAE tests in power plants, other arrangements are required. We chose a variation in

Table 2. Characteristics of Retired CrMoV Rotors						
Rotor	Class	Years in Service	Est. Temp at sample °C	Metalloid Content ppm		
				P	Sn	S
Mustang 1	C	29	850	320	110	270
Mustang 3	C	28	"	"	"	"
Buck 6	C	17	760	440	900	360
Riverbend 6 (Rings 7&8)	NR	NR	700	170	220	120
Wyman 1 (Cold End)	D	30	<400	280	100	310
Wyman 1 (ot End)	D	30	860	"	"	"
Riverside 5	C	34	≈800	350	NR	280
NR - Not Reported						

which etchant is pumped around a closed loop that contains a reservoir of temperature-controlled etching solution. The etchant is confined to the area of the etching head; a built-in valve opens when the head is pressed against the object to be etched and the solution then flows across the metallographically-prepared surface and back to the reservoir. This was achieved by modifying a commercially-available, portable apparatus made by Struers. The main modification was provision for active cooling of the pump motor. The resulting system was very well behaved; there was no leakage of etchant during 2.5-hour exposures, and spillage accompanying disconnect was typically one drop.

Replication. Early in the investigation a complicated procedure for making replicas was devised, which involved three layers of cellulose acetate each 0.1mm thick [7]. It later turned out that equivalent results could be obtained with Struers Transcopy replicas (self-adhesive with aluminum backing), provided that the replica was applied with about 14kg force normal to the surface, without sliding or rotation, and the force maintained for at least one minute.

Three replicas were prepared for each specimen and examined in the SEM. Heights of individual features on the replica were determined by reference to glass or polystyrene spheres (5μm) placed on the replica surfaces before a 20-50nm sputtered coating was applied. A typical micrograph is shown in Figure 1.

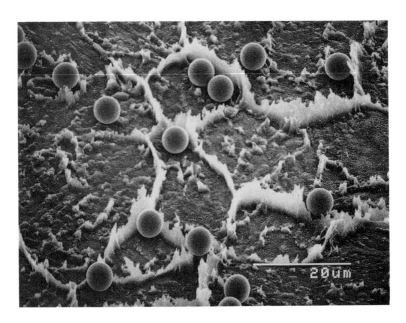

Figure 1.   Replica from preferentially-etched sample of aged NiCrMoV Steel (Heat 045).

RESULTS

Laboratory NiCrMoV Steels. Table 3 summarizes the AES results [7] and the maximum etch depths determined by the PAE-Replica procedure. Although phosphorus segregation increased consistently with aging time at 480°C for each of the heats, there was considerable variability in the monolayer coverage within each specimen. Since we always sought the maximum etch depths (protrusion heights) on the replicas, the maximum of each percent-coverage range is plotted against etch depth in Figure 2. These results show excellent correspondence between etch depth and phosphorus segregation, which can be expressed by the following linear relation:

$$\text{Max.EtchDepth, } \mu m = 5.28 + 0.64(\text{Max.P, \% Monolayer}) \quad (1)$$

with a correlation coefficient, r, of 0.994. Moreover, the etch-depth measurements from replicas compare very well with those from incremental polishing; the correspondence between 100% IG depths [7] and replica depths is shown in Figure 3. Note that the positive intercept of the least-squares line is attributable to differences in the way the specimens were etched in the two methods, about which we will say more in the Discussion.

Table 3. Summary of Auger-Electron Spectroscopy and Etch Depths for Laboratory Heats of NiCrMoV Steels.

Heat	Aging Treatment, °C/hours	P Coverage, % Monolayer	Maximum Etch Depth from Replics, μm
10	As Tempered	2.1 - 5.3	8.1
	480/100	8.9 - 18.0	16.2
	480/5000	8.4 - 26.5	22.0
13	As Tempered	3.2 - 7.1	10.9
	480/100	5.0 - 15.8	15.2
	480/5000	14.0 - 23.3	20.9
14	As Tempered	—	5.7
	480/100	0.8 - 1.8	6.7
	480/5000	2.0 - 5.4	8.2

Commercial NiCrMoV Steels. An important objective of this work was to confirm that PAE-Replica tests can consistently indicate embrittlement in rotor steels. Since AES determinations of grain-boundary segregation were not performed for the commercial NiCrMoV steels,

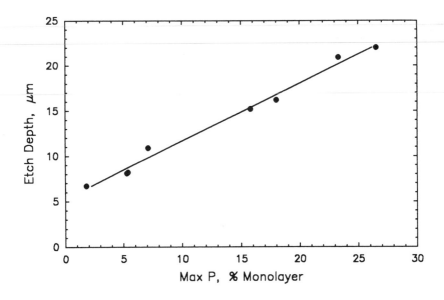

Figure 2.   Correspondence between phosphorous segregation to prior-austenite boundaries and depth of attack in PAE test for laboratory heats of NiCrMoV steels.

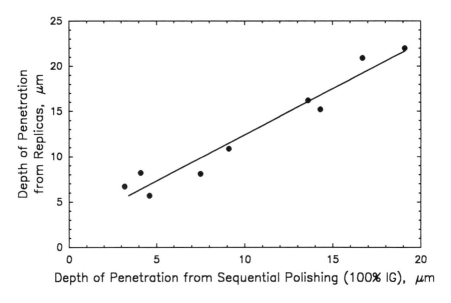

Figure 3.   Comparison between etch depths measured by sequential polishing and by replicas for laboratory heats of NiCrMoV steels.

we relied on measurements of the shift in FATT (ΔFATT) between the original manufactured condition and the aged condition. Table 4 presents the aging treatments, current FATTs, ΔFATTs, and etch depths from replicas. The correspondence between ΔFATT and etch depth is shown in Figure 4. Except for one determination (Heat 055-A), there appears to be a close relationship.

Service-Embrittled 1CrMoV Steels. The response of these specimens to PAE was unexpected, because past trials with 1CrMoV rotor samples had produced only modest etch depths

	Table 4. PAE-Replica Results for Aged NiCrMoV Steels			
Specimen	Aging Treatment, °C/hours	FATT, °C	ΔFATT, °C	Etch Depth from Replics, μm
045	343/16K	-17	11	9.5 - 9.8
055-A	343/39K	27	17	13.7 - 14.0
618-A	343/39K	27	42	13.7 - 14.9
653-B	371/37K	99	64	18.4 - 19.6
659-A	343/34K	43	8	7.8 - 8.0
691 (5)	343/79K	36	41	13.1 - 13.3
691-B	371/32K	71	78	23.3 - 24.2
693-B	371/32K	35	53	15.0 - 16.0

[7]. In contrast, a typical micrograph of an embrittled rotor sample in the present study is shown in Figure 5. Measurements are summarized in Table 5 and plotted in Figure 6. The line drawn in Figure 5 refers to C-class rotors, which were austenitized at 1010°C; measurements of two locations on the D-class Wyman rotor (austenitized at 954°C) clearly belong to a different population.

DISCUSSION

The meaning of the results reported here hinges on deciding correctly about what the PAE method actually measures. At this juncture, it seems reasonable to state that the method reflects the amount of phosphorus at prior-austenite grain boundaries. That information was not available for commercial NiCrMoV and 1CrMoV steels. Therefore, etch depths were correlated with FATT or ΔFATT, although that step certainly would have been taken anyway.

Several factors other than phosphorus segregation (e.g., strength level, grain size, uniformity and distribution of microstructural constituents, and other segregants) contribute to FATT of rotor steels. Thus, consistent correlation of PAE depths to FATT or ΔFATT should not be expected. The fairly regular dependence of etch depth on ΔFATT of commercial NiCrMoV steels can probably be attributed to a uniformity of initial strengths and microstructures in this

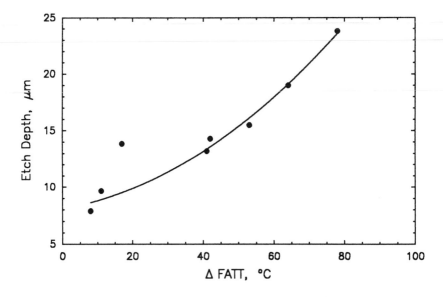

Figure 4.   Correlation between etch depths and ΔFATT of commercial NiCrMoV steels.

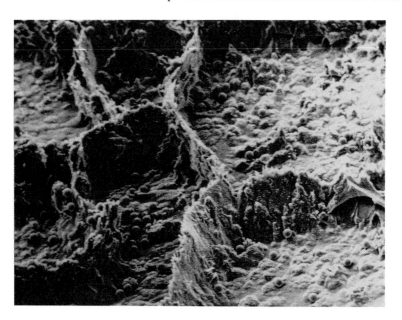

Figure 5.   Replica from preferentially-etched sample of service-embrittled 1CrMoV steel (Riverside).

Table 5. PAE-Replica Results for Service-Embrittled 1CrMoV Steels

Rotor	Current FATT, °C	Etch Depth, μm
Mustang 1	246	22.7 ± 1.0
Mustang 3	202	17.4 ± 1.2
Buck 6	315	41.2 ± 2.0
Riverbend 6-R7	207	23.3 ± 1.2
Riverbend 6-R8	210	21.3 ± 1.0
Wyman 1-CE	99	29.4 ± 1.4
Wyman 1-HE	232	50.1 ± 1.7
Riverside 5	382	56.5 ± 1.5

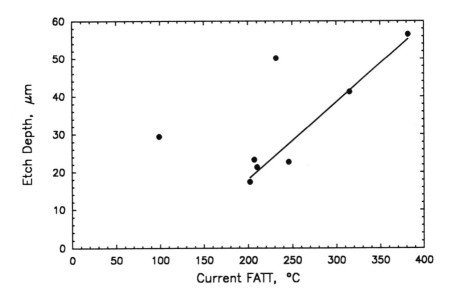

Figure 6.   Relation between etch depth and current FATT of service-embrittled 1CrMoV rotors.

class of rotor materials. 1CrMoV microstructures are more variable and so the etch depths are less consistent. On the other hand, consistent results might be expected from the FATT difference between aged and thermally de-embrittled (denoted $\Delta FATT_Q$) specimens, since those changes would be a consequence primarily of phosphorus segregation.

A particular source of frustration has been the lack of agreement between results of sequential polishing and results from replication of embrittled 1CrMoV steels, summarized in Table 6. There appears to be no correlation between 90% IG depths and replica depths, and only slight correlation between 100% IG and replica results. We suspect that the way specimens were etched is the main cause of disagreement.

Initial feasibility of the PAE-Replica method was established by immersing specimens in small quantities of etchant and stirring vigorously. A dark flocculent was always generated in the solution by the more embrittled (deeper etching) specimens. This observation implies that there is a tendency for prior-austenite boundaries to be occluded by corrosion products during the long etching exposures. As it happens, the etching procedure for sequential polishing involved only "light" stirring of picric-acid solution.

If there is a tendency to occlude more embrittled boundaries, differences between etch depths from full-flow impingement (for which occlusion effects would be much smaller) and etch depths from immersion should depend on the degree of embrittlement (i.e., of etch depth). This is demonstrated in Figure 7. Note that replica depths (with the exception of Riverbend 6 - Ring 8) are now highly correlated with the <u>differences</u> between replica depths and sequential-polishing depths, either 90% IG (r=0.993) or 100% IG (r=0.994).

Table 6. Results of Two Different PAE Methods		
	Etch Depth by Sequential Polishing, μm	Etch Depth by Replicas, μm
Rotor	Avg. 90% IG Max 100% IG	
Mustang 1	10.4 18.8	22.7
Mustang 3	10.3 17.9	17.4
Buck 6	12.0 21.9	41.2
Riverbend 6-R7	13.0 18.3	23.3
Riverbend 6-R8	16.1 24.3	21.3
Wyman 1-CE	7.5 18.0	29.4
Wyman 1-HE	12.5 21.2	50.1
Riverside 5	12.8 26.2	56.5

CONCLUSIONS

PAE-Replica tests, in which etching is performed by full-flow impingement, can assess nondestructively the degree of phosphorus segregation to prior-austenite boundaries. Although characterization of embrittlement is not nearly as secure as evaluation of phosphorus segregation, good correlation is obtained between etch depth and ΔFATT of aged NiCrMoV steels, and between etch depth and current FATT of C-class 1CrMoV steels. It is felt that determination of embrittlement would be improved by subsequent thermal de-embrittlement of embrittled samples and re-measurement of etch depths to establish baseline differences in etch depths that should reflect more accurately the degree of phosphorus segregation.

Acknowledgements. Most of this work was conducted under contracts RP2426-7 and RP1957-7 to the Electric Power Research Institute. The author is grateful for that support and

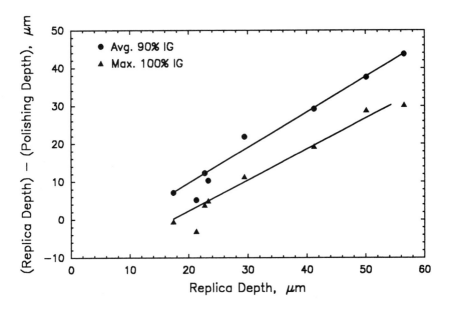

Figure 7.   Dependence of the difference between etch depths measured by replicas and by sequential polishing on the degree of embrittlement (replica depths) of 1 CrMoV rotor samples.

for the guidance provided by the EPRI project managers, Drs. R. Viswanathan and S.M. Gehl, respectively. Special thanks are also due to J.A. Maasberg for electron microscopy, to R.L. Cargill for development of the procedures and apparatus, and to Dr. S.M. Bruemmer for numerous discussions during the course of the investigation.

REFERENCES

1) R. Viswanathan, "Temper Embrittlment in a Ni-Cr Steel Containing Phosphorus as Impurity", **Metall. Trans.**, Vol. 2, p. 809, 1971.

2) G.C. Gould, "Long Time Isothermal Embrittlement in 3.5Ni, 1.75Cr, 0.50Mo, 0.20C Steel", **Temper Embrittlement in Steel, ASTM STP 407**, Philadelphia, PA: American Society for Testing and Materials, p. 90, 1968.

3) R. Viswanathan and R.I. Jaffee, "Toughness of Cr-Mo-V Steels for Steam Turbine Rotors", **ASME Joint Power Generation Conf.**, Paper No. 82-JPGC-Pwr-35, 1982.

4) A.H. Ucisik, C.J. McMahon, Jr., and H.C. Feng, "The Influence of Intercritical Heat Treatment on the Temper Embrittlement Susceptibility of a P-Doped Ni-Cr Steel", **Metall. Trans. A**, Vol. 9A, p. 321, 1978.

5) T. Ogura, A. Makino, and T. Masumoto, "A Grain Boundary Etching Method for the Analysis of Intergranular P-Segregation in Iron-Based Alloys", **Metall. Trans. A**, Vol. 15A, p. 1563, 1984.

6) S.M. Bruemmer, L.A. Charlot, and B.W. Arey, **Grain Boundary Composition and Integranular Fracture of Steels. Vol. 2: Evaluation of Segregation and Its Synergism with Stress Corrosion and Hydrogen Embrittlement**, Palo Alto, CA: Electric Power Research Institute, EPRI RD-3859, 1986.

7) R. Viswanathan, S.M. Bruemmer, and R.H. Richman, "Etching Technique for Assessing Toughness Degradation of In-Service Components", **ASME Journ. Engng. Mater. Technol.**, Vol. 110, p. 313, 1988.

ANALYSIS OF MERCURY DIFFUSION PUMPS

Kerry A. Dunn[1]

ABSTRACT

Several mercury diffusion pump stages in the Tritium Purification process at the Savannah River Site (SRS) have been removed from service for scheduled preventive maintenance. These stages have been examined to determine if failure has occurred. Evidence of fatigue around the flange portion of the pump has been seen. In addition, erosion and cavitation inside the throat of the venturi tube and corrosion on the other surface of the venturi tube has been observed. Several measures are being examined in an attempt to improve the performance of these pumps. These measures, as well as the noted observations, are described.

[1]Westinghouse Savannah River Company, Savannah River Laboratory, Aiken, SC 29802

SUMMARY

Six stages [two machined (MP) and four electron beam (EB) welded] from the mercury diffusion pumps operating in the Tritium Purification process at SRS have been analyzed to determine their condition after nine months of usage. Several cracks were found around the necked region of the two MP stages. The EB welded stages, however, seemed to perform better in service—two of four stages showed cracking. The cracking is caused by fatigue that has been enhanced by high stresses and tritium in the flange area.

The EB welded stage appears to be a step in the right direction. Since the EB weld is a shrink fit, the surface is in compression, thereby eliminating crack propagation. In addition, shot peening has been employed to produce a compressive material surface since fatigue usually originates at the surface.

Pitting was observed down the throat of the venturi. This pitting was caused by cavitation and erosion along the length of the venturi tube.

Corrosion and pitting was seen on the exterior walls of the diffuser tubes. Stress–corrosion cracks were observed emanating from these corrosion pits. The corrosion likely occurred from the chloride ions present in the process cooling water. Shot peening is now being used in an attempt to place the outside of the diffuser tube in compression to eliminate the stress–corrosion cracking.

INTRODUCTION

The mercury diffusion pump in tritium is a three–stage, booster–ejector pump originally manufactured by Consolidated Vacuum Corporation (CVC). This pump pulls gas from the furnaces through a uranium decomposer bed to the Sprengel forepumps for further purification. The unit can increase gas pressure from microns up to approximately 30 kPa, with a maximum capacity of approximately 8000 cm^3/minute (similar to most Hg vacuum pumps). Mercury is vaporized in the boiler and pushed to the top of the pump where it condenses and entrains incoming gas and flows downward through a diffuser tube or venturi. The diffuser tubes are cooled by the surrounding cooling water jackets that initially condense the mercury as it flows down the venturi [1]. The mercury vapor temperature inside the diffuser tube is around 573K and the outside temperature is about 293K, which creates a tremendous thermal gradient between the walls. Figure 1 represents one stage of the mercury diffusion booster–ejector pump.

There are two mercury diffusion pumps (237 and 232) in the Tritium Purification process at SRS. The 237 pump requires a more rigorous operation than the 232 pump because it sees tritium gas prior to purification. Once the gas reaches the 232 pump, it has already been through the uranium decomposer bed and the palladium/silver diffuser for purification. Originally, the position of the stages and their performance were thought to be independent of one another. However, because of insufficient evidence for that hypothesis, stage performance in reference to position is now being monitored.

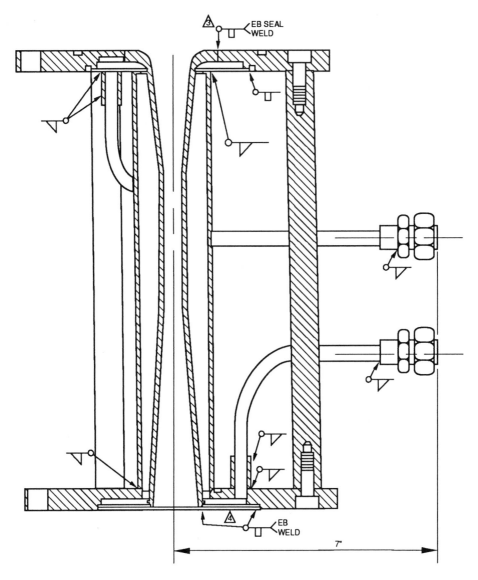

Figure 1. Schematic of CVC Third Stage

The original mercury diffusion pump stage design incorporated a tungsten inert gas (TIG) weld to interface the flange with the diffuser tube. Failures with these stages, at the TIG weld, occurred in 1986 and 1987 after only three months of service. An investigation was initiated to determine how the life of the diffuser stages could be extended.

Table 1. Mercury Diffusion Pump Stage Information

Name	Stage	Start Date Examine	How Fabricated	Pump Position
1988 EB	3rd	10/88	EB welded	—
1988 MP	3rd	10/88	machined from single block	—
1989 EB (cracked)	3rd	10/89	EB welded and shot peened from weld to outer flange	—
1989 EB	3rd	10/89	EB welded and shot peened from weld to outer flange	—
1990 EB	3rd	7/90	EB welded and shot peened on all flange and outer venturi	237
1990 MP	3rd	7/90	machined from single block	232

Finite element analyses were employed to examine the stress states that were present due to thermal stress, residual weld stress, shrink–fit weld stress, and the combination of thermal and residual weld or shrink–fit weld stresses on the stage of the mercury diffusion pump diffusers [2]. The analyses showed that the presence of principal stresses, as a result of high bending stresses, were located in the venturi tube area and the necked region for the thermal case. However, the stress analysis for the residual welding stress case predicted tensile stresses in the horizontal and necked regions of the mercury diffusion pump stage. According to the study, neither of the stress–related problems would, separately, cause any cracking through the wall. However, combining the two stresses would result in high–tensile stresses located in the necked region. From this finite element analysis, several modifications to the previous design were made. The two most recent designs included a stage machined out of a single block of steel, while the other design incorporated a shrink–fit EB weld between the top flange and the diffuser tube.

In an attempt to maintain failure-free pumps, the current preventive maintenance schedule requires the removal of the third stages after nine months of service and the first and second stages after 18 months of service. Although no service failures have occurred while using this preventive maintenance regimen, cracking has been observed. Table 1 shows the third stages examined and their properties.

PROCEDURE

All stages removed from service were tested with a dye penetrant to determine if any cracking was visible in the necked region of the stages. Once tested, cracked stages were stored in an air hood until the material had sufficiently outgassed. This was done so that sectioning could be performed without tritium contamination. Most of the flange was cut away to expose the diffuser tube region. The diffuser tubes were sectioned along their length adjacent to cracks. Once the pieces were of the workable size, stereomicroscopy and scanning electron microscopy (SEM) were used to examine the cracks in the necked region of the venturi. In addition, metallography was used to examine both crack morphology and crack depth.

Energy dispersive spectroscopy (EDS) was used to examine the extraneous debris on the inner surface of the diffuser tube as well as the corrosion product on the water side. SEM was employed to examine pitting of both the inner and outer surfaces of the stage prior to metallographic examination of the pits and cracking.

RESULTS AND DISCUSSION

Dye-penetrant checks of the first and second stages showed no cracking in the flange. Since these stages see a smaller velocity of gas and mercury flowing through them, their cyclic and tensile stresses are lower, therefore decreasing the probability of fatigue failure.

One of the more recent designs has involved machining the stages out of a single block of 304L stainless steel in an attempt to eliminate flaws that may be introduced by welding processes. Two machined third stages were removed from service, one in 1988 and one in 1990. Dye–penetrant testing revealed approximately 12 cracks around the necked portion of each stage (Figures 2 and 3). A full destructive examination of both the 1988 MP and the 1990 MP stages has been conducted.

Figure 2. Dye–Penetrant Test from 1988 MP

Another design that has been examined recently is the shrink–fit, EB weld technique to connect the diffuser tube to the top flange. The shrink fit was used to place the origin of cracking (inside surface of necked region) in compression.

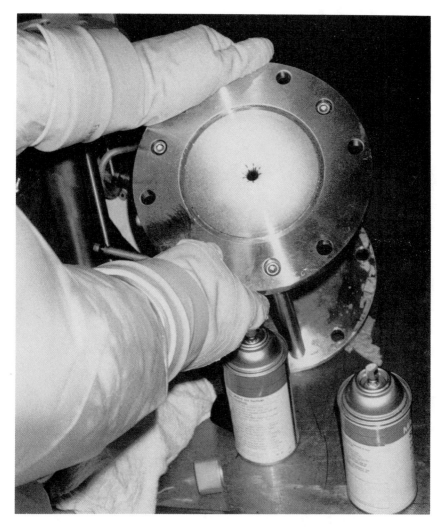

Figure 3. Dye–Penetrant Test from 1990 MP

In two of the four EB welded third stages removed from service since 1988, dye–penetrant testing revealed cracking (Figures 4 and 5).

Previous investigations by Ehrhart and Eberhard [1, 3] have attributed the mercury diffusion pump failures to liquid metal embrittlement and stress–induced cracking. They also claim that mercury globules were present in the cracks of the metallographic specimen. It is not unusual, however, that mercury droplets were found in the cracks since mercury flows through the tubing. Furthermore, any type

179

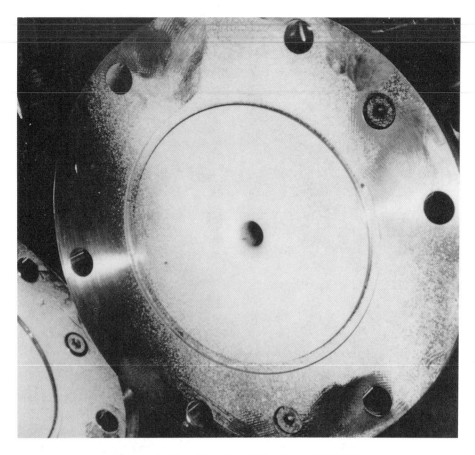

Figure 4. Dye–Penetrant Test from 1988 EB

of debris present on the parts will tend to congregate at a crack site. Upon examination of liquid metal embrittled samples, one should see a complete coating of the liquid metal. In fact, the liquid metal is in such close contact with the solid that it is generally difficult to remove. Therefore, the presence of mercury droplets in the crack is not sufficient justification for concluding liquid metal embrittlement. The present studies found no coating of mercury, but a few mercury droplets were observed. Liquid metal embrittlement has, for the most part, been ruled out as the failure mechanism.

Following the sectioning of the stages, it was observed that the cracks in all samples were approximately 2–4 cm along the length of the diffuser tube. The cracks were examined by SEM (Figure 6). It was unclear from these micrographs whether the cracking occurred intergranularly or transgranularly. In addition, the SEM of opened cracks did not reveal a crack origin or reason for fracture. Several particles were observed on the interior surface of the stages near to and around the cracks (Figure 7). The particles were rich in silicon (Si), aluminum (Al), and potassium (K), as shown by the EDS spectrum in Figure 8. Although great in number, the particles were probably not a factor in the failure of the mercury diffusion pumps. There are two possible explanations for the presence of the particles. First, Al and Si were observed during an elemental examination of a

Figure 5. Dye–Penetrant Test from 1989 EB

spot–check penetrant used in the dye–penetrant testing (Figure 9). The second explanation deals with the presence of zeolite beds upstream in the process, which may provide a source of Al, Si, and K.

Cross–sections of the top portion of four mercury diffusion pump stages were cut to conduct metallographic examinations. Optical microscopy on the mercury

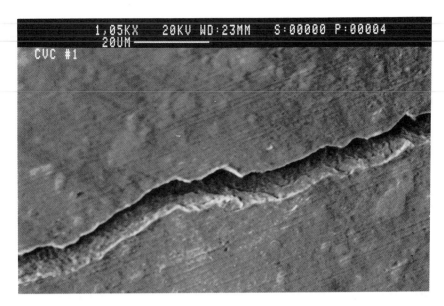

Figure 6. SEM Micrograph of Opened Crack in 1989 EB Welded Stage

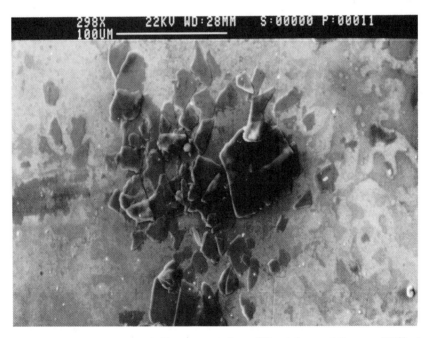

Figure 7. SEM Micrograph Representative of Particles on Mercury Diffusion
Pump Stage

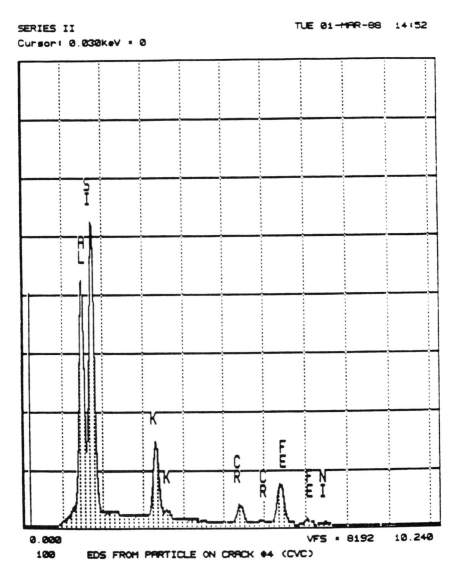

0.000   VFS = 8192   10.240
100   EDS FROM PARTICLE ON CRACK #4 (CVC)

Figure 8. EDS of Particles on Mercury Diffusion Pump Stage

diffusion pump stage pieces was employed to observe the cracking in the necked regions (Figures 10–12). As can be clearly seen, the cracks are mostly transgranular in nature. Propagation of the crack has occurred from the inside of the necked region toward the outside wall and is not indicative of stress–corrosion cracking. In addition, no pitting is visible at the origin of the cracking. The cracks

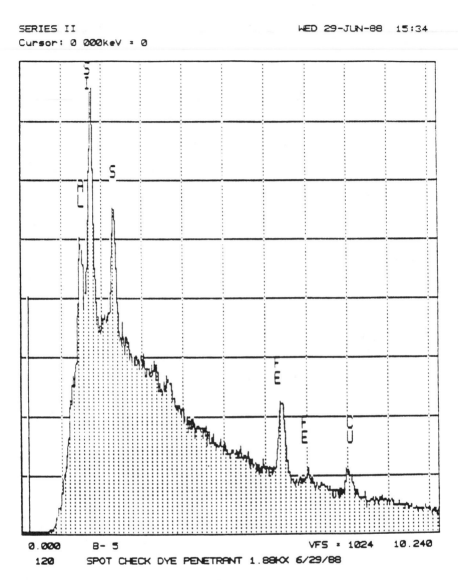

Figure 9. EDS of Dye-Penetrant Check

appear to be mainly related to fatigue. They are clean, there is not much branching, and they are fairly straight. The cracks extended, at the most, one-third of the way into the sections and no through-wall cracks were observed. This is not unusual since the cyclic stress needed to cause fatigue cracking is less than the yield stress of the material. Therefore, a through-wall crack may not occur at all.

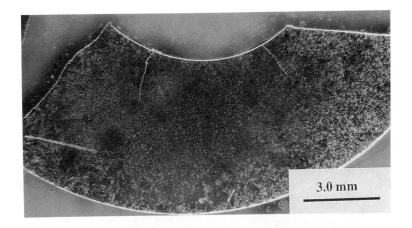

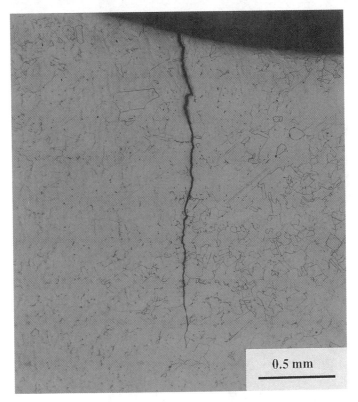

Figure 10. (a) Flange Region of 1988 MP
(b) Higher Magnification of Crack in (a)

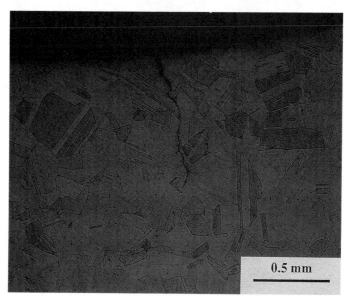

Figure 11. (a) Flange Region of 1988 EB Welded Stage
(b) Higher Magnification of Crack in (a)

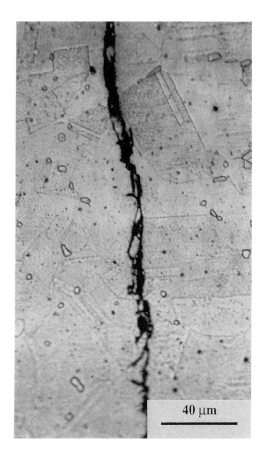
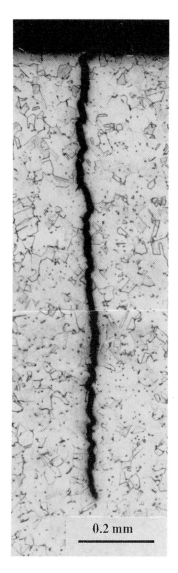

40 μm

0.2 mm

Figure 12. Transverse Section of Crack in 1989 EB Sample
(Note the outlined artifacts in the material.)

There were unusual regions observed in the metallographic section of the 1989 EB welded stage shown in Figure 12. They are small, approximately 5 μm in diameter, and appear to be ferrite. Figure 13, a longitudinal section from this same area, shows the presence of stringers (probably ferrite). Transmission

Figure 13. Longitudinal Section of Crack in 1899 EB Stage

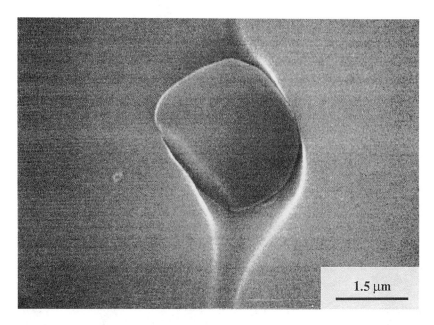

Figure 14. Cr and Mn Rich Particle in Matrix

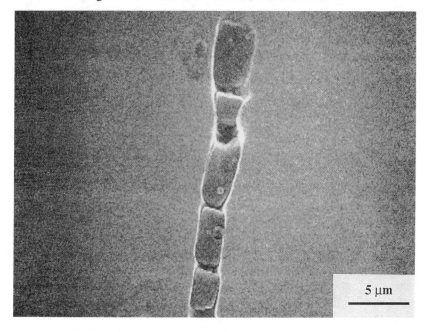

Figure 15. MnS Inclusions in Matrix

electron microscopy (TEM) was employed to identify these artifacts. Particles found in the matrix were either chromium (Cr) and manganese (Mn) rich or manganese sulfide (MnS) inclusions (Figures 14 and 15). Because of problems with thinning, the particles in Figure 14 could not be identified with diffraction patterns. Since Cr is a ferrite stabilizer, it is likely that the particles are ferrite. However, because of the lack of diffraction information and the presence of Mn, ferrite cannot be confirmed.

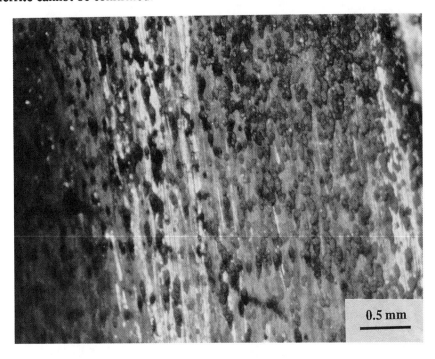

Figure 16. Pits Inside Diffuser Tube from 1988 EB

A large degree of pitting was observed down the length of the venturi tube, primarily in the 1988 stages that were examined. The 1988 EB welded piece was pitted far more than the remaining stages. Macrographs representative of the pits that formed on the interior wall of the 1988 EB welded stage are shown in Figure 16, while SEM micrographs of the pitting are shown in Figure 17. The appearance of the pits is typical of cavitation, which involves a material buildup along a surface. The pressure is increased at the surface into a range such that a cavity is formed from the force of the eventual pressure release. As indicated by the micrographs, the pits are approximately 62.5 μm deep and numerous.

SEM micrographs representative of the pitting in the 1988 MP piece can be seen in Figure 18. The pits in the MP piece typify erosion pitting, where erosion

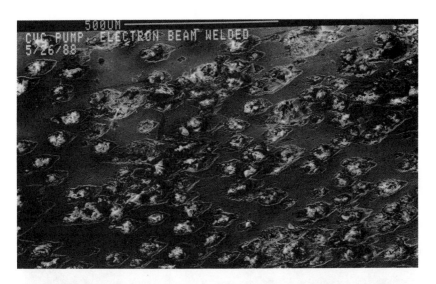

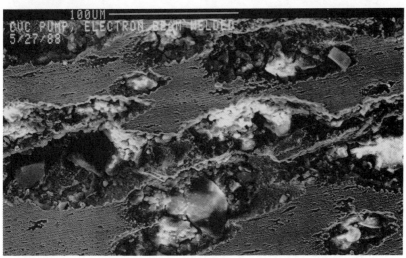

Figure 17. SEM Micrographs of Pits from Inside 1988 EB

is defined as the removal of surface material due to numerous impacts of the surface from solid or liquid particles.

The pits in both the 1988 EB piece and the 1988 MP piece had several particles in them that were rich in Al, Si, and K. However, the particles are not attributed to the cracking in the pump or the pitting. They are believed to be from the process system.

It is evident from the metallography performed on the cross section of the 1988 EB part that, although the pits were very prominent and visible, none of the interior wall pits propagated cracks (Figure 19). This fact is encouraging since the interior of the mercury diffusion pump stage should be in compression from the shrink-fit weld. The pits on the 1988 MP stage (Figure 20) are not as noticeable by metallography as those from the 1988 EB welded stage.

The fact that no cracks were found propagating from the cavitation/erosion pits in 1988 MP and 1988 EB and that pitting was not evident of the interior walls

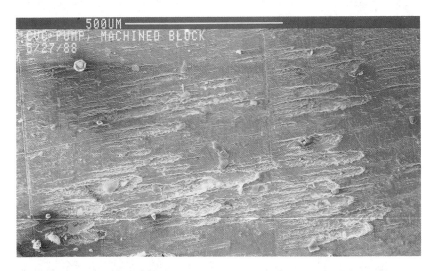

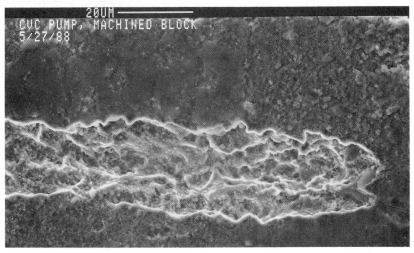

Figure 18. SEM Micrographs of Pits Inside 1988 MP

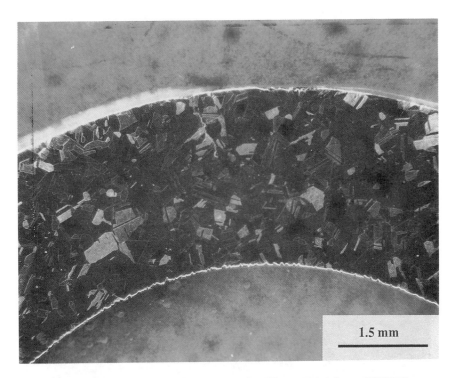

Figure 19. Metallographic Section of Diffuser Tube from 1988 EB

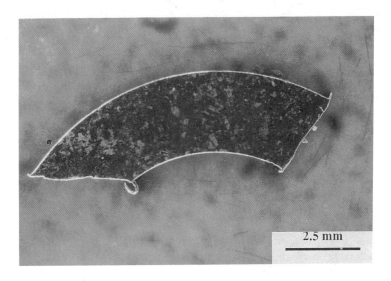

Figure 20. Metallographic Section of Diffuser Tube from 1988 MP

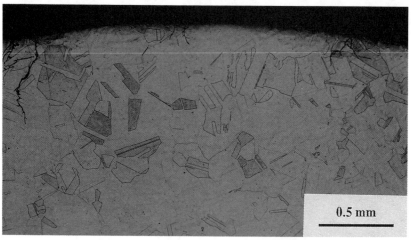

Figure 21. Stress Corrosion Cracks Propagating from
Exterior Pits from 1988 EB

of the two 1989 EB welded stages provides sound reasoning to assume that the
pitting will not contribute to the failure of the pump.

There were, however, several corrosion pits observed on the exterior wall of
most diffuser tubes examined. The pits acted as crack initiators in the 1988 EB and

1988 MP stages (Figures 19–22). The cracking that was observed typifies stress–corrosion cracking. Chloride ions present in the cooling water are suspected to be the corrosive medium. To combat this problem, the outer walls are now shot peened to place the surface in compression. Stage 1990 EB was shot peened on the outside of the diffuser tube and it did not exhibit any corrosion.

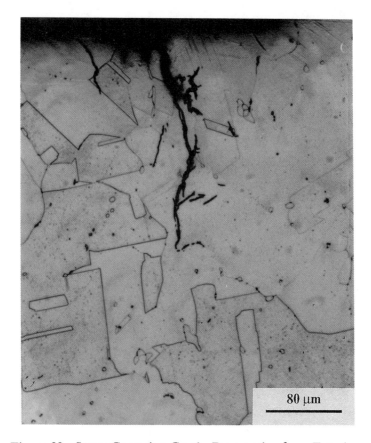

Figure 22. Stress Corrosion Cracks Propagating from Exterior Pits from 1988 MP

CONCLUSIONS

Table 2 shows the results from the stage analyses. From the course of these analyses, several conclusions can be made. The EB shrink–fit weld procedure is

Table 2. Results from Mercury Diffusion Pump Stage Analyses

Name	Cracks in Flange	Pits Inside Diffuser	Pits Outside Diffuser	Shot Peen
1988 EB	1	yes—cavitation	yes—corrosion and crack propagation	no
1988 MP	= 12	yes—erosion	yes—corrosion and crack propagation	no
1989 EB (cracked)	= 6	no	corrosion product/no cracking	yes—EB weld out on flange
1989 EB	no	no	corrosion product/no cracking	yes—EB weld out on flange
1990 EB	no	no	no pits or corrosion product	yes—all flange surface and outside diffuser
1990 MP	= 12	no	some corrosion/no cracking	no

necessary since it appears to have eliminated much of the tensile stress associated with the mercury diffusion pumps. A decreased number of cracks present around the necked portion of the diffuser tube indicates that the design and strength of the material is a step in the right direction. However, it is important to note that material quality is a necessity to having failure-free pumps. Machining the stages out of a single piece of metal will be eliminated because of the presence of numerous cracks observed in the examined stages.

Shot peening the entire surface of the flange also seems to have an advantageous effect toward eliminating the cracks in the mercury diffusion pump stages. However, the shot peening must be conducted on the entire flange rather than from the EB weld outward, or not at all. Stage 1990 EB, the only stage that was shot peened on the entire flange, did not have any cracks in the necked region of the diffuser tube. However, of the three stages that were not shot peened on the entire flange, 1988 EB, 1989 EB, and 1989 EB (cracked), two had cracks in the necked region of the diffuser.

Shot peening produces compressive residual stresses on the surface by the introduction of a high-velocity stream of shot to the surface. With the material surface in compression, the cracks are inhibited. However, the cracks will propagate once the stress is overcome. Future analyses of third stages will continue to provide information on the success of the shot peening.

No cracks were observed propagating from the erosion/cavitation pits in the 1988 MP stage and the 1988 EB stage. In addition, no significant cavitation or erosion was observed along the inside length of the diffuser tubes from the remaining four stages examined. Therefore, neither the cavitation nor the erosion should be deleterious to the life of the mercury diffusion pump stages.

Finally, shot peening the outside of the diffuser tube has been used to eliminate the stress-corrosion pitting and crack propagation associated with the chloride ions. Since shot peening has been shown to provide protection against corrosion and crack propagation, it will continue to be used on future mercury diffusion pump stages.

REFERENCES

1)   W. S. Ehrhart and B. A. Eberhard, EED850081, "Failed CVC-237 Mercury Pump Stage", 1985.

2)   R. J. Thomas and J. M. Cahill, 863807, "Savannah River Plant 232-H Tritium Facilities Booster Pump Diffuser Stress Analysis", 1985.

3)   W. S. Ehrhart and B. A. Eberhard, EED850343, "Cracked CVC-232-F Mercury Pump - 3rd Stage Diffuser", 1985,

4)   R. J. Thomas and J. M. Cahill, WR864295, "Savannah River Plant – Tritium Facilities – 232-H CVC Pump Repair Assistance Three Stage Diffuser Stress Analysis Interference Fit", 1987.

ACKNOWLEDGMENT

The information contained in this paper was developed during the course of work done under Contract No. DE–AC09–89 SR18035 with the U.S. Department of Energy.

METALLOGRAPHIC PREPARATION TECHNIQUES

FOR MICROMECHANICAL DEVICES

D. Clausen, H. Hansen and J.W. Yardy[1]

ABSTRACT

A method for the preparation of metallographic sections of micromechanical devices etched in silicon wafers is described. Because of the small size and complex shape of these devices, in this case 1,5 μm thick underetched beams and thin diaphragms, it was necessary to modify conventional metallographic preparation techniques. By paying particular attention to casting viscosity and curing rate of the mounting resin, it was possible to obtain metallographic sections without entrapped air pockets and with good edge retention. Specimen sectioning, grinding, and polishing techinques were modified to avoid damaging the brittle silicon material.
Metallographic sections of micromechanical devices prepared in this manner were quite suitable for examination by optical microscopy. As a result, the shape of these devices was determined accurately and manufacturing processes were then changed to achieve optimal geometry.

INTRODUCTION

Micromechanical sensors and actuators are produced by micromachining silicon wafers with a combination of doping and etching techniques. A new method has been developed [1], which exploits the sharp selectivity of HF

---

[1] Danfoss A/S, DK-6430 Nordborg Denmark

anodic etching between   p-Si and n-Si, using masked implantation of phosphorous in a p-type silicon wafer for geometry definition. This technique offers new opportunities in the field of micromachining of silicon and has been used to manufacture suspended beams 3 mm long, 1,5 μm thick, 15 μm wide with only 20 μm clearance between beams and 10 μm clearance under the beam.

During development of the techinque it became necessary to determine the effect of etching upon the geometry of the silicon wafer. Initially, micromechanical devices were examined by scanning electron microscopy. But it was decided that metallographic sections would give a more direct determination of etching geometry. Having no previous experience with such small specimens it was decided to modify conventional metallographic preparation techinques to suit the small size and complex shape of these devices.

PROCEDURE

The etched wafers were embedded in epoxy resin to avoid damaging the fragile beams during sectioning and subsequent operations. Prior to embedding, pieces of wafer were supported above the bottom of a casting mould on small spacers to allow free circulation of the epoxy resin. To allow the two component epoxy resin (Araldit D) to flow more readily, the resin component was heated to 305K to reduce viscosity, mixed with the hardening component, and degassed under vacuum to release entrapped air. The casting mould was placed in a Struers Epovac vacuum embedding apparatus and filled under vacuum with the degassed resin. Evacuation was continued until all air bubbles were removed, this was ascertained with the help of a stereomicroscope.

The resin was then cured slowly in a refrigerator at 280K for 70 hours to avoid the build-up of internal stress, thus ensuring good edge retention. After a period at room temperature the resin was given a final hardening cure at 350K for 90 mins. As the cured specimens were only 10 mm high they were sectioned on a Struers Accutom low-speed saw using a diamond cut-off wheel and recast in a mould using cold-mounting resin Demotec 35 to make handling easier. Both 15° taper sections and sections perpendicular to the wafer surface were prepared.

The grinding and polishing procedures were carried out manually, the specimens being successively rough ground on silicon carbide paper from 220 down to 4000-grit. Best edge retention was achieved when grinding towards the silicon edge to be observed. Polishing was also carried out in the same direction using 3-μm diamond paste on Pan-W polishing cloth. Final

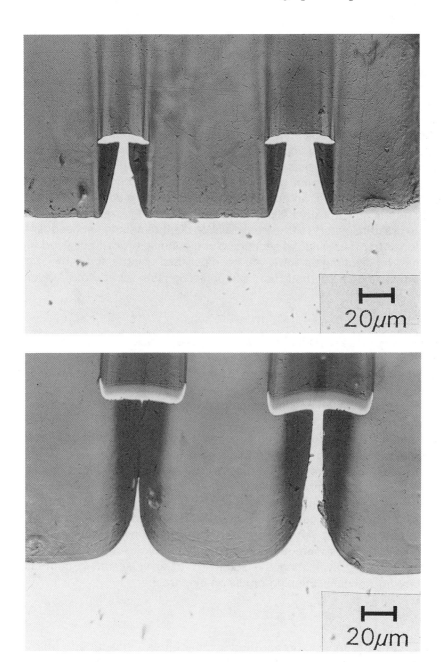

Figure 1. Taper sections of beams partially etched(A) and undercut(B) showing good edge retention between epoxy and silicon.

polishing, with 1-μm diamond on nap polishing cloth, was found to give the best result when the specimen was rotated continuously.

The silicon diaphragms were received bonded to a glass plate, except in areas where a gold electrode was plated on the glass. In these areas the diaphragm bows slightly away from the electrode due to degassing of the glass during previous operations. As one of the aims of the investigation was to ascertain the range of diaphragm displacement, a number of diaphragms were placed in moulds and subjected to vacuum in the vacuum embedding chamber to accentuate vertical displacement. The degassed epoxy resin was then filled into the moulds and alllowed to cure slowly under vacuum for 18 hours. Prior to sectioning, an opening was made in the diaphragm and the gap between it and the gold electrode was impregnated with resin. This resulted in greater support for the fragile silicon diaphram during subsequent preparation. After following the sectioning and grinding procedures used for the beams, the diaphragms were vibratory polished using a slurry of 1-μm alumina abrasive on a synthetic soft nap polishing cloth for approx. 4 hour.

RESULTS

By following these preparation techniques, well-defined cross-sections of beams in etched silicon wafers with good edge retention and free from entrapped air pockets can be achieved. Transparent light-brown areas were observed on the top surface of the beams, indicating that these areas are amorphous due to a high concentration of phosphorous used to produce n-type regions.

As a result of these observations, doping and etching methods were adjusted to produce beams with optimal geometry.

Due to their transparency, the thin silicon diaphragms were initially examined by using a combination of transmitted and reflected illumination during microscopy. The gap between the diaphragm and gold electrode produces interference fringes which can be used as contour lines to measure diaphragm displacement. After filling the gap and sectioning, the shape and thickness of diaphrams were measured directly by microscopy.

We are currently evaluating the suitability of laser scanning confocal microscopy for optical sectioning of silicon diaphragms.

CONCLUSIONS

The metallographic preparation techniques described allow micromechanical devices, in this case beams and diaphragms, to be sectioned and examined by microscopy in order to define the geometry achieved by isotropic anodic etching when micromachining silicon. Metallograghic sections were produced with good edge retention and freedom from entrapped air pockets by paying particular attention to the viscosity and curing rate of the epoxy embedding resin.

ACKNOWLEDGMENT

The authors are grateful to Hanne Søllingvraa for preparing the manuscript.

REFERENCES

1) J. Branebjerg, C.J.M Eijkel, J.G.E Carderiers and F.C.M. Van de Pol "Dopant Selective HF Anodic Etching of
Silicon", **Proc. IEEE conf. MEMS-91**, Japan, Feb. 1991, p.221.

JOB TRACKING SOFTWARE FOR THE METALLOGRAPHIC LABORATORY

J.C. Grande[1]

ABSTRACT

A job tracking system consisting of a microcomputer in conjunction with database management software offers an inexpensive yet custom alternative to Laboratory Information Management Systems (LIMS). A computerized job tracking system offers several advantages over manual information tracking. These include: consistent and informative final reports, consumable tracking, fast and flexible searches, backlog information, and automatic backup.

The system described utilizes a Macintosh® computer system with a program called Foxbase+/Mac® for database handling. The various software modules are described and lend themselves to simple modification for other analytical services such as elemental analysis and spectroscopy labs. The software has been extended into other areas of the metallographic lab including a computerized etchant filing system (EtchFind) and grain size determination.

---

[1] General Electric Corporate Research & Development, P.O. Box 8, Schenectady, New York 12301
© A trademark of Apple Computer, Inc.
© A trademark of Fox Software, Inc.

Various issues are also discussed regarding security, database backup, user-friendliness, and compatibility with other computer systems.

INTRODUCTION

A metallography laboratory typically handles hundreds of specimens and thousands of negatives a year with a commensurate amount of information which is needed for identification purposes. In the past, this light microscopy laboratory kept both a notebook and a job card which were used to track information regarding each metallography job. A computerized filing system was desired to automate several tasks so that metallographers could spend more time doing metallography, not bookkeeping.

The components of the job tracking system are a Macintosh® computer configured with a hard disk and a laserwriter printer. The software is a custom program which uses the database program called Foxbase+/Mac® to handle the database manipulations. The program not only tracks metallography information so that final reports can be generated but also provides a tool with which consumable purchases can be anticipated, backlogs can be listed, and general workload information can be easily generated.

HARDWARE AND SOFTWARE REQUIREMENTS

The job tracking system requires at least a Macintosh® Classic computer with a hard disk. However, information processing is performed at a much faster rate with any of the faster Macintosh® computers, i.e., Macintosh® LC, Si, SE/30, II, IIx. The built-in monitor of the Classic is sufficient for all tasks; however, a full page monochrome monitor makes editing of the final report much easier. A postscript printer is desired for the professional looking reports generated, but any Macintosh® compatible printer will suffice. A computer with 2 megabytes (Mbytes) of RAM can be dedicated to job tracking. A 4-5 Mbyte machine is recommended if any word processing or graphing programs are used at the same time as job tracking. A 20 Mbyte hard disk is the minimum recommended hard disk capacity.

The software required is the Foxbase+/Mac® database handling program and the job tracking software custom designed for the metallographic laboratory.

JOB TRACKING CAPABILITES

The job tracking program for the metallography laboratory is called MICROLog (fig. 1) and is always available on a Macintosh® SE/30 computer conspicuously located in the microscopy lab.

Figure 1. Main menu screen for MICROLog job tracking software.

Scientists at the Lab submit specimens for preparation and evaluation to the metallographic laboratory. Either a customer or a metallographer enters the appropriate information into the program. New customers who submit specimens for preparation, enter lab location, phone numbers, billing information, specimen material and comments in addition to the specimen identification and required analysis. The information is automatically stored so that future specimen submissions by a repeat customer only require the new specimen identification.

A metallographer selects the job he/she is about to prepare from the new job queue (fig.2) and inputs his/her name.

JOB	CUSTOMER	RECEIVED	MATERIAL
71524	CASEY	07/03/91	Ti 6-4
71527	POTTER	07/03/91	Ti
71530	LAY	07/03/91	silver
71531	LEBLANC	07/03/91	RTG
71532	CASEY	07/03/91	Ti-Al-Nb
71533	CASEY	07/03/91	Ti-Al-Nb
71534	HEDENGREN	07/03/91	I.C.'s
71535	TODT	07/08/91	PBT / blass fiber composite
71536	CASEY	07/08/91	Ti-Al-Nb rolled plate coupons
71538	GREY	07/08/91	Nb3Sn foils
71539	BARREN	07/08/91	PET DBKS discs
71541	KAMBOUR	07/08/91	lexan sheet

SELECT JOBS TO BEGIN

Figure 2. New job queue for metallographers to select which metallography jobs to begin.

A customer can quickly determine which metallographer has his/her job and when it was started using this information. The metallographer completes a job information query, fig.3, when the metallography job is completed. In this way an overview of the consumables utilized, the type of work, and the equipment used can be tracked. Once this job completion query has been filled out a final report can be generated. The final report is automatically generated after answering some basic questions as to the type of illumination, print designations, type of metallography request, and if applicable, the etchant used.

The tracking program is integrated with the computerized etchant filing system, ETCHFIND [1] so that etchants used during the course of a metallography request can easily be selected and automatically incorporated into the final report (fig. 4)

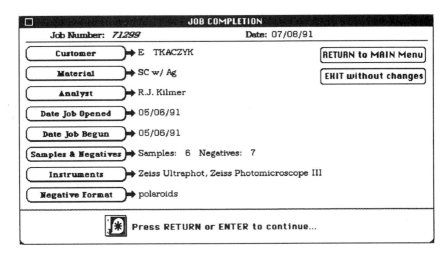

Figure 3. Job completion screen where information is entered into the database before a final report is generated.

One of the most utilized capabilities is backlog tracking. An overall backlog sheet which lists all unfinished metallography requests is automatically generated and printed on Fridays. The supervisor can then easily track which requests might need greater customer interaction and ascertain expected turn-around times. The software also lists each metallographers backlog.

There are basically two types of searches which are built into the tracking program: metallography job or customer information and overall workload information. The database can be searched to locate a previous request armed only with the date of possible submission and the name of the person who submitted it. Often customers want a printout of requests they have submitted during particular time frames. The overall workload information searches print out a month by month tally of negatives, specimens, and requests. The actual type of negative, i.e., 4 x5, 35mm, or Polaroid, can even be broken down on the printout. This information has proved invaluable for keeping track of consumables and tracking overall trends.

A browse feature for displaying all pertinent information in a spreadsheet type format is available through the menu bar located at the top of the Macintosh® screen. This format permits the quick visual scan of information without a formal search routine.

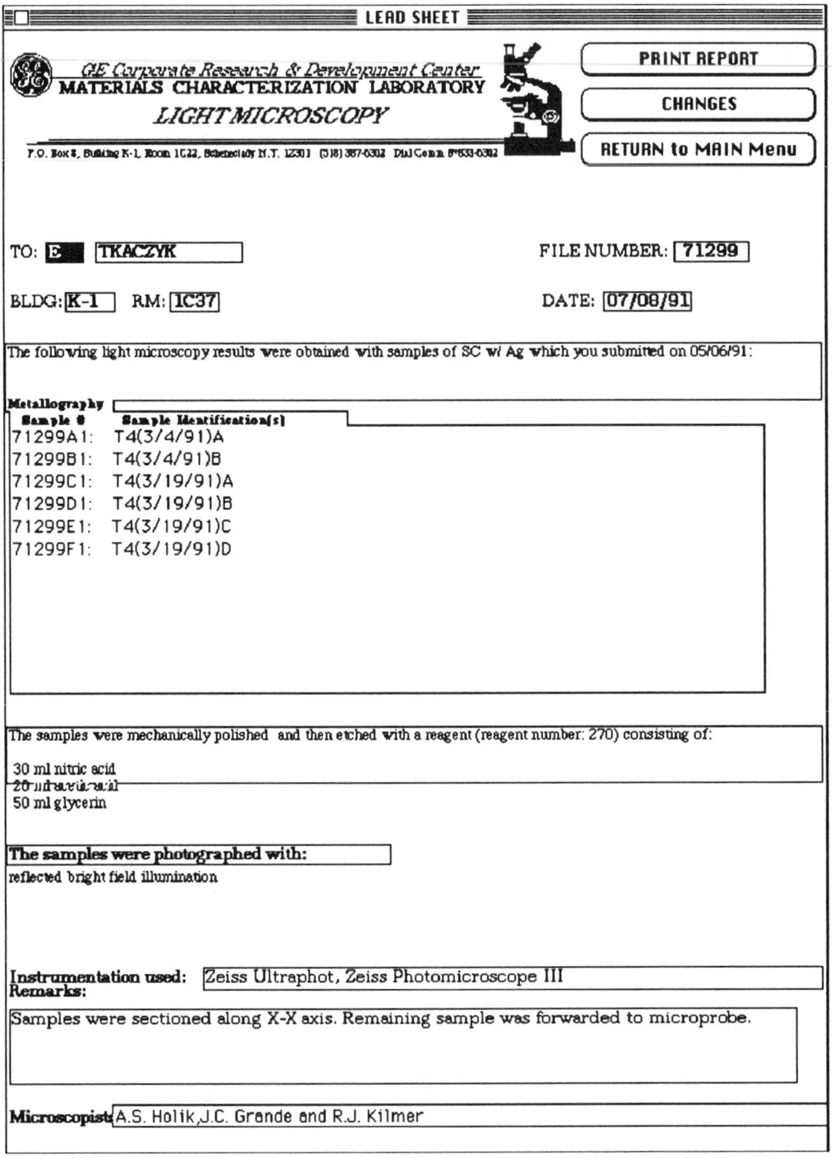

LEAD SHEET

PRINT REPORT

CHANGES

RETURN to MAIN Menu

GE Corporate Research & Development Center
MATERIALS CHARACTERIZATION LABORATORY
LIGHT MICROSCOPY

P.O. Box 8, Building K-1, Room 1C22, Schenectady N.Y. 12301 (518) 387-0302 Dial Comm 8*833-0302

TO: E TKACZYK

FILE NUMBER: 71299

BLDG: K-1 RM: 1C37

DATE: 07/08/91

The following light microscopy results were obtained with samples of SC w/ Ag which you submitted on 05/06/91:

Metallography

Sample #	Sample Identification(s)
71299A1:	T4(3/4/91)A
71299B1:	T4(3/4/91)B
71299C1:	T4(3/19/91)A
71299D1:	T4(3/19/91)B
71299E1:	T4(3/19/91)C
71299F1:	T4(3/19/91)D

The samples were mechanically polished and then etched with a reagent (reagent number: 270) consisting of:

30 ml nitric acid
20 ml acetic acid
50 ml glycerin

The samples were photographed with:
reflected bright field illumination

Instrumentation used: Zeiss Ultraphot, Zeiss Photomicroscope III

Remarks:
Samples were sectioned along X-X axis. Remaining sample was forwarded to microprobe.

Microscopist: A.S. Holik, J.C. Grande and R.J. Kilmer

Figure 4. Preview of final report which is automatically generated after job completion information is input into the program.

DATABASE BACKUP AND SECURITY

All the Macintosh® computers in the research center are linked through the phone system to form a local area network (LAN). The network also supports many other computer platforms. This LAN permits the workload information to be automatically transferred to the supervisor/manager computer and allows the routine database backup necessary for secure information filing. Using off-the-shelf software, a remote professionally maintained and backed-up mainframe computer is automatically connected to the server Macintosh® used for job tracking. The job tracking software stores a copy of the database at the end of the day onto this remote computer just as if it was a floppy disk inserted into the Macintosh®.

Access to the database is password protected so that only those in the metallography lab can have editing capabilities to the stored information. Encrypting schemes have not been employed due to the unclassified nature of the information stored.

SUPERVISORY UTILITIES

Some of the tasks of any supervisor is to track the type of work coming in and going out, turn-around time, customers, and resource allocation. Aside from the automatically generated backlog information mentioned previously, the program offers several routines which aides the supervisor in determining where future resources should be focused and whether existing equipment is meeting metallography needs. For example, we had used the workload by month breakdown of type of negative format to ascertain whether customer were actually utilizing the 35mm capabilities of a new metallograph. The data showed that 35mm usage had dramatically increased over a 1.5 year period by 100% and was continuing an upward trend. This information not only confirmed our belief of increased throughput with 35mm film but will also be used when purchasing our next metallograph.

The software also generates a list of the top ten customers, based on number of specimens and negatives, so that a guide as to the type of customers utilizing our services can be generated. This information is useful when trying to determine future metallography usage from various research groups within our laboratory.

211

An overall summary, including number of jobs entered and closed out for the previous week, is also automatically generated every Monday morning and stored into the supervisor's computer.

SOFTWARE DESIGN

The two most important aspects of any program which tracks information and which many people use are: user friendliness and speed. The Macintosh® platform used in conjunction with the Foxbase+/Mac® database software affords many features which make MICROLog and the other analytical laboratory tracking programs I have written very easy to use, even for the computer novice. Generous use of help screens are available when needed in addition to an animated face which uses my digitized voice to assist users through initial metallography requests.

The speed of the system is dependent on many factors, e.g., type of computer system, database handling software used, and size of the database. We are limited in which computer system we can use for job tracking so speed improvements there are limited. The database handling program is recognized as being one of, if not the fastest database program on the market. Size of the database is one of the biggest factors in program speed. Two job tracking databases are utilized in order to lesson the impact of database size and to permit a consistent apparent speed to the system during use. One database is an archived database (dating from 1987 and containing ~6000 requests) and has all the job tracking information for jobs which have been closed out and are older than 60 days. The other database, which I refer to as the active database (having ~500 requests), contains the most current metallography tracking information. The increase in speed arises from the fact that most information that is required occurs within ~60 days of a metallography request. The active database is searched first so that if the information is found, then the much larger archived database is not required. Information is automatically transferred from the active database to the archived database on a weekly basis. It is transparent to the user that two databases are effectively being used during all tasks.

DISCUSSION

The system was initially developed in 1987 for the IBM PC-AT®
computer using Dbase III+® database software. The programs and
databases were transferred to the Macintosh® environment in 1989 so
that more capability could be built into the system, especially using the
well established networking system at the research center. The
Foxbase+/Mac® software permitted instant utilization of the programs
and database without modification while the programs were adapted to
provide more user friendly features. This was a tremendous advantage
since it provided a smooth transition to a more feature packed
environment while maintaining compatibility with older files and
programs.

The utilization of a full page monitor has also been a great time
savings device for several reasons including: 1) much more displayed
information on one page, 2) the final report can be displayed in its
entirety in a WYSIWYG format so that any changes can be readily
displayed.

The use of Foxbase+/Mac® software offers the ability to
simultaneously share databases across a network and in a mixed IBM
PC® -Macintosh® environment. This is quite useful when integrating
job tracking among ever changing computers.

CONCLUSION

The use of a computerized job tracking system has successfully been
implemented in the metallography laboratory. While offering many of
the advantages of expensive and cumbersome LIMS tracking systems,
the use of this customized software is quite inexpensive and is flexible
enough to grow with the laboratory. The main disadvantage of using
this type of system is that it does not track the actual measurement data
or micrographs for metallography requests. However, it does remove
the cumbersome manual logging and tracking of jobs which was at
times very time consuming. The software has raised the professional
quality of our final reports and permitted the quick retrieval of
information.

© A trademark of Aston-Tate Corporation
© A trademark of International Business Machines Corporation

The other successful feature of this software is that over 300 scientists and technicians with various degrees of computer expertise have effectively utilized the software without assistance.

REFERENCES
1) A. S. Holik and J. C. Grande, "An Easy Access Computerized Etchant File System", Microstructural Science, Vol. 15, ASM International, pp. 489-495, ed. by M. E. Blum, P. M. French, R. M. Middleton, and G. F. VanderVoort (1987).

COMPUTERS REPLACE FILM:

THE MACINTOSH AS A METALLOGRAPHIC TOOL

K.S. Fischer and W.D. Forgeng, Jr.

California Polytechnic State University
San Luis Obispo, CA 93401

ABSTRACT

The Materials Engineering department at the California Polytechnic State University is replacing traditional film photography with a computer imaging system. The computer system saves lab time and reduces photographic material costs. At one to three dollars per image recorded, instant sheet film expenditures approached ten thousand dollars annually. By implementing imaging computers, laser prints cost only a few cents each. Multiple copies of prints are fast and easy. Images can be stored permanently on high capacity media. The cost of the system is recovered in only a few years. Applications for the system are found anywhere traditional photography is used, including failure analysis of fracture surfaces by scanning electron microscopy, as well as microstructural analysis by light microscopy.

Apple Macintosh™ computers are used with video scanning hardware and software to record images from a video source rather than using traditional photographic techniques. The imaging workstations were installed in conjunction with light microscopes and a scanning electron microscope.

215

By using the computer, many time consuming photographic techniques can be replaced with simple image manipulation. Multiple images from light or scanning electron microscopes can be combined to map a large area of the specimen at high resolution or show enlarged detail insets. Colorizing of images improves the researcher's ability to communicate findings through enhanced illustrations.

INTRODUCTION

For more than 100 years, the social and technical fields have been permeated with an invention popularized by George Eastman. This invention was the photographic camera. Through the years, this invention has been refined and improved. The process has become simple, easy, and relatively cheap. It is, however, not perfect. The human factor is still present. Photos must have the correct exposure, the correct focus, and the photographer must operate the camera properly. Any mistake, and a piece of film is wasted. This waste manifests itself in time and money.

The film used in the Materials Engineering department at Cal Poly is primarily Polaroid™ instant film. While this film is expensive, it gives instant results. This is beneficial in case the photo needs to be retaken. The photos are also instantly available for the students to continue with their lab work. Unfortunately, mistakes made in the process of taking a photo result in high costs associated with film that is thrown away. For the photos that are good, the students must still spend time processing the negative, drying it, and later making additional prints of it for lab partners and the professor.

In today's era of computers, there is a solution to this problem. The Macintosh computer can be used as a high tech camera. This process, which will be referred to as 'imaging', 'scanning', or 'digitizing' in this paper, can use many different types of sources. The most common is the 'flat bed scanner', which works in a manner similar to a copying machine where the copy is an image in the computer. The second is a 'video scanner'. Video scanning brings in an analog signal, usually from a common television (NTSC) or home video camera, and converts it to a digital representation that is understood by the computer.[1]

By combining the microscopy equipment already in use in most labs with a simple video camera, a video scanner, and a Macintosh system, digitized images are only a few keystrokes away. Black and white prints are easily made by a laser printer.[2]

The scope of this project was to provide a computer imaging system to reduce consumable costs within the department. Images were to be captured from a scanning electron microscope and light microscopes. Images were to be printed with two primary considerations: extremely low cost and sufficient quality for student lab notebooks.

EXPERIMENTAL PROCEDURES

The hardware and software in the system are all commercially available. Research in this project involved combining the existing hardware to create new capabilities. This system is based on a standard video signal, known as NTSC, which is used by most video cameras and monitors.

The NTSC video signal is converted into a binary code by a computer peripheral known as a digitizer, or scanner. In technical terms, the NTSC signal is analog. The scanner is an "analog to digital converter". The computer, in effect, lays a grid over a continuous tone image (Figure 1A) and then assigns each square a shade of gray corresponding to the average value within the square (Figure 1B). These squares are known as pixels. A computer image can be thought of as a three dimensional mathematical array, where the first two dimensions locate a pixel in the X-Y plane, and the third dimension determines the color or shade of gray of the pixel.[3]

While there are many different video scanners on the market with a wide range of capabilities, the MacVision™ scanner manufactured by Koala Technologies was selected for its simplicity and low cost.[4] The computer used in this project is an Apple Macintosh IIx with a 19-inch color monitor, and an Apple LaserWriter IInt printer. Many other system combinations would work equally well and this should be considered as just one possible combination. A typical system is shown in Figure 2.

While the basic use of the system consists of scanning and printing images, the real power stems from further manipulation of the images by special image processing software.[5] This is what separates the imaging system from computer image analyzers and video printers currently available on the market. While there are many software packages available on the market, *Photoshop*, published by Adobe Systems, is the most powerful and is considered the standard in the industry.

A classic example in metallography that benefits from this manipulation is galvanized steel. As etchant is applied to a polished

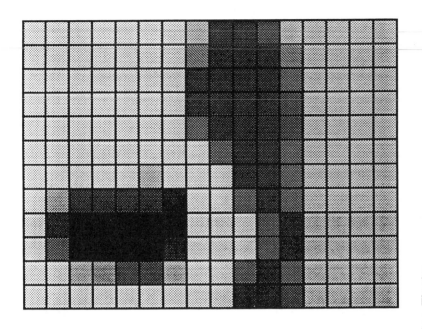

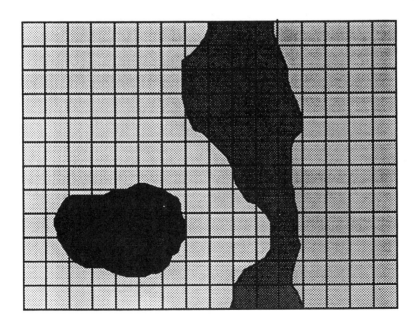

Figure 1: How Digitizing Works.

(A) Computer grid placed over image. (B) Computer representation of image

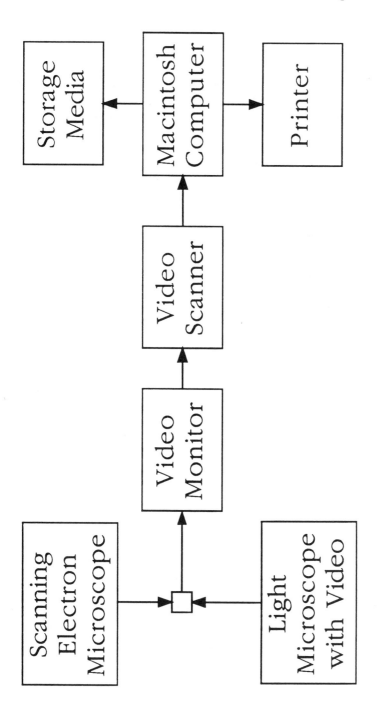

Figure 2: A Typical Imaging System.

cross section, the zinc coating attempts to protect the steel by acting as a sacrificial anode, just as it should.

Unfortunately, this makes it difficult to photographically illustrate the microstructure of the steel and zinc simultaneously. The following series of images shows the power of the computer. The first image is of the unetched microstructure (Figure 3). Note the clarity of the zinc plating. The next image shows the sample after five seconds of etching in 2% nital (Figure 4), The steel is underetched and the zinc has been darkened. After another five seconds of etching (Figure 5), the steel is properly etched, but the zinc is severely corroded. By using Adobe Photoshop, the zinc plating in Figure 3 can be combined with the base metal from Figure 5. This results in an image which illustrates the zinc plating on top of the etched metal (Figure 6). While this combination never actually existed under the microscope, it is more effective in communicating the microstructure than any of the previous images. This technique has been used for years, but the method has been to splice two negatives and print them together, a much more time consuming venture with a less convincing end result.

Sometimes a photo at low magnification fails to show the detail needed to communicate a concept while a photo at high magnification loses its sense of direction - the reader does not know how it fits into the overall scheme. This example of a copper braze on steel (Figure 7) would benefit from more magnification. By placing inset images at higher magnification (Figure 8), the metallographer better communicates both the detail and the setting of the microstructure.

This concept also applies to microstructures that change over a large distance at low magnification. This raw image of a copper-zinc diffusion couple (Figure 9) hints that there is more to see, both beyond the field of view of the image, and at higher magnification. To better communicate the overall structure, another image is made at the same magnification and fused to the first image. High magnification insets are placed on the image to add detail. Finally, the image is colored similar to the actual diffusion couple (Figure 10). The color effect could also be captured by a color scanner, although that increases cost and complexity.

The scanning electron microscope provides excellent images to be scanned because it is a fully electronic system. The image as seen by the operator and the film is already a video image and the imaging system has been specifically engineered by the manufacturer. Examples are shown in Appendix 1.

Printing is still the weak point in the system. Most prints from the laser printer are of sufficient quality only for class and notebook work.

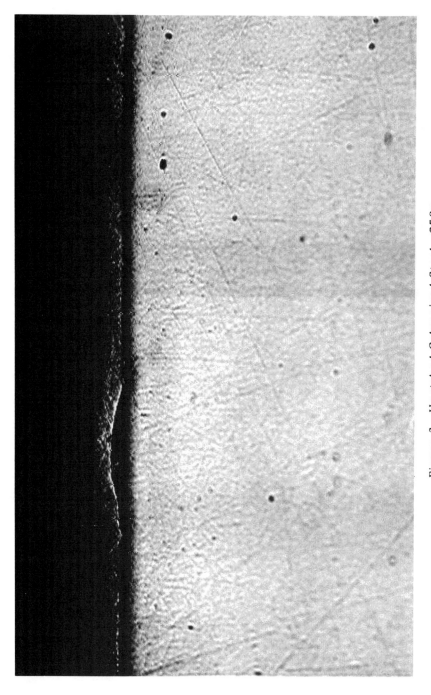

Figure 3: Unetched Galvanized Steel, 250x.

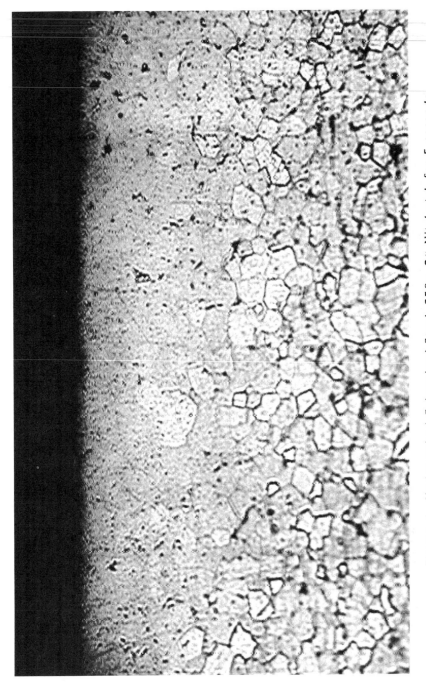

Figure 4: Underetched Galvanized Steel, 250x, 2% Nital etch for 5 seconds.

Figure 5: Properly Etched Galvanized Steel, 250x, 2% Nital etch for 10 seconds.

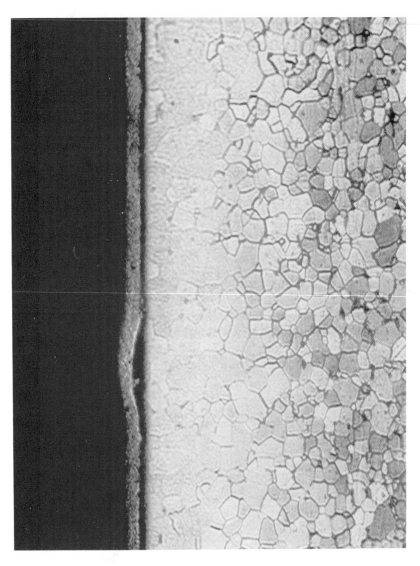

Figure 6: Superimposed Zinc Microstructure on an Etched Galvanized Steel Microstructure.

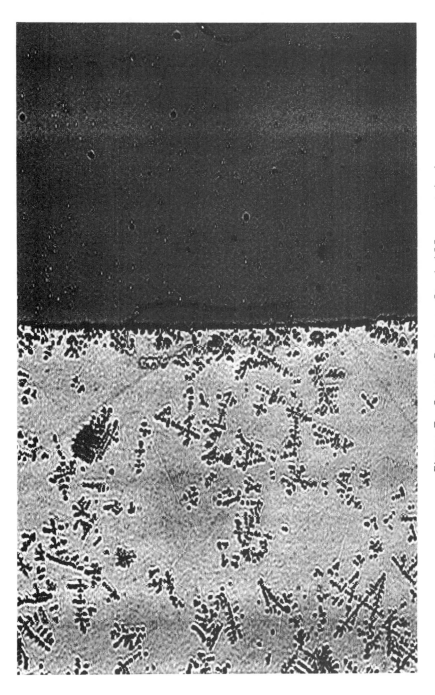

Figure 7: Copper Braze on Steel, 125x, unetched.

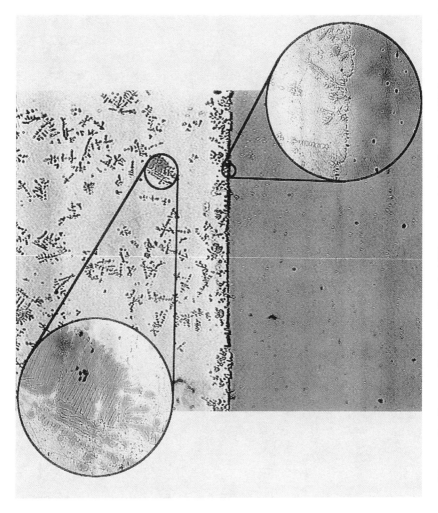

Figure 8: Copper Braze on Steel, 125x, unetched. Insets: 1250x, unetched.

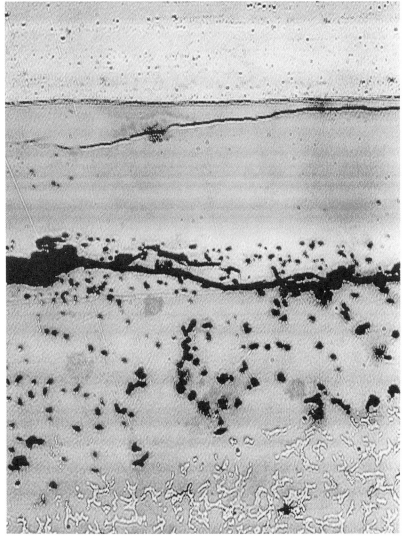

Figure 9: Copper-Zinc Diffusion Couple, 125x, unetched.

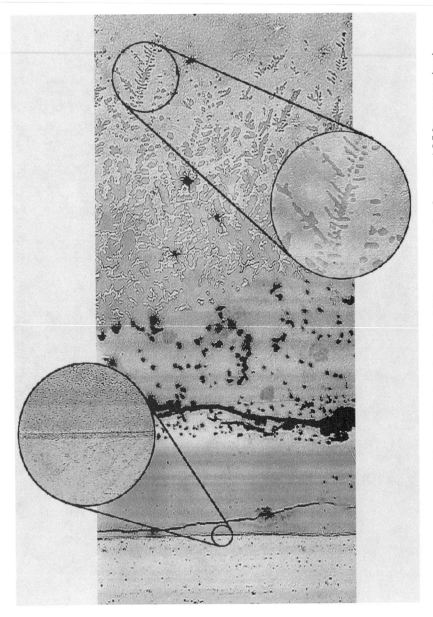

Figure 10: Copper-Zinc Diffusion Couple, 125x, unetched.   Insets: 1250x, unetched.

Several printing alternatives exist to provide high-quality images. One uses a high resolution professional printing system. Another is based on the new Canon color copier.

The high resolution printer improves the images dramatically because it simulates a gray by halftoning (Figures 3, 4, 5, 7 and A1 to A6). A halftone is a fine grid of black dots with white spaces. The denser the dots, the darker the simulated gray. The result is a stunning print of a quality on par with a good photograph. Prints can be made on paper or transparency.

The Canon color copier has a much lower resolution, but is capable of printing true shades of gray(Figures 6, 8 & 9) or color if the image is in color (Figure 10). This greatly improves the quality of the prints. A special postscript processor connects the copier to a Macintosh computer.

These printing processes represent the two basic forms of printing, high resolution halftones like the laser printer and high resolution professional printers, and true shades of gray from color printers as from the Canon color copier or other personal color printers.

RESULTS AND DISCUSSION

The Macintosh imaging station has proved to be quite valuable in a year of tight university budgets. Film has been severely rationed, and yet the students were able to use the computer to produce even more images than the previous year's class did with a normal supply of film. The students have used the system for several months and find it easy and convenient. It has become the normal way of doing routine metallographic work.

The future of video and the Macintosh is evolving rapidly. Apple Computer has just announced a new extension to its operating system called QuickTime. QuickTime will bring new standards in video integration and will allow better image compression and use of video within documents. One scenario provided by Apple envisions a short movie to be viewed within a word processing document. This is one more step to the removal of paper and hardcopy manuals and documents from our everyday lives.[6]

SUMMARY

A new method of capturing images has been implemented which uses computer imaging, rather than traditional film photography. The computer eliminates the messy chemicals, darkroom, and wasted film associated with traditional photography. While film photography still has a place in metallography, the majority of its current use can be replaced with computer imaging, resulting in a great savings in photographic consumables. As the technology progresses, and as metallographers further explore the possibilities of computer imaging, the computer is expected to completely replace film.

REFERENCES

1. A. Naiman, **The Macintosh Bible**, Publishers Group West, Emerville, p 273, 1988
2. R.F. Wickham, **MacVision User's Manual Version 2.0**, Koala Technologies, San Jose, 1988.
3. J. Morton, *"Shades of Gray,* **MacWorld**, Vol. 5, No. 1, p 110, 1988
4. F. Tessler, *"Review - MacVision 2.0"*, **MacWorld**, Vol. 6, No. 7, p 155, 1989
5. S. Roth, *"Grade-A Gray Scale"*, **MacWorld**, Vol. 7, No. 10, p 202, 1990
6. L. Poole, *"QuickTime in Motion"*, **MacWorld**, Vol. 8, No. 9, p 154, 1991

Appendix 1
Scanning Electron Microscope
Image Gallery

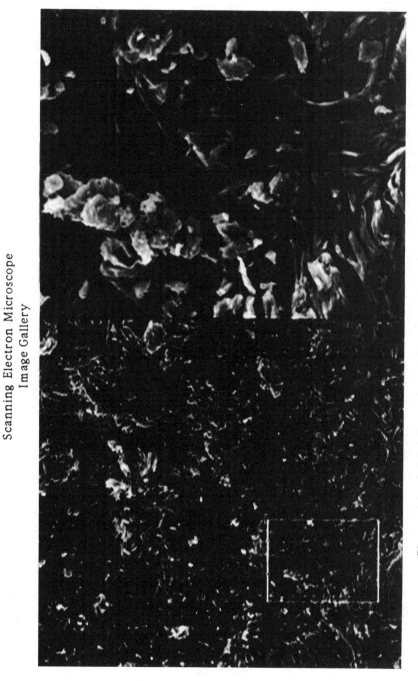

Figure A1: Initial Corrosion of Zinc Layer. Left: 200x Right: 600x

231

Figure A2: Complete Corrosion of Zinc Layer. Left: 100x Right: 630x

Figure A3: Zinc and Iron Oxides, 1000x.

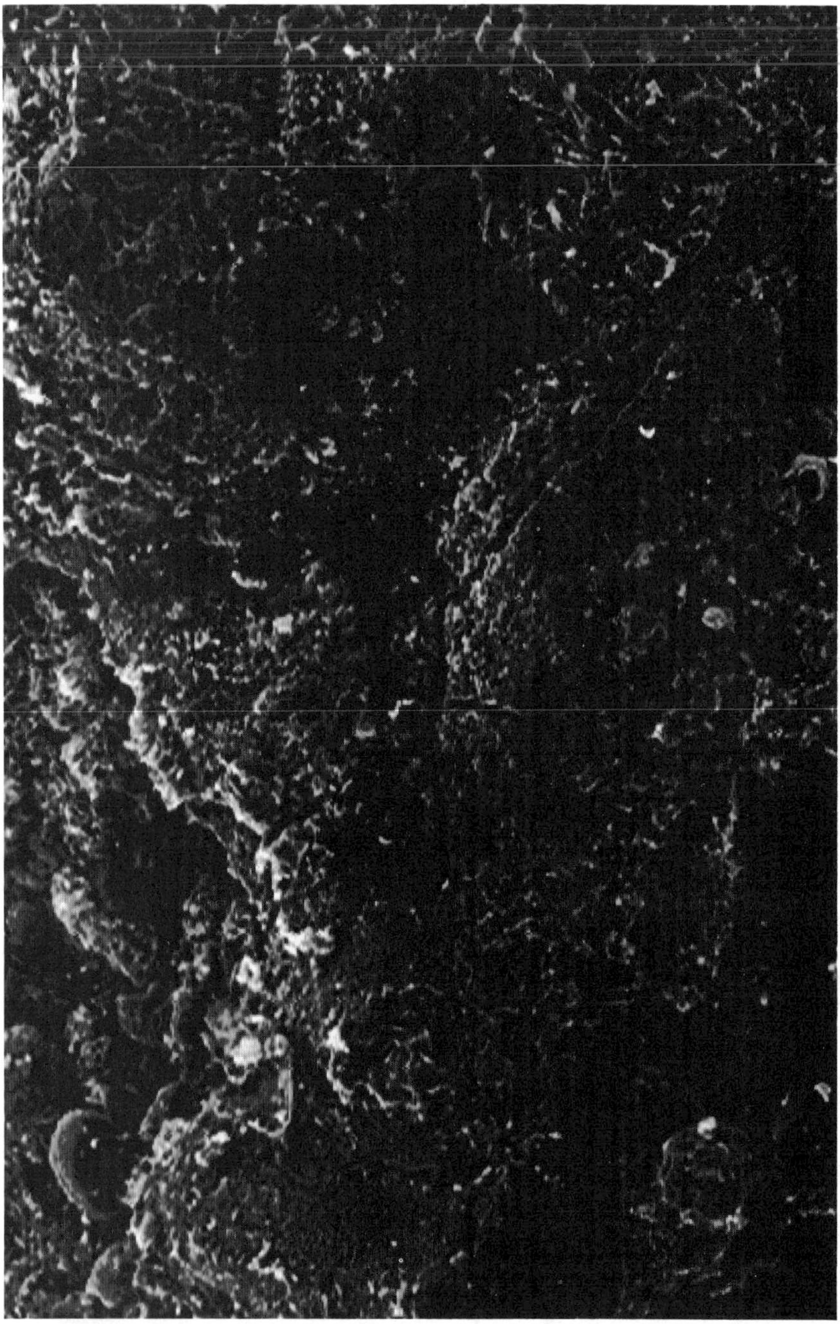

Figure A4: Coherent Iron Oxide Coating, 270x.

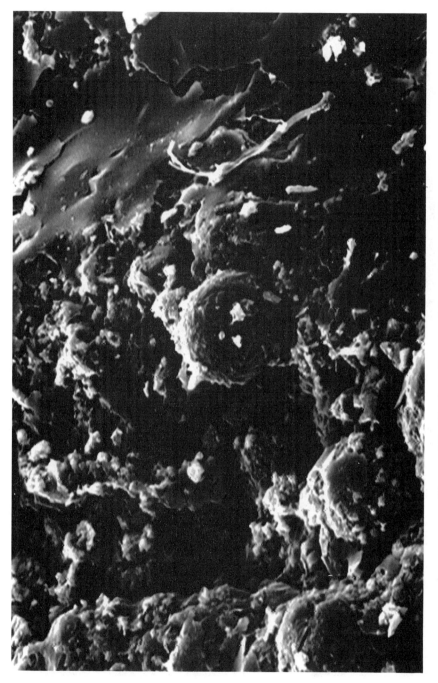

Figure A5: Various Forms of Iron Oxide, 1000x.

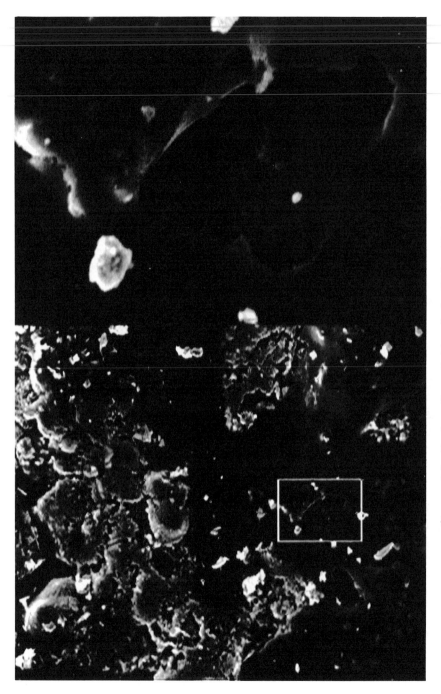

Figure A6: Broken Iron Oxide. Left: 300x Right 1800x

A STUDY IN THE DYNAMICS AND WEAR

OF RIGID DISC SYSTEMS

Stephen D. Glancy* and Morten J. Damgaard*

ABSTRACT

Rigid grinding disc (RGD) technology was introduced into the metallographic laboratory on a commercial level during the early 1980's. This came when conventional metallographic preparation methods could no longer satisfy the needs of the materials laboratory. The ability of RGD to produce high quality "true microstructures" was excellent. The RGD technology revolves around specially designed discs of various compositions that perform abrasion of specimens using two common modes of operation, grinding and lapping. These discs are particularly effective for the "high-tech" composites that are now becoming common-place. However, like the simple silicon carbide grinding paper, this technology has limitations. RGD techniques must be specifically designed for the material being prepared, rather than just substituted into a procedure that calls for the use of a conventional grinding format. The RGD technique demands continual maintenance to insure operating efficiency. If the operation and application of these discs are not fully understood and controlled, specimen quality and disc life will be greatly reduced. Therefore, the operation (dynamics) and the maintenance (wear) of Rigid Grinding Disc systems must be studied.

INTRODUCTION

Rigid Grinding Disc (RGD) technology was introduced on a commercial level during the early 1980's. RGD come in a variety of sizes and compositions. Unlike traditional grinding formats, RGD require external application of the abrasive. The

---

*Struers, Incorporated, Westlake, Ohio

most common abrasive is diamond. However, they may be used with other abrasives such as SiC and Al_2O_3. Material is removed through two modes of abrasion, grinding and lapping.

Grinding is a two-body abrasive technique in which the grains are affixed to a substrate such as paper or plastic [1]. These grains remove material in a manner similar to a lathe in which the abrasive particles act as small cutting tools. Grinding is characterized by excellent material removal rates, minimal sub-surface deformation, and, in most cases, acceptable flatness. This method is used primarily for the preparation of metallic specimens and is generally done with conventional SiC paper. Because the grains are fixed, most of the force is directed parallel to the specimen surface. A ductile specimen which has been subjected to a grinding process can be identified by the presence of uniform scratches on the surface (Figure 1).

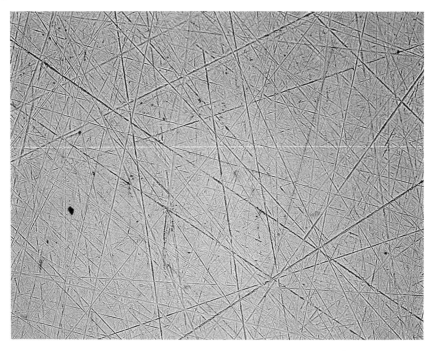

Figure 1: Carbon steel, ground on 1200 grit SiC paper.

Lapping is a three-body abrasive technique in which loose or free grains roll between the specimen and disc. As a result, little or no material is removed except in the case of brittle materials [2]. This method of abrasion is used predominantly in the preparation of brittle materials, such as ceramics and minerals. Lapping discs come in a wide variety of compositions. Examples include cast iron, plastic, paraffin and glass [3]. Lapping is characterized by extremely low material removal (except for brittle materials), moderate sub-surface damage, and excellent flatness. A ductile specimen which has been lapped can be identified by its matte or dull, hammered appearance (Figure 2).

Figure 2: Carbon steel, lapped on a cast iron disc with 1200 grit SiC.

Application

The best method of abrasion is determined by the material that is to be prepared. Metals or ductile materials are usually prepared by grinding. Ceramics or brittle materials are most effectively prepared by lapping. Attempts to prepare metals with lapping or brittle materials with grinding often result in unacceptable microstructures.

The introduction of "high-tech" composites into the laboratory has presented the metallographer with a difficult task. A good example is the metal-matrix composite of SiC fibers in a titanium matrix. Grinding the titanium matrix with SiC paper presents no problem. However, the paper is quite ineffective against the SiC fibers. Grinding with SiC paper results in extensive damage to the fibers. Lapping is very effective for the SiC fibers, but not for the titanium matrix. The ideal method should include a system that performs both grinding and lapping. This is the idea behind today's RGD systems. The degree of grinding or lapping is controlled by the hardness (composition) of the disc. RGD that are hard abrade material primarily by lapping, since the abrasive particles roll on the disc. RGD that are soft perform abrasion primarily by grinding, since the abrasive particles embed in the disc and become fixed. By choosing the correct disc(s), it is possible to prepare effectively almost any composite.

As with most tools, proper operation and maintenance are critical if the desired results are to be obtained. RGD should only be used on automatic preparation equipment because it is necessary to control the applied force and specimen movement across the RGD. Two important areas determining RGD effectiveness are *dynamics* and *wear.*

Dynamics

The dynamics, or motion, of the specimen across the RGD is of critical importance. There are many variables associated with the dynamics of these discs. Following is a list of the most important variables.

- ▸ RPM (RGD)
- ▸ RPM (specimen holder)
- ▸ Rotation direction (RGD)
- ▸ Rotation direction (specimen holder)
- ▸ RGD diameter
- ▸ Specimen size
- ▸ Specimen holder diameter
- ▸ Specimen holder off-set

Each of these variables plays an important role in the production of high quality "true microstructures." All of these variables are also directly linked together. Theoretically, it would be advantageous to maintain a constant non-directional relative velocity ($V_{b/a}$) between the specimen and the RGD, regardless of the specimen's position on the disc. This would eliminate any adverse effects associated with directional abrasion. The $V_{b/a}$ between a point on the specimen and a point on the RGD can be expressed as:

$$V_{b/a} = \sqrt{b^2(\omega_a{}^2 + \omega_b{}^2 - 2\omega_a\omega_b) + c^2\omega_b{}^2 + 2cb\cos A\ (\omega_a\omega_b - \omega_b{}^2}$$

$V_{b/a}$ =   velocity of the specimen in relation to the Rigid Grinding Disc (RGD) in meters/second [m/s]

ω_A =   angular velocity of the specimen holder in [rad/s]

ω_B =   angular velocity of the RGD in [rad/s]

b   =   distance from the center of the sample holder to the point on the specimen [m]

c   =   distance between the center of the specimen holder and the center of the RGD [m]

A   =   angle between velocity vector V_a and V_b [rad] (Figure 3)

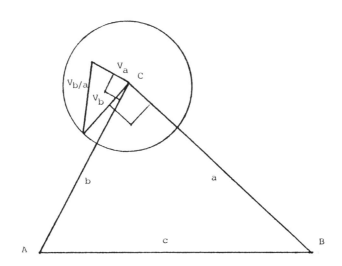

Figure 3: Vector diagram for RGD System.

By varying A from 0 to 2π, $V_{b/a}$ can be graphed. Figures 4 and 5 show the $V_{b/a}$ graphs, using polar coordinates, for several available grinding systems with 160mm specimen holders and 300mm discs.

The graphs clearly illustrate that, under these conditions, $V_{b/a}$ cycles dramatically during one revolution of the specimen holder. In Figure 5, the velocity of the specimen cycles from 0.05 to 3.82 m/s, 50 times a minute. This cycling effect can cause preparation artifacts like preferential abrasion, smearing and pullout.

241

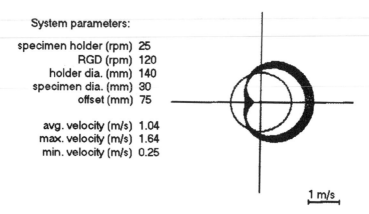

System parameters:

specimen holder (rpm)	25
RGD (rpm)	120
holder dia. (mm)	140
specimen dia. (mm)	30
offset (mm)	75
avg. velocity (m/s)	1.04
max. velocity (m/s)	1.64
min. velocity (m/s)	0.25

1 m/s

Figure 4: 25 rpm specimen holder, 120 rpm disc, co-rotation

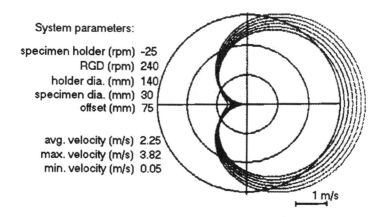

System parameters:

specimen holder (rpm)	-25
RGD (rpm)	240
holder dia. (mm)	140
specimen dia. (mm)	30
offset (mm)	75
avg. velocity (m/s)	2.25
max. velocity (m/s)	3.82
min. velocity (m/s)	0.05

1 m/s

Figure 5: 25 rpm specimen holder, 240 rpm disc, contra-rotation.

There is also a problem with varying velocities across the face of the specimen. Graph #5 shows the difference in velocity from inside to outside of the specimen is almost 1m/s. This will cause preferential abrasion of the inside section of the specimen.

In addition, Formula 1 shows that if ω_A is equal to ω_B, then the $V_{b/a}$ is constant, regardless of **A**.

With the specimen holder speed equal to the RGD speed, the constant relative velocity can be varied by changing the specimen holder/RGD speed and/or **c,** the distance between the center of the specimen holder and the center of the RGD. This is especially important when working with materials that are sensitive to directional abrasion. These materials often require a reduction in the relative velocity to prevent preparation artifacts. A reduction in either the specimen holder or RGD RPM would lower the $V_{b/a}$ of the system, but would increase preferential abrasion.

In order to maintain a constant non-directional $V_{b/a}$, the following conditions must be met.

1. Rotate the specimen holder and RGD in the same direction.

2. Rotate the specimen holder and RGD at *approximately* (see section on wear) the same RPM.

With these conditions there is less chance of introducing preparation artifacts. If these conditions cannot be met, then the additional induced damage must be removed in the polishing steps. This is usually done by extending the polishing time. However, long polishing times should be avoided because of induced relief and rounding.

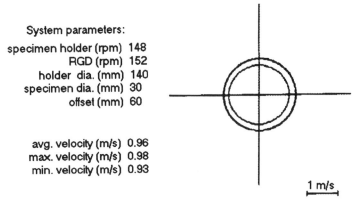

System parameters:

specimen holder (rpm)	148
RGD (rpm)	152
holder dia. (mm)	140
specimen dia. (mm)	30
offset (mm)	60
avg. velocity (m/s)	0.96
max. velocity (m/s)	0.98
min. velocity (m/s)	0.93

1 m/s

Figure 6 graphs the $V_{b/a}$ for an "ideal" dynamic setup, with the specimen holder and the RGD rotating in the same direction at approximately the same RPM.

Wear

Even if all the optimum dynamic conditions are satisfied, another important aspect of RGD procedures must be addressed. This often overlooked problem is RGD flatness. These discs will not produce acceptable results if they are not kept extremely flat. The effects of an unplane disc usually show up as deep scratches and rounding on the outer periphery of the specimen. Another associated problem is the dramatic increase in preparation time required to achieve plane specimens. Since most RGD systems use diamond as an abrasive, this can become very expensive. Figure 7 shows a computer-generated cross-section of a 290mm RGD that was used under the ideal dynamic conditions described earlier. Note that this an accelerated projection of wear. The area of highest wear is concentrated in the center of the disc, with little wear at the outer area of the disc. For this reason the center section of the disc is removed during manufacturing.

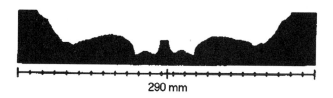

290 mm

Figure 7: Computer generated cross-section of RGD under "ideal" dynamic conditions.

A second area of concern is the outer-most area of the disc. Wear of this area depends on the size ratio of the specimen holder to RGD. Figure 8.1 shows the wear pattern for a 290mm RGD used with a 160mm specimen holder, with 25mm specimens. The disc wears only where the specimens track. If a set of 30mm specimens were to be run on the same disc, the outer portion of the specimen would track over the high section of the disc as in Figure 8.2. This would result in deep scratches and rounding of the specimen as illustrated in Figure 8.3.

Therefore, it is critical to use the proper diameter combination in the RGD system. The general rule is to select the size of RGD so the specimen holder will rotate over the center of the disc and then off the outside edge. Figure 9 shows the wear pattern for the correct sizing of RGD and specimen holder, i.e., a 230mm RGD with the 160mm specimen holder, under optimum dynamic conditions. This results in a much more uniform wear pattern than that illustrated in Figure 7.

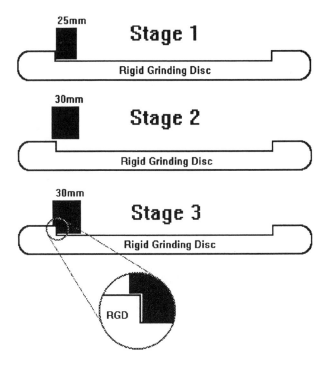

Figure 8:

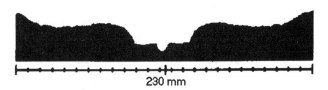

Figure 9: Wear pattern for correct sizing of RGD and specimen holder under optimum conditions.

245

It is also very important to maintain the peak cutting rates for RGD. The material abraded from the specimens (swarf) will often cause the disc to "load-up" and quit grinding. This swarf must be removed by dressing (grinding) with an aluminum oxide stick. This procedure is also used to flatten the disc. In severe cases, flatness can only be restored with a lathe or surface grinder. Running a holder loaded with specimens of a hard material, such as tungsten carbide, is ineffective for flattening the disc. This will, however, help to restore an effective cutting surface.

There are several brands of RGD systems available for metallographic preparation. Please refer to the manufacturer's instructions for proper disc maintenance.

Wear Warning!

It is important that the specimen holder and RGD RPM do not match exactly! If both rotate at exactly the same RPM, the specimens will track in an undesirable pattern on the RGD (Figure 10). This pattern will cover a limited area of the disc and will result in unacceptable wear of the disc. A variation of *one* RPM is sufficient to prevent this condition.

Figure 10: Unacceptable wear pattern created when specimen holder and disc rotate at identical rpm.

Conclusion

While Rigid Grinding Disc systems are effective in preparing difficult composite materials, caution must be exercised when establishing dynamic parameters. Important parameters include the following:

▸ Specimen holder and RGD rotation direction

-AND-

▸ Specimen holder and RGD RPM.

In situations where the relative velocity ($V_{b/a}$) of the system must be adjusted to minimize preparation artifacts, the following guidelines should be observed.

▸ Adjust both the specimen holder and RGD RPM to achieve the desired velocity. Changing just one of these will result in preferential abrasion.

-OR-

▸ Adjust the variable "c" (the distance between the center of the specimen holder and the center of the RGD).

Besides meeting the optimum conditions, it is critical to establish and follow a disc maintenance program. This will insure the discs are kept flat. *Flat discs produce flat specimens.*

References:
[1] L.E. Samuels, Metallographic Polishing by Mechanical Methods, 3rd ed., American Society for Metals, p 32, 1982.

[2] Ibid., p 32

[3] G. F. Vander Voort, Metallography Principles and Practice, McGraw-Hill Inc., p 100, 1982.

Environmentally-Induced Cracking

STRESS CORROSION CRACKING FRACTOGRAPHY OF A

HIGH STRENGTH LOW ALLOY CASING STEEL EXPOSED

TO A SOUR GAS ENVIRONMENT

D.A. Diakow[1] and W.J.D. Shaw[2]

ABSTRACT

A high strength low alloy tubular steel, Nippon NT-90SS grade (690 MPa UTS), was exposed to a NACE TM0177-90 environment. Tensile, double cantilever beam and compact tension fracture toughness specimens were exposed and tested under various environmental and loading conditions. Prior immersion or pre-conditioning unloaded components in a sour solution allowed the separation of transient effects of hydrogen charging and hydrogen embrittlement from stress corrosion cracking effects. Testing of fracture toughness specimens using a type of slow strain rate method - a slow constant crack opening displacement rate, allowed determination of the amount of stress corrosion cracking occurring without the influence of hydrogen embrittlement. Fractographic analysis at the different testing conditions allowed the identification of various mechanisms due either to stress corrosion or hydrogen embrittlement. The material exhibited a marginal decrease in toughness due to hydrogen charging but a substantial decrease as a result of stress corrosion effects. Reduced ductility regions around microvoids and secondary cracking were typical characteristics of hydrogen embrittlement while extensive fissures and crevices, secondary intergranular cracking and transgranular fracture were characteristic of stress corrosion cracking for this steel.

[1] NOVA Corporation of Alberta, P.O. Box 2535, Station 'M', Calgary, Alberta, Canada T2P 2N6.
[2] Department of Mechanical Engineering, University of Calgary, Calgary, Alberta, Canada T2N 1N4

INTRODUCTION

Stress corrosion cracking, SCC, of high strength steels is recognized as an important design consideration. As a result of the increasing demand for low weight, high strength products, there has been a considerable amount of SCC research performed on these materials. Much of this work has been reviewed in detail by Phelps et al [1]. Through such research it is generally recognized that steels with high yield strengths, in particular, are very susceptible to SCC in environments containing H_2S (hydrogen sulfide) [1-3]. This is of particular interest in the oil industry where H_2S is commonly found in subterranean hydrocarbon reservoirs. As a result, high strength SCC-resistant tubulars have been developed [4,5] for use in such environments.

Materials which cannot resist H_2S environments have only limited, if any, application for oil and gas production. Consequently, there is a need for test methods to determine if a material is susceptible to SCC. Many test methods have been developed and include constant load, constant displacement and, more recently, slow strain-rate, SSR, tests [6]. These tests normally employ the use of smooth, round tensile specimens but since cracks, notches and other defects or imperfections are often present in materials, application of pre-cracked fracture mechanics specimens in SCC testing are also used. The stress intensity factor is used to characterize the mechanical driving force in SCC [7] and in this manner a practical measure of fracture toughness of a material in an environment can be determined. This type of specimen can be used in constant displacement tests as a crack-arrest type of specimen to determine the threshold stress intensity value of SCC, K_{ISCC} [8].

It has been suggested that pre-cracked specimens be used in SSR testing to achieve a constant strain rate throughout the test [6]. Constant load and displacement tests are in essence, variable strain rate tests. A constant crack-opening displacement, COD, rate closely simulates a constant strain rate at the crack tip. Such constant COD rate tests have been previously employed [9] in an attempt to screen the SCC susceptibility of aluminum alloys. It has also been confirmed that the effect of a decreasing applied strain rate increases the severity of hydrogen damage [10, 11] in SCC tests. This would suggest that a value of K_{ISCC} may be obtained by testing fracture mechanics specimens at low COD rates and may result in a new method of determining SCC susceptibility. Using a standard K_{ISCC} test [8] for comparison purposes, fractography will be used to verify the constant COD rate test method results in a characteristic SCC test.

There are two different types of mechanisms by which stress corrosion is believed to occur: active path corrosion and hydrogen embrittlement, HE. However, there has been no general agreement on whether the mechanism of SCC in high strength steels is one of active path corrosion, HE or possibly a combination of both [1,12,13,14]. An attempt is made in this study to separate and thus identify using fractography, the effects and characteristics of HE and SCC.

EXPERIMENTAL PROCEDURE

The material used in this study was a Nippon NT-90SS (sulphide service) tubular steel with the chemical composition as shown in Table 1.

Table 1. NT-90SS Chemical Composition
(Weight percent)

C	Si	Mn	P	S	Ni	Cr	Mo	Nb	Ti	B	Al
0.18	0.12	1.31	0.006	0.002	0.02	0.02	0.12	0.02	0.016	0.001	0.020
Balance Fe											

This is a low alloy, modified AISI type 4130 steel developed for enhanced SCC resistance used primarily in oil and gas applications in sour environments.

Specimens were cut from a piece of 244.5 mm diameter tubular, with a wall thickness of 12.5 mm.

The test solution used was standard NACE aqueous solution containing 0.5% (by weight) CH_3COOH (glacial acetic acid) and 5% NaCl saturated with 1 atm H_2S (referred to as *NACE solution* in this paper). Solution pH varies from 2.7 to 3.5 over a 14 day test period [4].

Fracture mechanics properties were evaluated using compact tension, CT, specimens, as defined in ASTM E399, Appendix A4. Specimens were 32 mm long by 30 mm high and 9.53 mm thick. Specimens tested were machined in the T-L configuration. The CT specimens were fatigue pre-cracked to a crack length-to-width ratio of a/w = 0.35. The fracture toughness, K_C, was determined according to ASTM E561.

Tensile properties were determined using round tensile bars, 9.53 mm in diameter and 50.8 mm gage length, machined longitudinally from the tubular walls. HE effects on tensile properties were evaluated by testing specimens which were pre-immersed in an unloaded condition in the NACE solution for a specific period of time [15,16].

HE effects on K_C were also determined by 'pre-conditioning' or pre-immersion of unloaded CT specimens in the NACE solution to obtain hydrogen saturation of the steel. Once saturated, the specimens were removed from solution and immediately tested according to ASTM E561 to obtain K_C .

SCC tests were performed according to NACE Standard TM0177-90 [8]. 'Pre-conditioned' double cantilever beam, DCB, specimens 102 mm long by 25 mm high and 9.53 mm thick were tested to obtain a value of K_{ISCC}. The specimens were fatigue pre-cracked to a crack length-to-half height ratio of a/h = 3.0. Wedge-opening displacement was 0.84 mm giving an initial stress intensity of 65 $MPa*m^{1/2}$. A minimum specific volume, which is solution volume per unit specimen surface area, of 15 ml/cm^2 was used. Previous testing [3,17] indicated this specific volume must be maintained to minimize pH drift with time throughout the test.

SSR or COD tests were conducted using the identical DCB specimens as used in the SCC tests. The load pins were isolated from the specimen and solution by coating with heat shrink plastic tubing. The load grips were Teflon coated to isolate them from the solution. A plexiglass container holding 2.0 litres of solution (giving a specific volume of 30 ml/cm^2) was used for the test cell. Continuous bubbling of H_2S into the solution and throughout the test maintained a slight positive pressure in the test cell, ensuring H_2S saturation and eliminating the possibility of air/oxygen contamination of the solution.

All constant rate pin-loading displacement rate tests resulted in constant COD rates which were varied between 10^{-4} to 10^{-9} m/s. One specimen was tested at each condition. Again, all specimens were 'pre-conditioned' for a set period of time to ensure hydrogen saturation after which time the test began. As a means of control, a combination of the AWS Standard A4.3-86 [18] and the Barnacle test methods [19,20] were used to the determine diffusible hydrogen concentration of test specimens. This was done to ensure similar 'pre-conditioning' and thus hydrogen saturation conditions existed prior to SCC testing.

After testing, the CT fracture toughness specimens were completely fractured and stored in a desiccator until subsequent examination. The DCB specimens were unloaded, the solution purged with nitrogen, the specimens removed from solution, chilled in liquid nitrogen, fractured to reveal the failure surfaces and then placed in the desiccator.

Fractography was conducted using a Cambridge S100 scanning electron microscope, SEM, after ultrasonic cleaning of the surfaces in 5% citric acid for 10 minutes [21]. Microstructure evaluation was investigated using a Zeiss ICM-405 metallograph in addition to the SEM.

DISCUSSIONS

Microstructure

The NT-90SS steel microstructure is shown in Figures 1(a) and 1(b). Etchant used was 5% nital with specimens additionally sputter coated with gold for SEM analysis. The microstructure is fully tempered martensite, with an acicular non-directional grain structure and fine carbide precipitates [4] within the grains and along the grain boundaries as shown in Figure 1 (a). Some segregation of P and Mn [5] in the rolling direction is observed with larger spherical precipitates, identified by EDS analysis as being high in Ca, Si and S content randomly dispersed throughout, Figure 1 (b).

Tensile behaviour

The ambient temperature tensile properties of the NT-90SS steel are shown in Table 2. The material possesses high strength with very good ductility. It also displays unusual behaviour due to strain aging effects. A large amount of ductility and work softening was exhibited prior to work hardening.

Pre-Conditioning Effects

An attempt was thus made in this study to separate or identify the effects of SCC and HE. Tensile specimens 'pre-conditioned' in the NACE solution were found to be 'saturated' with hydrogen generated by the anodic reaction of the steel in the solution, after a period of 48 hours. In this manner, transient effects of hydrogen charging [17] of the steel were eliminated from the SCC tests. It also provided a common basis for testing and allows for a more 'true' evaluation of SCC characteristics. This saturation time was determined by observing total elongation and reduction of area, RA, decreasing with immersion time to a point where no further degradation in these properties was found, Figure 2. A decrease of 39% and 31% was observed in elongation and RA, respectively. Similar effects have been reported previously

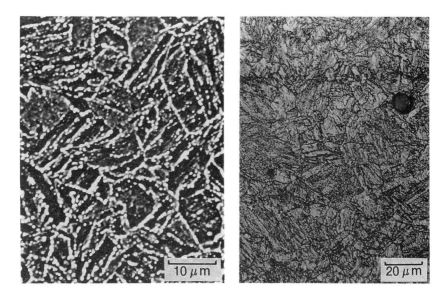

Figure 1(a) (left). SEM Micrograph of NT-90SS Microstructure.
Figure 1(b) (right). Light Micrograph of NT-90SS Microstructure.

Table 2. Tensile Properties: NT-90SS Steel

Property	As-Received Condition	Pre-conditioned in NACE Solution
Ultimate Tensile Stress, MPa	694	698
Yield Stress, MPa	671	671
Elongation, m/m	0.248	0.152
Reduction of Area, percent	73.4	50.5

[22]. The effect of hydrogen saturation on these mechanical properties is summarized and compared to the as-received values in Table 2.

Fracture Toughness/ Hydrogen Embrittlement Effects

CT specimens tested in the as-received and 'pre-conditioned' H_2 embrittled conditions gave K_c values of 231 Mpa*m$^{1/2}$ and 206 Mpa*m$^{1/2}$, respectively resulting in a decrease in toughness due to hydrogen embrittlement of 11%. This slight decrease was somewhat

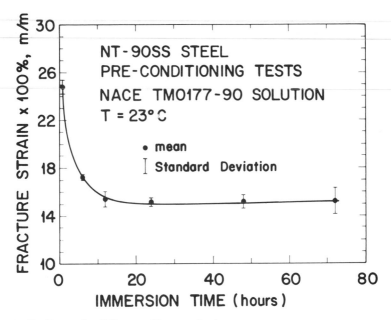

Figure 2. Pre-Immersion Effects on Fracture Strain.

unexpected due to the significant reduction in ductility observed with the tensile specimens.

As shown in Figures 3(a) and 3(b), fractographic analysis of the as-received specimens indicated fracture to be almost entirely one of microvoid coalescence with the microvoids being initiated by carbide precipitates or sulfide inclusions [23].

The 'pre-conditioned' CT specimens, Figure 3(c), indicated a completely different fracture morphology when compared to that of Figure 3(a). At higher magnifications, however, as shown in Figure 3(d), microvoid coalescence is still the predominant fracture mechanism with the voids being significantly more shallow, indicative of a very much refined fracture mechanism. Areas of low ductility within and surrounding the microvoids are also observed. It appears that the decreased ductility regions have minimal effect on fracture toughness whereas microvoid coalescence dominates the fracture mechanism.

Figure 3(c) also shows extensive secondary cracking occurring at the fatigue/overload transition zone. This is thought to be caused by diffusion of hydrogen to the plastic zone upon loading. It has been determined that martensitic structures, such as those found in high strength steels, absorb atomic hydrogen at an increasing rate with increasing stress. As well, this atomic hydrogen diffuses to and concentrates at regions of high triaxial stress [12,13,22,24]. This would tend to increase the embrittling effect ahead of the crack tip and possibly account for the occurrence of secondary cracking. Figure 3(d) also reveals some minor microscopic cracking in the area surrounding the particles at the bottom of the microvoids. This secondary cracking as a result of mechanical separation is a typical HE characteristic found in high strength steels [1].

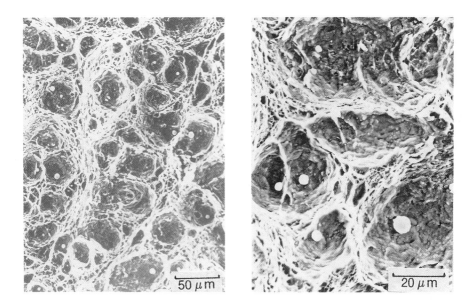

Figure 3(a) (left). SEM Fractograph of CT Specimen.
Figure 3(b) (right). Same as 3(a) Showing Microvoid Coalescence.

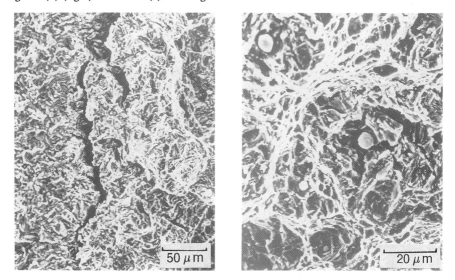

Figure 3(c) (left). SEM Fractograph of Pre-Immersion CT Specimen.
Figure 3(d) (right). Same as 3(c) Showing Reduced Ductility and Microvoid Coalescence.

Threshold Fracture Toughness Tests, K_{ISCC}

NACE TM0177-90 tests with DCB fracture toughness specimens yielded a K_{ISCC} value of 30 Mpa*m$^{1/2}$. This is a substantial decrease in toughness and indicates a significant susceptibility to SCC.

Since this test is a constant displacement or K-decreasing test, different fracture morphologies were observed along the SCC crack length. These different zones are shown at various locations along the crack front in Figures 4(a) through 4(d). Crack propagation is from left to right.

The fracture surface at the fatigue/overload transition zone or high stress intensity area resembles that of the 'pre-conditioned' CT specimen. One major difference is the secondary cracking is more extensive and not limited to the transition zone. Upon detailed examination, the cracks appear as wide, deep fissures or crevices which suggests a dissolution-type feature more typical of a SCC mechanism rather than HE.

Further along the crack length as shown in Figure 4(b), it is interesting to note that the secondary cracking becomes more intensive and severe. The lower intensity of the secondary cracking of Figure 4(a) as compared to that in Figure 4(b) is thought to occur as a result of rapid crack propagation due to wedge insertion prior to the SCC testing. The additional effects of the environment would be non-existent in this zone at that time. The secondary cracking decreases, Figure 4(c), as the crack length progresses due to the decreasing stress intensity and is at a minimum at the crack arrest region, Figure 4(d).

To examine the fracture morphology , the zones described in Figures 4(a) through 4(d) were observed in greater detail in Figures 5(a) to 5(d). In the region of high stress intensity, Figure 5(a), microvoid coalescence is dominant in combination with small amounts of less ductile transgranular fracture. As the crack progresses, the microvoids are almost indistinguishable with transgranular tearing becoming the dominant fracture mechanism, Figure 5(b). In Figure 5(c), the fracture changes to one of intergranular secondary cracking with transgranular tearing. Some microvoid coalescence is still evident at this stage. At the crack arrest region, the fracture mode appears to have changed again to one absent of microvoid coalescence and consisting of intergranular secondary cracking and approaching a quasi-cleavage mechanism.

Constant COD Rate Tests

The results of the constant COD rate tests are given in Figure 6. The critical stress intensity, K_{SCC} decreases with decreasing COD rates and approaches K_{ISCC} at loading rates in the order of 10^{-9} m/s. Machine limitations did not allow testing at slower rates. It is possible, therefore, that the critical SCC threshold stress intensity may be obtainable through application of a slow constant COD rate to a fracture mechanics specimen.

Since the NACE TM0177-90 test is a K-decreasing test (constant displacement) while the constant COD rate test is that of K-increasing, it was assumed that some similar fracture morphologies would be observed. Fractographs at COD rates ranging from 3 x 10^{-6} to 1.0 x 10^{-9} m/s are shown in Figures 7(a) to 7(d).

Fracture morphology at the higher COD rates, Figure 7(a), exhibits microvoid coalescence and extensive ductile rupture in the regions between. At medium COD rates, a more brittle

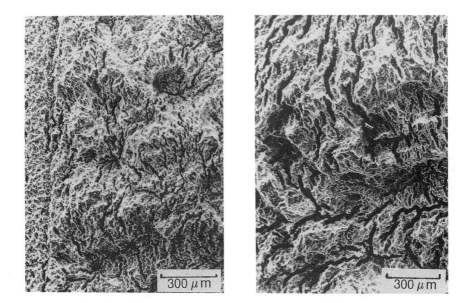

Figure 4(a) (left). SCC Fracture Surface at Fatigue/Overload Transition Region.
Figure 4(b) (right). SCC Fracture Surface at 1/3 Distance Along Crack Length.

Figure 4(c) (left). SCC Fracture Surface at 2/3 Distance Along Crack Length.
Figure 4(d) (right). SCC Fracture Surface at Crack Arrest Region.

259

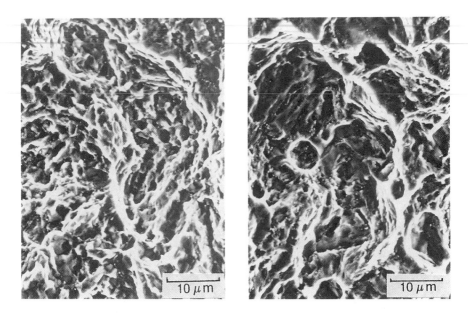

Figure 5(a) (left). SCC Fracture Surface at Fatigue/Overload Transition Region.
Figure 5(b) (right). SCC Fracture Surface at 1/3 Distance Along Crack Length.

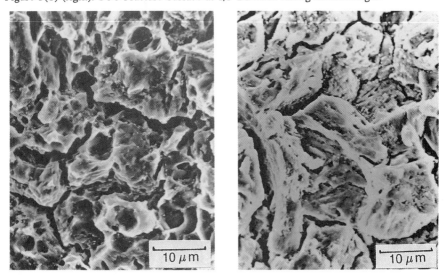

Figure 5(c) (left). SCC Fracture Surface at 2/3 Distance Along Crack Length.
Figure 5(d) (right). SCC Fracture Surface at Crack Arrest Region.

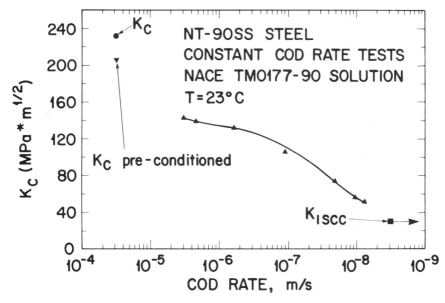

Figure 6. Variation of K_C With COD Rate.

type of appearance is observed with reduced microvoid coalescence and extensive transgranular tearing occurring throughout as shown in Figure 7(b). The fracture at low COD rates, is similar to that observed in the crack arrest region in the K-increasing tests, Figure 5(d). That is, the fracture morphology is one predominated by intergranular secondary cracking and extensive transgranular tearing as shown in Figures 7(c) and 7(d).

In comparison to the K-decreasing tests, the extensive secondary cracking is not observed in the COD rate SCC tests. A possible explanation could be that the faster strain rates promote faster increases in the stress intensity causing a more ductile crack propagation prior to adequate hydrogen transport and absorbtion to the plastic zone region [25]. As a result, full embrittlement effects causing secondary cracking resulting from mechanical separation is not observed.

SCC characteristics are also absent as the intensive fissures and crevices are not evident at any COD rate tested. This would suggest that the COD rates used do not allow sufficient time for SCC mechanisms to be operative to a great degree. This is substantiated by analysis of the COD tests. The lowest critical stress intensity obtained was 52 MPA*m$^{1/2}$ as compared to the K_{ISCC} value of 30 MPA*m$^{1/2}$. Since the initial stress intensity imparted by the wedge was approximately 65 MPA*m$^{1/2}$, only COD rate tests less than 5x10^{-8} m/s, as shown by Figure 6, would be expected to exhibit features similar to the K_{ISCC} tests. This appears to be the case as only Figure 7(d) exhibits the intergranular secondary cracking and transgranular tearing of the K_{ISCC} tests and compares well with Figure 5(c). Therefore, COD rates in the order of 10^{-8} to 10^{-10} m/s could be expected to exhibit features of SCC, HE and a possible value of K_{ISCC}.

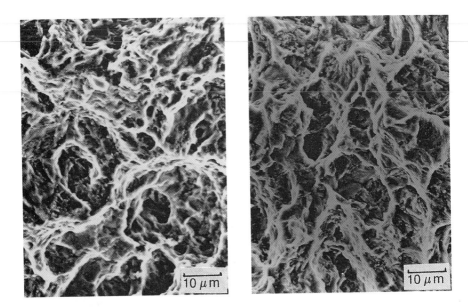

Figure 7(a) (left). Fracture Surface at COD Rate of 3.0 x 10⁻⁶ m/s.
Figure 7(b) (right). Fracture Surface at COD Rate of 1.0 x 10⁻⁷ m/s.

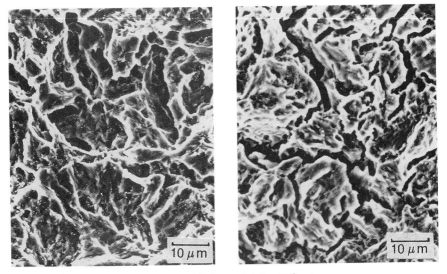

Figure 7(c) (left). Fracture Surface at COD Rate of 1.0 x 10⁻⁸ m/s.
Figure 7(d) (right). Fracture Surface at COD Rate of 4.0 x 10⁻⁹ m/s.

CONCLUSIONS

• Hydrogen embrittlement, HE, fracture morphology for NT-90SS steel is typified by extensive secondary cracking indicative of an embrittling mechanism coupled with a large degree of mechanical separation.

• Stress corrosion cracking, SCC, fracture morphology is characterized by extensive fissures and crevices typical of non-mechanical separation. The combination of the applied stress and the corrosion reaction at the crack tip to supply hydrogen to the plastic zone ahead of the crack, appears to intensify the embrittling process resulting in intergranular secondary cracking and transgranular fracture as additional SCC characteristics.

• Pre-immersion of specimens to avoid transient hydrogen charging effects enabled the identification of SCC and HE mechanisms.

• SEM analysis of fracture morphologies allowed the identification of HE and SCC characteristics.

• Constant crack opening displacement, COD, rate tests appear to be a viable test to obtain K_{ISCC} at very low COD rates. The critical stress intensity was found to decrease with decreasing COD rate for NT-90SS Steel.

ACKNOWLEDGEMENTS

The authors wish to acknowledge Shell Canada Limited, the Alberta Oil Sands Technology and Research Authority and the Natural Sciences and Research Council of Canada for financial support of this research.

REFERENCES

1) E.H. Phelps, "A Review of the Stress Corrosion Behaviour of Steels with High Yield Strength", **Proceedings of Conference, Fundamental Aspects of SCC**, National Association of Corrosion Engineers, p. 398, 1969.

2) B.G. Pound, G.A. Wright and R.M. Sharp, "The Anodic Behaviour of Iron in Hydrogen Sulfide Solutions", **Corrosion**, Vol. 45, No. 5, p. 386, 1989.

3) R.B. Heady, "Evaluation of Sulfide Corrosion Cracking Resistance in Low Alloy Steels", **Corrosion**, Vol. 33, No. 3, p. 98, 1977.

4) H. Asahi, Y. Sogo, M. Ueno and H. Higasiyama, "Metallurgical Factors Controlling SCC Resistance of High Strength, Low-Alloy Steels", **Corrosion**, Vol. 45, No. 6, p. 519, 1989.

5) T. Sato, H. Higashiyama, K. Yamamoto and T.Inoue, "Development of OCTG Resistant to Sour Environments", **14th Annual Offshore Technology Conference**, OTC 4331, p.379, 1982.

6) R.N. Parkins, "Development of Slow Strain-Rate Testing and its Implications", **ASTM STP 665**, p. 5, 1979.

7) R.P. Wei, S.R. Novak and D.P. Williams, "Some Important Considerations in the Development of Stress Corrosion Cracking Test Methods", **Materials Research and Standards**, Vol. 12, No. 9, p. 25, 1972.

8) NACE Standard TM0177-90, "Laboratory Testing of Metals for Resistance to Sulphide

Stress Cracking in H$_2$S Environments", **National Association of Corrosion Engineers (NACE)**, p. 17, 1990.

9) W.J.D. Shaw, "Aspects of Accelerated SCC Tests on Aluminum Alloys in 3.5% NaCl Solution", **Microstructural Science**, Vol. 13, p. 565, 1986.

10) C.D. Kim and B.E. Wilde, "A Review of the Constant Strain Rate Corrosion Cracking Test", **ASTM STP 665**, p. 97, 1979.

11) D.R. McIntyre, R.D. Kane and S.M. Wilhelm, "Slow Strain Rate Testing for Materials Evaluation in High Pressure H$_2$S Environments", **Corrosion**, Vol. 44, No. 12, p. 920, 1988.

12) A.A. Sheinker and J.D. Wood, "Stress Corrosion Cracking of a High Strength Steel", **Stress Corrosion Cracking of Metals - A State of the Art**, ASTM STP 518, p. 16, 1971.

13) H.H. Johnson, "On Hydrogen Brittleness in High Strength Steels", **Proceedings of Conference, Fundamental Aspects of SCC**, p. 439, 1969.

14) G.V. Karpenko and I.I. Vasilenko, **Stress Corrosion Cracking of Steels**, Freund Publishing House, Tel-Aviv, Israel, p. 72, 1977.

15) Huang H. and W.J.D. Shaw, "Effect of Cold Working on the Fracture Characteristics of Mild Steel Exposed to a Sour Gas Environment", **The 23rd International Metallographic Technical Meeting**, July 22-25, Cincinati, Ohio, 1990.

16) Huang H. and W.J.D. Shaw, "Microstructural Corrosion Behaviour of Cold Worked Steel in a Sour Environment", **The 23rd International Metallographic Technical Meeting**, July 22-25, Cincinati, Ohio, 1990.

17) H. Asahi, Y. Sogo and H. Higashiyama, "Effects of Test Conditions on K_{ISCC} Values Influenced by SCC Susceptibility of Materials", **Corrosion/87**, Paper 290, 1987.

18) ANSI/AWS A4.3-86,"Standard Methods for Determination of the Diffusible Hydrogen Content of Martensitic, Bainitic and Ferritic Weld Metal Produced by Arc Welding", **American Welding Society, Inc.**, 1985.

19) J.J. DaLuccia and D.A. Berman, "An Electrochemical Technique to Measure Diffusible Hydrogen in Metals (Barnacle Electrode)", **Electrochemical Corrosion Testing**, ASTM STP 727, p. 256, 1981.

20) D.A. Berman and V.S. Agarwala, "The Barnacle Electrode to Determine Diffusible Hydrogen in Steels", **Hydrogen Embrittlement: Prevention and Control**, ASTM STP 962, p. 98, 1988.

21) S.A. Bashu and S.V. Reddy, " Preparation of Elevated Temperature Fracture Surfaces for SEM Studies", **Short Communications; Metallography**, Vol. 22, p. 275, 1989.

22) M. Smialowski, **Hydrogen in Steel**, Permagon Press Ltd., New York, p. 224, 1962.

23) S.W. Ciarldi, "Microstructural Observations on the Sulfide Stress Cracking of Low Aby Steel Tubulars", **Corrosion**, Vol. 40, No. 2, p. 77, 1984.

24) D.I. Phalen and D.A. Vaughan, "The Role of Surface Stress on Hydrogen Absorption by 4340 Steel", **Corrosion**, Vol. 24, No. 8, p. 243, 1968.

25) J.R. Scully and P.J. Moran, "Influence of Strain on Hydrogen Assisted Cracking of Cathodically Polarized High-Strength Steel", **ASTM STP 1049**, p. 5, 1990.

FRACTOGRAPHIC ANALYSIS: SULFIDE STRESS CRACKING

OF MILD STEEL EXPOSED TO A SOUR GAS ENVIRONMENT

H. Huang and W.J.D. Shaw

ABSTRACT

Sulfide stress cracking damage of mild steel in a complex brine solution containing a high concentration of hydrogen sulfide in conjunction with carbon dioxide was investigated. Slow strain rate tests using tensile specimens were conducted with respect to the interactive effect of the sour gas environment and the material including both crack initiation and propagation. Multiple cracks were found on the outside surface of some components, indicating typical anodic stress corrosion cracking. Fractographic analysis indicated that crack initiation as a result of hydrogen embrittlement was located at the enter of the component due to the enhanced triaxial stress condition. Slow strain rates ($1x10^{-4}$ -$5x10^{-6}$/s) resulted in outside surface cracks indicating a dominance of the stress corrosion cracking mechanism. However very slow strain rates ($<10^{-7}$/s) changed the crack initiation to the center of the component and resulted in nearly 100% brittle failure. Similarly the occurrence of crack initiation shifted to the center of the component when there was an absence of carbon dioxide in the environment, which was representative of a hydrogen embrittlement mechanism.

Department of Mechanical Engineering, The University of Calgary
Calgary, Alberta, Canada T2N 1N4

INTRODUCTION

There has been an increase in use of low-to-medium strength steel in pipeline application dealing with deep sour wells. Sulfide stress cracking, SSC, damage of low strength steel in sour gas environments has been occurring [1-2]. When failure of low strength steel in corrosive environments occurs it is usually by anodic stress corrosion cracking, SCC, or hydrogen embrittlement.

SSC damage of high strength steel exposed to sour gas environments is believed to be associated with hydrogen embrittlement [3-4]. SSC requires the following conditions; tensile loading, a sour environment and a susceptible steel, usually thought to be a material having a hardness greater than HRC 22 (HB=235) [5]. However, Treseder et al [6] have reported SSC damage of low strength steel (HB=135) in an H_2S (hydrogen sulfide) environment. Failure of low strength steel in H_2S environments was considered to be a matter of environment severity, applied stress, and not the material's tensile properties (or hardness) [2].

The mechanism of SSC damage of low strength steel is not totally understood. For low strength, low nickel steel, Dunlop [7] pointed out that the mechanism of SSC is an anodic driven reaction, which was supported by other investigators [8]. On the other hand, some results indicate that the hydrogen embrittlement mechanism is responsible for SSC damage of low strength steel [2,9].

The purpose of this study is to investigate the susceptibility of low strength steel to SSC damage under slow strain rate loading and identify the SSC mechanism with the help of fractographic analysis. The main program objective is to gain an understanding of effects of strain rate on SSC damage.

EXPERIMENTAL

The material used in this study was a ferritic/pearlitic AISI 1020 steel in the hot rolled condition. The aqueous and gaseous environment used for this project is based upon analysis of some problem gas wells in Alberta [11]. The aqueous salt water as used in this study consisted of the following: 48,500 mg/L Na^+, 14,250 mg/L Ca^{2+}, 1,045 mg/L Mg^{2+}, 91,500 mg/L Cl^-,180 mg/L HCO_3^- and 150 mg/L SO_4^{2-}. The gaseous constituents consist of 34% hydrogen sulfide, 10% carbon dioxide with the balance being methane. This gas mixture was bubbled through the solution at a rate of 0.4 L/min. The tensile specimens were 3.2 mm in diameter with a gage length of 27 mm, however all elongation data has been adjusted to a standard gage length of 50 mm according to Barba's law.

The mechanical properties were determined after various lengths of time of exposure in an unloaded condition to the sour gas environment. This approach was used to identify the effect of hydrogen embrittlement. Slow strain rate tests were conducted with respect to the interactive effect of the sour gas environment and the material's crack initiation and propagation. The effects of strain rate and environment were studied using fractographic examination of the components. Fractography was conducted on a Cambridge Scanning Electron Microscope, SEM,

S100 after cleaning the fracture surface of the component using Clark' solution [11] and then sputter coating with gold.

EXPERIMENTAL RESULTS

PRE-EXPOSURE TEST

The tensile behaviors of mild steel after 0, 24, 72 and 216 hours of unstressed immersion in the sour environment were measured. The yield and ultimate strengths indicated an insensitivity to the effect of hydrogen accumulated in the mild steel during pre-immersion. A change of ductility reflects the extent of hydrogen embrittlement [12]. The ductility decreased with increasing exposure time as shown in Figure 1.

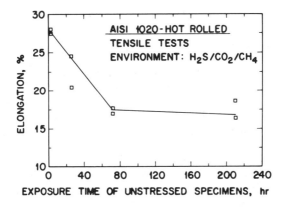

Figure 1. Relationship Between Elongation and Exposure Time
of Tensile Specimens

SLOW STRAIN RATE TEST

Slow strain rate tests were conducted using tensile specimens for evaluating SCC behavior. The results of slow strain rate tests using stain rates ranging from $1x10^{-4}$ to $1x10^{-7}$/s indicated that the ductility decreases with a decrease in strain rate as can be seen in Figure 2.

In the absence of H_2S, the mild steel tested in the brine solution containing CO_2/CH_4 gases at the strain rate of $1x10^{-5}$/s has a greater amount of ductility compared to that tested in the sour gas environment as can be seen in Figure 2. However the CO_2/CH_4 brine environment does cause some embrittlement of the mild steel. The results of the slow strain rate test exposed to this brine solution containing only H_2S gas is also found on Figure 2. This condition indicated a greater decrease in ductility as compared to that tested in the mixed sour gas environment.

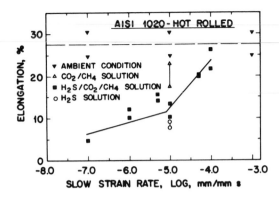

Figure 2. Relationship Between Elongation and Slow Strain Rate

FRACTOGRAPHIC ANALYSIS

The fractographic features of tensile specimens after a 216 hour unloaded pre-exposure in the sour gas environment indicates mainly microvoid coalescence, Figure 3. It is found that the fracture features are no different to those tested under ambient condition [12]. The slight change in the ductility and no apparent difference in the fractography shows that mild steel is not very sensitive to

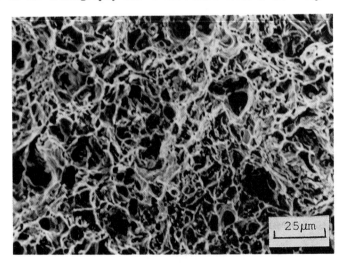

Figure 3. Fractographic of Tensile Specimen after 216 Hr Immersion in a Sour Environment

hydrogen embrittlement as evaluated from unstressed prior immersion specimens in the sour gas environment. In comparison the cold worked steel is considerably more sensitive to hydrogen embrittlement and indicates a considerable decrease in ductility after 72 hour pre-exposure in the sour gas environment [12]. Previous information [5] indicates that the susceptibility of carbon steel and low alloy steels to SSC decreases with decreasing strength level and becomes insignificant at hardness values below HRC 22 (HB=235). However hydrogen embrittlement of low strength steel does occur in sour environments or under cathodically charged conditions [1-2].

Examination of the outside surface of components under slow strain rate loading shows that multiple surface cracks occurred when components were tested at strain rates from 1×10^{-4} to 1×10^{-6}/s. Multiple cracks appear, prior to the domination of one crack and its subsequent propagation through the material to failure. Typical multiple cracks on the outside surface of a component are shown in Figure 4. The maximum occurrence of multiple cracking was found at a strain rate of 1×10^{-5}/s, while no multiple cracks occurred at 1×10^{-7}/s. The multiple cracks were also found on components tested in the CO_2/CH_4 brine solution. There was an absence of multiple crack on components tested in 100% H_2S brine environment.

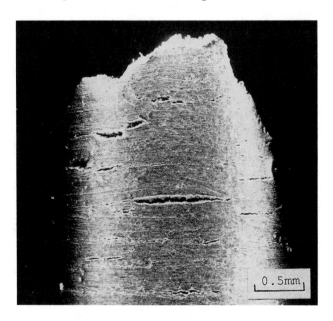

Figure 4. Typical Multiple Cracks on the Outside Surface of SCC Specimen
Tested in a Sour Environment at a Strain Rate of 10^{-5}/s

Figure 5 shows the fractography of mild steel under slow strain rate loading at a strain rate of 1×10^{-4}/s. The initiation of cracks occurs on the outer edge. The crack propagation then moved inward to the center of the component. The edge initiation is likely the result of a corrosion reaction of material with the environment while under applied loading [7]. The overall fracture surface can be seen in Figure 5a. The various regions can be seen in Figures 5b and 5c. The fracture feature at the center of the component is characterized by the ductile microvoid coalescence fracture, as shown in Figure 5c. This indicates that the failure process at the center of the component is controlled mainly by pure mechanical loading due to the limited testing time.

The fractography of mild steel under slow strain rate loading in the sour environment at a strain rate of 1×10^{-5}/s is shown in Figure 6. This figure indicates that the fracture features are somewhat similar to those tested at a strain rate of 1×10^{-4}/s. Again the overall fracture surface is shown in Figure 6a and is comparable to the previous case, Figure 5a. The propagation of the multiple cracks begins from the outside surface and moves inward, Figure 6b. The difference between these two rates lies in the number of multiple cracks and the deeper initiation region for the components tested at the slower strain rate of 1×10^{-5}/s. Ductile microvoid coalescence at the center of the component can be seen in Figure 6c. However the depth of the ductile dimples decreases for the slower strain rate condition.

Once a strain rate is so slow that there is lots of time for the interactive reaction of material with the environment, SCC cracks may be developed and maintained [13]. For anodic SCC to occur it is necessary that the electrochemical conditions for active corrosion be maintained at the crack tip. Cracking of the mild steel in the CO_2/CH_4 brine solution was found to result in a failure mechanism of anodic SCC. The anodic SCC mechanism is present in the mixed sour gas environmental condition due to the contribution of the CO_2/CH_4 gas components under slow strain rate loading when strain rate is faster than 1×10^{-6}/s. The hydrogen sulfide in the solution enhances the acidity of the solution and increases the reduction of hydrogen. As a result of the reduction of hydrogen a steady source of monotonic hydrogen occurs on the surface of the specimen, and subsequently diffuses into the steel. The hydrogen diffusion mechanism is enhanced by the presence of strain and results in a hydrogen embrittlement failure. Therefore the mechanism of SSC is a mixture of anodic SCC and hydrogen embrittlement in low strength mild steel.

Although strength level is not expected to have a significant effect on anodic stress corrosion cracking, it has been pointed out that hydrogen embrittlement can be related to the strength level when sour environments are present [5]. Once the strength level of a steel is low, and the H_2S environment contains components causing anodic SCC, such as CO_2, the contribution from an anodic SCC mechanism will likely be contributing to the failure.

Further decrease in the strain rate to 1×10^{-7}/s results in almost completely brittle fracture as shown in Figure 7. There is an absence of multiple cracks on the outer specimen surface. The overall fracture surface is shown in Figure 7a. The fracture is now brittle with small amounts of ductile coalescence at the outer edge can be seen in Figure 7b. A very small shear lip occurs on the outer edge of the fracture surface, Figure 7b. Figure 7c shows almost complete cleavage failure

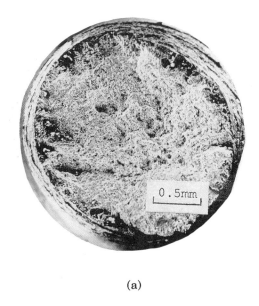

(a)

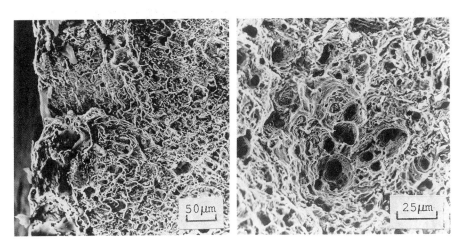

(b)   (c)

Figure 5. Fractography of Tensile Specimen at a Strain Rate of 10^{-4}/s
in a Sour Environment
a) Overview of Fracture Surface
b) At the Edge   c) At the Center

271

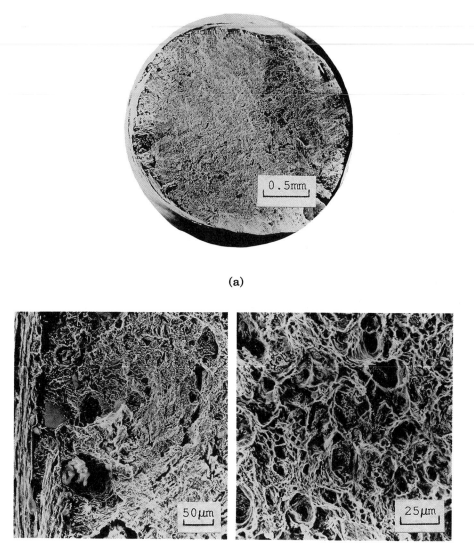

(a)

(b)   (c)
Figure 6. Fractography of Tensile Specimen at a Strain Rate of 10^{-5}/s
in a Sour Environment
a) Overview of Fracture Surface
b) At the Edge   c) At the Center

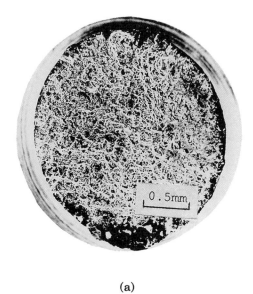

(a)

(b)   (c)

Figure 7. Fractography of Tensile Specimen at a Strain Rate of 10^{-7}/s
in a Sour Environment
a) Overview of Fracture Surface
b) At the Edge   c) At the Center

occurring at the center of the specimen. This arrangement indicates that initiation occurs at the center of the specimen and propagation occurs outwards to the exterior surface similar to regular tensile testing behavior. This is in contrast to the behavior of components tested at strain rates ranging from $1x10^{-4}$/s to $1x10^{-6}$/s.

At low strain rates (less than 10^{-6}/s), the rate of film rupture on the surface diminishes. When the rate of film repairing exceeds that of film rupture, the electrochemical condition for an anodic SCC mechanism is severely reduced and in some cases eliminated [13]. This is likely the reason why multiple cracks do not form on the outside surface of the component at a strain rate of 10^{-7}/s. Rather cracks initiate in the interior and propagate outward due to accumulated hydrogen acting with triaxial stress. Cracks that initiate at the center of tensile specimens affected by H_2S environments are typical hydrogen embrittlement behavior [12,14]. An increase in testing time due to the slower strain rate allows the hydrogen atom from the surface reaction to easily diffuse into the material. Hydrogen embrittlement is promoted under slow strain rate loading conditions. The results of this study show that the failure mechanism of mild steel is one of hydrogen embrittlement under slow strain rate loading when the strain rate is in the order of 10^{-7}/s.

According to the above discussion, the results in this study show that the slow strain rate technique, supplemented by the fractographic examination, is an effective method to classify the mechanism between anodic SCC and hydrogen embrittlement.

CONCLUSIONS

1. Sulfide stress cracking of mild steel exposed to the sour gas environments results in a mixture of anodic stress corrosion cracking and hydrogen embrittlement when tested under a slow strain rate conditions greater than 10^{-6}/s.

2. Under regular tensile testing, mild steel is quite insensitive to hydrogen embrittlement as determined from prior immersion testing in the sour gas environment.

3. Sulfide stress cracking damage of mild steel exposed to concentrated H_2S solution results primarily from hydrogen embrittlement.

4. Slow strain rate loading promotes susceptibility of mild steel to hydrogen embrittlement.

5. A low strength steel exposed to H_2S environments containing components causing anodic SCC, such as CO_2, is susceptible to a failure mechanism of anodic SCC.

6. The slow strain rate test technique, supplemented by fractographic examination, is an effective method to classify the contribution between anodic stress corrosion cracking and hydrogen embrittlement.

ACKNOWLEDGEMENT

The authors wish to thank the Petroleum Graduate Research Program and AOSTRA for financial support of this work. Thanks are also given to Bob Konzuk of the Department of Mining, Metallurgy and Petroleum Engineering at the University of Alberta for his efforts in cold rolling the material.

REFERENCE

1. J.G. Erlings, H.W. De Groot and J. Nauta, "The Effect of Slow Plastic and Elastic Straining on Sulfide Stress Cracking and Hydrogen Embrittlement of 3.5% Ni Steel and API %L X60 Pipeline steel", **Corrosion Science**, Vol. 27, p 1153, 1987
2. J.L. Turn. Jr., B.E. Wilde and C.A. Troianos, "On the Sulfide Stress Cracking of Line Pipe Steel", **Corrosion**, Vol. 39, p 364, 1983
3. K. Matsumoto, Y. Kobashi, K. Ume, K. Murakami, K. Taira and K. Arikata, "Hydrogen Induced Cracking Susceptibility of High Strength Line Pipe Steels", **Corrosion**, Vol. 42, p 337, 1986
4. T. Taira, K. Tsukada, Y. Kobayashi, H. Inagaki and T. Watanabe, "Sulfide Corrosion Cracking of Linepipe for Sour Gas Service", **Corrosion**, Vol. 37, p 5, 1981
5. National Association of Corrosion Engineers Standard MR-01-75, 1978 Revision
6. R. S. Treseder and T.M. Swanson, "Factors in Sulfide Corrosion Cracking of High Strength Steels", **Corrosion**, Vol. 24, p 31, 1968
7. A.K. Dunlop, "Stress Corrosion Cracking of Low Strength, Low Alloy Nickel Steels in Sulfide Environments", **Corrosion**, Vol. 34, p 88, 1978
8. K. Motoda, Y. Yamane, K. Uesugi and Y. Nakai, "Effect of the Potential on the sulfide Stress Cracking of Steel", **Iron &Steel Institute of Japan**, Vol. 67, p S478, 1981
9. E. Snape, "Sulfide Stress Corrosion of Some Medium and Low Alloy Steel", **Corrosion**, Vol. 23, p 154, 1967
10. G.I. Ogundele, "An Electrochemical Study of Corrosion of Carbon Steel in Aqueous Sour Gas Media", Ph.D. Thesis, The University of Calgary, 1984
11. "Preparing, Cleaning, and Evaluation Corrosion Test Specimens", Annual Book of ASTM Standards, ASTM G1-81, Vol. 03.02, p 89, 1987
12. H. Huang and W.J.D. Shaw, "Effect of Cold Working on the Fracture Characteristics of Mild Steel Exposed to a Sour Gas Environment", **Microstructural Science**, Vol. 19, 1991 in press
13. R.B. Diegle and W.K. Boyd, in Stress Corrosion Cracking -- The Slow Strain Rate Technique, ASTM STP 665, G.M. Ugiansky and J.H. Payer, Eds., American Society for Testing and Materials, Philadelphia, p 5, 1979
14. S.W. Ciaraldi, "Microstructural Observation on the Sulfide Stress Cracking of Low Alloy Steel Tubular", **Corrosion**, Vol. 71-81, p 77, 1984

HYDROGEN INDUCED FRACTURE IN PWA 1480E AT 1144K

P.S. Chen[1] and R.C. Wilcox[2]

ABSTRACT

The high temperature fracture behavior of single crystals of the PWA 1480E nickel-base superalloy was studied. Notched crystals with three different crystal growth orientations near [100], [110], and [111] were tensile tested at 1144K in both helium and hydrogen atmospheres at 34 MPa. A stereoscopic SEM technique and planar γ' morphologies were applied to identify the orientation of cleavage planes. Specimens tested in helium failed predominately by {111}-type cleavage. The notched tensile strength in helium was orientation dependent. In hydrogen, all the specimens had very nearly the same tensile strength and failed predominately by {111}-type cleavage. Hydrogen effects appeared to be minimized at 1144K although some degradation was detected.

INTRODUCTION

The deformation behavior of single crystals of various nickel-base superalloys has been studied [1-6]. PWA 1480 was found to have an increase in strength with increasing testing temperature [3]. This unusual behavior was due

---

[1]P.S. Chen, Associate Engineer, IIT Research Institute, Huntsville, AL 35812.
[2]R.C. Wilcox, Professor, Materials Engineering Program, Department of Mechanical Engineering, Auburn University, Auburn, AL 36849.

to a thermal activated cube cross-slip process inhibiting dislocation glide. In the alloy Rene N4, specimens with [100] and [110] orientations exhibited octahedral slip at room temperature, at 1033K and at 1253K [4]. However, primary cube slip occurred for the [112] oriented specimen at 1033 and 1253K.

Nickel-base superalloys have been shown to be susceptible to hydrogen embrittlement [1,7-9]. Moreover, strength degradation of single crystals in an hydrogen environment was found to be orientation dependent at room temperature [1,9]. However, hydrogen effects at high temperature (1144K) on single crystals of nickel-base superalloys have not been reported. In order to determine the role of hydrogen, fracture surfaces of notched tensile specimens of single crystals of PWA 1480E with three different crystal growth directions were examined. A comparative analysis of the fracture behavior of these three crystals was made to understand the susceptibility of PWA 1480E to hydrogen embrittlement.

EXPERIMENTAL PROCEDURE

Notched PWA 1480E single crystals with three different orientations, [100], [110], and [111], were furnished by NASA, Marshall Space Flight Center, Huntsville, AL. The composition of the PWA 1480E alloy is given in Table 1 [10]. Cast single crystal slabs with growth orientations of [110], [111], and [100] were produced by the Howmet Turbine Components Corporation, Whitehall, MI [10]. These slabs were solution heat treated at 1561K and then at 1569K for proprietary lengths of time. Blanks were cut from the slabs and reheated to 1353K for 4 hrs in vacuum, then gas quenched and aged for 32 hrs in air at 1144K. Notched tensile samples were machined from these blanks by Williams International, Walled Lake, MI and were tensile tested by Pratt and Whitney, East Harford, CN [10]. Specimens with orientations parallel to the [110], [111], and [100] tensile axes were machined from the [110], [111], [100] slabs, respectively. Tensile specimens were 8.5mm in diameter by 50mm long with a 6mm notch diameter. The notch angle was 60°C with a 0.0508mm radius. After machining, the crystals were stress relieved at 1123K for

Table I.   Spectrographic analysis of PWA 1480E [10].

Element	Wt Percent*	Element	Wt Percent*
Carbon	0.004	Sulfur	0.0009
Phosphorus	<0.01	Manganese	<0.10
Silicon	<0.10	Chromium	10.0
Nickel	Balance	Tungsten	4.10
Cobalt	5.23	Titanium	1.35
Copper	<0.10	Lead	<0.0001
Bismuth	<0.00003	Aluminum	4.88
Iron	0.06	Columbium and	
Zirconium	0.001	Tantalum	11.78
Boron	0.001	Hafnium	0.003
Selenium	0.00005	Tellurium	<0.00005
Thallium	<0.00005		

* Howmet Turbine Components Corporation,
  Whitehall, MI.

8 hrs in vacuum. Tensile testing was performed at 1144K (crosshead speed of 0.127mm/min.) in both helium and hydrogen atmospheres at 34 MPa to simulate conditions in the space shuttle engine [10]. Helium was considered to be a hydrogen-free atmosphere. The total exposure time for each environment was up to 30 min. due to heating of the specimens.

A JEOL 840 SEM operating at 20 kV was utilized to study the fracture behavior. Stereo photo pairs 10-15° apart were taken from various cleavage planes to calculate the angle between the crystal axis and the cleavage step normal. The procedure involved the use of a reference point and three other points in the stereo pair [11] and was based on a stereographic technique developed for chemical analysis of fracture surfaces [12]. A Microcomp Image Analysis System was used to determine the area fraction of {111}-type fracture planes with linear dimensions of larger than $10\mu m$. Cleavage plane orientations near (within $\approx 0.5mm$ of the notch root) and outside of the notch region as determined by trace analysis and the stereoscopic technique are given in Table 2 for the six crystals studied.

Microstructural Science, Volume 19

Table 2. Cleavage plane orientations near the notch area and the area fraction of {111} planes found in PWA 1480E single crystals at 1144K.

CRYSTAL ORIENT.	ORIENT. DEVIATION (DEG.)[10]	NEAR NOTCH CLEAVAGE PLANE		ANGLE*	AREA % OF {111} CLEAVAGE	
		He	H$_2$		He	H$_2$
[100]	1.5	{111}	{111} (100)	55	54	42
[110]	1.0	{111}	{111}	35	58	45
[111]	5.0	{111}	{111}	70	60	38

* Angle between the cleavage plane normal and the crystal axis.

RESULTS

The microstructure of the PWA 1480E crystals consisted of about 40% γ and 60% γ' with some carbides and less than 2% γ/γ' eutectic and 0.4-0.5% porosity. The γ' particles had a cube shape (average size of 0.6-0.7μm) in which the [100] direction of the γ' was parallel to the [100] direction in the γ matrix [10,13]. This γ' morphology and orientation relationship to the γ matrix produced a square γ' network on {100} planes and a triangular network on {111} planes [13]. Carbides were present on all fracture surfaces but did not influence fracture. All crystals had an orientation deviation of 5° or less from the desired tensile axis (Table 2) [10].

Table 3 gives the notched tensile strengths of the crystals tested in helium and hydrogen environments at 1144K. Notched tensile strengths at 295K are provided for comparison purposes. All tensile strengths are the average of two tests. Because of the use of notched specimens, yield strength and elongation were not measured [10]. In helium, the [100] specimen was the strongest while the [110] crystal was the weakest. In hydrogen, the [111] specimen was slightly stronger than the other

280

Table 3. Degradation of tensile strength of PWA 1480E single crystals at 1144K.

CRYSTAL ORIENT	NOTCHED TENSILE STRENGTH (MPa) [10]		
	He		% DEGRADATION
	295K	1144K	From 295 to 1144K in He
[100]	1534.1	1441.9	- 6.0
[110]	1692.1	1168.5	-30.9
[111]	1566.2	1308.5	-16.5

| | H_2 | | % DEGRADATION | |
| | | | From 295K to | From He to |
	295K	1144K	1144K in H_2	to H_2 at 1144K
[100]	876.8	1296.5	+47.9	-10.1
[110]	1335.0	1254.0	- 6.1	+ 7.3
[111]	1377.6	1297.5	- 5.8	- 0.8

crystals, while the [110] orientation was the weakest. The difference in strength for the helium-charged crystals was about 273 MPa while for the hydrogen-charged crystals the difference was 44 MPa. The notched tensile strength of single crystals tested in helium at 1144K [10] was found to be orientation dependent. The strength of the crystals tested in hydrogen was not orientation dependent.

Fracture surfaces of [100], [110] and [111] crystals tested in helium are shown in Figure 1. In all cases, examination at high magnification in the region of the notch root (within ≈0.05mm of the notch) revealed a jagged appearance which was the result of cleavage along crystallographic planes (Figure 2). Stereoscopic measurements of the normals to the cleavage planes near the notch gave an angle of about 55°, 35°, and 70° with respect to the [100], [110], and [111] crystal axes, respectively. Cleavage planes with these angular

Figure 1. SEM profiles of the fracture surfaces of crystals tested in helium at 1144K; (a) [100]. (b) [110]. (c) [111].

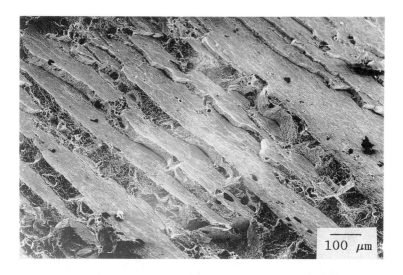

100 μm

Figure 2. SEM fractograph of the [100] specimen tested in helium at 1144K showing the {111}-type cleavage planes found near the notch area.

relationships were then determined by trace analysis to be {111}-type planes. An equilateral triangular type network morphology of γ' was found on these cleavage planes (Figure 3). This morphology was typical of cleavage fracture on {111}-type planes in the cube shaped γ' particles [6] and confirmed the {111}-type cleavage planes. In all crystals, {111}-type cleavage planes also were found outside the notch region toward the specimen center.

The overall area fraction of {111}-type cleavage planes in the [100], [110], and [111] crystals were determined to be about 54, 58, and 60%, respectively (Table 2). In all three crystals about 50% of these {111} planes were located in the notch region. Therefore, the fracture behavior of the crystals tested in helium at 1144K can be summarized as fracture mainly on the {111}-planes in the notch region followed by {111}-type cleavage outside of the notch area.

The fracture behavior of notched [100], [110] and [111] single crystals of PWA 1480E at room temperature has been studied [6,13]. At room temperature, the crystal with the

Figure 3. Equilateral triangular γ' network on {111}-type cleavage planes found near the notch area in crystals tested in helium at 1144K; (a) [100] crystal. (b) [110] crystal. (c) [111] crystal.

[100] orientation was the weakest while the [110] crystal was the strongest (Table 3). However, at 1144K the [100] specimen was the strongest of the three crystals and the [110] crystal was still the weakest. The [111] specimens exhibited a tensile strength between these two extremes at both room or at high temperature.

Cleavage of all specimens tested in helium occurred primarily on {111} planes despite the single crystal orientation. No obvious macroscopic {100} cleavage was observed on the fracture surface in any of the crystals. In addition, the area fraction of {111}-type cleavage planes was found to increase from 25% at room temperature [6,13] to about 55% at 1144K. This was strong evidence that high temperature promoted {111} octahedral slip and resulted in a large increase in the area fraction of {111} cleavage planes.

In PWA 1480E, cleavage planes originated at the notch and propagated along crystallographic planes in all specimens. The observed cleavage planes were either a large single {111}-type plane or a combination of smaller {111} planes. The single crystal orientation did not influence the initial cleavage plane orientation in the notch area.

Fracture surfaces of notched [100], [110] and [111] orientated crystals tensile tested in hydrogen are shown in Figure 4. As in the helium tested specimens a jagged appearance of the cleavage planes again was observed. Measurements of the cleavage planes normals near the notch gave angles of about 55°, 35° and 70° with respect to the [100], [110] and [111] axes, respectively. Cleavage planes with these angular relationships were determined to be of the {111}-type. Further confirmation of the nature of the cleavage planes was made by the equilateral triangular type γ' network (Figure 5).

In addition to the {111}-type cleavage planes found in the [100] crystal, cleavage on (100) planes near the notch was observed. The square γ' morphology (Figure 6) identifies these planes as {100}-type. The (100) cleavage plane was not as smooth as those found at room temperature in previous studies [6,14]. The rougher (100) plane implied that a greater amount of plastic deformation occurred at 1144K before final fracture than at room temperature.

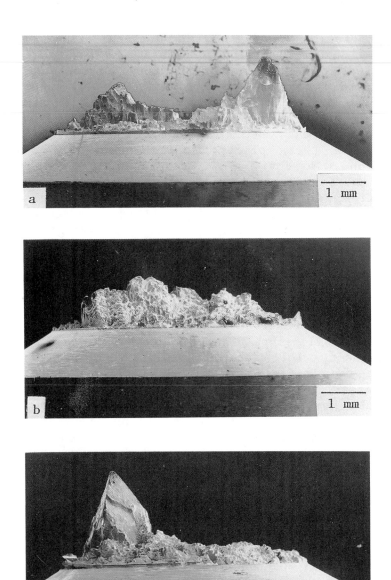

Figure 4. SEM profiles of the fracture surfaces of crystals tested in hydrogen at 1144K; (a) [100]. (b) [110]. (c) [111].

Cleavage plane orientations near the notch and the overall area fraction of {111}-type cleavage planes in the crystals tested in hydrogen at 1144K are given in Table 2. Cleavage on the (100) plane in the [100] crystal only was found close to the notch root. Outside the notch region toward the specimen center, {111}-type cleavage planes were detected and further confirmed from the triangular morphology of the γ'. The overall area fraction of {111}-type cleavage planes in the [100] specimen was determined to be about 42%. Approximate 50% of the {111} cleavage was located near the notch region. Fracture of the [100] specimen tested in hydrogen occurred mainly on the {111} planes with some {100} cleavage in the notch region followed by {111}-type cleavage outside of the notch area.

The area fraction of the {111}-cleavage planes in the [110] crystal were determined to be close to 45% and about 38% in the [111] specimen. About 50% of these cleavage planes were located near the notch area. Cleavage planes of the {111}-type also were found outside the notch area in both of these crystals. Therefore, in hydrogen at 1144K the [110] and the [111] specimens failed primarily by {111}-type cleavage near the notch followed by {111}-type cleavage outside of the notch area.

The hydrogen-induced fracture behavior of PWA 1480E single crystals at room temperature has been studied [6,14]. At room temperature, the specimen with the [100] orientation was the weakest while the [111] crystal was the strongest. At 1144K, the [111] specimen was the strongest while the [110] crystal was the weakest. The [100] specimen exhibited a notched tensile strength between these two extremes at high temperature. However, the greatest strength degradation occurred in the [100] specimen but not in the [110] crystal as shown in Table 3.

Examination of all the hydrogen charged specimens revealed that fracture occurred predominantly along {111}-type of planes near the notch region. The occurrence of {111} cleavage planes appeared to be independent of the single crystal orientation. Decohesion along {100} planes was not found in the crystals tested at high temperature except for the hydrogen tested [100] specimen. Also, the (100) cleavage planes were not as smooth as those found at

287

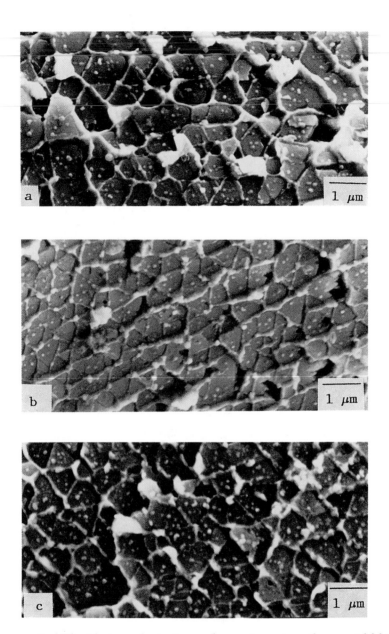

Figure 5. Equilateral triangular γ' network on {111}-type cleavage planes found near the notch area in crystals tested in hydrogen at 1144K; (a) [100] crystal. (b) [110] crystal. (c) [111] crystal.

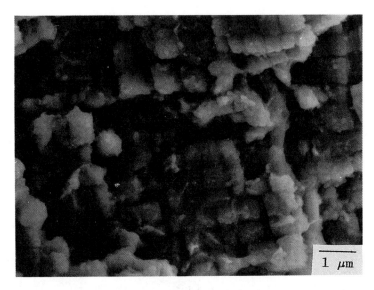

Figure 6. (100) cleavage plane with a square γ' network found near the notch root of the [100] crystal tested in hydrogen at 1144K.

room temperature [6,14]. Tearing of both γ and γ' phases was observed (Figure 6).

Unlike the large strength difference which occurred at 295K [6,14], all specimens tested in hydrogen at 1144K were very close in notched tensile strength (Table 3). This implied that the effects of hydrogen on orientation differences at high temperature were much less compared to those at 295K. Therefore, the lower differences in strength must be related to different fracture modes. At 295K, {100} type cleavage dominated in the three different orientations [6,14]. However, {111}-type cleavage was the controlling fracture mechanism at high temperature in that all crystals tested at 1144K failed primarily by {111}-type cleavage regardless of the single crystal orientation.

Although the influence of hydrogen was small, hydrogen effects still existed. The most obvious difference was the area fraction of the {111}-cleavage planes observed between the two testing atmospheres. For the helium tested crystals, the area fraction of large {111}-cleavage planes ranged from 50 to 60 %, while only about 38 to 45%

of the fracture surface in the hydrogen charged crystals were {111}-type cleavage planes. In addition, cleavage along the (100) γ/γ' interface was observed only in very limited areas and the cleavage planes were not as smooth as those found at room temperature [6]. At room temperature in hydrogen, only about 8 to 10% of the fracture surface was {111}-type cleavage planes. This was again evidence that the high temperature environment reduced hydrogen-induced embrittlement along the (100) γ/γ' interface and enhanced slip on {111} planes.

CONCLUSIONS

The following conclusions resulted from the investigation of hydrogen-induced fracture in single crystals of the nickel-base superalloy PWA 1480E at 1144K (871°C):

1. Failure occurred by {111}-cleavage in both hydrogen and helium atmospheres.
2. The occurrence of {111}-type cleavage was independent of the single crystal orientations.
3. The amount of {111} cleavage planes was greater in specimens tested in helium than those tested in hydrogen.
4. The notched tensile strength of PWA 1480E tested in helium was dependent on the crystal orientation, while those tested in hydrogen were not.

ACKNOWLEDGEMENT

This work was supported by NASA, Marshall Space Flight Center under contract no. NAS8-38184.

REFERENCES

1) R.L. Dreshfield and R.A. Parr, "Application of Single Crystal Superalloys for Earth-to Orbit Propulsion System," **NASA Memorandum 89877, ALAA-87-1976**

2) D.L. Anton, "Fracture of Nickel-Base Superalloy Single Crystals," **Materials Science and Engineering**, Vol. 57,p 97, 1983

3) D.M.Shah and D.N.Duhl,"The Effect of Orientation, Temperature and Gamma Prime Size on the Yield Strength of a Single Crystal Nickel Base Superalloy," **Superalloys 1984**, M. Gell, C.S. Kortovich, R.H. Bricknell, W.B. Kent, and J.F. Radavich, eds., AIME, Warrendale, PA, p 105, 1984

4) R.V. Miner, R.C. Voigt, J. Gayda, and T.P. Gabb, "Orientation and Temperature Dependence of Some Mechanical Properties of the Single-Crystal Nickel-Base Superalloy Rene N4: Part I. Tensile Behavior," **Met. Trans.**, Vol. 17A, p 491, 1986

5) R.V. Miner, T.P. Gabb, J. Gayda and K.J. Hemker, "Orientation and Temperature Dependence of Some Mechanical Properties of the Single-Crystal Nickel-Base Superalloy Rene N4: Part III. Tension-Compression Anisotropy," **Met. Trans.**, Vol. 17A, p 507, 1986

6) P.S.Chen, R.C.Wilcox, "Hydrogen Induced Cleavage in Single Crystals of the Ni-based Superalloy PWA 1480E," **Microstructural Science**, T.A. Place, J.D. Braun, W.E. White and G.F. Vander Voort, eds., ASM International, Materials Park, OH, Vol 18, p 443, 1990

7) M. Dollar and I.M. Bernstein, "The Effect of Hydrogen on Deformation Substructure, Flow and Fracture in a Nickel-Base Crystal Superalloy," **Acta Metall.**, Vol. 36, p 2369, 1988

8) C.L. Baker, J. Chene, I.M. Bernstein and J.C. Williams, "Hydrogen Effects in [001] Oriented Nickel-Base Superalloy Single Crystals," **Met. Trans.**, Vol. 19A, p 73, 1988

9) B. H. Kear and B. J. Piearcey, "Tensile and Creep Properties of Single Crystals of The Nickel-Base Superalloy Mar-M200," **Trans. AIME**, Vol. 239, p 1209, 1967

10) P. Nagy, K. Seitz and K. Bowen, Final report for contract NAS8-35915 by Williams International for Marshall Space Flight Center, Huntsville, AL, March 1986

11) P.S. Chen and R.C. Wilcox, "Stereographic Technique for Quantitative Analysis for Cleavage Plane Orientation," **Materials Characterization**, Vol. 26, p 9, 1991

12) J.L. Bomback, "Stereoscopic Techniques for Improved X-Ray Analysis of Rough SEM Specimens," **Scanning Electron Microscopy**, IIT Research Institute, Chicago, p 97, 1973

13) P.S. Chen and R.C. Wilcox, "Fracture of Single Crystals of the Nickel-Base Superalloy PWA 1480E in Helium at 22°C," **Metall. Trans. A**, Vol. 22A, p 731, 1991

14) P.S. Chen and R.C. Wilcox, "Fracture of Single Crystals of the Nickel-Base Superalloy PWA 1480E in Hydrogen at 22°C," **Metall. Trans.**, in press.

PRE-PLASTIC STRAIN EFFECT ON HIGH TEMPERATURE AGING

EMBRITTLEMENT OF TYPE 316 STAINLESS STEEL

W. Zhao and X. Mao

ABSTRACT

Type 316 austenitic stainless steel (with 0%, 13%, and 26% pre-plastic strain) was aged at a temperature of 650°C for 750 hours. The microstructures of these pre-deformed samples were examined by optical and scanning electron microscope (SEM). The tensile property of these samples was measured by normal mechanical tensile test. The fracture surfaces were examined by SEM. It has been found that a large amount of chromium carbides are precipitated in deformation lines introduced by pre-plastic deformation and then chromium carbide precipitation in grain boundary is reduced. The final ductility of the pre-deformed and subsequent aging treated 316 stainless steel is increased. Plastic deformation pretreatments and subsequent aging treatments of the stainless steel can improve its mechanical property.

Department of Mechanical Engineering, The University of Calgary
Calgary, Alberta, Canada T2N 1N4

293

INTRODUCTION

Type 316 Austenitc stainless steel has found wide application in steam generating plants as piping and superheater tube material as well as structural material in nuclear reactor. Long-time exposure of Type 316 stainless steel to elevated temperature (400°C ~ 900°C) is known to course high temperature embrittlement due to chromium carbides and σ phase precipitating in grain boundaries.

Numerous investigations have been published on the mechanical properties and microstructure changes occurring during exposure to high temperatures [1-13]. All the investigations generally can be divided into two areas: (1) mechanical property loss such as ductility, creep resistance and fracture toughness of the aged material and (2) phase instabilities during high temperature exposure such as time-temperature-precipitation (TTP) diagrams. The effect of cold work on aging is to shift the TTP diagram to shorter times and lower temperatures. The formation of $M_{23}C_6$ and σ is markedly accelerated [1, 2]. Creep behavior of several austenitic stainless steels has been studied as a function of prior cold work. In the case of Type 316 stainless steel, Donati et al [13] showed that improvements in creep resistance increased with degree of cold work. However, no investigations on pre-plastic-deformation effect on chromium carbide precipitation in grain matrix and grain boundary during high temperatures aging of Type 316 stainless steel and then its effects on room temperature tensile property. Since the stainless steel sometimes is deformed before servicing in high temperatures, it is necessary to study the pre-plastic strain effect of the stainless steel on the microstructure change and mechanical property change during high temperature exposure. Therefore, the purpose of the present investigation was i) to study the chromium carbide precipitation in grain boundary and grain matrix of pre-deformed Type 316 stainless steel; ii) to determine the pre-deformation effects on mechanical property change.

EXPERIMENTAL PROCEDURE

Commercial Type 316 austenitic stainless steel was employed for this study. The chemical composition of this material is listed in Table 1. Cylindrical samples of

Table.1   Chemical composition of Type 316 stainless steel

Composition,			Wt Pct			
C	Cr	Ni	Mo	Mn	Si	Fe
0.066	17.4	12.3	2.05	1.57	0.21	balance

approximately 0.5 in. diameter were solution heat-treated at 1050°C in an electric furnace in stagnant air for 20 minutes. After solution treatment, conventional tensile specimen were machined from the cylindrical samples. Thirteen and twenty-six percent

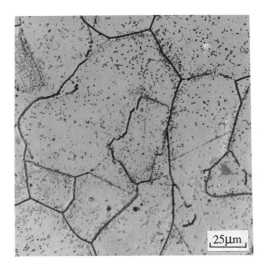

a)

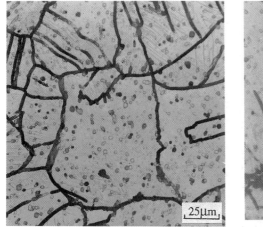

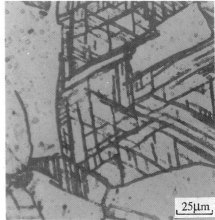

b)   c)

Figure 1   Pre-plastic strain effect on chromium carbide precipitation after aging
                at 650°C for 750 hours.
        a) 0% pre-strain, solution treated. b) 13% pre-strained and aged at 650°C,
        750hrs.
        c) 26% pre-strain and aged at 650°C, 750 hrs.

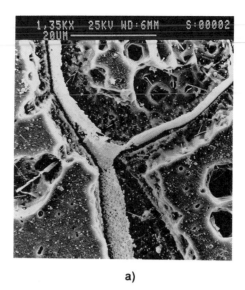

a)

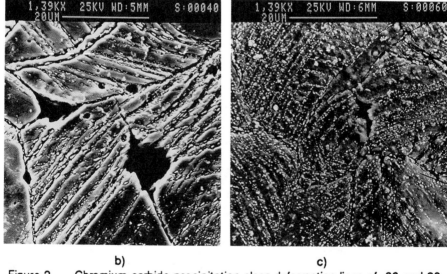

b)   c)

Figure 2   Chromium carbide precipitation along deformation lines of 26 and 39
percent pre-strained samples after aging at 650°C for 750 hours.
a) 0% pre-strain, solution treated. b) 26% pre-strained and aged at
650°C,750hrs.
c) 39% pre-strained and aged at 650°C, 750 hrs.

of pre-plastic strain was introduced in some of the tensile test specimens by controlled elongation in conventional tensile test equipment. Then aging treatments were performed in an electric furnace in stagnant air at temperature of 650°C for a period of 750 hours. The aging temperature is so controlled at 650°C [13] that the pre-plastic strained material was kept in the recovery region during aging treatment. The specimens were rapidly cooled after the aging treatment.

Thin sections were cut from the aged material and mechanically grounded to remove the oxidized surface layers, conventionally mounted and polished. An etching solution of 45 pct $FeCl_3$ as well as an electrolyte of 10 pct oxalic acid were used to reveal the microstructure of chromium carbides. Observation of the aged material microstructure was done using both optical and scanning electron microscope. The mechanical tensile tests were carried out in air in accordance with ASTM E8-85. The strain rate used was 0.005/min. After tensile test, the fracture surfaces of specimens were examined under a scanning electron microscope.

RESULTS AND DISCUSSION

It is well known that high temperature exposure of Type 316 stainless steel results in the chromium carbide precipitation in grain boundary. The grain boundary carbide is of the $M_{23}C_6$ type, as determined by electron diffraction in a transmission electron microscope and by a Deby-Scherrer pattern of extracted residuals [11]. Figures 1(a), (b), and (c) show optical microstructures of the pre-deformed 316 stainless steel after 750 hours aging at temperature of 650°C (using an electrolyte of 10 pct oxalic acid). Figure 1(a) is microstructure of no plastic strain introduced sample. It can be seen that only grain boundary is heavily etched after using etchant of solution. Then chromium carbides were identified based on etched area by the etchant. However, with the introduction of plastic deformation before high temperature aging, the microstructure of the aged materials changed. The 13 and 26 pre-strain resulted in the formation of needle-shaped features observable in the Figure 1(c) and (b). The microstructure with further magnification of the needle-shaped features (26% pre-plastic strain) was taken by use of scanning electron microscope as shown in Figure 2. In Figure 2, fine particals (chromium carbides) along these deformation lines were identified by using etchent of $FeCl_3$ solution. It can be found that less chromium carbides were precipitated in grain boundaries than those in grain matrix. It also can be found that with the increasing of pre-strain, more fine chromium carbides precipitated in deformation lines as shown in Fig.2 (C). In the pre-plastic deformed and subsequently aged sample, the deformation lines in grain matrix acted as nucleation sites for numerous small chromium carbide particals. It seems that the higher the amount of pre-strain introduced the less amount of chromium carbide precipitated in the grain boundaries. Due to chromium carbide precipitation both in grain boundaries and in grain matrix, the mechanical property of pre-strained and unstrained 316 stainless steel after aging become different. A mechanical property test of pre-strained and unstrained materials was done. First, plastic strain was introduced in the solution-treated specimen by controlled strain in conventional tensile test machine and then the tensile specimen was unloaded. After being aged at temperature of 650°C for 750

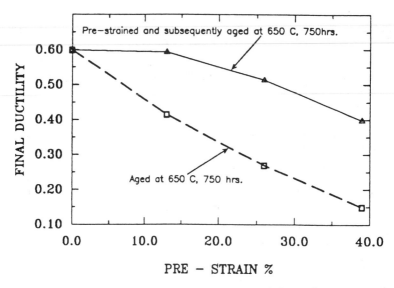

Figure 3 -   Final ductility relationship with pre-deformation treatment.

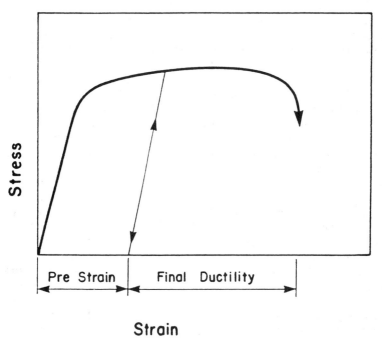

Figure 4 - Illustration of pre-strain and final ductility in stress vs strain curve.

a)

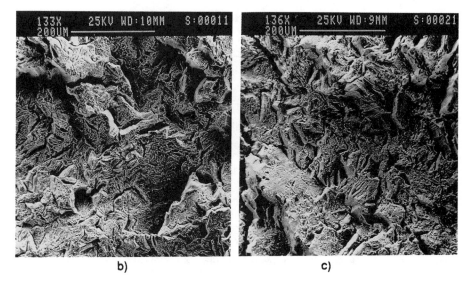

b)   c)

Figure 5   SEM fractographes showing secondary cracking on the fracture
                  surface.
        a) 0% pre-strain, solution treated. b) 13% pre-strained and aged at 650°C,
        750hrs.
        c) 26% pre-strain and aged at 650°C, 750 hrs.

299

Engineering Research Council for financial support for this work.

REFERENCES

1. B. Weiss and R. Stickler, "Phase Instabilities During High Temperature Exposure of 316 Austenitic Stainless Steel", **Metallurgical Transactions**, Vol. 3, pp. 851-866.(1972)

2. H.J. Bolton, J.E. Cordwell, A.J. Hooper, P. Marshall, and A. Wickens, "Preliminary examination of creep failures in Type 316 stainless steel", **CEGB**, Report No. RD/B/N3991.(1977)

3. X. Chen, R. Caretta, W. Zielinski and W.W. Gerberich,"Carbon/Oxygen Synergism During Elevated Temperature Sustained Load Cracking", **Acta metall.**, Vol. 38, No. 9, pp. 1719-1730.(1990)

4. S.M. Bruemmer and L.A. Charlot,"Development of Grain Boundary Chromium Depletion in Type 304 and 316 Stainless steels", **Scripta Metallurgica**, Vol. 20, pp. 1029-1024.(1986)

5. S.M. Bruemmer and L.A. Charlot,"Influence of grain Boundary Carbides and Phosphorous Segregation on The Low-temperature Intergranular Embrittlement of Type 316 stainless steel", **Scripta Metallurgica**, Vol. 23, pp. 1549-1554.(1989)

6. Charlie R. Brooks and Ji-Peng Zhou,"Microstructural Analysis of An Embrittled 422 Stainless Steel Stud Bolt after Approximately 30 Years Service in A Fossil Power Plant", **Metallography**, Vol. 23, pp. 27-55.(1989)

7. Yoshitaka Iwabuchi,"Temper Embrittlement of Typr 13Cr-4Ni Cast Steel", **Transactions ISIJ**, Vol. 27, pp. 210-217.(1987)

8. Ernest L. Hall and Clyde L. Briant,"Chromium Depletion in The Vicinity of Carbides in Sensitized Austenitic Stainless Steels", **Metallurgical Transactions A**, Vol. 15A, pp. 793-811.(1984)

9. J.K. Lai and A. Wickens,"Microstructural Changes and Variations in Creep Ductility of 3 Casts of Type 316 Stainless Steel", **Scripta Metallurgica**, Vol. 27, pp. 217-230.(1979)

10. J.K. Lai and D.J. Chastell and P.E.J. Flewitt,"Precipitate Phases in Type 316 Austenitic Stainless Steel Resulting from Long-term High Temperature Service", **Material Science & Engineering**, Vol. 49, pp. 19-29.(1981)

11. F.B. Pickering,"Physical Metallurgy of Stainless Steel Developments",

International Metals Reviews, pp. 227-268.(1976)

12.	D. Dan,"Tensile and Fracture Properties of Type 316 Stainless Steel After Creep", **Metall. Trans. A**. pp. 2155.(1982)

13.	P. Marshall,"Austenitic Stainless Steels - Microstructure and Mechanical Properties", Elsevier Applied Science Publisher Ltd.pp.262. (1984)

Fatigue

THE INFLUENCE OF WEAR ON THE FATIGUE

FAILURE OF A WIRE ROPE

Karol K. Schrems[1]

ABSTRACT

The fatigue failure of a wire rope used on a skip hoist in an underground mine has been studied as part of the ongoing research by the Bureau of Mines into haulage and materials handling hazards in mines. Macroscopic correlation of individual wire failures with wear patterns, fractography, and microhardness testing were used to gain an understanding of the failure mechanism. Wire failures occurred predominantly at characteristic wear sites between strands. These wear sites are identifiable by a large reduction in diameter; however, reduction in area was not responsible for the location of failure. Fractography revealed multiple crack initiation sites to be located at other less noticeable wear sites or opposite the characteristic wear site. Microhardness testing revealed hardening, and some softening, at wear sites.

INTRODUCTION

Wire ropes are used in many heavy industrial applications, including mines, offshore oil rigs, and barges. Premature failure of ropes can be costly in any application. In mining applications, failures become not only costly in dollars due to replacement or shaft damage, but can be devastating in its human costs. The Bureau of Mines, through its interest in the health and safety of mine workers, has been studying the failure mechanisms of wire ropes.

---

[1]Materials Engineer, Albany Research Center, U.S. Bureau of Mines, Albany, OR.

The ideal solution is to remove the rope when its useful life has been expended; and to remove it based upon an objective "retirement criteria". The current retirement criteria for wire ropes depend on the country. In U.S. mining applications, retirement criteria is based on the number and distribution of broken wires, loss of diameter of outside wires, reduction of rope diameter, loss of strength, corrosion, distortion, or heat damage.[1] Retirement criteria set by the Ontario Ministry of Labor in Canada includes loss of breaking strength, decrease in rope elongation, corrosion, decrease in wire torsions, and broken wires.[2] The complexity of objectively determining the end of the useful life of a wire rope is reflected in both the number of criteria examined and the differences for just these two sets of mining regulations.

Studies have been performed to understand the mechanisms involved with wire rope failure. Weischedel[3] reports two principal modes of deterioration of a wire rope: 1) Loss of metallic area (LMA) caused by external and internal abrasion, and corrosion, and 2) localized faults (LF) such as broken wires. Electromagnetic instruments have been designed for LMA and LF testing, but have a finite spacial resolving power.

By examining broken wires, the parameters for retirement indirectly address fatigue, which is a serious factor in the useful life of a rope. In 1945, after studying 20 years worth of data on ropes running over sheaves, Drucker and Tachau identified fatigue as being the primary cause of failure.[4] Savill, a ropemaker, attributes the majority of fatigue breaks to initiation at surface defects caused by corrosion, damage, martensite, or other defects.[5] Wiek states that in light of the strong interaction among wear, corrosion, and fatigue, OIPEEC, the International Organization for the Study of the Endurance of Cables, uses a broader definition for wire rope fatigue that includes these interactions.[6]

In mine hoisting, wear is present in three broad categories: 1) Crown wear on the outside of the rope due to motion over a sheave or drum, 2) wear from line contact between wires, and 3) wear from point contact between wires. It is highly conceivable that wear has an influence on the fatigue life of a wire rope. Wear by itself can reduce the maximum load capable of being carried by the rope, create nicks (the effects of point contact) which act as stress risers, and produce metallurgical changes in the wire.[7] To determine if wear affects fatigue, a failure of a mine hoisting rope was examined in three areas: 1) Correlation of individual wire fatigue failures with point contact between strands, 2) fractography of selected wire failures, and 3) micro-hardness testing of wire subsurface layers.

EXPERIMENTAL PROCEDURE

Wire ropes consist of multiple strands, which in turn consist of multiple wires. Many different constructions devised by changing the arrangement of the wires and the strands have been developed over the years to work in different applications. Names such as Seale, Warrington, and filler wire describe particular wire arrangements used in strand constructions. The wires within the strand are wound in a helix. If the helix of the wires is opposite the helix of the strand in the rope, it is referred to as regular lay. If the helix is in the same direction, this produces a lang lay. The use of a lang lay rope provides for more resistance to abrasion and bending fatigue. The strands can be laid in a left or right rotating direction, which distinguishes a left lang lay from a right lang lay. The center of a rope is referred to as the core, which distinguishes it from the center of a strand. Lay length is defined as the length in which it takes a wire or a strand to make one complete revolution.[8]

The rope obtained, Figure 1, was a non-rotating rope, 2 inches in diameter, with a nominal breaking strength of 455 kips. The nineteen strands in the rope are arranged in three layers. The center layer is a strand core of Seale left lay construction. The middle layer consists of six flattened strands of 29 wires wound in a left lang lay direction. The outside layer consists of twelve flattened strands wound in a right lang lay direction. The counter torques developed by the outside and middle layers provide for the rotation-resistant properties.

Wire ropes are so constructed that the failure of an individual wire does not affect the load bearing capacity of the rope. The number, distribution, and type of wire breaks within a rope affect the final failure. The failure of a rope, whether caught when one strand or multiple strands sever, contains information that can be misleading to a wire-by-wire failure analysis approach. Of critical importance is the knowledge of whether wire breaks identified as overload or fatigue were present prior to rope failure, or if they were a result of the rope failure.

Ropes operating over sheaves are subject to bending stresses; the ratio of the sheave diameter (D) to the rope diameter (d), plays a key role in the useful life of a rope, with a smaller ratio indicating a shorter life. The failed rope being examined is an underground mine hoist rope with a generous D/d ratio of 84.

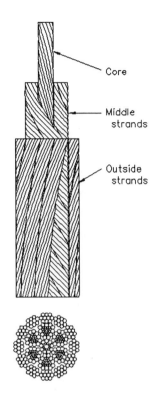

Figure 1. Construction of Strands to Form the Non-rotating Rope. The middle strands and outside strands are laid in opposite directions to produce the rotation resistant properties.

The failed segment was that part of the rope that rested on the head sheave when the skip was at the bottom and being loaded.

Two segments of rope were chosen for examination. The first, color-coded orange, was slightly over four feet long and included the final rope break. The core and middle layer were recovered intact; five of the original 12 outside strands were separately recovered. Because of the desire to study the mechanism leading up to the failure of the rope, the area of study in the failed segment was chosen to be 30 inches of complete strand length, not including the area of the final rope break. The second segment, color-coded blue, was from a location 49 feet from the break; it was also cut to a length of 30 inches. This segment was chosen as a comparison rope based on its distance from the break and the external visual condition of the rope.

The ropes were disassembled and cleaned in solvent. Each strand and wire within the strand was labeled for identification. The small inner wires of the middle strands were by far the most prone to failure. Since the large outside wires of the strands are considered to be of primary importance in the life of the rope, and are considered to carry the majority of the load, this study will concentrate the outside wires.

After disassembly, the outside wires were examined for location of failures. These occurred largely at characteristic wear patterns formed by the interaction of the outside wires of a strand with an identical strand in the same layer, or with the outside wires of a strand in an adjacent layer, Figure 2. The five possible wear interactions are: A) outside layer strand to strand, B) outside layer strand to middle layer strand (or middle layer strand to outside layer strand), C) middle layer strand

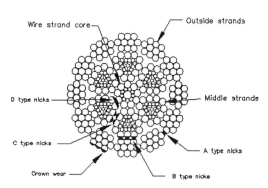

Figure 2. Cross-sectional View of the Rope Construction Illustrating the Location of the Characteristic Wear Patterns.

to strand, D) middle layer strand to core (or core to middle layer strand), and E) all other failures, including crown wear of the outside strands. The lack of any visible patterns identifiable as A type nicks indicates that the outside strands did not touch each other, and therefore no failures could be attributed to this wear pattern.

The characteristic wear patterns of the outside wires appear as deep saddle-shaped nicks at an angle to the wire axis. The remaining diameter, measured with a knife-edged micrometer at the deepest section of the nick and parallel to the nick direction, was used as an indication of loss of cross-sectional area. The wires also experienced wear from adjacent wires within the strand. Since the wear patterns from adjacent wires are very fine and can extend several inches down the side of a wire, these areas are referred to as general wear. The category of general wear can be further defined with the identification of line contact and point contact sites. Since the initial contact of adjacent outside wires is a Hertzian line contact, and the initial contact between the outside wires and inner wires is a Hertzian point contact, these will be referred to as line contact and point contact, respectively. Measurements of remaining diameter at the nick site and general wear were made on selected wires of each strand. The remaining diameter at the failure was also measured.

Several fracture samples were chosen and prepared for examination on a scanning electron microscope (SEM). The samples were cleaned in a commercial industrial strength cleaner and degreaser, rinsed in water and isopropyl alcohol. The samples were viewed using an Amray 1000A[*] scanning electron microscope, using 10 kV excitation voltage, 100 micron aperture, an 8 mm working distance, and the secondary electron imaging mode.

Microhardness samples of 1) areas of general wear and 2) wear patterns corresponding to C type nicks (interaction of two middle strands) were metallographically prepared for comparison between the orange segment and the blue segment. These were mounted in iron powder filled epoxy for a transverse view. The samples were prepared on an automatic polisher by: (i) grinding sequentially on 120, 220, 500, 1000 grit SiC wet grinding paper (FEPA Standard P-Series); (ii) polishing with 6 micron diamond particles on a hard paper cloth with an alcohol-based lubricant; (iii) polishing with 1 micron diamond particles on a medium nap cloth with the same lubricant; and (iv) final polishing with a proprietary aqueous colloidal suspension of 0.04 micron silica particles (OP-U from Struers, Inc.[*]) on a medium nap chemical resistant cloth. The samples were etched using a 1% Nital solution (1 ml HNO_3 + 100 ml ethanol or methanol) for approximately 10 seconds.

Microhardness testing was performed on a Wilson Tukon MO at 100 grams with a 15 second dwell time. Due to the light loads used, a Knoop indentor was chosen over diamond pyramid hardness because of the longer length of the diagonal, and the increased sensitivity to surface effects.[9] Calibrated indentations were read and test indentations were made. Both were found to be within specifications. The long diagonal of the indentor was oriented parallel to a tangent to the surface of the wire. Lengths were measured by two different people to agreement within 4 filar units and then averaged. The distance from the long diagonal to the edge of the wire was between 20 and 30 microns. With a short diagonal less than

---

[*] Reference to specific products does not imply endorsement by the Bureau of Mines

10 microns, this is approximately the minimum distance to edge spacing of 2.5 diagonals reported by Samuels to be conservative.[10] After microhardness readings were taken, 200x photographs of the circumference of the wire were taped together. The circumference of the wire and location of the hardness indentations in the resulting montage were entered into a computer by tracing with a digitizer, whereupon the hardness readings were added.

RESULTS

Results of failure correlation

Failures of the outside wires of the outside strands, Figure 3, could fall into two categories: B, wear of the outside strands with those in the middle layer; and E, other, predominantly crown wear. For comparative purposes, the number of wire failures for the segment was divided by the number of strands, since only 5 of the orange segment outside strands were recovered. The total number of failures per strand was 1.2 for the orange segment, and

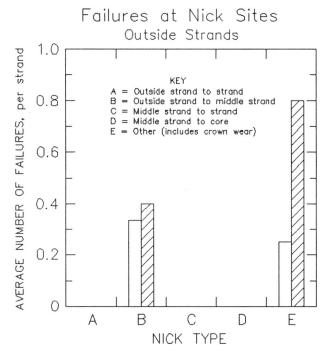

Figure 3. Failures of the Outside Strands at Characteristic Nick Sites. Blue segment - open bars. Orange segment - cross-hatched bars.

0.58 for the blue segment, with the difference primarily being the number of crown wear failures.

Failures of the middle strand outside wires could occur at one of four locations, Figure 4. These were: B, wear of the outside wires of the middle strands with those in the outside layer (also responsible for the B type nicks of the outside strands); C, wear of the outside wires of the adjacent middle strands; D, wear of the outside wires of the middle layer with those of the core; and E, all others. Due to the construction of the rope, B and C type nicks occurred twice as often as D type nicks. Wire breaks at all four types of nicks were seen; however, failures at E type sites were the least common. The orange segment averaged 14.7 failures per strand compared to 7.17 for the blue segment. The middle strands had approximately 10 times as many failures as the outside strands.

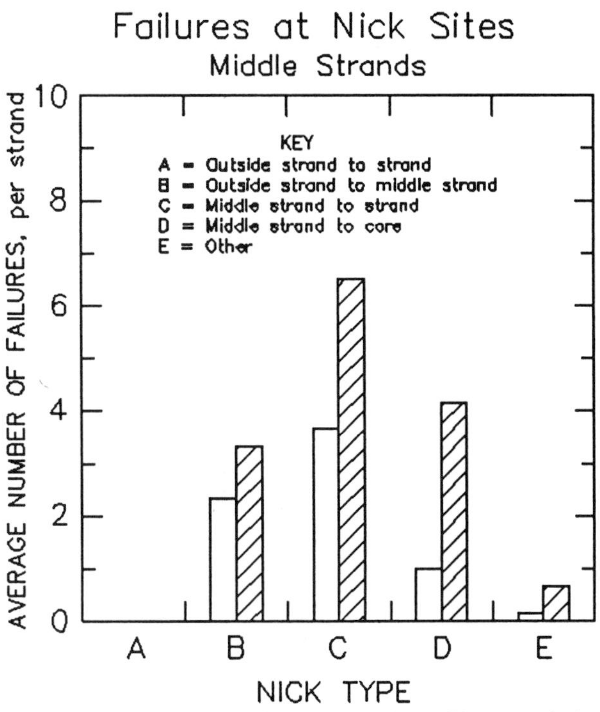

Failures at Nick Sites
Middle Strands

KEY
A = Outside strand to strand
B = Outside strand to middle strand
C = Middle strand to strand
D = Middle strand to core
E = Other

Figure 4. Failures of Middle Strands at Characteristic Nick Sites. Blue rope segment - open bars. Orange rope segment - cross-hatched bars.

The remaining diameter at the nick sites was measured for selected outside wires in the outside, middle, and core layers. These were averaged for single values for general wear and nick sites. The remaining diameter at the failures was measured and averaged for comparison. No failures were discovered in either strand core, and are therefore not reported. The results are presented in Tables 1, 2, and 3.

Results of fractographic examination

The results of the failure correlation with nick patterns revealed that the largest number of failures of outside wires occurred within the middle layer of the rope segments and at the point contact with the adjacent strands in the layer, a C type nick. Three samples of this type, two from the blue segment and one from the orange, were selected for fractographic examination. One sample of a failure at

Table 1 - Remaining diameter of general wear and at characteristic nick sites of selected wires from the outside strands

	General Wear	B
Blue	0.097	0.084
Orange	0.097	0.085
Blue failures	0.100	0.088 (range 0.003)
Orange failures	0.099	0.082 (range 0.024)

Table 2 - Remaining diameter of general wear and at characteristic nick sites of selected wires from the middle strands

	General Wear	B	C	D
Blue	0.109	0.097	0.100	0.096
Orange	0.108	0.098	0.101	0.098
Blue failures	0.109	0.099	0.102	0.099
Orange failures	0.109	0.103	0.101	0.101

Table 3 - Remaining diameter of general wear and at characteristic nick sites of selected wires from the core

	General Wear	D
Blue	0.106	0.101
Orange	0.107	0.103

a site of interaction between the middle strand and the core (D type nick) was also examined. A variety of sources on fractography ([11], [12], [13], [14], [15], [16]) were consulted in order to understand the failure mechanism of these wires.

The first sample, from the blue rope segment, is shown in Figure 5. The C type nick is to the left of the photograph. A dark spot in the lower right hand corner across from the large nick is the crack initiation site, which is not associated with any wear site. The line in the upper right hand corner marks the top of a plateau bounded on one side by a crack located at the edge of a line contact wear site. The ductile overload fracture is located at the other side of this plateau, adjacent to the nick in the upper left hand corner.

The second sample of a C type nick from the blue segment, is shown in Figure 6. In the upper right hand corner is the C type nick. The left side reveals a crack initiation site at the corner of a point contact site between this wire and one of the middle wires of the same strand. The wear site at the bottom of the photograph is the line contact between the outside wires in the strand. The final fast fracture is to the right of the photograph.

The third sample, Figure 7, a C type nick from the orange segment, indicates several possible initiation sites. The C type nick is at the lower right hand corner associated with the plateau just barely visible in the photograph. The initiation site at the top of the photograph is not associated with a wear site, although it is opposite the C type nick. The line contact with the adjacent wires in the strand is visible in the lower left hand corner. The point contact with the inner wires is located to the left, and contains a crack initiation site at one corner. The final fracture follows a longitudinal crack through the length of the nick.

The fourth sample, Figure 8, shows a D type nick from the blue segment at the top of the photograph. A clear-cut crack initiation site is difficult to discern, although several possibilities are evident. One is the recessed semicircle located opposite the nick at the bottom of the photograph. Another is the line contact wear area located to the right, which reveals a plateau the width of the wear area, a crack close to one edge, and a stair-step appearance to the fatigue surface. The overload fracture is located near the right corner of the large D type nick.

Results of microhardness testing

Four samples were chosen for microhardness testing; a site of general wear and C type nick of both the blue and the orange segments. The general wear sample of the blue rope, Figure 9, was used as a comparison for the other samples. This appears to be a reasonable assumption, as the wire appears round, with only one line contact area, and with Knoop microhardness values that consistently fall in the range of the mid to upper 400's. The only anomaly, a softening, appears beside the line contact wear site.

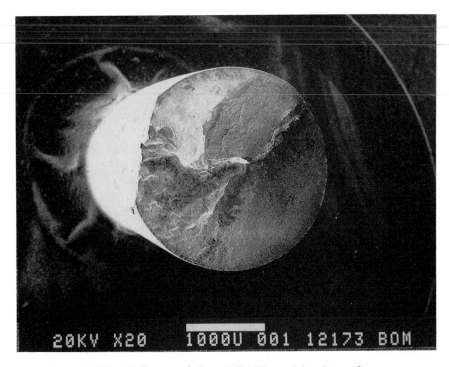

Figure 5. Wire Failure at C Type Nick From Blue Rope Segment.

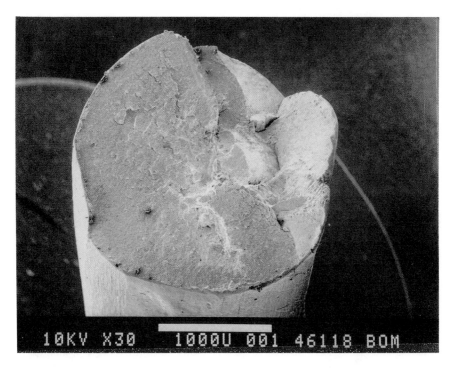

Figure 6. Wire Failure at C Type Nick From Blue Rope Segment.

Figure 7. Wire Failure at C Type Nick From Orange Rope Segment.

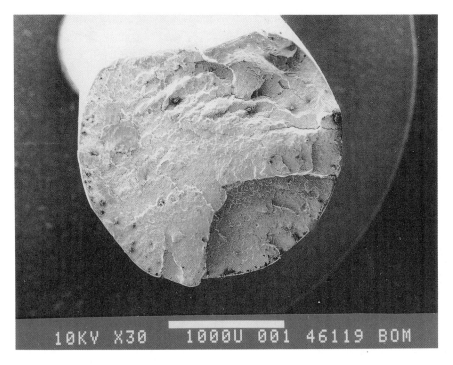

Figure 8. Wire Failure at D Type Nick From Blue Rope Segment.

The other sample of general wear, Figure 10, is from the orange segment and reveals a number of wear areas; a general flattening, two line contacts, and a point contact. General flattening was associated on the wires with B and C type nicks, and could extend into areas of general wear. The Knoop microhardness values contain considerably more scatter than the baseline blue general wear sample.

The region of general flattening appears to be slightly softer than the baseline. One of the line contact sites shows an increase in hardness across the extent of the wear site. Next to this site is localized softening. The microhardness reading of the other line contact and the point contact are similar to the baseline data. Microhardness readings across the wire diameter from both line contacts are above the baseline data.

The previous two wire samples contained areas of general wear. These were rarely associated with the failures found in the disassembled rope. The C type nicks were associated with failures the most, and the orange segment more so than the blue. A wire with a C type nick from a blue segment, Figure 11, shows multiple wear areas. An area of general flattening is associated with the nick; following around the wire reveals a line contact, a point contact, possibly the end of

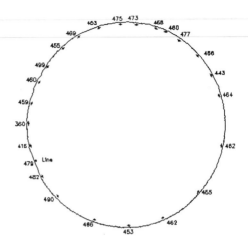

Figure 9. Subsurface Knoop (100 gram) Microhardness Readings of General Wear Wire Sample Removed From the Blue Rope Segment.

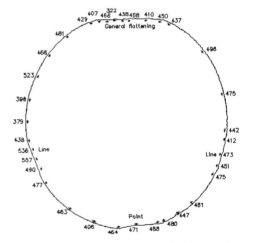

Figure 10. Subsurface Knoop (100 gram) Microhardness Readings of General Wear Wire Sample Removed From the Orange Rope Segment.

318

a point contact, and another line contact. Again, the microhardness readings were more erratic than the blue general wear sample. On this sample, both the general flattening and the C type nick experienced some hardening. Similar to the orange general wear sample, one of the line contacts exhibited the most hardening of the wire, and the readings at the other line contact and the point contact are similar to the baseline. An area of softening appears between these last two.

The wire with a C type nick from the orange segment, Figure 12, again reveals multiple wear areas. The nick is associated with a large area of general flattening. Following around the edge is a line contact, a point contact, either a shift in the point contact or the end of another point, and the end of a line contact. On this sample, the nick itself was the hardest area. The general flattening experienced

some softening; the line, point, shifted point, and end of line all showed a general increase in hardness.

DISCUSSION

Nick sites or notches have long been recognized as failure sites in wire ropes [17], [18], [19], although they have generally been reported as being

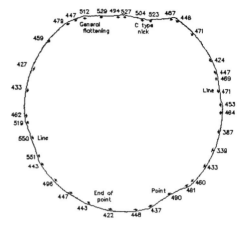

Figure 11. Subsurface Knoop (100 gram) Microhardness Readings of C Type Nick Wire Sample Removed From the Blue Rope Segment.

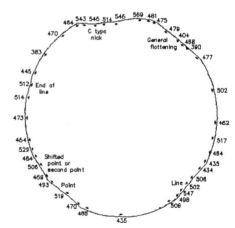

Figure 12. Subsurface Knoop (100 gram) Microhardness Readings of C Type Nick Wire Sample Removed From the Orange Rope Segments.

the origin of the crack. [20],[21]   One explanation involves the formation of martensite due to the localized heating of the wires resulting from the friction of the wear process. The over-heated metal appearance of the C and D type nicks in this rope, the most prevalent failure site, initially suggested martensite as an initiation site for the fatigue cracks. This does not appear to be the cause for two reasons. First, optical microscopy of the samples reveals a highly wrought pearlitic structure to the edge. Secondly, the microhardness values, when roughly converted to the Rockwell C scale, fall in the range of 35-45, considerably less than that expected for a brittle martensite. In addition, with a similar alloying elements, Table 4, the microhardness values compare favorably to the values of 441 to 527 HK reported by Lowery [22] for hard drawn pearlitic wires from a wire rope.

The results of the comparison of the remaining diameter of the wire at the failure site and the corresponding nick pattern, Tables 1-3, indicate that the reduction in area effect is not the reason for the failure at the nick sites. Consistently, the remaining diameter at the failure is greater than the minimum remaining diameter at the nick site. This could be because it failed before the nick had reached the maximum loss; however, the almost identical values between the blue segment and the orange segment for remaining diameter does not support this. In addition, a visual examination of the wires reveals that the failures are not at the maximum loss, but more towards the end of the wear site, near the top of the saddle-shaped area.

The SEM fractography work illuminates the failure mechanism of the wires. All four samples had the final overload fracture occur in proximity to the large nick site. The C type nicks had at least one initiation site, which generally fell into one of two predictable categories: 1) at a site of no wear opposite the major nick, or 2) at the corner of a point contact wear site. This is in agreement with Smith, Stonesifer, and Seibert [23], who found that the fatigue crack originated opposite the characteristic wear site, and terminated at the characteristic wear site.

The line contact wear site appears to be the initiation site in the D type nick examined. For the C type nicks, it contributed to the failure by the initiation of secondary cracks or longitudinal cracks.

Much work [24], [25], [26] has been done on various models for the distribution of stresses in ropes; however, the complexity of the interactions within a single rope construction, and the large number of rope constructions severely limits the applicability of the models. Wiek reports that friction can change the tension stress considerably within one turn of a wire's helix around the strand

Table 4 - Results of analysis of selected wires, weight %

Wire Location						
Layer	Wire	C	Si	Mn	S	P
Outside	Outside	0.676	0.34	0.79	0.014	0.015
Outside	Center	0.827	0.21	0.66	0.018	0.017
Middle	Outside	0.731	0.26	0.82	0.013	0.011
Middle	Inner	0.695	0.20	0.70	0.018	0.016
Core	Outside	0.853	0.24	0.55	0.010	0.018
Core	Inner	0.576	0.22	0.75	0.014	0.012
Core	Center	0.795	0.19	0.57	0.016	0.016

axis.[27]   The semi-circular cliffs present on the fracture in Figure 7 resemble the cracks formed by high compressive forces in samples by Pantucek.[28] Phillips, Miller, and Costello developed theoretical expressions for the stresses caused by the contact forces between wires in adjacent layers.[29]   They determined that the contact stresses can be significant, even though the axial load is relatively small. These contact stresses contribute to the surface hardening seen at the wear sites of the wires.

The comparison sample, general wear of the blue rope segment, shows a consistent hardness, but only one wear area. The general wear sample of the orange segment has four wear areas with minimal hardening. The area labeled general flattening may have been softened by frictional heat. A comparison of the two nick samples shows the orange segment to have a larger number of and more extensive wear areas. The hardness readings from the orange segment are also slightly higher in nature than those from the blue sample. It appears that the localized hardening caused by contact pressure and wear, and the number and extent of wear sites, combined to make the orange segment more prone to failure.

CONCLUSIONS

1. A macroscopic correlation of failures with distinguishing features revealed that the majority of failures occurred at characteristic nick sites. A comparison of the remaining diameter at these sites with the diameter at the failure sites, reveals that a reduction in area affect is not responsible for this correlation.

2. Fractographic examination reveals that crack initiation occurs at other, less noticeable wear sites such as point contact with inner wires or line contact with adjacent outside wires, or at sites opposite the characteristic nick.

3. Microhardness testing along the circumference of wires at areas most prone to failure reveals hardening at the characteristic nick, at sites of line contact with adjacent wires, and occasionally at areas without wear opposite a wear site.

REFERENCES

1. U.S. Code of Federal Regulations. Title 30--Mineral Resources; Chapter 1--Mine Safety and Health Administration, Department of Labor; Subchapter N--Metal and Nonmetallic Mine Safety; Part 56, Subpart R, and Part 57, Subpart R; Subchapter O-Coal Mine Safety and Health; Part 75, Subpart O, and Part 77, Subpart O; July 1, 1989.

2. The Occupational Health and Safety Act (R.S.O. 1980, c321 as amended) Bill 208 and Regulations for Mines and Mining Plants (R.R.O. 1980 Reg. 694 as amended) Jan. 1990 Edition.

3.Herbert R. Weischedel, "The Inspection of Wire Ropes in Service", Wire Journal International, Vol. 18, no. 9, p. 180, 1985.

4. D.C. Drucker, and H. Tachau, "A New Design Criterion for Wire Rope", Journal of Applied Mechanics, March 1945, p. A-33.

5. L. Peter Savill, "Parameter's which affect the endurance of wire ropes: A ropemaker's viewpoint", Wire Industry, Vol. 47, no. 554, p. 105, 1980.

6. I. L. Wiek, "Calculation of rope endurance whether the cause may be wear or fatigue - part 1", Wire, Vol. 38, no. 2, p. 218, 1988.

7.Russell A. Lund, Gary J. Fowler, and Donald O. Cox, "Failure Analysis of Wire Rope", Proceedings First Annual Wire Rope Symposium, p. 139, 1980.

8. American Iron and Steel Institute, <u>Wire Rope Users Manual</u>, p. 10, 1979.

9. Peter J. Blau, "The Use of Knoop Indentations for Measuring Microhardness near Worn Metal Surfaces", Scripta Metallurgica, Vol. 13, p. 95, 1979.

10. Leonard E. Samuels, "Microindentations in Metals", Microindentation Techniques in Materials Science and Engineering, ASTM STP 889, P.J. Blau and B.R. Lawn, Eds., American Society for Testing and Materials, Philadelphia, 1986. p. 5.

11. Metals Handbook, Ninth Edition, Volume 12, Fractography, ASM International, 1987.

12. Metals Handbook, Eighth Edition, Vol. 9, Fractography and Atlas of Fractographs, American Society for Metals, 1974.

13. W.R. Warke, N.A. Nielsen, R.W. Hertzberg, M.S. Hunter, and M. Hill, "Techniques for Electron Microscopic Fractography", Electron Fractography, ASTM STP 436, American Society for Testing and Materials, p. 212, 1968.

14. A. Phillips, V. Kerlins, R.A. Rawe, and B.V. Whiteson, <u>Electron Fractography Handbook</u>, Metals and Ceramics Information Center, Columbus, Ohio, 1976.

15. T.S. Sudarshan, T.A. Place, M.R. Louthan, Jr., and H.H. Mabie, "Torsional Fatigue Fractures of Aged Aluminum Alloy 2024 in Dry Argon", Microstructural Science, Volume 13, Corrosion, Failure Analysis, and Metallography, American Society for Metals, p. 111, 1986.

16. Marion Russo, "Analysis of Fractures Utilizing the SEM", Metallography in Failure Analysis, edited by James L. McCall and P.M. French, American Society for Metals, 1978.

17. American Iron and Steel Institute, <u>Wire Rope Users Manual</u>, p. 57, 1979.

18. E.h.E. Bahke, "150 Years of Wire Rope Research, Part II", Wire, Vol. 35, p. 203, 1985.

19. H.L. Smith, F.R. Stonesifer, E.R. Seibert, "Increased Fatigue Life of Wire Rope Through Periodic Overloads", 10th Annual Offshore Technology Conference, Houston, TX, May 8-11, 1978, OTC 3256.

20. Philip T. Gibson, "Wire Rope Behavior in Tension and Bending", Proceedings, First Annual Wire Rope Symposium, p. 3, 1980.

21. G. H. Beeman, "Factors Affecting the Service Life of Large-Diameter Wire Rope", NTIS 23111 02321, 1978.

22. R.R. Lowery and G.L. Anderson, "An Analysis of 6X19 Classification Wire Hoist Rope", RI 8817, Bureau of Mines Report of Investigations, p. 1, 1983.

23. H.L. Smith, F.R. Stonesifer, E.R. Seibert, "Increased Fatigue Life of Wire Rope Through Periodic Overloads", 10th Annual Offshore Technology Conference, Houston, TX, May 8-11, 1978, OTC 3256.

24. I. L. Wiek, "Calculation of rope endurance whether the cause may be wear or fatigue - part 1", Wire, Vol. 38, no. 2, p. 218, 1988.

25. George A. Costello, "Analytical Investigation of Wire Rope", Applied Mechanics Reviews, Vol. 31, no. 7, p. 897, 1978.

26. George A. Costello and Gary J. Butson, "A Simplified Bending Theory for Wire Rope", **ASCE J. Eng. Mech. Div.**, Vol. 108, no. EM2, p. 219, 1982.

27. I. L. Wiek, "Calculation of rope endurance whether the cause may be wear or fatigue - Part II", Wire, Vol. 38, no. 3, p. 321, 1988.

28. E.h.E. Bahke, "150 Years of Wire Rope Research, Part II", Wire, Vol. 35, p. 203, 1985.

29. James W. Phillips, Robert E. Miller, and George A. Costello, "Contact Stresses in a Straight Cross-Lay Wire Rope", Proceedings, First Annual Wire Rope Symposium, Denver, CO, March 18-20, 1980, p. 177.

THE INFLUENCE OF SULFUR ON THE ROLLING CONTACT
FATIGUE LIFE OF S.A.E. 52100 BEARING QUALITY STEEL

Nicholas J. Ruffer
Federal Mogul Research
Ann Arbor, MI

ABSTRACT

The size and frequency of manganese sulfide inclusions in a steel are directly related to the sulfur content. Manganese sulfide inclusions have been considered to be the least detrimental type of inclusion and have even been considered benign with respect to the fatigue life. The benign tendency has been thoroughly documented but has always been assessed while other types of inclusions were present. Because these other types of inclusions were more detrimental to the fatigue life than the manganese sulfide inclusions the adverse effects of the manganese sulfide inclusions may have been overlooked. Therefore, the effects of manganese sulfide inclusions were determined when other types of inclusions were virtually absent. This study shows that the rolling contact fatigue life of an S.A.E. 52100 steel containing 0.004% sulfur was reduced by 37% at the L10 life and by 50% at the L50 life compared to an S.A.E. 52100 steel that contained 0.025% sulfur. These reductions are attributed to the presence of manganese sulfide inclusions. These results demonstrate that manganese sulfide inclusions were detrimental to the rolling contact fatigue life at a contact stress level of 5.42 GPa. This detrimental effect on the rolling contact fatigue life at a contact stress of 5.42 GPa was not present when the contact stress was reduced to 4.83 GPa.

INTRODUCTION

Rolling contact fatigue (RCF) is fatigue under conditions where there is rolling contact between two objects which commonly occurs in bearings. RCF life is a measure of a materials ability to carry a cyclic load for a given period of time. An RCF test is an accelerated test used to determine the fatigue life of a given steel or iron [1]. The tester allows the fatigue life of various materials to be evaluated in a short time and in a cost-effective manner [2]. The RCF tester operates at very high contact stresses which cause the test material to fail earlier than it would under normal loading conditions. In a ball/rod type RCF tester the standard contact stress is 5.42 GPa [1]. The high contact stress is the result of a small spring load (25.9 Kilograms) applied to a very small elliptical contact area.

The objective of this investigation was to determine the influence that varying amounts of sulfur have on the RCF life of S.A.E. 52100 bearing quality steel. Ball bearings are cyclically loaded at low stresses, relative to the tensile and yield strength, and when failure occurs it is often a result of fatigue. The fatigue life of a given ball bearing is usually determined by full scale bearing life tests or vehicle testing. Data from bearing tests always have a certain amount of scatter [2].

INCLUSIONS AND FATIGUE

Inclusions are always present in steels because of the melting practice and the raw materials used in the making of steel. The fatigue life of a steel is usually directly related to the cleanliness of the material. With all other factors being held constant, the cleaner the steel, the longer the fatigue life [3,4,5,6,7]. The procedure for rating the cleanliness of a material has been outlined by ASTM E 45, Standard Practice for Determining the Inclusion Content of Steel [8]. The present rating charts are perhaps out of date [4]. Some types of inclusions are not rated on these charts and the frequency of inclusions on the plates are not typical of what is seen in steels today.

In the past it was concluded that soft inclusions, rather than hard, would lead to a longer fatigue life [3,5,9,10,11,12]. Hard inclusions typically are highly angular and act as a preformed crack that can begin to propagate immediately. Whether the particles are hard or soft, the fatigue life of a material is affected by the amount of non-metallic inclusions present [3,6], but the relationship between the cleanliness rating of many materials and the fatigue life remains poor

[13]. Manganese sulfide inclusions, the softest inclusions rated in steels, are for all practical purposes, the only type of inclusion found in large quantities in bearing quality S.A.E. 52100. The maximum amount of sulfur allowed in S.A.E. 52100 is 0.025 weight percent [14]. The average sulfur level in vacuum degassed bearing quality S.A.E. 52100 material is typically in the range from 0.010 to 0.020 weight percent. Past research has indicated that manganese sulfide inclusions have little or no detrimental effect on the the fatigue life [3,4,5,12,15,16]. In some instances, it was concluded that the presence of manganese sulfide inclusions in a steel was actually advantageous [5,16]. Manganese sulfide inclusions have not been reported to contain stress induced microstructure (butterflies) present around the inclusion, while other types of inclusions have [16,17]. A butterfly is the result of fatigue and is always found surrounding an inclusion. When these butterflies are present, they are often found to contain small cracks that can propagate and cause failure. The type of tester, loads used or the presence of other types of inclusions in the steel may be why past research has not found butterfly(s) around manganese sulfide inclusions. The effect of manganese sulfide inclusions on the fatigue life can only be determined when other types of inclusions are absent. Low sulfur levels can be obtained in both high and low carbon steels through vacuum degassing or ladle additions of rare earth elements when the steel is in the molten stage. Calcium is added for shape control of manganese sulfide to improve machinability and castability but can lead to a reduced fatigue life [3].

EXPERIMENTAL PROCEDURE

The test bars were made from tube samples of two different heats of vacuum degassed S.A.E. 52100 steel, both heats of steel came from the same melt shop. The chemistry of the two heats are given in Table 1 and are within the specification limits for S.A.E. 52100 [14]. The sulfur content (weight percent) of the steels used in this investigation was 0.004 and 0.023 and were not attained by calcium additions. All of the test bars were hardened and tempered together. The bars were austenitized for twenty minutes at 843° C in an atmosphere controlled furnace with a carbon potential of 1.0 and then quenched into a moderately agitated oil bath held at 71-82° C. The bars were then tempered to Rc 62 to 64. The test bars were centerless ground to a surface finish of 3.5-4.5 Ra. The test bars were typically 10 cm long allowing 10 tests per bar. The test balls were new 1.27 cm diameter grade 25 balls from the same lot made from S.A.E. 52100 steel. The surface

roughness of the balls ranged from 0.3 to 0.4 Ra.   The hardness range
of the balls were between Rc 64-65.   The oil used for lubrication was
Exxon Teresstic 68.   This oil chosen was based on; the speed of rotation
of the test machine, load, operating temperature and lubrication factor.
The average operating temperature was 85° C.   The lubrication factor
required is 1.0 to ensure that the fatigue life results are not a result of
oil starvation or over lubrication.   Too thin of an oil will produce a lu-

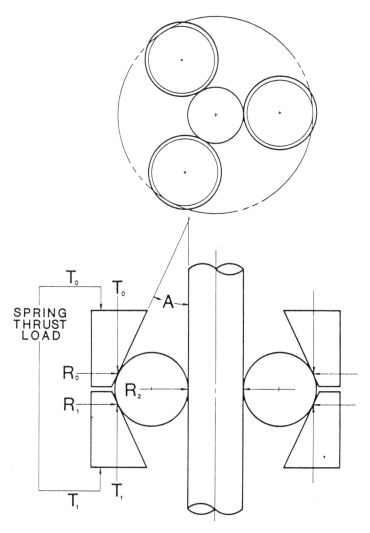

Figure   1   Schematic of rolling contact fatigue test fixture
Copied from ASTM STP 771, Douglas Glover, Ref 1.

brication factor that is less than 1.0 resulting in a reduced fatigue life due to inadequate lubrication. A lubrication factor greater then 1.0 is produced by having too thick of an oil resulting in an exaggerated fatigue life [3].

The fatigue testing was done on RCF ball/rod type test machines located at the Federal Mogul Bearing Research and Technical Center in Ann Arbor, Michigan. The tester consists of three radially loaded balls separated by a brass retainer with the load being transferred to a rotating 0.9525 cm test bar, Figure 1. The test fixture has three springs in compression; these springs apply a load to three balls, which in turn transfer the load to the rotating bar. The springs can be adjusted to different loads to allow testing at different contact stresses. Two different spring loads were used in this investigation that resulted in contact stresses of 4.83 GPa and 5.42 GPa. The test bar is held firmly in a collet and is rotated at 3600 RPM. An accelerometer is attached to all test heads that will automatically shut off the test machine when damage occurs. If failure did not occur by 300 hours, 150 million cycles, the test was suspended.

RESULTS AND DISCUSSION

Rolling contact fatigue life tests were carried out for two different contact stresses, 4.83 GPa and 5.42 GPa, on test bars having two different sulfur levels, 0.004 and 0.023 weight percent. For simplicity, the 0.023 weight percent sulfur test bars will be identified as the high sulfur test bars and the 0.004 weight percent sulfur test bars as low sulfur test bars. A total of 139 tests were completed on 19 RCF test bars, 101 of those being run at 5.42 GPa and the remainder (38) at 4.83 GPa. There were 80 tests completed on the high sulfur test bars and the remainder (59) on low sulfur bars. Following testing, the data was entered into a standard weibull analysis program to determine the estimated L10 and L50 fatigue lives of the bars. It has been shown that the L10 life is more sensitive to random accidental factors that can lead to unusually short life test results [18]. The estimated fatigue lives from the weibull analysis program can be found in Table 2 .

The weibull estimated fatigue life (L10 and L50) of the bars tested at 5.42 GPa show a significant difference between the high and low sulfur test bars, Table 2. This table shows that the low sulfur test bars have a significantly higher fatigue life (L10 and L50) than the high sulfur test bars. The L10 and L50 lives of low sulfur test bars were 18.5 and 119.8 hours respectively and the L10 and L50 life for the high sulfur test bars were 11.1 and 60.6 hours respectively. The L10 life of the low sulfur test bars is 37.1% greater than that of the high sulfur test bars while the L50 life of the low sulfur material is 49.4% greater than the high sulfur test bars. The correlation coefficient of the data for the low sulfur test bars was 0.98 and 0.99 for the high sulfur test bars, which indicates that the data fits closely to the line connecting the estimated fatigue lives (L10 and L50). The slope of the weibull plot for the low sulfur test bars is 1.03 and 1.14 for the high sulfur test bars. Both of these slopes indicate the failures were random and independent of time. These slopes further indicate that the amount of scatter is low, the distribution is symmetrical and thereby an acceptable weibull analysis. To determine if the greater L10 and L50 life of the low sulfur material was significant, a test of significant difference was done. This test indicated that the low sulfur material is superior to the high sulfur material with a 75% confidence level. The graph for the high and low sulfur material tested at 5.42 GPa can be found in Figure 2. As expected from the high correlation coefficient, the test data falls very close to the line drawn on the graph from the estimated L10 to the L50 life. It can be seen that the low sulfur test bars have a higher estimated fatigue life (L10 and L50) than the high sulfur test bars.

The difference in the fatigue life between the high and low sulfur bars tested at 5.42 GPa was not present in the tests run at 4.83 GPa, Table 2. The weibull estimated fatigue life, L10 and L50, for both the high and low sulfur test bars are almost the same. Experimentally, the high sulfur material had a L10 and L50 fatigue life of 33.6 hours and 142.7 hours respectively while the low sulfur had an L10 life of 22.1 hours and an L50 life of 150.9 hours. The correlation coefficients of the high and low sulfur test bars were 0.98 and 0.96 respectively, which indicates that the data fits very closely to the line estimated by the weibull program. The slope of the weibull line for the high sulfur was 1.41 and 1.04 for the low sulfur test data. Both of these slopes indicate that the failures were random and not dependent of time. The slope indicates that the amount of scatter is low and that the distribution is symmetrical and an acceptable weibull analysis. The data points fall very near the estimated weibull life line and there appears to be virtually no difference in fatigue life for the two different sulfur levels, Figure 3. The L10 and L50 lives can not be significantly different when the estimated life lines cross as shown in Figure 3.

The estimated L10 and L50 lives for both the high and low sulfur bars tested at 5.42 GPa and 4.83 GPa are plotted in Figure 4. This figure shows that the L10 and L50 lives of the high and low sulfur test bars from the 4.83 GPa tests and the low sulfur bars tested at 5.42 GPa are nearly the same and that the L10 and L50 life of the high sulfur test bars from the 5.42 GPa test is lower. The test results for the low

TABLE 1 Chemical Composition of Test Material

| | Composition in weight percent | | | | | | | |
	C	S	P	Si	Mn	Cr	Ni	Mo
High Sulfur	0.99	0.023	0.008	0.26	0.27	1.37	0.14	0.02
Low Sulfur	1.04	0.004	0.01	0.24	0.31	1.44	0.08	0.02

TABLE 2 Weibull Analysis of Test Results

Sulfur Level (Wt. %)	Test Load (k.s.i.)	Number of Tests	Number of Susp.	Life Cycles (Million)		Slope	Corr. Coef.
				L10	L50		
.025	786	55	8	5.80	30.32	1.14	.99
.025	700	25	11	11.07	75.47	1.04	.96
.004	786	46	13	9.23	59.90	1.03	.98
.004	700	13	5	16.79	71.33	1.41	.98

TABLE 3   J K CLEANLINESS RATINGS OF TEST BARS

Inclusion Rating Number	Manganese Sulfide Type A		Oxides Type D	
	Thin	Heavy	Thin	Heavy
High Sulfur Test Bars				
0.5	7.7	7	6.3	4.3
1	10	12	0.3	0
1.5	6	4	0	0
Low Sulfur Test Bars				
0.5	6.5	0.5	2.5	0.5
1	0	0	0	0
1.5	0	0	0	0

* No Type B or C inclusions present

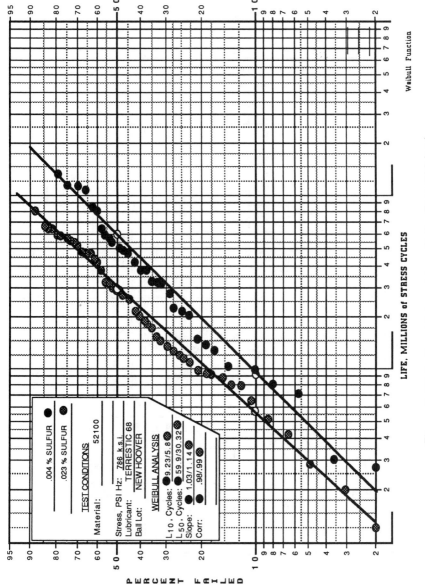

Figure 2 Weibull on High and Low Sulfur Bars 786 k.s.i.

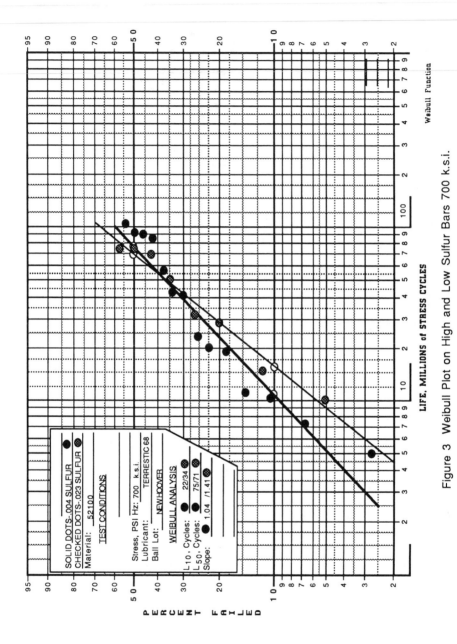

Figure 3 Weibull Plot on High and Low Sulfur Bars 700 k.s.i.

sulfur material did not change when the test load changed but this was not true for the high sulfur material. The high sulfur test bars did experience an increase in the fatigue life due to a 10% decrease in the test load. The data seems to indicate that the failure mechanism for the high sulfur test bars tested at 5.42 GPa was different than that on the high sulfur test bars tested at 4.83 GPa. The fatigue life of S.A.E. 52100 tested at 5.42 GPa is dependent on the sulfur level present. This load to sulfur dependency appears to be true only on the 5.42 GPa tests and not on the bars tested at 4.83 GPa.

MICROSCOPIC EXAMINATION

Sections were taken both parallel and perpendicular to the length of the bar. The longitudinal sections were examined in an unetched condition to determine cleanliness ratings while the transverse sections were etched with 2% nital for microstructural evaluation.

All cleanliness ratings were evaluated in accordance with ASTM E - 45 method D [8]. The cleanliness rating for the low sulfur material was substantially lower than the high sulfur material, Table 3. The low sulfur material had an inclusion rating of 0.5 maximum for all four types of inclusions rated. The high sulfur test bars had inclusion ratings as high as 1.5 and the frequency of inclusions, all types, were higher than the low sulfur test bars. There were no type B or C inclusions present in either heat.

The cross sections of the bars were mounted in a clear lucite mount to allow visual determination of when the sample had been ground down to where the ball tracked. The microstructure of the test bars consisted of spheroidal carbides dispersed throughout a tempered martensitic matrix.

The high sulfur bars tested at 5.42 GPa had butterflies around the manganese sulfide inclusions, Figure 5. These butterflies were 0.05 to 0.30 mm in from the running surface. Some of the manganese sulfide inclusions had large amounts of butterflies present around them and others had small butterflies and some had no butterflies present. Not all of the large manganese sulfide inclusions had these butterflies nor were all the small manganese sulfide inclusions free from butterflies. In many instances the microstructure contained subsurface cracks propagating from a manganese sulfide inclusion surrounded by butterflies [16,17]. The subsurface cracking always ran parallel to the rolling contact surface. All of the butterflies viewed had a manganese sulfide inclusion at the center.

The low sulfur material tested at 5.42 GPa had microstructural attributes similar to the high sulfur bars tested at the same load. The

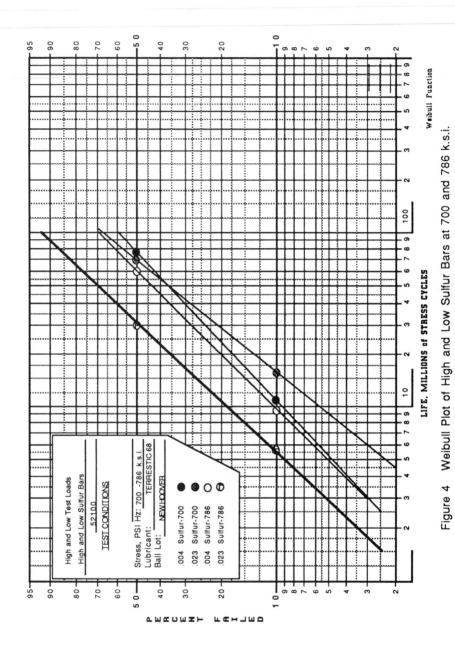

Figure 4   Weibull Plot of High and Low Sulfur Bars at 700 and 786 k.s.i.

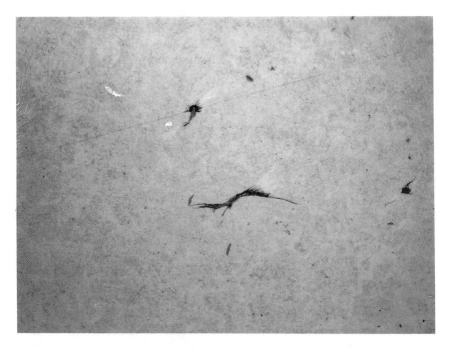

Figure 5   Butterfly surrounding Manganese Sulfide inclusion

depth of the subsurface cracking and butterflies were 0.05 to 0.46 mm from the running surface. The low sulfur bars had fewer manganese sulfide inclusions and a smaller frequency of butterflies around the inclusions as compared to the high sulfur bars tested at 5.42 GPa. The butterflies present were smaller than those viewed in the high sulfur bars tested at 5.42 GPa. There were subsurface cracks propagating from butterfly(s) parallel to the rolling contact surface.

The high sulfur bars tested at 4.83 GPa had microstructural attributes similar to that observed in the high sulfur bars tested at 5.42 GPa. Butterflies and cracks were present around some of the manganese sulfide inclusions but not as frequent or as large as those seen on the high sulfur bars in the 5.42 GPa tests. The depth of the butterflies and sub surface cracking ranged from 0.05 to 0.36 mm below the running surface.

The low sulfur bars tested at 4.83 GPa had microstructural attributes similar to that observed in the high sulfur test bars tested at 4.83 GPa and the low sulfur bars tested at 5.42 GPa. There were far fewer manganese sulfide inclusions present in the low sulfur material than in the high sulfur test bars. The depth of the butterflies and sub surface cracking ranged from 0.05 to 0.36 mm from the running surface. Some of the small manganese sulfide inclusions had butterflies present around them while other large inclusions did not. The butterflies always contained a manganese sulfide inclusion at its center. The subsurface cracking originated at a butterfly and propagated parallel to the running surface.

CONCLUSIONS

1. The fatigue life of the high sulfur (0.023%) test bars tested at 5.42 GPa have a significantly lower L10 (37%) and L50 (50%) fatigue lives than low sulfur (0.004%) test bars.

2. There is no significant difference in the rolling contact fatigue life of a SA.E. 52100 steel at contact stresses of 4.83 GPa, due to fluctuations of the Sulfur content within the range of 0.004 - 0.023 weight percent.

3. Butterflies form around manganese sulfide inclusions in which the frequency of such is dependent upon the stress level and the sulfur content of the steel. All butterflies contained a manganese sulfide inclusion at the center.

4. Subsurface cracking is directly related to butterflies, subsurface cracks were not present without the presence of butterflies.

REFERENCE

1. Douglas Glover, A Ball-Rod Rolling Contact Fatigue Tester, ASTM Spec. Tech. Publ. 771, 1982 pp 107-121.
2. Johnson, Sewell, Robinson, Relevance of Steel Cleanness and Fatigue Tests to Bearing Performance, ASTM Spec. Tech. Publ. 575, 1975 pp 114-137.
3. J. Monnot and B. Heritier and J. Cogne, Relationship of Melting Practice, Inclusion Type and Size with Fatigue Resistance in Bearing Steels, ASTM Spec. Tech. Publ. 987, 1988 pp 149-165.
4. Cogne, Heritier and Monnot, Cleanness and Fatigue Life of Bearing Steels, Inst. of Metals, 1987 pp 26-31.
5. Brooksbank, Andrews, Stress Fields Around Inclusions and Their Relation to Mechanical Properties,Production and Application of Clean Steels, Iron and Steel Institute, 1972 pp 186-198.
6. Kumagai, Takata, Yamada, Mori, Fatigue Life of High Carbon Chromium Ball Bearing Steel Produced by Electric Furnace-Vacuum Slag Cleaner-Ladle Furnace-RH Degassing-Curved Continuous Caster, ASTM Spec. Tech. Publ. 987, 1975 pp 248-359.
7. Tsushima, Maeda, Nakashima, Rolling Contact Fatigue Life of Various Kinds of High-Hardness Steels and Influence of Material Factors on Rolling Contact Fatigue Life, ASTM STP 987, 1988 pp 132-148.
8. ASTM, Annual Book of Standards, Volume 3.01, E 45-87, 1990 pp 221-231
9. J. Hampshire and E. King, Quantitative Inclusion Ratings and Continuous Casting: User Experience and Relationships with Rolling Contact Fatigue Life, ASTM Spec. Tech. Publ. 987, 1988 pp 61-81.
10. Ohshiro, Ikeda, Matsuyma, Improvement of Fatigue Life of Valve Spring Wire by Morphology of Nonmetallic Inclusions, ASM, Aug 1987 pp 36-40.
11. Zelbet, Laposhko, Voinov, Effect of Non-Metallic Inclusions on the Fatigue Strength of Rolling Bearings, Steel in USSR, Oct 1971 pp 798-799
12. Okamoto, Kazuo, Shikoh, Saburo, Effect of Non-Metallic Inclusions on the Rolling Fatigue Life of Ball Bearing Steel, Nippon Steel Tech. Report Overseas, January 1973 pp 49-57.
13. C. A. Stickles, Carbide Refining Heat Treatments for 52100 Bearing Steel,Metall. Trans., 5(1974) 865-874.
14. American Society for Metals, Metals Hand Book Ninth Edition, Volume 1., 1978
15. Kinoshi, Massao, Koyanagi, Akira, Effect of Nonmetallic Inclusions on Rolling Contact Fatigue life on Bearing Steels, ASTM Spec. Tech. Publ. 575, 1975 pp 138-149.
16. R. Tricot, Relative Detrimental Effects of Inclusions on Service Properties of Bearing Steels, Production and Application of Clean Steels, Iron and Steel Institute, 1972 pp 186-198.
17. Meredith, Sewell, Assessment of Nonmetallic Inclusion Content and the Relationship to Fatigue Life of 1 percent C-Cr Steels, ASTM Spec. Tech. Publ. 575, 1975 pp 66-95.
18. Stickels, Janotik, Heat Treatment for Ball Bearing Steel to Improve the Resistance to Rolling Contact Fatigue, 1977 Patent Application 4,023,988.

EFFECT OF PRIOR CREEP DAMAGE ON FATIGUE AND

CREEP-FATIGUE PROPERTIES OF 1Cr-1/2Mo STEEL

F. V. Ellis*

ABSTRACT

The effect of prior creep damage on the remaining life of base material and HAZ material subjected to pure fatigue and creep-fatigue loading was determined. To produce creep damaged material, an interrupted creep testing program was conducted on a single heat of 1Cr-1/2Mo steel. Metallographic examination of the pre-crept material revealed the primary creep damage mechanism was thermal softening for the base material and creep cavitation for the HAZ material. The pure fatigue life was the same for the pre-crept base and 0.2 life fraction pre-crept HAZ material as for the virgin material while the 0.6 life fraction pre-crept HAZ material had a greatly reduced life. Creep-fatigue tests showed that the as-received HAZ material exhibited a lower creep-fatigue life than the as-received base material. This was due to the lower creep ductility associated with the HAZ material. Thermal softening of the pre-crept base material increased the base material ductility and increased the creep-fatigue life compared to the as-received base material. The cavitation creep damage formed during prior creep reduced the HAZ material ductility and caused a corresponding reduction in creep-fatigue life.

---

*   Tordonato Energy Consultants, Inc.
    4156 S. Creek Road, Chattanooga, TN 37406

INTRODUCTION

The remaining life of high temperature boiler pressure parts can be divided into two distinct load regimes: steady creep and cyclic. Metallographic techniques have been developed for the evaluation of steady creep load damage accumulation and life assessment [1,2]. However, there are several factors which require life assessment techniques applicable to the second operating condition of cyclic loads. First, the incidence of thermal fatigue cracking in conventional plants is now becoming more frequent due to the increasing need to two-shift (two shifts on, one shift off) or load-follow these plants as new nuclear units become available. Second, there is an economic incentive to develop methods that extend plant life and increase availability. The cyclic duty of a plant subjected to two shifting conditions can give rise to extreme temperature changes during both start up and shut down procedures. These two operating conditions, combined with a steady running conditions during which creep damage can form, can result in thermal fatigue or creep-fatigue damage accumulation and failure. An additional mitigating factor which needs to be considered is that, in most cases, the plants subjected to two-shifting conditions have already been operating for a considerable fraction of design life under base load (steady creep loading) conditions. Based on these concerns, a preliminary study was undertaken to examine the influence of prior creep deformation on the fatigue and creep-fatigue behavior of 1Cr-1/2Mo base and HAZ material.

EXPERIMENTAL PROCEDURE

Pure fatigue tests were conducted the virgin base and HAZ material for comparisons with the pre-crept material. The test specimen geometry used for the pure fatigue and creep fatigue tests had a 6mm gage diameter and a 12mm gage length established by machined ridges for extensometer attachment. Tensile dwells of sixteen hours were employed to simulate two-shifting conditions. The creep-fatigue tests were conducted on a servo electric machine. Strain control was facilitated via extensometry attached to the ridges of the specimen and displacement was monitored by averaging two LVDT transducers. All tests were conducted at 535°C, temperature being monitored by three thermocouples attached to the specimen gauge length. Stress-strain hysteresis loops were monitored at regular intervals and the stress relaxation behavior monitored continuously. Specimen failure was defined as a 20 percent reduction of the peak tensile load from the saturated plateau region. The total strain rate during the cyclic part of the test was always $10^{-3}s^{-1}$.

After fatigue or creep-fatigue testing, specimens were sectioned longitudinally and examined metallographically in order to determine the failure mode. In addition, limited transmission electron microscopy (TEM) using carbon extraction replicas was conducted to examine the effect of prior creep on the microstructure.

MATERIAL

The base and simulated HAZ specimens were produced from the same heat of 1Cr-1/2Mo. The heat treatment used for the base material was: normalize 1 hour at 590°C, temper 3 hours at 640°C, air cool. The simulated HAZ heat treatment consisted of the following: grain coarsen for 5 minutes at 1300°C, oil quench, temper for 2 hours at 650°C, air cool. The base material microstructure consisted of a mixed ferrite-bainite structure with a Vickers hardness (30 Kg load) of 144, while the simulated HAZ was fully bainitic with a Vickers hardness of 234. The room temperature ultimate tensile strengths for the base and HAZ material were 465 MPa and 708 MPa, respectively. The tensile elongations were 33.3 percent for the base and 13.5 percent for the HAZ material.

INTERRUPTED CREEP TESTS

A program consisting of interrupted creep tests was performed to produce creep damaged or pre-crept material for the pure fatigue and creep fatigue test program. There were 17 base metal and 14 HAZ tests in this program. These interrupted creep tests were conducted at accelerated temperature (compared to service) and constant load conditions. It was found that these tests were subject to a systematic error in the temperature measurement such that some tests operated at a higher temperature than desired, i.e., the desired temperature of 590°C was actually 594.4°C. The corrected test durations for exposure at the intended test temperature were calculated using time-temperature parametric methods and both planned and actual life fraction values are given in Table 1. Typical hardness valves and average strain valves for the multiple tests performed at common conditions are also given in Table 1.

TABLE 1. Interrupted Creep Test Results for 1Cr-1/2Mo
Base and Simulated HAZ Material

Stress (MPa)	Temp.-°C Plan	Temp.-°C Actual	Life Fraction Plan	Life Fraction Actual	Vickers Hardness	Average Strain - %
			Base			
80	590	594.4	.6	.82	140	9.6
42	630	635	.2	.24		1.8
42	630	635	.6	.6	126	12.2
			HAZ			
58	605	609.6	.2	.27	203	.35
58	605	609.6	.6	.5	200	.63

The creep damage mechanism (microstructural or grain boundary cavitation) developed during the interrupted creep test was determined by metallographically examining either spare pre-crept specimens or the non-gauge sections of the creep-fatigue specimens. Examination of the interrupted HAZ creep specimens revealed that the major component of damage was the formation of creep cavitation shown in Figure 1. At 0.2 t_r, this damage was present in the form of discrete cavities, whereas at 0.6 t_r the cavities had interlinked to give rise to grain boundary cracking. Evidence of thermal softening damage occurring in the HAZ structure at 605°C is shown by the Vickers hardness values of 203 and 200 for 0.2 and 0.6 life fraction. However, it would appear that the creep cavitation dominate subsequent creep behavior at 535°C.

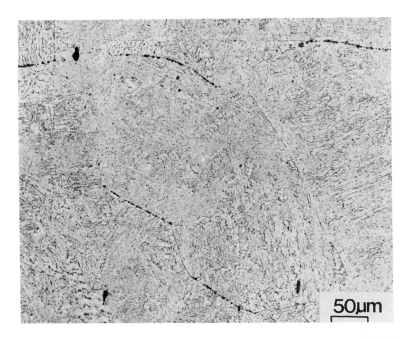

Figure 1. Creep cavitation damage in HAZ material pre-crept at 605°C and 58 MPa to a life fraction of 0.6.

For the base material, metallographic examination revealed that tests carried out at 590°C and 80 MPa tended to induce more creep cavitation damage than those using the higher temperature test conditions of 630°C and 42 MPa. Although creep cavitation was still observed at the higher temperature conditions, it was only in the form of discrete cavities. In addition to this creep cavitation, microstructural coarsening of the carbides was also found to have taken place as shown by comparison of the nature of the carbide distribution found in carbon extraction replicas for the as-received base material and material tested at 590°C and 80 MPa and at 630°C and 42 MPa. The microstructure in the higher temperature specimen had changed considerably compared to the lower temperature condition and extensive coarsening of the carbides had taken place in both the bainitic and ferritic regions. The extent of carbide coarsening for both prior creep conditions is exhibited in the hardness measurements in Table 1, which show a greater decrease in hardness at 630°C compared to 590°C. The hardness variation agrees well with measurements made in the previous study [2].

In summary, metallographic examination of the interrupted creep specimens has revealed two forms of damage: grain boundary cavitation and coarsening of carbides. The grain boundary cavitation is the dominant damage mechanism in the HAZ material, whereas carbide coarsening is the dominating mechanism in the base material.

EFFECT OF PRIOR CREEP ON PURE FATIGUE

The fatigue properties for the pre-crept base and HAZ material are shown in Figure 2 on a plot of plastic strain range versus number of cycles to failure. Data for the as-received material are also shown for comparison. Except for the 0.6 t_r pre-crept HAZ material, the fatigue properties are nominally the same as those for the virgin or as-received material.

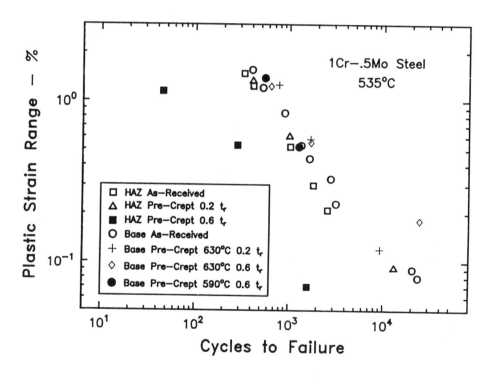

Figure 2. Effect of prior creep on pure-fatigue life of 1Cr-1/2Mo base and HAZ material at 535°C.

The base material exhibited cyclic hardening while the HAZ material cyclically softened. It is evident that, after initial hardening or softening, a plateau is reached as the material becomes cyclically stable. This stability is followed by a rapid decrease in load as a fatigue crack forms and propagates through a specimen. The effect of the prior creep on the peak stress in the fatigue cycle was substantial. Comparing the pre-crept to virgin material condition, large decreases in the peak stress were found for both the base and HAZ material. This decrease was greater for the base material than for the HAZ with typical base material peak stress differences greater than 100 MPa.

Metallographic examination of the as-received fatigue specimens revealed failure to be due to the nucleation and growth of a transgranular fatigue crack which propagated from the surface into the specimen. In the HAZ material pre-crept at 605°C and 58 MPa to 0.2 life fraction, transgranular crack propagation was predominant with some intergranular cracking in the material ahead of the major fatigue crack. However, when the life fraction was extended to 0.6, intergranular cracking was the major crack mode and intergranular cracks were uniformly distributed throughout the sample volume. The base material also exhibited intergranular cracking after being pre-crept at 590°C and 80 MPa to 0.6 life fraction.

EFFECT OF PRIOR CREEP ON CREEP-FATIGUE

The effects of adding hold times to the fatigue cycle are discussed in this section. Metallographic examination of the specimens was performed and interpreted using a failure mode characterization scheme discussed in a recent review on the creep-fatigue behavior of engineering alloys [3]. In general, the failure mode under creep-fatigue conditions can be categorized into three regions: fatigue dominated, creep-fatigue interaction, and creep dominated. Fatigue dominated failures are characterized by transgranular surface connected cracks, while the creep dominated failures have volumetric damage consisting of intergranular creep voids and cracks. The fatigue-creep interaction failure mode occurs only at specific strain ranges and cyclic values and has a mixture of the other two modes. The interrelationships between the failure mode and creep-fatigue life is important in understanding creep-fatigue and the effect of prior creep.

347

Virgin Base and HAZ Material

From the discussion in the previous sections, it is evident that the pure fatigue tests conducted on the as-received base and HAZ materials are compatible with the failure mode categorization described above. The pure fatigue tests on the virgin base and HAZ materials all exhibited fatigue dominated failure due to the initiation and propagation of predominately transgranular surface cracks and no significant difference was observed between the lives of the materials. However, as shown in the data presented in Figure 3, a large reduction in the life of the base material occurred with a 16-hour tensile dwell was introduced into the fatigue cycle.

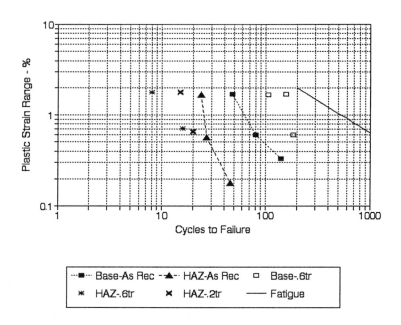

Figure 3. Effect of prior creep on creep-fatigue life (16 hour tensile dwell) of 1Cr-1/2Mo base and HAZ at 535°C.

Metallographic examination of the specimens revealed that, coincident with the reduction in life, there was a change from fatigue to creep dominated failure mode. This failure mode was exhibited at all strain ranges examined and was a consequence of creep damage accumulation during the tensile dwell period. Creep cavitation nucleated and grew under relaxing stress conditions and was found throughout the specimen gauge length. This mode of damage dominated in each cycle and the total fatigue damage was limited because of the large reduction in the number of cycles available for nucleation and growth of fatigue cracks. Fatigue damage was therefore limited to minor transgranular surface cracks which did not propagate into the specimens and no interaction between the two damage modes was observed.

The data presented in Figure 3 show similar trends for the virgin HAZ material compared to the base material. However, the introduction of a tensile dwell into the fatigue cycle resulted in a larger reduction in life for the HAZ material compared to the base material. This was a result of the lower creep ductility exhibited by this material. Examination of failed specimens revealed fracture to be a consequence of creep damage formation throughout gauge length, giving rise to creep dominated fracture mode as shown in Figure 4. In these tensile dwell tests, the peak stress decreased continuously during the test, showing cyclic softening behavior similar to that found in the pure fatigue tests on this material.

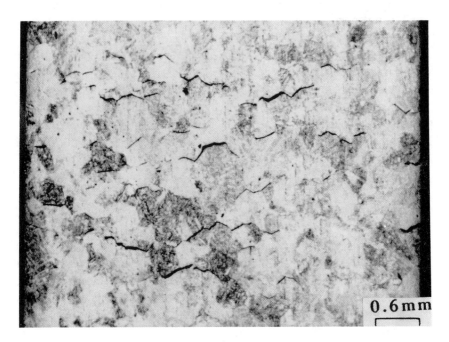

0.6mm

Figure 4. Creep dominated failure mode for creep-fatigue test on as-received HAZ material.

Comparisons of creep-fatigue data generated on 1Cr-1/2Mo at 535°C with other creep-fatigue data on ferritic steels were made by Miller et al. [3]. It is evident that the pure fatigue data on all low alloy ferritic steels examined fall within a narrow scatterband. Furthermore, the influence of creep ductility on the life behavior and failure mode was clearly demonstrated; the lower the creep ductility the lower the creep-fatigue life. Creep dominated failure is favored by long dwell periods and low ductility while intermediate strain ranges, short dwell periods, and high creep ductility favor creep-fatigue interaction. Therefore, it is clear that the HAZ material has a lower life because the ductility is approximately five times lower than that observed in the base material [1,2].

Pre-Crept HAZ Material

The effects of prior creep at 605°C and 58 MPa to life fraction of 0.2 and 0.6 t_r on the creep-fatigue of the HAZ material are depicted graphically in Figure 3. The prior creep resulted in the subsequent creep-fatigue lies at 535°C being shorter than those of the as-received material and this reduction continued as the amount of prior creep increased.

The reduction in life was a consequence of the prior creep cavitation damage present in the material. This cavitation resulted in a lower creep ductility at 535°C compared to the as-received material, and this reduced the creep-fatigue lives. The prior creep cavitation eliminated the time required to cause cavity nucleation, which was necessary in the as-received material. This reduced the number of cycles to failure because it was only necessary to grow the creep cavities. The failure mode was creep dominated with no evidence of any contribution from fatigue.

Pre-Crept Base Material

In the base material, prior creep at 590°C, 80 MPa, and 0.6 t_r and at 630°C, 42 MPa and 0.6 t_r increased the creep-fatigue lives, as shown in Figure 3. The stress relaxation behavior of the pre-crept material was such that the total stress drop was comparable to that found in the as-received material. However, the prior creep produced a significant drop in the peak stress of the cyclic stress-strain behavior.

Examination of the specimen pre-crept at 630°C, 42 MPa, and 0.6 t_r revealed that fracture had occurred due to creep-fatigue interaction as shown in Figure 5. The specimen fractured due to the growth of a fatigue crack interacting with creep cavitation. The material pre-crept at 590°C, 80 MPa, and 0.6 t_r showed similar behavior in that it also exhibited a longer life compared to that of the as-received material. However, the increase in life was less than that observed in the material pre-crept at the higher temperature. Metallographic examination of the specimens showed the failures to be creep dominated with extensive creep cavitation throughout the gauge length. Small fatigue cracks were observed on the specimen surface, but growth was not extensive.

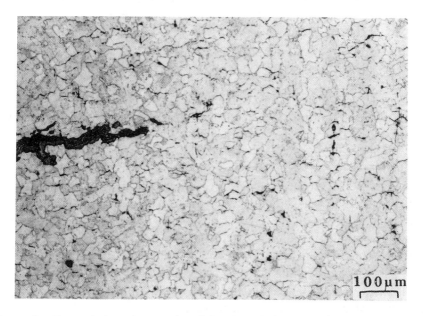

Figure 5. Creep-fatigue interaction failure mode in creep-fatigue test on pre-crept (630°C, 42 MPa, 0.6 t_r) base material.

These results may seem surprising at first and are the reverse of the effect caused by prior creep in the HAZ material. It was shown that, in base material, prior creep produced carbide coarsening as the dominant damage mechanism. This resulted in the ductility of the pre-crept material at 535°C being higher than that of the as-received base material. The strong influence that ductility has on creep-fatigue life has been found in other studies. Therefore, due to the thermal softening and higher ductility arising from the accelerated temperature prior creep conditions, the creep-fatigue lives at 535°C were longer than for the as-received material.

This is further supported by the data obtained from the two different prior creep conditions. It was found that the creep-fatigue life after prior creep at 630°C, 42 MPa, and 0.6 t_r was greater than that after 590°C, 80 MPa, and 0.6 t_r. This result can be explained in terms of the associated amount of thermal softening which is more extensive for the higher temperature and longer time condition. Therefore, the ductility at 535°C was ranked according to the temperature of the prior creep conditions with the ductility after prior creep at 630°C being the highest and that of the as-received material the lowest. This ductility ranking gave rise to an identical ordering of the creep-fatigue lives.

Structure-Property Correlation

The general effects of prior creep on the creep-fatigue behavior of 1Cr-1/2Mo material are shown schematically in Figure 6. It has been found that prior creep induces two damage processes: grain boundary creep cavitation and carbide coarsening. The extent of these two processes will be strongly dependent on the prior creep conditions used. The formation of creep cavitation will exhaust some of the material ductility, provided microstructural damage is limited, as is the case for the HAZ material. This leads to a reduction in the creep-fatigue life compared to that of the as-received material. If the prior creep conditions result in carbide coarsening being the dominant mechanism, then creep ductility and creep-fatigue life will be increased, as is the case for the base material. If other prior creep conditions were used for the base material which limited carbide coarsening, but enhanced creep cavitation, then behavior similar to that found in the HAZ material would be expected.

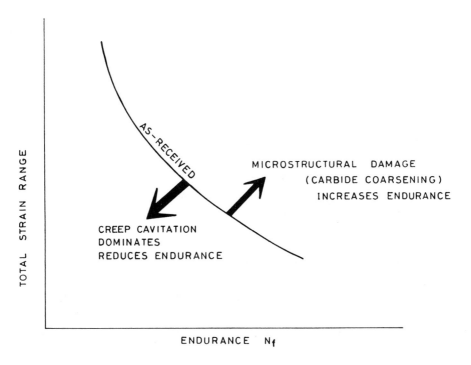

Figure 6. Schematic illustration of the influence of prior creep on creep-fatigue life of 1Cr-/2Mo steel.

CONCLUSIONS

Conclusions reached based on the current study of the effect of prior creep on fatigue and creep-fatigue of 1Cr-1/2Mo base and HAZ include:

1. The prior creep conditions used in this study resulted in the dominant mechanism being creep cavitation in the HAZ structure and thermal softening from carbide coarsening in the base material.

2. The introduction of a sixteen hour tensile dwell into a fatigue cycle significantly reduces the life of 1Cr-1/2Mo base and HAZ materials. Consistent with this reduction in life is a change in failure mechanism from fatigue to creep dominated failure mode.

3. The creep-fatigue life of HAZ material is less than that exhibited by base material due to the lower creep ductility exhibited by the HAZ.

4. Prior creep of the HAZ material resulted in a reduction in pure fatigue and creep-fatigue life compared to the as-received material life. For creep fatigue, the reduction in life increased with increasing prior creep life fraction. This reduction in life was the result of creep cavitation which reduced the creep ductility.

5. Prior creep of the base material resulted in increasing the creep-fatigue life compared to that of the as-received material. This increase in life was due to thermal softening which increased the base material creep ductility.

ACKNOWLEDGEMENTS

Financial support was provided by the Electric Power Research Institute. The guidance and encouragement of Dr. R. Viswanathan is gratefully acknowledged. I also wish to acknowledge the major contribution of Dr. D. A. Miller and Mr. D. Gladwin of National Power (U.K.) in the conduct and understanding of the creep-fatigue phase of the project.

REFERENCES

1. M. S. Shammas, F. V. Ellis, and J. F. Henry, "Remaining-Life Estimation of Boiler Pressure Parts, Volume 4: Metallographic Models for Weld-Heat-Affected Zone," EPRI CS-5588, Volume 4, Electric Power Research Institute, Palo Alto, CA, Nov. 1989.

2. M. C. Askins, "Remaining-Life Estimation of Boiler Pressure Parts, Volume 3: Base Metal Model," EPRI CS-5588, Volume 3, Electric Power Research Institute, Palo Alto, CA, Nov. 1989.

3. D. A. Miller, R. H. Priest, and E. G. Ellison, "A Review of Material Response and Life Prediction Techniques under Creep-Fatigue Loading Conditions," High Temperature Materials and Processes, Volume 6, p. 155, 1984.

THE FATIGUE CRACK GROWTH BEHAVIOR OF NOTCHED NICKEL

C.E. Price[1] and Yiao-Teng Ho[1]

ABSTRACT

Slip band growth, the initiation of microcracks and the transition of microcracks to macrocracks were monitored, using Nomarski interference contrast microscopy, on specimens of coarse grain size nickel. The testing was in reversed bending, with the cracks initiating at notches. Coarse slip bands, originated away from the grain boundaries and grew rapidly at first and then slowly, as a grain boundary was approached. The slip band cracks, likewise, grew rapidly at first and then more slowly. After the initial upsurge, crack growth rates in the first grain were irregular but often approximated a parabolic decay. The same irregularities in growth rate persisted across the second and third grains by which time the life was largely over. The irregular crack growth profiles arose because of the initial and continued crystallographic nature of the cracking. Both the orientation of slip bands and intensity of slip bands ahead of the cracks influenced the growth pattern.

NTRODUCTION

The fatigue behavior of nickel (N02200) in reversed bending was tudied previously using radiused specimens [1-4]. It was found that a) racks initiate at slip bands early in the fatigue life, after appreciable yclic hardening has occurred, b) the majority of the fatigue life is spent vhile the cracks traverse three grains, and c) the cracking continues to be

Oklahoma State University, School of Mechanical and Aerospace Engineering, Stillwater, OK 74078 USA.

crystallographic. Because the cracks originate at many locations over the cross section and crack linkage may occur, reversed bending on radiused specimens is not the most convenient test mode to study crack growth behavior. Tensile fatigue testing on notched specimens would seem preferable. However, there is a problem in that the fatigue cracks originate as many individual slip band cracks in the thickness dimension of the notch root [5]. There is no single short fatigue crack. Accordingly, the authors reverted to the reversed bending test mode for the present study, using a coarse grain size to facilitate observations but with double edged notched specimens that defined the crack initiation sites. The features of interest were the slip band growth and growth rates, the initiation of microcracks, the transition of microcracks to macrocracks and the subsequent growth of the macrocracks. It is usually found that the initial crack growth rate across the first grain occurs at a decreasing rate [6-8]. However, Baxter and McKinney [9,10] concluded that in 6061-T6 aluminum, under constant amplitude loading, the persistent slip bands elongate rapidly at first and then more slowly, following a parabolic dependence on the number of fatigue cycles in a small grain, whereas in a large grain specimen the rate remained constant. To study nickel is pertinent to the many nickel based commercial alloys, for their fatigue crack propagation is often crystallographic, too [11,12].

PROCEDURE

Specimens of width 12.5 mm were cut from 2.3 mm thick sheet. Two opposite edge notches, each 1 mm deep and 1 mm wide were cut with a low speed diamond saw. The specimens were annealed at 1050°C for 7 hours to give a coarse grain size of 400-500 µm and slow cooled. They were chemically polished at 85-90°C for two minutes in a mixed acid solution [13]. This also had the effect of smoothing the notch tip contours. The fatigue testing was in fully reversed bending at a frequency of 35 Hz. The specimens were removed frequently for observations, utilizing Nomarski interference contrast microscopy. In a test lasting about 500,000 cycles, there were approximately 50 removals for inspection, with the interruptions being more frequent early in the life. As previously, the interruptions had no discernable effect on the fatigue lives or surface appearance [1,4]. A detailed photographic record was kept and measurements of slip bands and cracks were made on the photographs. Tests were run on samples of lifetimes 14,000 to 500,000 cycles approximately. For short life specimens, the quantity of slip markings in the notch and crack tip vicinities became too great to conveniently monitor individual features. Accordingly, the most detailed measurements, reported here, were concentrated on specimens of lifetimes approaching 500,000 cycles.

RESULTS

Slip markings were visible under Nomarski contrast after 1 cycle. As cycling continued, the slip became concentrated in coarse bands which extended across the width of the grains. Later, some fine slip initiated between the coarse bands. These coarse bands developed after 2000-4000 cycles, within the first 1% of the lifetime. This appearance was as described previously [4]. The total length and number of all the slip bands in grains at a notch root, where a crack eventually developed, were measured, Figure1, for example. At first, the total length of slip bands grew rapidly, then slowed down with the increase in band width and quantity. The growth rate was rapid initially and then decreased, Figure 2. In this instance, a crack initiated after 25,000 cycles in a zone of intense and intersective slip. By this time, the slip band formation and growth were largely complete.

A complicating factor in the measurement of individual slip band and short crack characteristics was the large number of bands that appeared in the notch tip vicinity. Some interacted with each other, sometimes linking up. Grains containing several slip bands, that were not involved in such interactions, were chosen for individual measurements. Sample slip band length and growth rate plots are shown in Figures 3 and 4. The growth rate increased initially and then decreased as the slip bands reached 20%-50% of the grain width. Eventually, the slip bands were arrested at the grain boundaries.

Of the order of a quarter of the lifetime was spent while a crack traversed the first grain, Figure 5. As with slip band growth, the crack growth rate was irregular, high initially and then decreasing, Figure 6. A parabolic decay of the crack growth rate was usually a reasonable fit to the data. The crack growth characteristics were largely unchanged once the first grain had been crossed; indeed, the number of cycles to cross the second grain was usually higher than for the first grain, because the initial high growth rate segment was not repeated, Figure 7. Only after the crack had grown beyond about 3 grains, after about 300,000 cycles, was a marked general acceleration obvious, Figure 8.

The growth of cracks was crystallographic, in an irregular zig-zag pattern. When a growing encountered a coarse slip band, it often spread along the slip band for some distance and then connected with a neighboring

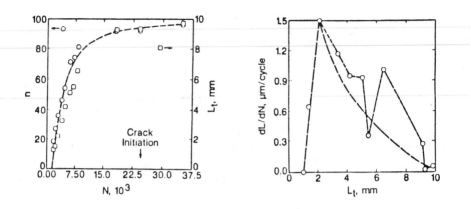

Figure 1. (left)   The growth in total number (n) and total length (L_t) of sli bands in a grain at the notch root, shown versus the number of cycles (n where a crack originated after 25,000 cycles.

Figure 2. (right)   The total growth rate (dL/dN) of slip bands in a notch roc grain, as a function of total slip band length (L).

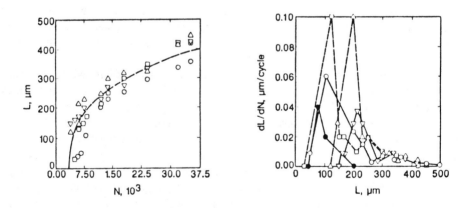

Figure 3. (left)   The length (L) of 4 slip bands in a grain at the notch root, a a function of number of fatigue cycles (N).

Figure 4. (right)   The growth rate (dL/dN) of 5 slip bands across a grain, o width 500 μm, at the notch root.

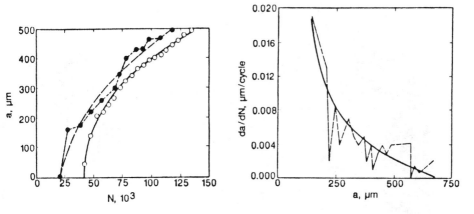

Figure 5. (left) The increase in crack lengths (a) of 2 cracks growing from notch roots, in specimens lasting about 500,000 cycles.

Figure 6. (right) An example of initial crack growth rate that reasonably fits parabolic decay.

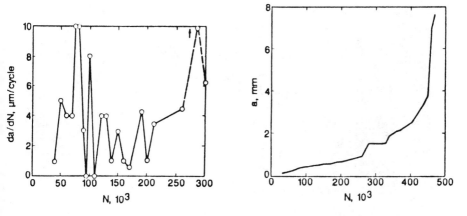

Figure 7. (left) Crack growth rates across the first three grains approximately.

Figure 8. (right) A plot of crack length (a) versus fatigue cycles (N) over the entire life of a specimen.

Figure 9. Cracks growing from notches in two specimens each lasting abou 500,000 cycles. The numbers indicate the crack position after that numbe of cycles.

Figure 10. The early growth of one of the cracks from Figure 9, showing how the growth is contrary to the dominant slip system.

slip band after intensive general slip in the intervening zone. As a crack approached a grain boundary, retardation was usual and intense localized slip occurred in the next grain. The crack continuation path was rarely obvious beforehand. After a crack grew over a distance of 3-4 grains, it sometimes connected with another crack that had initiated at coarse slip bands, ahead of the main crack. In this way several short cracks were often left hanging. Occasionally the crack bifurcated, with maybe one arm propagating across several grains before petering out, or rejoining the other crack. Cracks grew very occasionally along grain boundaries or twin boundaries.

In these edge notched specimens, crack initiation took anywhere from 25,000 to 50,000 cycles for a life of about 500,000 cycles. With essentially two or three grains, at the crack root, the orientation of the slip bands mattered much; although, sometimes the crack began at an unexpected location or grew initially in an unexpected direction. Two examples of early crack growth are shown in Figure 9, illustrating expected and unexpected crack initiation directions. A point here is that if there is an "obvious" slip

Figure 11. A crack tip late in the fatigue life. It continues to grow by jogging sideways to the cracked slip band indicated.

band crack growth path across the first grain and it is followed, the crack may not grow more rapidly, but the growth is more regular. This contrast can be seen in the two curves of Figure 5. A detail of a difficult crack path is shown in Figure 10, showing the zig-zag growth between two slip systems. It can be seen, too, that many of the slip bands are cracked apart in their mid sections. In this instance, the slip bands, at the crack root have broadened through cross slip, rather than cracked. The crack is heading directly for a band that has cracked rather than broadened. Similar situations were noted late in the fatigue life, Figure 11. It may also be recognized from Figures 9 and 10 that the quantity of slip at the notch root is such that the instant of crack initiation cannot be discerned and, indeed, several cracks may start. Only when a crack has traveled 50 - 100 μm can it be identified unambiguously. The fields of view of Figure9 were selected to encompass 3-4 grains in width. The number of cycles to traverse portions of the field of view are indicated. Two features are that firstly the field of view encompasses most of the fatigue life in terms of crack growth, and secondly, the irregularity of the actual crack path can be discerned. It may be noticed, as mentioned previously, that the crack growth across the second grain rate takes longer than across the first grain.

DISCUSSION

The fatigue crack initiation at the notch root is unchanged from that in the radiused specimens, insofar as coarse slip bands develop within the first one percent of the life that shortly become cracked in many instances. With few grains at the notch root, the details of the slip bands matter. Obviously orientation is one factor, but the interference is that the extent to which the slip bands broaden also matters. That, in the situation illustrated in Fig. 10, the crack did not initiate at the obvious location and propagate in the direct line can be explained by the pertinent slip bands broadening, through cross slip instead of cracking. The crack, when it initiates heads towards the nearest conspicuously cracked slip band, one that has not broadened. There are, therefore, three variables inherent in the initial fatigue stage, the initiation step which depends on the slip character, and the first growth, that is governed both by crystallographic orientation and the nature of the prior 'damage' in the zone ahead of the embryo crack. The irregular crack propagation rate is a consequence.

In single crystals of nickel base superalloys, where the cracking is also crystallographic, there is no decrease in crack propagation rate after the commencement of growth [14]. Why, therefore, the fall off in this study? Usually the fall off is attributed to the forthcoming barrier of the grain boundary [6-8]. However, in the present case there is no following upsurge and it is a three grains distance rather than one that is critical. Note that growth over three grains dominated the life in the radiused specimens that also had a much finer, about 120 µm, grain size [4]. The likely explanation is that the plastic zone ahead of the notch (or a fatigue crack) initially extends across these grains. It would be large because of the high cyclic stress level necessary to get fatigue, above the nominal yield stress, as stated earlier, and because of the appreciable cyclic hardening capacity of nickel. A further point is that the slip band cracking in the plastic zone ahead of the crack tip facilitates crack branching. In a nickel base alloy, Waspaloy, crack branching was found to reduce the effective stress intensity [15]. Branching would be most important in short cracks because of the lower stress intensity and the preponderance of the life spent in this situation.

363

CONCLUSIONS

The conclusions to this study are as follows:

• While coarse slip bands develop within the first one percent of the fatigue life and individual bands may crack promptly, the ultimately damaging notch root cracks take 5-10 percent of the life to develop.

• The notch root cracks initiate preferentially at slip bands showing minimal wavy slip.

• Once initiated, notch root cracks grow rapidly through the first part of the first grain and then the growth rate decreases, but irregularly. A parabolic decay is an approximation.

• There is no increase in crack growth rate beyond the first grain; more cycles may be taken to traverse the second and third grains than the first grain because of the absence of the initial acceleration.

• The fatigue life is largely spent while the crack traverses 3 grains.

• The crack growth path is mostly along slip bands. Often, the crack moves sideways to link up with slip band cracks that have generated ahead of the main crack. This, also, leads to frequent crack branching.

REFERENCES

1. Price, C.E. and Fila, L.J., "Surface Zone Hardening During the Bending Fatigue of Nickel," **Metall. Trans.,** Vol. 12A p 623, 1981.

2. Price, C.E. and Fila, L.J., "Observations on the Surface Zone Hardening Rate During the Bending Fatigue of Nickel," **Scr. Metall.,** Vol. 16 p 1157, 1982.

3. Price, C.E. and Kunc, R., "Occurrences of Faceted Fatigue Fractures in Nickel," **Metallography,** Vol. 19 p 317, 1986.

4. Price, C.E., "The Progression of Bending Fatigue in Nickel," **Fatigue Frac., Engng. Mater. Struct.,** Vol. 11 p 483, 1988.

5. Price, C.E. and Everhart, L.G., "Characterizing Plastic Deformation Around Fatigue Cracks in Nickel," **Microstructural Sci.,** Vol. 18 p 315, 1990.

6.   Lankford, J., "The Growth of Small Fatigue Cracks in 7075-T6 Aluminum," **Fatigue Fract. Engng. Mater. Struct.,** Vol. 5 p 233, 1982.

7.   Miller, K.J., Mohamed, H.J. and de los Rios, E.R., "Fatigue Damage Accumulation Above and Below the Fatigue :Limit," in K.J. Miller and E.R. de los Rios (eds) **The Behavior of Short Fatigue Cracks** (EFG MEP) p 491, 1986.

8.   Kendall, J.M. and King, J.E., "Short Fatigue Crack Growth Behavior: Data Analysis Effects," **Int. J. Fatigue,** Vol. 10 p 163, 1988.

9.   Baxter, W.J., "The Growth of Persistent Slip Bands During Fatigue," in K.J. Miller and E.R. de los Rios (eds) **"The Behavior of Short Fatigue Cracks** (EFG MEP) p 193, 1986.

10.   Baxter, W.J. and McKinney, T.R., "Growth of Slip Bands During Fatigue of 6061-T6 Aluminum," T.R., **Metall. Trans.,** Vol. 18A, p 83, 1988.

11.   Gell, M. and Leverant, G.R., "The Characteristics of Stage 1 Fatigue Fracture in a High Strength Nickel Alloy," **Acta. Metall.,** Vol. 16, p 553, 1968.

12.   Price, C.E., "Observations on the Faceted Fatigue Fracture of a SEL Superalloy Turbine Blade," **Metallography,** Vol. 17, p 469, 1984.

13.   Tegart, W.J.M., **The Electrolytic and Chemical Polishing of Metals,** (Pergamon Press Oxford, U. K., 2nd edn.) p 102, 1959.

14.   Howland, C., "The Growth of Fatigue Cracks in a Nickel Base Single Crystal," in K.J. Miller and E.R. de los Rios (eds) **The Behavior of Short Fatigue Cracks** (EFG MEP) p 229, 1986.

15.   Hussey, I.W., Byrne, J. and Duggan, T.V., "Behavior of Small Fatigue Cracks at Blunt Notches in Aero Engine Alloys," in K.J. Miller and E.R. de los Rios (eds) **The Behavior of Short Fatigue Cracks** (EFG MEP) p 337, 1986.

MICROSTRUCTURAL INFLUENCE ON FATIGUE CRACK GROWTH BEHAVIOR OF HIGH STRENGTH LOW ALLOY STEEL WELDMENTS

Abhijit Sengupta[1], Jack Schaefer[2] and Susil K. Putatunda[3]

ABSTRACT

The influence of microstructure on the fatigue crack growth behavior of high strength low alloy (HSLA) steel welments was investigated. Compact tension specimens were prepared from ASTM grade A514 in such a way that the cracks propagated through the heat affected zone, base metal and weldments. This arrangement permitted the influence of the heat affected zone on fatigue crack growth rate to be determined in both the linear and threshold regions of Paris curves in a room temperature air environment. The influence of microstructure on the fatigue threshold and the micromechanism of crack growth in the threshold and linear region were also investigated. Welding was accomplished by the gas metal arc welding (MIG) process.

[1] Graduate Student, Department of Materials Science and Engineering, Wayne State University, Detroit, MI 48202

[2] Metallurgical Research Engineer, Detroit Edison, Detroit, MI 48226

[3] Associate Professor, Department of Materials Science and Engineering, Wayne State University, Detroit, MI 48202

The results of the present investigation indicates that the near-threshold fatigue crack growth rate was higher at the same load ratio when the crack propagated through the heat affected zones. Consequently, the fatigue threshold was lowest in these specimens. The crack growth process was found to be predominantly by striations. Extensive crack branchings and secondary cracking were observed in the high ΔK region but were not generally seen in the threshold region. A linear relationship was observed between log C and m (Paris constants).

INTRODUCTION

The fatigue crack growth rate da/dN, has been related to the stress intensity range, ΔK and the Paris equation [1] has been found to be extremely useful for characterizing the fatigue crack growth rate. The Paris equation [1] relates the crack growth rate da/dN with ΔK in the form of a power law,

$$da/dN = C \cdot (\Delta K)^m$$

Here C and m are material constants and $\Delta K = K_{max} - K_{min}$ = the difference between maximum and minimum stress intensity factors.

However, when the experimental crack growth rate data is plotted in terms of a log da/dN versus log ΔK, it shows a sigmoidal curve with varrying slopes (Figure 1) instead of a single straight line, as one would expect from such a plot. There are three distinct regions in this plot. In region I or the slow crack growth region, the crack growth rate deviates from the Paris equation and the crack growth rates are very slow. In region II or the linear region, the crack growth rate increases linearly with ΔK and the Paris equation [1] is usually obeyed by most materials. In region III or the fast fracture region, the crack growth rate is very rapid and again deviates from the Paris equation.

In addition to the above three regions, there is a threshold stress intensity factor, ΔK_{th}, below which the crack growth rate approaches a zero value. According to ASTM standard E 647 [2], threshold is defined as the value of the stress intensity factor at which the crack growth rate is of the order 10^{-10} m/cycle.

The threshold region is of great importance, since a significant portion of the life of the structural components is spent in this region. The threshold stress intensity factor ΔK_{th} is also a very important parameter for structural design, since structural components designed on the basis of fatigue threshold are expected to have infinite lives or at least last for a very long period of time. Microstructural parameters can significantly affect the fatigue crack growth rate both in linear and threshold region and fatigue threshold. Hence microstructural modifications appears to be the only possible way to improve the fatigue threshold of metals and alloys.

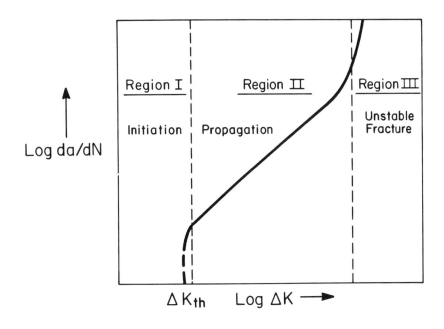

Figure 1: Schematic Representation of Fatigue Crack Growth
According to Paris Equation

High strength low alloy steels, such as ASTM grade A514 are used as blade materials in both forced and induced draft fans in utility steam generators. These materials have been used in Detroit Edison's various plants. Fatigue failure of these fan blades have been observed in past in Detroit Edison's Belle River plant at welded joints in forced draft fans. Hence the fatigue crack growth data of these weldment materials are extremely important and needed for life prediction of the fans. Moreover, in order to improve the overall fatigue life of these components, it is essential to establish the influence of microstructural parameters on fatigue crack growth rate and fatigue threshold for these weldments.

Based on the above facts, an investigation was undertaken to examine the influence of microstructure on fatigue crack growth behavior of high strength low alloy steel weldments (ASTM Grade A514). The earlier study [3] examined the effect of microstructure on fatigue crack growth behavior where the welding was done by Submerged Arc Welding (SMAW). The present investigation is a continuation of the above study where the welding was done by Gas Metal Arc Welding process (MIG).

The objective of the present investigation was to examine the influence of microstructure on the fatigue threshold and fatigue crack growth rate of the high strength low alloy steel weldment ASTM grade A514 both in linear and threshold region and to establish the influence of heat effected zone on fatigue crack growth rate and fatigue threshold.

EXPERIMENTAL PROCEDURE

MATERIAL

2(T) Compact tension specimens were prepared from ASTM grade A514 high strength low alloy steel weldment. The specimens were designed in three different ways. In the first series of specimens, the crack tip was located at the mid-section of weld metal. These specimens are designated as C1. In the second series of specimens (designated as C2), the notch tip was located in such a way that the crack would propagate through the heat affected zone. In the third series of specimens, the notch of the crack was oriented in a manner such that the crack would propagate through the base metal. These specimens are designated as the C3 series. The welding in all these specimens were done by Gas Metal Arc Welding (MIG) process as per American Welding Society Standard AWS A5.18-79 and ASME SFA-5.18 Standard. Table 1 shows the designation of the specimens. Chemical composition of the base metal as well as the welding electrodes, and weld deposits are reported in Tables 2a, 2b and 2c, respectively. The mechanical properties of base metals and weld deposits are reported in Table 3.

The Compact Tension specimens were prepared as per ATM standard E-647 [2]. The thickness of the specimens was kept about 6 mm and the width was 50.8 mm. The initial aspect ratio was kept at $(a/W) = 0.30$. A schematic of the compact tension specimens used in this study is shown in Figure 2.

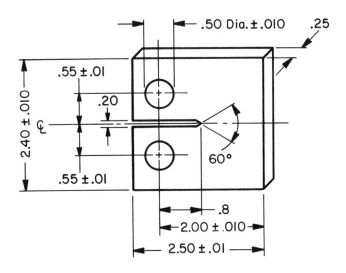

Figure 2: Compact Tension Specimen (2T)
Scale: Full Size: All Dimensions in Inches

371

FATIGUE TESTING

The specimens were ground on both surfaces and then polished with 600 grit level emery paper. This was found extremely useful in locating the crack tip during the fatigue testing. The specimens were initially precracked in fatigue for 2 mm at a ΔK level of 15 $MPa\sqrt{m}$. After fatigue precracking, the fatigue testing was carried out in a servohydraulic MTS test machine under load control mode. All testing was carried out in tension tension mode in room temperature ambient atmosphere. A constant amplitude sinusoidal wave form was applied and tests were carried out at constant load. The load ratio R was kept constant at R = 0.1.

Table 1

Specimen Designations

C1:   Crack notch oriented through the weld metal.

C2:   Crack notch oriented through the heat affected zone.

C3:   Crack notch oriented through the base metal.

Table 2

2(a): Chemical Composition of the Electrodes in Weight Percentage

C	-	0.10
Mn	-	1.20
Si	-	0.55
S	-	0.015
P	-	0.008

2(b): Chemical Composition of the Base Material

C	-	0.18
Mn	-	1.02
Si	-	0.60
S	-	0.010
P	-	0.005
Cr	-	0.65
Mo	-	0.21

Table 2

2(c): Chemical Composition of Weld Deposit

C	-	0.088
Mn	-	0.72
Si	-	0.33
S	-	0.014
P	-	0.007
Cu	-	0.108

Table 3

Mechanical Properties of the Material

Material	Y.S. MPa	UTS (MPa)	% Elongation	Hardness
Weld deposit	430	535	29	88 RB
Base Material ASTM Grade A514	690	830	18	28 (RC)

The crack lengths were monitored with the help of an optical travelling microscope. The crack lengths and number of cycles were measured continuously. The crack growth rates were determined as per ASTM standard E-647 [2] test procedures.

The threshold was determined using the load shedding techniques. The threshold was determined by slowly decreasing the load values and recording the crack growth rate. This reduction of load at any stage was done maximum up to 5% and that also only done after the crack has propagated by at least 1.00 mm at the previous ΔK level. The load shedding done in this manner was sufficient to avoid any retardation phenomenon due to prior ΔK levels. The threshold was identified as the ΔK level at which the crack growth rate was of the order of 10^{-10} m/cycle as per ASTM standard E-647. At least 4 specimens from each series were tested and average values were taken as the representative of crack growth rate and fatigue threshold.

RESULTS

The microstructure of the material in C1, C2 and C3 conditions are shown in Figures 3a, 3b and 3c, respectively. The hardness values of the materials are reported in Table 4.

Figure 4 reports the fatigue crack growth behavior of C1 specimens in both linear and threshold regions. Figures 5 and 6 report the fatigue crack growth behavior of C2 and C3 specimens in linear as well as threshold regions. In Figure 7, the fatigue crack growth behavior of the materials C1, C2 and C3 are compared at same load ratio R = 0.1. Paris constants C and m of the materials and the fatigue threshold values are reported in Table 5. It is evident from all these figures (Figures 4 through 7) that the near threshold crack growth rate is highest in C2 material where the crack propagated through the heat affected zone. This higher near threshold crack growth rate in C2 material has resulted in the lowest fatigue threshold in this material. Comparing the results in Figure 7, one can also observe that this the near threshold crack growth rate is lowest in C3 material (i.e., the base material) and consequently the fatigue threshold is highest in this series of specimens.

Figure 3(a) : Microstructure of weld metals

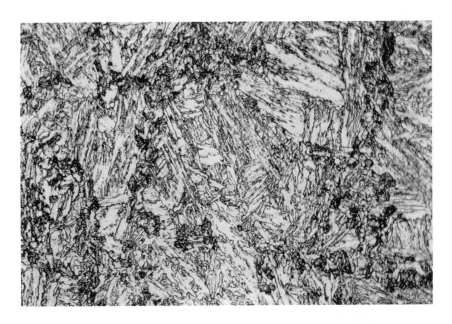

Figure 3(b): Microstructure of HAZ Material

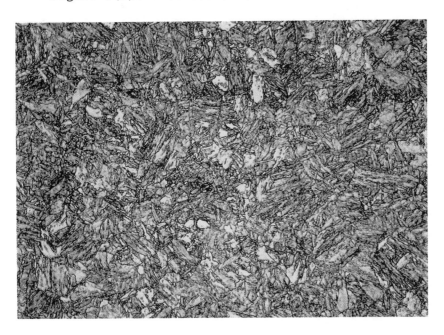

Figure3(c): Microstructure of base metal

Table 4

Hardness of the Materials

Specimen Series	Hardness
C1	28
C2	27
C3	25

Table 5

Paris Constants and Fatigue Threshold Values of the Material

Paris Constants

Specimen Type	C	m	ΔK_{th} $MAa\sqrt{m}$
C1	3.0×10^{-13}	3.98	8.83
C2	3.2×10^{-13}	3.47	6.38
C3	1.5×10^{-12}	3.42	6.74

One interesting aspect of these results is the fact that around $\Delta K \approx 45MPa\sqrt{m}$ or so one can observe a transition ΔK_T, at which the crack growth rate in C1 series of specimens becomes significantly higher than either C2 or C3 series of specimens. Moreover at this ΔK level we also observe that the crack growth rate in C2 specimens becomes lowest among all these specimens. Such transition behavior has been reported in the past in several other materials by earlier workers [4-6] and it has been suggested [4-6] that this transition occurs due to a change in crack growth mechanism in fatigue. Though such behavior has not been reported in weldments, our earlier studies indicated [3] that this transition occurs in weldments also. This transition stress intensity factor appears to be related to the value of ΔK at which the cyclic plastic zone size becomes equal to the (prior austenitic) grain size of the material.

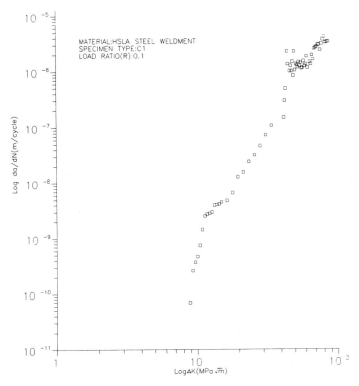

Figure 4: Fatigue crack growth behaviour of weld metal

DISCUSSION

The higher fatigue crack growth rate in the threshold region in the C2 series of specimens appears to be related to its microstructural features. The heat affected zone had a coarse grain structure with extensive martensite structure present in it. These martensites are more susceptible to fatigue cracking during cyclic loading. As a result of the presence of martensite in HAZ material, fatigue crack growth rate accelerates in the threshold region and reduces the fatigue threshold for these series of specimens. Furthermore, formation of these martensites in HAZ material would increase the yield strength of this material and as a result the cyclic plasticity induced crack closure [7] K_{oP} is lower in this material. Since effective stress intensity factor $\Delta K_{eff} = K_{max} - K_{oP}$ and since the crack closure

377

stress intensity factor K_{oP} is lower in HAZ material (C2 series of specimens) the effective stress intensity factort is significantly higher for the same applied ΔK in this material. Hence the crack growth rate in this material (C2) is higher than either C1 or C3 series of specimens.

In addition to above, there exists residual tensile stresses [8] in the heat affected zones as a result of uneven cooling and heating process during the welding cycle. This presence of tensile residual stresses in the HAZ material would increase the effective stress intensity factors or in other words the mechanical drying force for propagation of a crack. Hence the crack growth rate in the C2 series of specimens increases in the threshold region and consequently fatigue threshold decreases in this series of specimens.

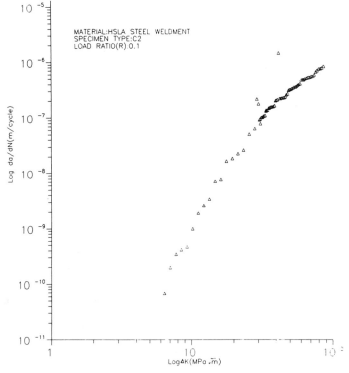

Figure 5: Fatigue crack growth behaviour of HAZ material

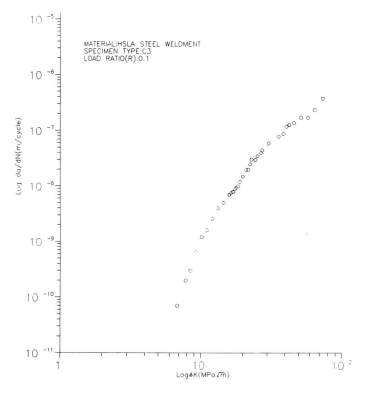

MATERIAL:HSLA STEEL WELDMENT
SPECIMEN TYPE:C3
LOAD RATIO(R):0.1

Figure 6: fatigue crack growth behaviour of base metal

At higher ΔK level, the effect of crack closure is greatly diminished and hence we do not observe any significant difference in crack growth rate behavior C1, C2 and C3 series of specimens. As mentioned earlier at or around $\Delta K \approx 45\ MPa\ \sqrt{m}$, fracture mode transition occurs and as a result the crack growth rate becomes lower in C2 series of specimens from that ΔK level onwards.

RELATIONSHIP BETWEEN C AND m

Many earlier workers [9-11] have found and reported that there exists a linear relationship between Paris constants C and m in the form $m = x\ log\ C + y$.

McCarteney are Irving [9] have shown on the basis of dimensional analysis that the linear relationship between C and m in the above equation is a necessary condition. Picoll

[10] analyzed the data from several materials and has shown that in a wide range of materials this relationship holds true.

Our test results were analyzed based on the above observation. Figure 8 reports the plot of log C and m for the materials in C1, C2 and C3 series of specimens. It is evident that a linear relationship exists between log C and m in these specimens. The values of x and y obtained from this plot were different than reported by other workers. Apparently, this is due to the different micromechanism of crack growth in fatigue in different materials.

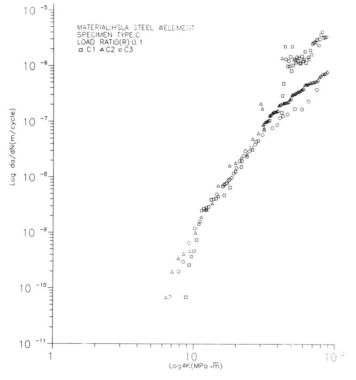

Figure 7: Comparison of fatigue crack growth beahvior of C1,C2 and C3 specimans.

380

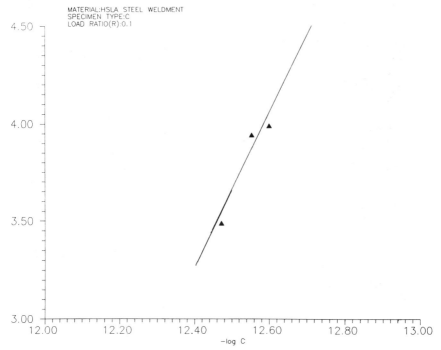

MATERIAL:HSLA STEEL WELDMENT
SPECIMEN TYPE:C
LOAD RATIO(R):0.1

Figure 8: Relation between Paris constants

FRACTOGRAPHY

The SEM fractographs of C1, C2 and C3 specimens in the threshold region is reported in Figures 9 (a), (b) and (c), respectively. The fractographs in the high ΔK region for all these specimens are reported in Figures 10 (a), (b) and (c).

The fractographs indicate that fatigue crack growth mechanism is mainly by ductile striation. Whereas in high ΔK region, one can observe extensive crack branching, we do not observe these crack branchings in low ΔK region for all these specimens. The fracture surface shows typical faceted regions in all these specimens.

381

Figure 9(a):Fractograph of weld metals in the threshold regio

Figure 9(b): Fractograph of HAZ material in threshold regior

Figure 9(c): Fractograph of base metal in threshold region

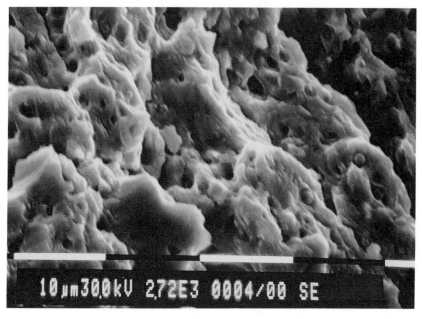

Figure 10(a): Fractograph of the weld metal in high K region

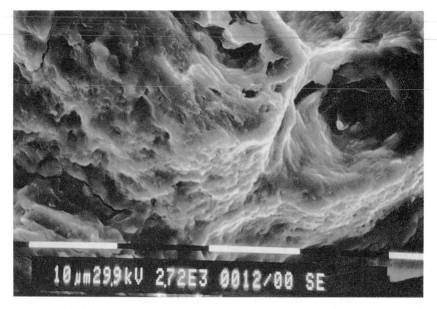

Figure 10(b): Fractograph of the HAZ material in the high
K region

CONCLUSIONS

1. The near threshold fatigue crack growth rate was found to be highest in C2 series of specimens where the crack propagated through heat affected zone.

2. Fatigue threshold was lowest in HAZ material. This is due to the lower cyclic plasticity induced crack closure in this material and presence of residual tensile stresses in the HAZ material.

3. A linear relationship was observed between log C and m.

4. The crack growth process was found to be predominantly by ductile striations. Extensive crack branchings and secondary crackings were observed in high ΔK region but were not seen in threshold region.

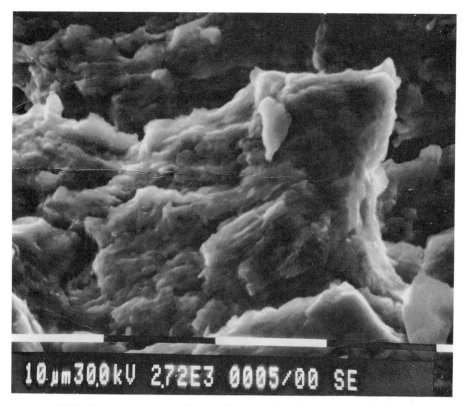

Figure 10(c): Fractograph of base metal in threshold region

REFERENCES

1) P.C. Paris and F. Erdogan, **Journal of Basic Engineering,** Vol. 85, p. 528, 1967.

2) A.S.T.M. E-647, **Annual Book of ASTM Standards,** Vol. 03.01, p. 647, 1988.

3) A. Sengupta, J. Schaefer and S.K. Putatunda, "Fatigue Crack Growth Behavior of HSLA Steel Weldments", **Microstructural Science,** Vol. 22, 1990.

4) G.R. Yoder, L.A. Cooley and T.W. Crooker, "Proceedings of 2nd International Conference on Mechanical Behavior of Materials," **ASM,** Metals Park, Ohio, p. 1010, 1976.

5) G.R. Yoder, L.A. Cooley and T.W. Crooker, **Journal of Engineering Materials Technology,** Vol. 99, p. 313, 1977.

6) G.R. Yoder, L.A. Cooley and T.W. Crooker, **Engineering Fracture Mechanics,** Vol. 11, p. 805, 1979.

7) W. Elber, **Engineering Fracture Mechanics,** Vol. 21, p. 323, 1970.

8) K. Easterling, "Introduction to Physical Metallurgy of Welding", **Butterworth's,** p. 104, 1983.

9) L.N. McCarteney and P.E. Irving, **Scripta Metallurgica,** Vol. 11, p. 181, 1977.

10) E.H. Nicoll, **Scripta Metallurgica,** Vol. 10, p. 295, 1976.

11) J.P. Benson and D.V. Edmonds, **Scripta Metallurgica,** Vol. 12, p. 645, 1978.

EFFECT OF GRAIN SIZE, MICROSTRUCTURE, TEST TEMPERATURE, AND FREQUENCY ON THE LOW CYCLE FATIGUE PROPERTIES OF INCONEL* ALLOY 617

S. K. MANNAN, G. D. SMITH AND R. K. WILSON
INCO ALLOYS INTERNATIONAL
HUNTINGTON, WEST VIRGINIA

ABSTRACT

Tension-tension axial stress-controlled Low Cycle Fatigue (LCF) tests were conducted in air on INCONEL* alloy 617 sheet specimens at 1255K, 1144K, 1033K, and 866K. The stress was varied in sinusoidal form at a rate of 50 cycles and 3.5×10^3 cycles per minute, maintaining a minimum stress of 34.5 MPa. At 866K and 1033K planar slip was the predominant mode of deformation, whereas at 1144K and 1255K both planar and wavy slip as well as grain boundary sliding contributed to the damage accumulation process. The fracture mode was mixed at 866K.

Triple point cavities and grain boundary cracks in the interior of test specimens and close to the surface suggest that at 1144K and 1255K the alloy failed by intergranular fracture. Transmission Electron Microscope (TEM) observation revealed tangled dislocations at 866K, uniform intragranular precipitates on dislocations at 1033K, partial to fully developed subgrains at 1144K, and well defined subgrains at 1255K. The wall thickness of subgrains was thicker for the specimen tested at 3.5×10^3

* INCONEL is a trademark of the Inco family of companies.

cycles per minute at 1144K than for specimens tested at 50 cycles per minute, tested at a comparable stress, whereas the fatigue life of the former was 28 times longer than the latter. The beneficial effects of higher frequency are related to reduced environmental interaction.

Fatigue resistance of alloy 617 is extremely sensitive to grain size. A fatigue life of 10^3 cycles at 866K at a stress of 500 MPa can be attained at a stress of 1000 MPa simply by decreasing the grain size from ASTM 1 to ASTM 9.5.

INTRODUCTION

INCONEL alloy 617 is a solid solution, nickel - chromium - cobalt - molybdenum alloy with a unique combination of high-temperature strength and corrosion resistance[1]. Due to exceptionally good high temperature properties up to 1273K, the alloy is being used in both aircraft and land based gas turbine components, and fossil-fueled and nuclear power plant components.

High temperature characteristics of alloy 617--microstructural stability [2,3], oxidation resistance [3,4], creep [5], and LCF [6,7] -- have been studied to understand the behavior of the alloy in service.

LCF is a rather complex phenomenon. Some of the factors which influence fatigue life include: grain size, strain rate or frequency of loading, shape of loading curve, test temperature, total strain, maximum stress amplitude and environment. Therefore, a large data base is required to understand behavior in service under different types of LCF loading conditions encountered in various applications.

Rao et al [6] have studied LCF of alloy 617 in the temperature range 1023-1223K at strain rates from $4 \times 10^{-3} s^{-1}$ to $4 \times 10^{-6} s^{-1}$. They observed dynamic strain aging at low temperature, high strain rate conditions. Burke and Beck [7] have reported LCF behavior of alloy 617 in a strain controlled mode at a frequency of 0.1 Hz at 1033K and 1144K. They suggested that grain boundary sliding was the primary mode of deformation at 1144K, while at 1033K the alloy deformed mainly by intragranular planar slip. In general, LCF resistance is dramatically influenced by grain size, and the effect of grain size on the LCF properties of alloy 617 has not been thoroughly investigated. Also, data on dislocation substructure strengthening under LCF conditions is scanty. In this investigation, LCF behavior of alloy 617 in the temperature range 866K - 1255K in the axial tension-tension mode is reported. Also, the effect of grain sizes in the range ASTM 1 to ASTM 9.5 on the LCF resistance at 866K and 1033K was investigated. Limited tests were conducted to determine the effect of frequencies (0.8 Hz and 59 Hz) on fatigue resistance at 1144K and 1255K. Further, the dislocation substructures pertaining to a large number of LCF conditions were characterized by TEM and microstructures were correlated with the observed fatigue properties.

EXPERIMENTAL PROCEDURE

The nominal composition of alloy 617 is given in Table I. LCF tests were carried out on annealed 5mm thick sheet specimens. Annealing temperatures, times, and cold rolling reductions were varied to obtain grain sizes in the range ASTM 1 to ASTM 9.5. The test samples were cut transverse to the rolling direction. Specimen dimensions are shown in Figure 1.

Table I. Chemical composition of alloy 617 (wt. %)

Ni	Cr	Co	Mo	Al	Fe	Ti	C
54	22.1	12.7	9.0	1.3	.085	0.06	0.3

Tension-tension axial stress-controlled LCF tests were carried out in air in an MTS closed loop servohydraulic system. The stress was varied in sinusoidal form at 0.83 Hz (50 cycles per minute), maintaining a minimum stress of 34.5 MPa to avoid buckling the sheet specimens. The tests were conducted in a resistance heated furnace. The temperature was measured by Type K thermocouples. The temperature variations were $\pm$ 10°C.

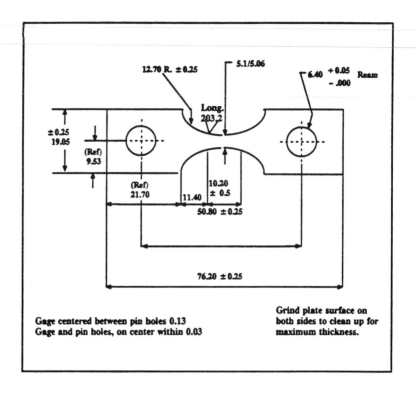

Figure 1 LCF test specimen (dimensions in mm).

Annealed and failed LCF test specimens were analyzed by optical microscopy, Scanning Electron Microscopy (SEM), and TEM. Transverse sections cut 3-5mm from the fracture surfaces were examined optically in the as-polished and two-stage etched conditions. Etching involved immersing a sample in concentrated HCl for 10 seconds followed by 5-10 seconds in 2% bromine in methanol. Specimens were not dried or washed with water after the first stage. Fracture surfaces and etched samples were examined by SEM to determine the fracture mode as well as the size and distribution of second phase particles. To understand the deformation mechanism, gage sections of the failed LCF samples were examined by TEM. Thin foils were prepared by grinding the samples down to 200-250μm followed by electropolishing in a solution of 10 volume percent perchloric acid in methanol using a twin jet apparatus at 233K at a voltage of 35V. The electron transparent regions were examined with a Phillips EM300 TEM at 100KV. The orientation of TEM foils was normal to the loading axis.

RESULTS

1. LCF PROPERTIES

A. LCF of alloy 617 for grain size ASTM 5

Fatigue lives at 1255K, 1144K, 1033K, and 866K as a function of maximum stress are presented in Figure 2. The plots show that, as

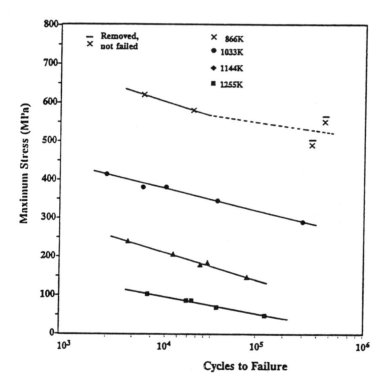

Figure 2   Maximum stress (σ_{max}) vs. cycles - to - failure (N_f) at 50 cycles per minute.

expected, fatigue lives decrease as the maximum stress is increased. Further, the slopes of S-N curves increase as the temperature is increased, indicating that the damage accumulation process is faster at higher temperatures. The damage accumulation process involves the combined effect of temperature, stress, and environment.

B. Effect of grain size

Fatigue lives at 866K as a function of maximum stress for grain sizes ASTM 1, ASTM 5, and ASTM 9.5 are shown in Figure 3. A fatigue life of 10^3 cycles is obtained at 500 MPa for grain size ASTM 1; interestingly, the same fatigue life can be obtained at 1000 MPa by decreasing the grain size to ASTM 9.5. The slopes of S-N curves increase as the grain size in increased, indicating that fine-grained material is more beneficial at

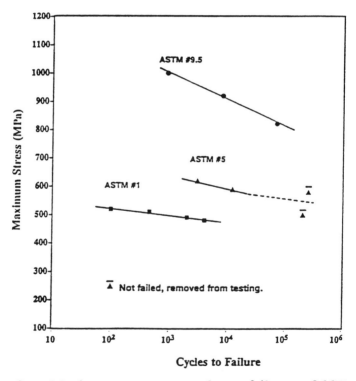

Figure 3   Maximum stress vs. cycles to failure at 866K.

higher stress. The degree of slip homogeneity is likely to be greater in fine-grained material, suggesting that crack initiation will be faster in coarse-grained material due to non-homogeneous deformation.

C. Effect of frequency

Table III summarizes the effect of frequency on fatigue lives at 1144K and 1255K. The data indicates that fatigue lives increase dramatically with increasing frequency at both test temperatures. These observations will be discussed later in term of the combined effect of environment and strain rate at these temperatures.

Table II

Effect of frequency on fatigue lives at 1144K and 1255K.

Temperature K	Maximum Stress MPa	Minimum Stress MPa	Frequency Hz	Cycles to Failures
1144K	241	35	59.0	342,377
	207	35	0.8	9,043
	172	35	59.0	2,269,573
	138	35	0.8	80,462
1255K	103	35	0.8	6,857
	86	35	59.0	1,108,670
	69	35	0.8	38,028

2. MICROSTRUCTURAL CHARACTERIZATION

A. Optical Microscopy

Transverse sections cut 3-5 mm from the fracture surfaces were optically examined. To avoid surface or near-surface effects, all the photomicrographs presented in this section are from the central part of the transverse section.

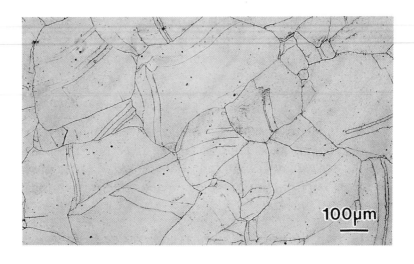

Figure 4a   Optical photomicrographs of specimens LCF tested at 866K, σ_{max}=414 MPa, N_f=13555, Magnification=80x.

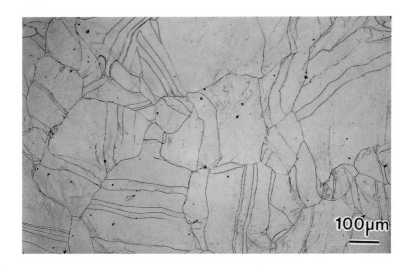

Figure 4b   Optical photomicrographs of specimens LCF tested at 866K, σ_{max}=517 MPa, N_f=182, Maginification=80x.

Coarse grained material (ASTM 1) LCF tested at 866K revealed both higher density and greater bending of twins when tested at maximum stress of 517 MPa as compared to 414 MPa (Figure 4). No inter- or intragranular cavities were observed. Figure 5 reveals the microstructural details of specimens tested at 1033K at various maximum stresses. A few triple-point cavities are observed only at the highest maximum stress (414 MPa). This does not rule out cavity formation at low stress levels because lower stress amplitudes tend to cavitate the material to a smaller thickness underneath the fracture surface. However, it does indicate that the contribution of the intergranular component to the fracture at a maximum stress of 414 MPa is relatively high. On the other hand, formation of a second phase (mainly $M_{23}C_6$ carbides) is more pronounced (Figure 5c) at the lowest stress due to longer fatigue lives.

Figure 5a   Optical photomicrographs of specimens LCF tested at 1033K, σ_{max}=414 MPa, N_f=2393, Magnification=400x.

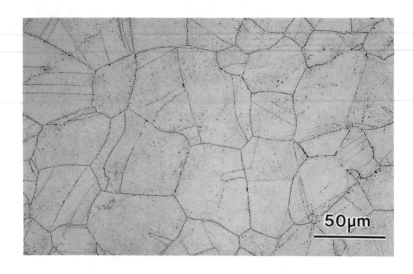

Figure 5b   Optical photomicrographs of specimens LCF tested at 1033K, σ_{max}=345 MPa, N_f=39462, Magnification=400x.

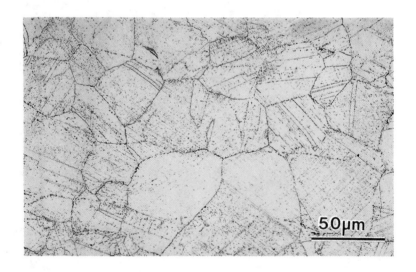

Figure 5c   Optical photomicrographs of specimens LCF tested at 1033K, σ_{max}=276 MPa, N_f=253503, Magnification=400x.

Figure 6a shows grain boundary cracks and triple-point cavities for a specimen LCF tested at 1144K. Extensive grain boundary cracks were observed at all of the test conditions at 1144K and 1255K (Figure 6). Since precipitation kinetics at 1144K and 1255K are quite fast, an attempt was made to reveal the carbide distribution at these temperatures.

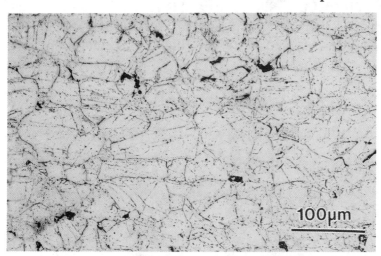

Figure 6a   Optical photomicrograph of LCF tested specimens at 1255, $\sigma_{max}=103$ MPa, $N_f=6857$, Magnification=200x.

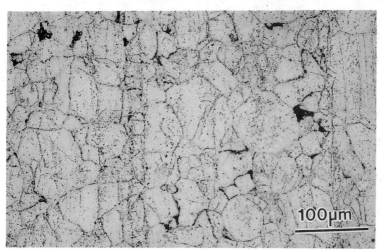

Figure 6b   Optical photomicrograph of LCF tested specimens at 1144K, $\sigma_{max}=138$ MPa, $N_f=80462$, Magnification=200x.

Mounted test specimens were ground down 2-3 mm away from the fracture surfaces to avoid heavily cracked and cavitated surfaces; they were then lightly etched to reveal carbides.   At 1255K, carbide precipitates form a continuous network at the grain boundaries (Figure 7a), whereas at 1144K only a few grain boundaries show the continuous network (Figure 7b).   Slip bands at 1144K are shown by arrow marks. Further, recrystallized grains and extensive distortion of prior grain boundaries are observed at 1255K (Figure 7a), indicating the onset of dynamic recrystallization.

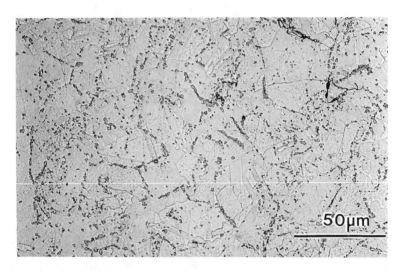

Figure 7a   Optical photomicrographs of LCF tested specimens at 1255K, $\sigma_{max} = 103$ MPa, $N_f = 6857$, Magnification = 500x.

B. Scanning Electron Microscopy

Figure 8 shows inter- and intragranular small cavities for specimens tested at 866K.   This indicates that under these test conditions, the fracture mode is mixed.   The precipitate shape, size, and distribution at 1144K are shown in Figure 9.

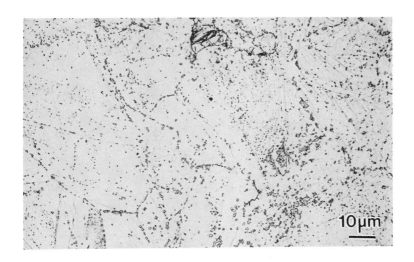

Figure 7b   Optical photomicrographs of LCF tested specimens at 1144K, σ_{max}=207 MPa, N_f=9043, Magnification=800x.

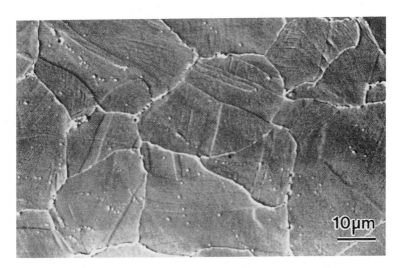

Figure 8   SEM photomicrograph of specimen LCF tested at 866K, σ_{max}=586, N_f=11043, Magnification=1000x.

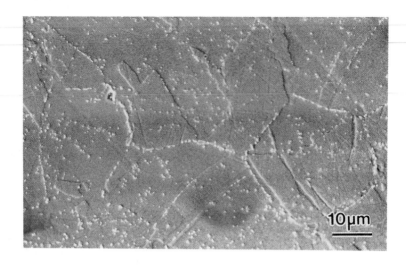

Figure 9a   SEM photomicrograph of specimen LCF tested at 1144K, $\sigma_{max} = 138$ MPa, $N_f = 80462$, Magnification $= 1000$x.

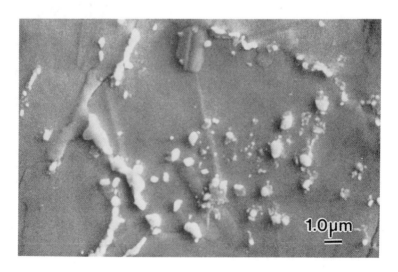

Figure 9b   SEM photomicrograph of specimen LCF tested at 1144k, $\sigma_{max} = 138$ MPa, $N_f = 80462$, Magnification $= 4800$x.

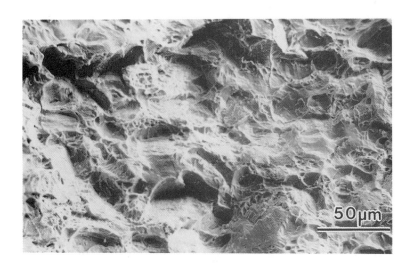

Figure 10a   Fracture surfaces of specimen for grain size ASTM #5 at
866K, σ_{max}=586 MPa, N_f=11043, Magnification=400x

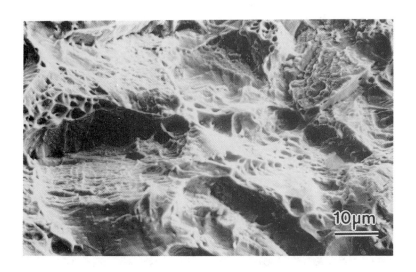

Figure 10b   Fracture surfaces of specimen for grain size ASTM #5 at
866K, σ_{max}=586 MPa, N_f=11043, Magnification=1200x

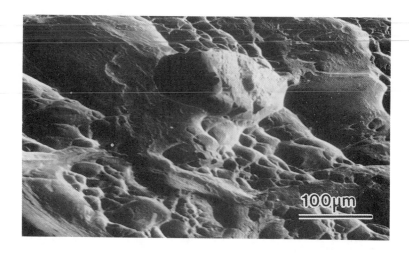

Figure 11a   Fracture surfaces of specimen for grain size ASTM #1 at 866K, σ_{max}=483 MPa, N_f=5019, Magnification=200x.

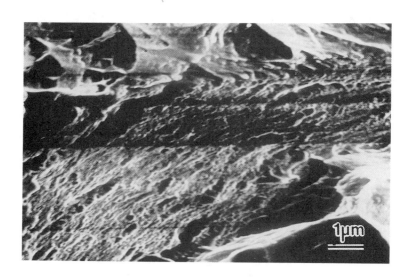

Figure 11b   Fracture surfaces of specimen for grain size ASTM #1 at 866K, σ_{max}=483 MPa, N_f=5019, Magnification=1000x.

Fracture surfaces of specimens for grain sizes ASTM 5 and ASTM 1, LCF tested at 866K indicate that the fracture modes in both samples are mixed inter- and intragranular (Figure 10 and 11). Both reveal striations at high magnifications, which are characteristics of planar slip. Well-defined slip bands are shown in Figure 11b. Higher test temperature fracture surfaces could not be analyzed since these were oxidized.

C. Transmission Electron Microscopy

Figure 12 shows alloy 617 in the as-annealed condition for grain size ASTM 5. It reveals geometric necessity dislocations around Ti(C,N) particles (Fig. 12a). The presence of Ti-rich particles in the as-annealed condition was confirmed by SEM/EDX analysis. Additionally, the annealed specimens have a few grown-in or accidentally introduced dislocations (Figure 12b). Figure 13a shows dense dislocation bands in LCF tested specimens at 866K and at 586 MPa. Similar observations are reported by Feltner and Laird on a Cu-Al alloy under LCF conditions [8]. The dislocation structures within these bands are shown clearly at higher magnification in Figures 13b. Dislocations tend to pile up on the slip bands. It appears that dislocation mobility is rather low at 866K.

Slip band spacings of three different magnitudes in the specimen subjected to LCF testing at 552 MPa are shown is Figure 14. This specimen survived 487, 925 cycles and was removed prior to fracture. The slip bands shown in Figure 14b were within the broad slip bands shown clearly in 14a. Similar observations are reported by Sanders et al [9] in alloy 718 under LCF conditions. Figure 14c shows slip band spacings finer than shown in Figures 14a and 14b in another grain. Wide variations in slip band spacing were observed from one grain to another in a thin foil and in different thin foils examined across the gage length. These observations indicate that planar slip is the primary mode of deformation at 866K under these LCF conditions. Different grains in polycrystalline material would deform to different extents depending upon the grain orientation with respect to the stress axis, resulting in wide variation in slip band spacings.

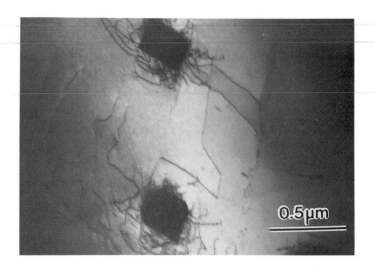

Figure 12a   TEM photomicrograph of the as annealed alloy 617. Note the dislocations around Ti(CN) particles.

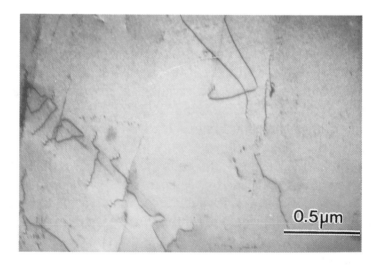

Figure 12b   TEM photomicrograph of the as annealed alloy 617.

Figure 13a   TEM photomicrograph of specimen LCF tested at 866K,
σ_{max}=586 MPa, and N_f=11043.

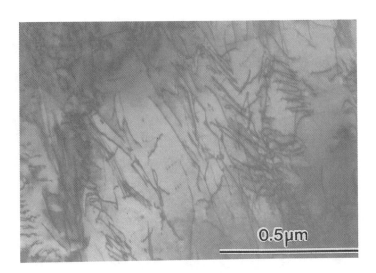

Figure 13b   TEM photomicrograph of specimen LCF tested at 866K,
σ_{max}=586 MPa, and N_f=11043.

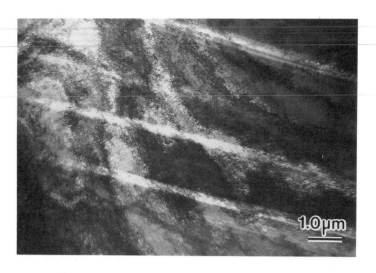

Figure 14a   TEM photomicrographs of specimen LCF tested at 866K
σ_{max}=552MPa, and N_f > 487925.

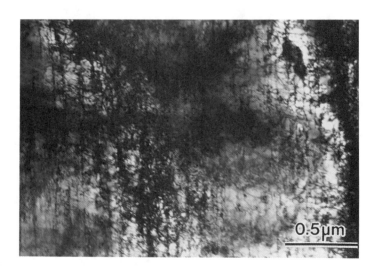

Figure 14b   TEM photomicrographs of specimen LCF tested at 866K
σ_{max}=552MPa, and N_f > 487925.

Figure 14c   TEM photomicrographs of specimen LCF tested at 866K
σ_{max}=552MPa, and N_f > 487925.

Figure 15a shows the dislocation tangles in coarse grained (ASTM 1) material tested at 866K. Rather closely spaced slip bands are shown at higher magnification (Figure 15b). Slip bands in this coarse grained material were not as prevalent as in fine grained (ASTM 5) material.

At 1033K, a LCF tested specimen revealed fine matrix precipitates on the dislocations (Figure 16a). These precipitates are likely to immobilize the moving dislocations. Figure 16b shows decoration of slip bands by fine precipitates and Figure 16c reveals the strain fields associated with dislocations. Lower stress amplitude LCF conditions at 1033K result in larger matrix precipitates (Figure 17a). Precipitates tend to form a continuous network on slip bands (Figure 17b). No subgrain formation was observed. Similar observations have been reported by Rao et al [6] under strain controlled LCF conditions at 1023K and at a strain rate of 6×10^{-5} s^{-1}. Earlier studies have shown the formation of $M_{23}C_6$ and M_6C type carbides under isothermal aging [3] and LCF conditions [6] in alloy 617 at 1033K and 1023K respectively.

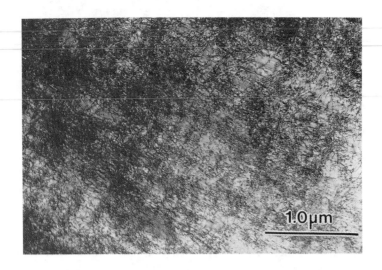

Figure 15a   TEM photomicrographs of specimen LCF tested at 866K,
σ_{max}=483 MPa, N_f=5019, and grain size ASTM #1.

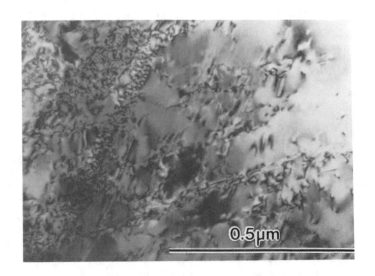

Figure 15b   TEM photomicrographs of specimen LCF tested at 866K,
σ_{max}=483 MPa, N_f=5019, and grain size ASTM #1.

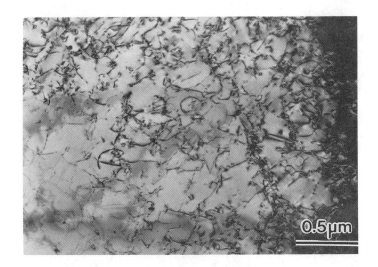

Figure 16a   TEM photomicrographs of specimen LCF tested at 1033K,
σ_{max}=414 MPa, and N_f=2393.

Figure 16b   TEM photomicrographs of specimen LCF tested at 1033K,
σ_{max}=414 MPa, and N_f=2393.

Figure 16c   TEM photomicrographs of specimen LCF tested at 1033K, σ_{max}=414 MPa, and N_f=2393.

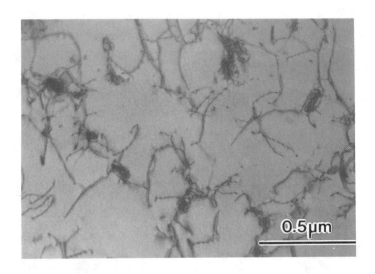

Figure 17a   TEM photomicrographs of specimen LCF tested at 1033K, σ_{max}=345 MPa, and N_f=39462.

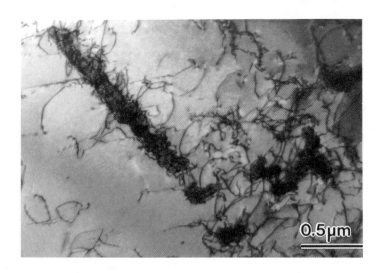

Figure 17b   TEM photomicrographs of specimen LCF tested at 1033K,
            σ_{max}=345 MPa, and N_f=39462.

Figures 18a and 18b show subgrain formation at 1144K. Subgrains
have high aspect ratios, suggesting that these may have been formed on
the slip bands. Carbide particles decorating a subgrain boundary and
subgrain formation within larger subgrains are shown in Figure 18b.
These photographs indicate that there are large variations in the subgrain
size, which is expected in a polycrystalline material under LCF conditions
due to the random grain orientation. Figure 18c shows a continuous rod-
type carbide particle on the slip band. Recrystallized nuclei at the
intersection of slip bands (arrow marks in Figure 18c) suggest the onset
of dynamic recrystallization. Optical and SEM observations did not
indicate the onset of dynamic recrystallization at 1144K due to the small
size of nuclei. Figures 18a, 18d, 18e and 18f indicate partially developed
subgrains, carbide precipitation, and wide variations in shape and size of
subgrains.

411

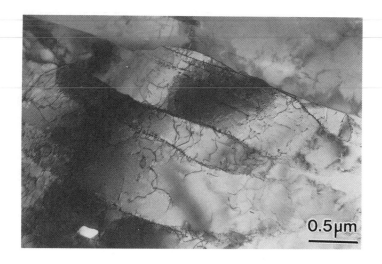

Figure 18a   TEM photomicrographs of specimen LCF tested at 1144K, σ_{max}=207 MPa, and N_f=9043.

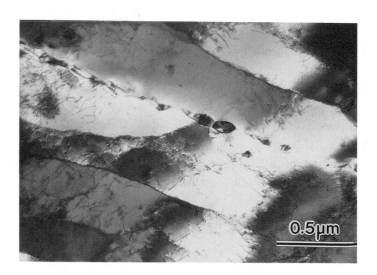

Figure 18b   TEM photomicrographs of specimen LCF tested at 1144K, σ_{max}=207 MPa, and N_f=9043.

Figure 18c   TEM photomicrographs of specimen LCF tested at 1144K, σ_{max}=207 MPa, and N_f=9043. Note the recrystallized nuclei.

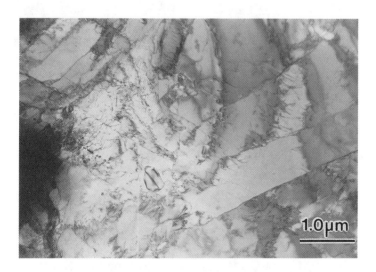

Figure 18d   TEM photomicrographs of specimen LCF tested at 1144K, σ_{max}=207 MPa, and N_f=9043.

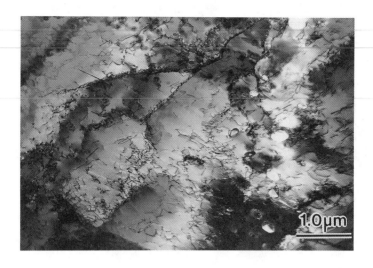

Figure 18e   TEM photomicrographs of specimen LCF tested at 1144K,
σ_{max}=207 MPa, and N_f=9043.

Figure 18f   TEM photomicrographs of specimen LCF tested at 1144K,
σ_{max}=207 MPa, and N_f=9043.

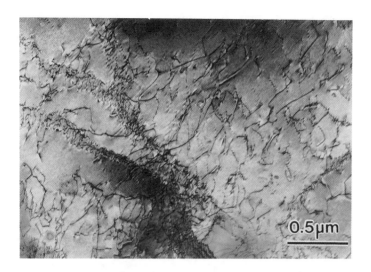

Figure 18g   TEM photomicrographs of specimen LCF tested at 1144K, σ_{max}=207 MPa, and N_f=9043.

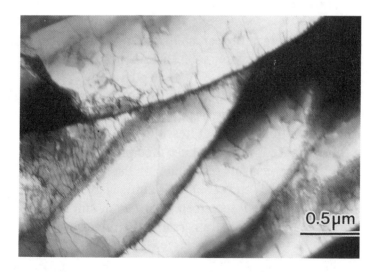

Figure 19a   TEM photomicrographs of specimen LCF tested at 1144K, σ_{max}=172 MPa, N_f=2269573, and frequency = 3.5 x 10^3 cycles per minute.

Figure 19b   TEM photomicrographs of specimen LCF tested at 1144K, σ_{max}=172 MPa, N_f=2269573, and frequency = 3.5 x 10^3 cycles per minute.

The activation energy for recrystallization generally increases with the alloy content due to dislocation solute interactions, and this phenomenon is reported to be true in the case of Fe-Cr-Ni alloys [10]. Atomic level compositional variations in heavily alloyed polycrystalline alloy 617, LCF tested in the carbide formation region, are likely to result in the wide variations in dislocation substructures. Dislocations at certain locations are likely to be pinned by the carbide precipitates, inhibiting dislocation mobility, and consequently resulting in the partial development of subgrains (Figures 18d and 18e). Figure 18g shows subgrain wall thickness in a LCF tested specimen. The wall thickness of subgrains depends on test temperature, strain rate and strain amplitude[10].

The change in test frequency of the sinusoidal wave in the axial tension-tension LCF conditions would change the effective strain rate, which in turn could change the dislocation substructure. Figure 19a shows subgrain wall thickness in a LCF tested specimen at 1144K at a frequency of 59 Hz. It should be mentioned that the rest of the data presented in the TEM characterization section pertains to a test frequency of 0.8 Hz. The wall thickness of subgrains in a specimen LCF tested at a frequency of 59 Hz is thicker than that of a specimen tested at 0.8 Hz at 1144K and at comparable stress amplitude. Further, higher frequency test specimens revealed more slip bands per unit area than for low frequency test specimens. Figure 19b shows a typical region for slip bands for a 59 Hz specimen. These observations indicate that, at 1144K under high frequency (59 Hz) LCF conditions, the deformation mode is mixed, consisting of planar and wavy slip. On decreasing the frequency to 0.8 Hz, the contribution of wavy slip to the damage accumulation process is increased.

Subgrain formation at 1255K at the maximum stress of 103 MPa is shown in Figure 20. Figure 20a shows slip bands on the subgrains, indicating that even at 1255K (highest test temperature) planar slip contributes to the damage accumulation process. A continuous network of carbides on the slip bands is shown in Figures 20b and 20c. A number of dislocations crossing a subgrain boundary are shown by the arrow marks in Figure 20d. McQueen and Hockett have reported that, in hot

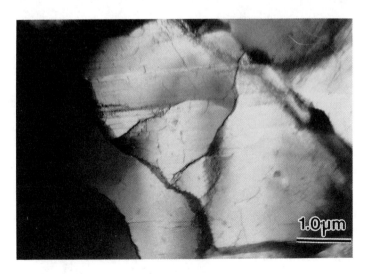

Figure 20a   TEM photomicrographs of specimen LCF test at 1255K, $\sigma_{max} = 103$ MPa, and $N_f = 6857$.

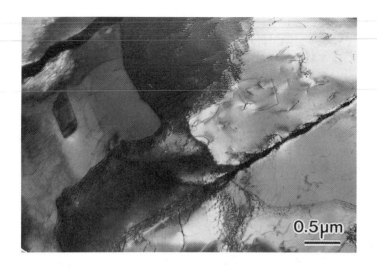

Figure 20b   TEM photomicrographs of specimen LCF test at 1255K,
σ_{max}=103 MPa, and N_f=6857.

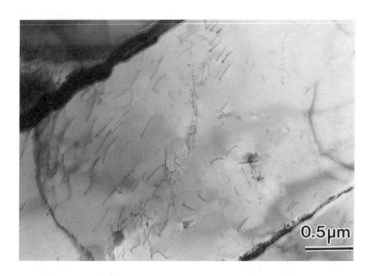

Figure 20c   TEM photomicrographs of specimen LCF test at 1255K,
σ_{max}=103 MPa, and N_f=6857.

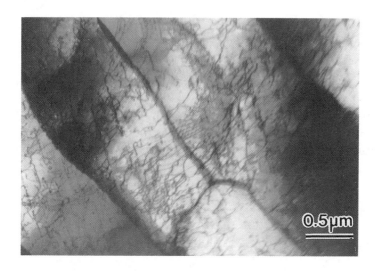

Figure 20d   TEM photomicrographs of specimen LCF test at 1255K, $\sigma_{max} = 103$ MPa, and $N_f = 6857$.

worked aluminum, increasing the temperature or decreasing the strain rate results in less dense and more regular cell walls [11]. Therefore, in general the sub-boundaries produced at higher temperatures and lower strain rates can thus be expected to have decreasing ability to impede dislocation motion [10].

DISCUSSION

The axial tension-tension S-N plots (Figure 2) indicate that fatigue resistance decreases as the temperature is increased. Test temperature is the most important variable since the contribution of various damage accumulation processes -- grain boundary sliding and diffusion assisted deformation, deformation induced by slip, and oxidation -- depends on temperature. Further, carbide precipitation, which indirectly affects the damage accumulation process, depends primarily on the test temperature. Therefore, the correlation between LCF behavior and microstructure will be discussed separately for different test temperatures.

LCF at 866K: Inter- and intragranular cavities in polished sections (Figure 8) as well as fracture surface analysis indicate that the fracture mode at 866K is mixed. Coarse grained (ASTM 1) material may have a larger intragranular component due to the deformation facilitated by twins. TEM observations reveal the presence of tangled dislocations and slip bands, indicating that planar slip is the predominant mode of deformation. Similar observations have been reported for alloy 718 under strain controlled high temperature LCF conditions [9]. No microstructural features could be detected to explain the large difference in fatigue life at 586 MPa and 552 MPa.

LCF at 1033K: R-type triple point cavities [12] were observed only at a maximum stress of 414 MPa. TEM analysis revealed the presence of carbides (mainly $M_{23}C_6$ and M_6C) in the matrix, slip bands, and grain boundaries. The precipitates are uniformly distributed in the matrix. Lipscomb et al [3] have reported a Time-Temperature-Transformation (TTT) diagram for alloy 617, indicating that precipitation kinetics are fastest at 1033K, i.e., 1033K corresponds to the nose of the C-curve for the TTT diagram. This implies that both nucleation and growth of precipitates are fastest at 1033K. In general for a typical C-type TTT diagram, nucleation is fast but growth is slow at temperatures lower than the nose temperature, while at temperatures higher than the nose temperature, nucleation is slower but growth is faster. A combination of rapid nucleation and growth at 1033K results in uniformly distributed intragranular precipitates. Further, under the influence of cyclic stress, growth kinetics of precipitates are likely to be enhanced due to the generation of heterogeneous nucleation sites, such as dislocations and faults, and perhaps also to the accelerated diffusion of solute atoms by the nonequilibrium vacancies produced.

The uniform intragranular precipitates on dislocations at 1033K are likely to pin dislocations, resulting in a reduction in dislocation pile-up at the grain boundaries. A negligible total elongation to fracture, no obvious necking, no surface cracking, and interganular R-type cavities all indicate that the material at 1033K may have failed by intergranular brittle fracture. The fracture mode could not be confirmed because the fracture surfaces were oxidized. Kimball et al [13] have reported the results of room temperature impact testing after thermal aging of alloy 617 in the range 866K to 1033K for durations up to 8000 h. They showed a minimum in impact energy for the samples aged at 983 to 1033K. Rao et al [6] have shown that alloy 617 fails by intergranular brittle fracture at 1023K under strain-controlled LCF conditions at a strain rate of 4×10^{-5} s^{-1}. The authors assumed that heavy strengthening of the matrix due to precipitation of fine $M_{23}C_6$ carbides on matrix dislocations causes the fracture of weak and less ductile grain boundary carbide film formed at 1023K, leading to brittle interganular fracture.

LCF at 1144K: Extensive grain boundary cavities and cracks at 1144K suggest that under the examined LCF conditions, creep-fatigue interactions are significant. At high temperature, cavity formation involves clustering of vacancies by diffusion to form a void cluster under the influence of stress. The cavities are likely to form at second-phase particles, in grain boundaries at the perimeter of contact between particles and grain boundaries, and at intragranular particle-matrix interfaces [12]. It was not possible to determine the mode of fracture due to oxidation of the fracture surfaces. Optical microscopy revealed slip bands, indicating that planar slip contributes to the damage accumulation process.

The presence of subgrains in specimens tested at 1144K indicates that creep is quite effective at this temperature. The dislocation substructure in specimens tested at 1144K was much cleaner than comparable specimens tested at 1033K due to rearrangement of dislocations in the form of subgrains, permitting fewer heterogeneous sites for nucleation of precipitates within the grains. On the other hand, the growth rate of precipitates at 1144K is expected to be higher than at 1033K. Therefore, at 1144K rod-type intragranular precipitates with a smaller number density but a larger size than those at 1033K were observed.

Rao et al [6] have reported that alloy 617 deformed by planar slip at 1123K under fully reversed axial strain controlled LCF conditions using symmetrical triangular strain wave cycles at strain rates of 4 x 10^{-3} s^{-1}, 4 x 10^{-4} s^{-1}, and 4 x 10^{-5} s^{-1}, but that sub-boundaries were observed only at a strain rate of 4 x 10^{-6} s^{-1}. Also, Burke and Beck [7] have shown that, under fully reversed constant strain amplitude LCF tests conducted at 0.1 Hz under the triangular wave form, sub-boundaries were observed in alloy 617 tested at 1144K but not subgrains. The former study was carried out on rod type specimens whereas the latter study was carried out on thin sheet specimens supported by guiding plates to prevent buckling. In the present study, carried out in the axial tension-tension load-controlled mode employing a sinusoidal wave form at a frequency of 0.8 Hz, subgrains of different size and shape were observed at 1144K. However, a minimum tensile stress of 34.5 MPa was maintained on the sheet specimen to prevent buckling. A constant minimum stress may have enhanced the creep contribution in the present study.

LCF at 1255K: Optical microscopy revealed extensive grain boundary cracks and cavities at 1255K. Recrystallized grains were observed, indicating the onset of dynamic recrystallization. A few small recrystallized grain nuclei were observed even at lower temperature (1144K) but only during TEM analysis. SEM analysis of 1255K tested specimens revealed fewer but larger intragranular carbides than at 1144K. Earlier studies of the effects on alloy 617 of isothermal aging at 1255K [3], creep at 1273K[5], and LCF at 1223K[6] have shown that chromium-rich $M_{23}C_6$ and molybdenum-rich M_6C type carbides co-exist at these temperatures. In the present study the existence of chromium-rich and molybdenum-rich precipitates was confirmed by SEM/EDX analysis, but their relative abundance at different maximum stress amplitudes could not be established. Figure 8a shows prior grain boundaries fully covered with a continuous network of coarse carbides, some grain boundaries shifted from their original positions and completely free from carbides and new recrystallized grains. Similar observations have been reported by Kihara et al for alloy 617 under creep conditions at 1273K [5]. The authors have proposed that grain boundary carbides migrate from grain boundaries that are under compressive stress to those under tensile stress.

This is accomplished by dissolution of carbides at the compressive boundary region followed by precipitation and growth of carbides at the tensile boundary region. When the population of carbides becomes low enough at a compressive boundary, as a result of grain boundary carbide dissolution, the grain boundary itself starts to migrate.

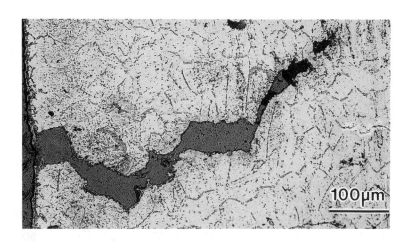

Figure 21a   Longitudinal cross section for LCF tested specimens at 1255K, σ_{max} = 103 MPa, N_f = 6857.

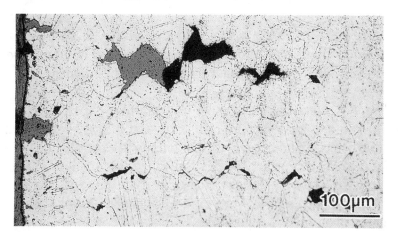

Figure 21b   Longitudinal cross section for LCF tested specimens at 1144K, σ_{max} = 138 MPa, N_f = 80462

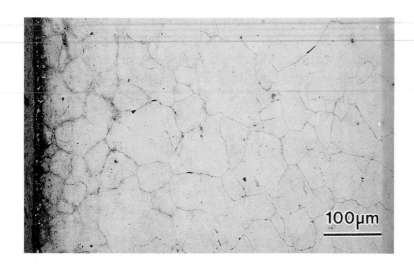

Figure 21c   Longitudinal cross section for LCF tested specimens at 1033K, σ_{max}=276 MPa, N_f=253503.

<u>Effect of Environment on LCF</u>:   To find out the effect of environment, longitudinal cross sections of failed LCF test specimens were polished and etched. Figure 21 shows grain boundary cracks at 1144K and 1255K, whereas at 1033K intergranular micro-cracks were observed. The presence of intergranular cracks near the surface and in the interior of LCF failed specimens at 1144K and 1255K suggests that the fracture mode at these temperatures is likely to be intergranular. Environmental contribution to intergranular fracture is more severe due to the existing interconnected path formed by grain boundaries.

Coffin has conducted extensive comparative experiments on LCF behavior in air and high vacuum (10^{-8} torr) --on A286, Nickel A and type 304 stainless steel -- and has shown that fatigue lives dramatically improve under vacuum [14]. Further, he has shown that the fracture mode at high temperature in some materials changes from intergranular in air to transgranular in vacuum. For pre-exposed Rene 80, Antolovich et al [15] have shown that high temperature LCF failure was related to grain boundary embrittlement due to oxidation. Since the LCF behavior of alloy 617 in vacuum is not known, the effect of environment can not be determined accurately. It is clear from Figure 25 that environment - associated damage increases as the test temperature is increased, which is expected due to increased oxidation at higher temperatures.

Effect of frequency on LCF: Table III indicates that, on increasing the frequency from 0.8 Hz to 59 Hz, fatigue lives at 1144K and 1255K increase dramatically at comparable stress amplitudes. Coffin has reported increases in fatigue lives for A286 and change in the fracture mode from intergranular to transgranular at higher test frequencies in high temperature LCF testing [15]. The beneficial effects of higher frequencies are explained based on the two factors [16]: 1) the assumption that frequency and strain rate are directly related, 2) at higher frequencies, environment-crack interaction time is reduced.

CONCLUSIONS

1. Under high temperature LCF conditions, planar slip is the predominant mode of deformation at 866K and 1033K.

2. At 1144K and 1255K both planar and wavy slip as well as grain boundary sliding contribute to the damage accumulation process. The crack initiation was assisted by grain boundary oxidation of the surface exposed to the air.

3. The fracture mode at 866K is mixed (inter- and intragranular), and 1033K, 1144K and 1244K is likely to be intergranular.

4. Necking and elongation to failure was lowest at 1033K, indicating that ductility is minimum at 1033K.

5. TEM analysis indicates the presence of fine uniformly distributed intragranular precipitates at 1033K; at higher temperatures (1144K and

1255K) a smaller number of coarse precipitates was seen; and at lower temperature (866K) a few isolated fine precipitates were observed under LCF conditions.

6.   Subgrains were not observed under LCF conditions at 866K and 1033K. At 1144K, partially to fully developed subgrains of wide variation in shape and size were present. At 1255K well defined subgrains were observed.

7.   High temperature LCF resistance of alloy 617 is extremely sensitive to grain size. A stress amplitude of 500 MPa for grain size ASTM #1 provides a fatigue life of 10^3 at cycles at 866K; the same fatigue life can be obtained at double the amount of stress, 1000 MPa, by refining the grain size to ASTM #9.5.

8.   Increasing the frequency of loading dramatically increases the LCF resistance of alloy 617 at 1144K and 1255K. A fatigue life of 80,464 cycles was obtained at 0.8 Hz at a stress amplitude of 138 MPa at 1144K. The fatigue life increased to 2,269,573 cycles at 59 Hz even at a higher stress amplitude (172 MPa).

REFERENCES

1.   INCONEL alloy 617 Data Sheet, Huntington Alloys, Huntington, WV, 1979.

2.   W. L. Mankins, J. C. Hosier, and T. H. Bassford, "Microstructure and Phase Stability of INCONEL Alloy 617", Metallurgical Transactions, Vol. 5, p. 2579, 1974.

3.   W. G. Lipscomb, J. R. Crum, and P. Ganesan, "Mechanical Properties and Corrosion Resistance of INCONEL Alloy 617 for Refinery Service", Paper No. 259 Corrosion '89, NACE Houston, Texas.

4.   P. Ganesan and G. D. Smith, "Oxide Scale Formation on Selected Candidate Combustor Alloys in Simulated Gas Turbine Environments", Journal of Materials Engineering, Vol 9 No. 4, p. 337, 1988.

5.   S. Kihara, J. B. Newkirk, A. Ohtomo, and Y. Saiga, "Morphological Changes of Carbides During Creep and Their Effects on the Creep Properties of INCONEL alloy 617 at 1000°C", Metallurgical Transactions, Vol. 11A, p. 1019, 1980.

6.   K. B. S. Rao, H. Schiffers, H. Schuster, and H. Nickel, "Influence of Time and Temperature Dependent Processes on Strain Controlled Low Cycle Fatigue Behavior of Alloy 617", Metallurgical Transactions, Vol. 19A, p. 359, 1988.

7.   M. A. Burke, and C. G. Beck, "The High Temperture Low Cycle Fatigue Behavior of Nickel Base Alloy IN-617", Metallurgical Transactions, Vol. 15A, p. 661, 1984.

8.   C. E. Feltner and C. Laird, "Cycle Stress-Strain Response of F.C.C. Metals and Alloys-II", Acta Metallurgica, Vol. 15, p. 1633, 1967.

9.   T. H. Sanders, Jr., R. E. Frishmuth and G. T. Embley, "Temperature Dependent Deformation Mechanisms of Alloy 718 in Low Cycle Fatigue", in Fatigue Environment and Temperature Effects , (Edited By J. J. Burke and V. Wiss), P. 163, Plenum Press, New York, 1980.

10.   H. J. McQueen, "The Production and Utility of Recovered Dislocation Substructures", Metallurgical Transactions, Vol. 8A, p. 807, 1977

11.   H. J. McQueen, and J. E. Hockett, "Microstructures of Aluminum Compressed at Various Rates and Temperatures" Metallurgical Transactions, Vol. 1A, P. 2887, 1970.

12.	R. Raj, "Mechanisms of Creep-Fatigue Interaction", in Flow and Fracture at Elevated Temperatures, (Edited by R. Raj), p. 215, American Society for Metals, Metals Park, Ohio, 1983.

13.	O. F. Kimball, G. Y. Lai, and G. H. Reynolds, "Effect of Thermal Aging on the Microstructure and Mechanical Properties of a Commercial Ni-Cr-Co-Mo Alloy (INCONEL 617), Metallurgical Transactions, Vol 7A, p. 1951, 1976.

14.	L. F. Coffin, Jr, "The Effect of High Vacuum on the Low Cycle Fatigue Law", Metallurgical Transactions, Vol. 3A, . 1777, 1971.

15.	S. D. Antolovich, P. Domas, J. L. Studel, "Low Cycle Fatigue of Rene 80 as Affected by Prior Exposure", Metallurgical Transactions, Vol. 10A, p. 1859, 1979.

16.	L. F. Coffin, "Overview of Temperature and Environmental Effect of Fatigue of Structural Metals", in Fatigue Environment and Temperature Effects, (Edited by J. J. Burke an V. Weiss), p. 13, Plenum Press, New York, 1980.

Wear

METALLOGRAPHY OF SLIDING WEAR IN SiC REINFORCED

ALUMINUM MATRIX COMPOSITES

A.T. Alpas[1]

ABSTRACT

Dry sliding wear properties of wrought (2014) and cast (A356) aluminum alloys reinforced with SiC particles were investigated by means of a block-on-disc type wear machine using normal loads of 0.9-150N and sliding velocities of 0.16 and 0.8 ms^{-1}. Metallographic studies were performed to determine the role of SiC particles on the wear mechanisms that control the wear resistance of composites. The results indicate that at applied normal loads, corresponding to stresses lower than their fracture strength, Si particles can act as load bearing elements and their abrasive action on the steel counterfaces causes transfer of iron oxide layers onto the contact surfaces of composites. Wear proceeds by spalling and reformation of these layers that also act as solid lubricants resulting in a tenfold increase in the wear resistance of the composites relative to unreinforced aluminum alloys. At higher loads (> 10N) SiC at the contact surfaces fracture and wear proceeds by a subsurface delamination process that involves decohesion of carbide matrix interfaces resulting in wear rates similar to those observed in the unreinforced alloys. Wear rates of unreinforced alloys increase by two orders of magnitude above a critical load (95N). SiC reinforcement impedes the transition to high wear regime.

---

[1]Engineering Materials Group, Department of Mechanical Engineering, University of Windsor, Windsor, Ontario, Canada N9B 3P4.

INTRODUCTION

Ceramic particulate reinforced metal matrix composites provide a possibility of designing lightweight structural materials with enhanced wear resistance. This is particularly applicable in the case of aluminum alloys reinforced by hard particles such as SiC and Al_2O_3 [1-3]. However, the micromechanisms that control the wear rates of composites have not been understood at a sufficient detail to predict the wear resistance of these materials under various operating conditions. In addition to the volume fraction, size, distribution of second phase particles the wear resistance of particulate composites is influenced by the ductility and the strength of the matrix phases as well as the mechanical properties of the interfaces. Depending on the magnitude and rate of applied load, particles may detach from the surfaces or may fracture [4,5]. Particles such as graphite or mica can act as solid lubricants [6,7] while hard constituents may enhance abrasive wear processes. The objective of the current study is to investigate the role of SiC particles on dry sliding wear behaviour of wrought and cast aluminum alloys under a broad range of applied loads. Sliding wear tests are performed in order to provide a direct comparison of wear mechanisms in carbide reinforced and unreinforced aluminum alloys.

EXPERIMENTAL DETAILS

Sliding wear tests were performed on wrought (2014) and cast (A356) aluminum matrix composites reinforced with 20 percent (by volume) SiC particles. Chemical compositions of the matrix alloys are given in Table 1. Both composites were hot extruded and heat treated to T6 condition prior to wear tests.

Microstructures of the composites are shown in Figure 1. The average SiC particle sizes in cast and wrought alloys were 10 ± 2 μm and 14 ± 3 μm respectively. A356-20%SiC contained also a fine dispersion (1.8 ± 0.9 μm) of silicon particles. The extrusion process induced directionality into the SiC particle distribution and carbide-rich bands were formed in the extrusion direction (Figure 1(a)). Wear tests were also performed on unreinforced 2024 and A356 alloys to clarify the effect of SiC reinforcement on the wear resistance of aluminum alloys.

A block-on-ring type wear machine was used for dry sliding wear tests under normal loads 0.9-150N and sliding velocities 0.16 and 0.8 ms^{-1}. Narrow surfaces of rectangular specimens (5 x 10 x 10 mm rectangular prisms) were worn against a SAE 52100 bearing steel slider so that the sliding direction was

Table 1 Chemical Composition (wt%) of the Matrices and Heat Treatment of Composites

Alloy	Si	Mg	Cu	Zn	Fe	Cr	Ti
A356	7.00	0.40	<0.01	<0.01	0.08	--	--
2014	1.00	0.6	4.00	0.25	0.7	0.1	0.15
2024	0.5	1.6	4.50	0.25	0.5	0.1	0.15

A356-20%SiC was solution treated at 540°C for 4 hours and age-hardened at 155°C for 9 hours; 2014-20%SiC was solution treated at 505°C for 1 hour and age-hardened at 175°C for 10 hours; 2024 was solution treated at 495°C for 1 hour and age-hardened at 190°C for 11 hours.

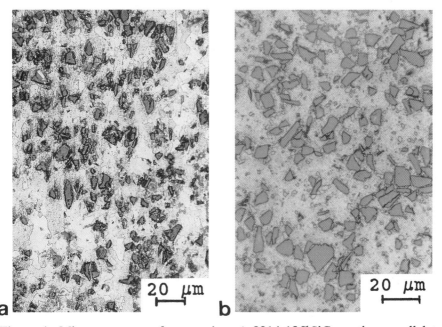

Figure 1. Microstructures of composites a) 2014-10%SiC, section parallel to extrusion direction; b) A356-20%SiC, section perpendicular to extrusion direction (silicon particles (1.8 μm diameter) are light-coloured small constituents in the matrix).

perpendicular to the extrusion direction of the composites. Prior to each test specimen surfaces were polished, ultrasonically cleaned and weighed to nearest 0.1 mg using an analytical balance. The weight losses due to wear were calculated from the differences in weight measured before and after each test.

X-ray diffraction and Energy Dispersive X-ray Analyses (EDXA) were performed to determine the changes in chemical composition of the worn surface layers and loose debris particles. Scanning electron microscopy (SEM) was used to study the microstructural, morphological features of the worn surfaces and material layers beneath these surfaces.

RESULTS

Wear tests were carried out at various constant load levels between 0.9-150N. At each load level the weight losses from the surfaces of SiC reinforced and unreinforced alloys were determined as a function of sliding distance. The wear rates were calculated from the slope of weight loss versus sliding distance curves. The results revealed that the effect of SiC particles on the wear rates varied with the applied load. The weight loss versus sliding distance curves for unreinforced and SiC reinforced alloys at 0.9N, 17N and 98N are shown in Figure 2. According to the differences in the wear rates of the unreinforced and reinforced alloys three distinct wear regimes were identified:

a) At low loads lower than about 10N the wear rates of the SiC reinforced composites were found to be an order of magnitude lower than those of the base alloys. The worn surfaces of composites that exhibited low wear rates ($\dot{W} \approx 10^{-7}$ g.m^{-1}) were characterized by the formation of iron oxide rich transfer layers. X-ray diffractometry indicated that about 80-85% of these layers consisted of a mixture of α-Fe$_2$O$_3$ and Fe$_3$O$_4$, the rest being γ-Al$_2$O$_3$ combined with silicon. The morphology of the wear scar in A356-20%SiC worn at a rate of 1.2×10^{-7} g.m^{-1} is shown in Figure 3. The thickness of iron rich layers was of the order of 50-20 μm. The underlying matrix structure was almost damage-free and the carbide particles on the contact surfaces remained unbroken during the sliding wear.

b) At 'intermediate' load levels between about 10-95N both the unreinforced and SiC reinforced alloys exhibited essentially similar wear rates (Figure 2(b)). For all the alloys wear rates increased slightly with the applied load (Table 2).

The worn surfaces of all the materials worn at rates 10^{-6} - 10^{-5} g.m.$^{-1}$ were characterized by extensive plastic deformation and damage in the form of cavitation and local grooving regardless of the presence of SiC particles. In fact, SiC particles were either pushed into the matrix or fractured during the wear process (Figure 4(a)). Within a subsurface region of about 40-50 μm deep, the microstructure was composed of deformed aluminum grains elongated in the sliding direction (Figure 4(b)) indicating accumulation of large strains and strain gradients on the layer of material near the worn surfaces. SEM metallography also indicated the presence of subsurface cracks that were nucleated around the SiC particles within the deformed zones (Figure 4(c)). These cracks propagated at a distance ~ 10 μm beneath the surfaces by following particle matrix interfaces. When these cracks reached the contact surfaces they produced delamination cavities as seen in Figure 5(a) and generated 'platelike' wear debris particles (Figure 5(b)). The thickness of these loose debris particles (Figure 5(c)) correlated very well with the loci of subsurface cracks.

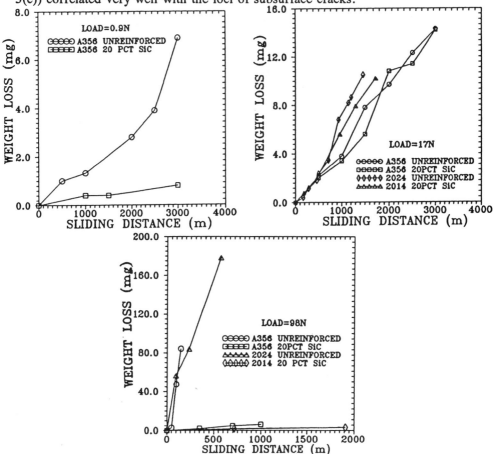

Figure 2. Weight loss versus sliding distance curves. a) 0.9N; b) 17N; c) 98N.

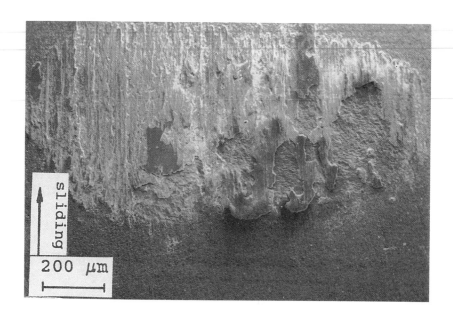

Figure 3. The worn surface morphology of A356-20%SiC (load = 0.9N).

Table 2 Dry Sliding Wear Rates of SiC Reinforced Composites
and Unreinforced Aluminum Alloys

LOAD (N)	WEAR RATE (x10^{-6} g.m^{-1})			
	A356-20%SiC	2014-20%SiC	A356	2024
0.9	0.1	---	2.0	---
1.7	0.3	---	3.2	---
9.3	2.1	3.7	3.7	6.5
17.2	4.9	6.0	4.9	6.8
42.0	8.3	7.3	6.0	11.0
56.0	7.7	---	7.4	29.0
98.0	7.4	7.6	595.0	340.0
147.0	8.2	8.4	850.0	---

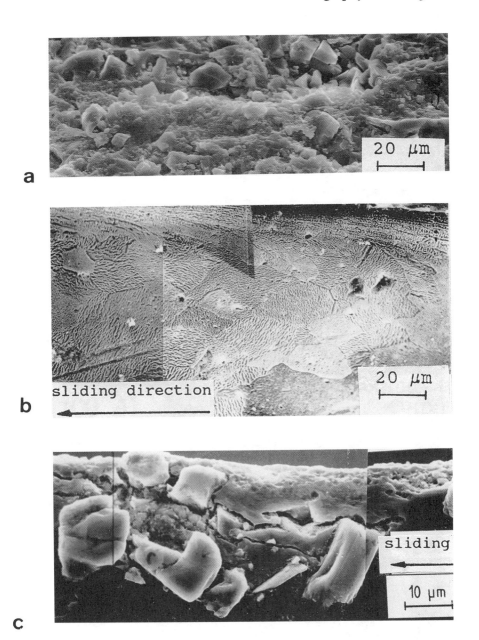

Figure 4. Wear processes leading to delamination of subsurface layers: a) Fractured SiC particles on the contact surfaces; b) deformation of aluminum matrix beneath the contact surfaces; c) subsurface crack propagation by decohesion of carbide/aluminum interfaces.

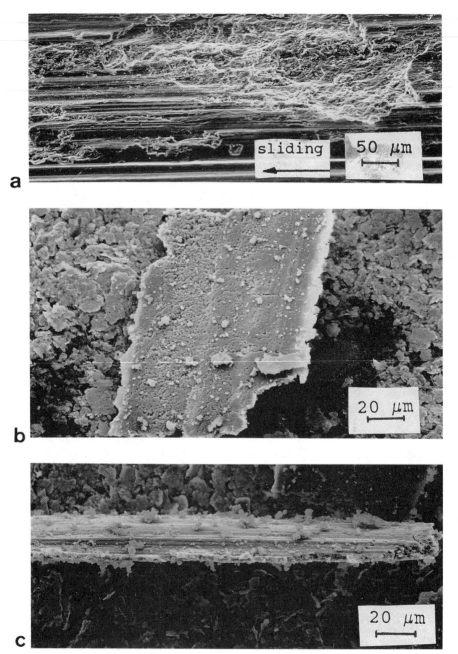

Figure 5. a) Delamination cavities formed on the worn surface of 2014-20%SiC; b) platelike debris particles (top view); c) platelike debris particles (side view).

Severe local damage and delamination of the subsurface layers were also the main processes of wear in unreinforced alloys worn at rates 10^{-6}-10^{-5} g.m^{-1}. As seen in Figure 6 in 356 alloy silicon particles acted as void nucleating agents and microcrack initiators.

c) The wear rates of unreinforced alloys accelerated by about two orders of magnitude at loads above 90-95N (Figure 2(c)). However, composites did not show such a transition to severe wear rate regime. In these materials delamination continued continued to be the dominant wear mechanism. The wear damage in the unreinforced alloys was much more extensive than the pre-transition (Figure 7(a)) and the size of deformed zones were ten times larger penetrating to depths about 400 μm beneath the surfaces. The debris particles were also coarser (200-500 μm thick) as shown in Figure 7(b).

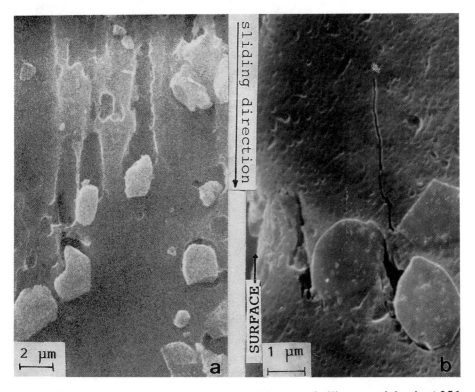

Figure 6. a) Nucleation of 'shear type' voids around silicon particles in A356 (section parallel to worn surface); b) Void nucleation between silicon particles and microcrack formation in A356 (section perpendicular to worn surface). Load = 27N.

439

Figure 7. a) The surface morphology of A356 worn at 147N ($\dot{W}$ = 8.5 x 10^{-4} g.m.$^{-1}$); b) Loose debris particles generated in the same material.

DISCUSSION

The low wear rates (10^{-7} g.m^{-1}) of the SiC reinforced composites are associated with the formation of iron-rich layers on the contact surfaces. Since there was little damage at the surfaces of composites but the surfaces of the steel sliders were characterized by the presence of longitudinal scratches or microgrooves it is concluded that material removal from the surface of the slider was due to the abrasive action of SiC particles. The detached wear fragments were subsequently oxidized and transferred to the surfaces of composites. Iron oxide layers (especially α-Fe_2O_3) have a low coefficient of friction and act as a solid lubricant [8]. Thus the formation of the iron oxide rich transfer layers is responsible for a tenfold drop in the wear rates of composites at low loads.

When the applied load produces stresses higher than the fracture strength of carbide particles, these particles lose their abilities to support the load and the aluminum matrix becomes in direct contact with the slider. Consequently, large local strains and strain gradients are generated in material layers beneath the contact surfaces. Using the methods developed to estimate the magnitude of subsurface strains from the distortion of the grain boundaries [9], it was found that within the deformed layers shear strains in excess of 5 were commonly reached. Clearly, SiC particles play a detrimental role by providing preferential nucleation and propagation sites for subsurface crack propagation. Thus, the lower ductility of the composites relative to the unreinforced aluminum alloys compensate the increased bulk hardness of these materials and therefore no significant increase in the wear resistance is observed for load ranges between ~ 10-95N.

At high loads (~ 80N for 2024, 95N for A356) the wear rates of the unreinforced aluminum alloys increase by a factor of 10^2. Such a drastic increase in the wear rates of aluminum-silicon alloys [10-12] is a practically important problem because the catastrophic nature of wear makes tribocomponents like cylinder pairs, piston rings made of aluminum alloys incapable of further use. Since none of the 20%SiC reinforced materials showed a transition to severe wear rates up to highest test load used (150N) SiC reinforcement is beneficial to improve the overall life of tribosystems. It has been observed that the temperatures on the surfaces of Al alloys reach in excess of 200°C at the onset of the transition. Consequently, material layers near the worn surfaces become susceptible to work softening and wear damage occurs at a massive scale (Figure 7).

It can be suggested that SiC reinforcement by improving the thermal stability of the aluminum alloys [13] enables the retention of room temperature strength during frictional heating and thus provide a way of suppressing the transition to severe wear rate regime.

CONCLUSIONS

The dry sliding wear behaviour of SiC reinforced composites varies with the applied load.

1. At low loads the SiC particles serve a load bearing elements, exposed portions of this particles create a local milling action on the slider such that the debris removed from the slider is transferred on to the surfaces of the composites. These layers act as solid lubricants resulting in a tenfold increase in the wear resistance of the composites ($\dot{W} \simeq 10^{-7}$ g.m^{-1}).

2. When the applied stress exceeds the fracture strength of the carbide particles normal compressive and shear forces are transmitted to the matrix causing accumulation of large strain gradients around the particles. Wear proceeds by subsurface crack nucleation and growth around SiC particles resulting in wear rates comparable to those observed in more ductile unreinforced alloys ($\dot{W} \simeq 10^{-6}$g.m^{-1}).

3. The wear rates of the unreinforced alloys accelerate by a factor of x100 above 80-95N. Incorporation of SiC particles into the matrix alloys impedes the transition to high wear regime.

REFERENCES

1) M.K. Surappa, S.V. Prasad and P.K. Rohatgi, "Wear and Abrasion of Cast Al-Alumina Particle Composites," Wear, Vol. 77, p. 295 (1982).
2) F.M. Hosking, F. Folgarportillo, R. Wunderling and R. Mehrabian, "Composites of Aluminum Alloys," J. of Mater. Sci., Vol. 17, p. 477 (1982).
3) A.G. Wang and I.M. Hutchings, "Wear of Alumina Fiber - Aluminum Metal Matrix Composites by Two Body Abrasion," Materials Science and Technology, Vol. 5, p. 71 (1985).
4) N. Saka and D.P. Karalekas, "Friction and Wear of Particle Reinforced Metal-Ceramic Composites," in Wear of Metals, Ed. by K.C. Ludema, ASME, New York, p. 784 (1985).
5) Y.M. Pan, M.E. Fine and H.S. Cheng, "Wear Mechanisms of Aluminum-Based Metal matrix Composites in Tribology of Composite of Materials," ed. by P.K. Rohatgi, P.J. Blair and C.S. Yust, ASM, Materials Park, p. 157 (1990).
6) D. Nath, S.K. Biswas and P. Rohatgi, "Wear Characteristics and Bearing Performance of Aluminum-Mica Particulate Composite Material," Wear, Vol. 60, p. 61 (1980).
7) P.K. Rohatgi, Y. Liu and T.L. Barr, "Tribological Behaviour and Surface Analysis of Tribodeformed Al Alloy-50 Pct. Graphite Particle Composites," Metallurgical Transactions, Vol. 22A, p. 1435 (1991).

8) K.C. Ludema, "A Review of Scuffing and Running in of Lubricated Surfaces with Asperities and Oxides in Perspective, **Wear**, Vol. 100, p. 315 (1984).

9) M.A. Moore and R.M. Douthwaite, "Plastic Deformation Below Worn Surfaces," **Metallurgical Transactions**, Vol. 7A, p. 1833 (1976).

10) A.D. Sarkar, "Wear of Aluminum-Silicon Alloys," **Wear**, Vol. 31, p. 331 (1975).

11) R. Shivanath, P.K. Sengupta, T.S. Eyre, "Wear of Aluminum-Silicon Alloys," **The British Foundrymen**, Vol. 70, p. 349 (1977).

12) V.K. Kanth, B.N.P. Bai and S.K. Biswas, "Wear Mechanisms in a Hypereutectic Aluminum Silicon Alloy Sliding Against Steel," **Scripta Metallurgica and Materialia**, Vol. 24, p. 267 (1990).

13) S.V. Nair, J.K. Tien and R.C. Bates, "SiC-Reinforced Aluminum Metal Matrix Composites," **Int. Metals Reviews**, Vol. 30, p. 275 (1985).

CAVITATION EROSION OF STRUCTURAL ALLOYS

R.H. Richman[1] and A.S. Rao[2]

ABSTRACT

The early stages of damage caused by cavitation collapse on both wrought and cast structural alloys were studied metallographically. First manifestations of damage varied widely with material. Normalized 1020 steel shows immediate, noncrystallographic rumpling with grains strongly delineated. Mass loss begins mostly in the ferrite phase at triple points and ferrite-pearlite boundaries. Martensite packets are defined first and material loss starts at lath boundaries in cast 13% Cr-4% Ni steel. In contrast, damage in Type 304 stainless steel begins with upheavals and subsequent mass loss at grain and twin boundaries. Weld-deposited Type 308 stainless evidences closely-spaced crystallographic features, but dendrites are delineated in preference to grain boundaries. There is no boundary definition at all in weld-deposited Stellite 21, only a subtle development of thin, linear, crystallographic traces. These observations are applied to the interpretation of cavitation damage in composite weldments. They are also shown to be consistent with fatigue as the main damage mechanism in cavitation erosion.

INTRODUCTION

Cavitation-erosion damage to hydraulic machinery is a common occurrence. One way of lessening such damage is to improve the cavitation-erosion resistance of the materials used for the original construction or for the repair of damaged equipment. Such improvements in the past have usually been guided by macroscopic measurements of volume or weight loss in laboratory or pilot-scale tests. A more modern approach would be to combine those macroscopic parameters with knowledge about the microscopic aspects of where the damage starts and how it propagates.

---

[1] Daedalus Associates, Inc., 1674 N. Shoreline Blvd., Mountain View, CA 94043.

[2] Powertech Labs, Inc., 12388 - 88th Ave., Surrey, BC, Canada V3W 7R7.

The damage induced by vibratory cavitation has been studied metallographically by others [1]. Typically, unalloyed metals and simple binary alloys served as exemplars for the effects of crystal structure [2], stacking-fault energy (SFE) [3], heat treatments [4], etc. As well, some industrial alloys have been examined for their pertinence to real machinery [5]. Almost all of the metallographic investigations, however, were conducted with wrought materials, with one notable exception [6]. Furthermore, many of the studies provide information only about surface appearance after significant exposure to cavitation; once the damage becomes widespread, the surface is so roughened that microstructural features are obscured. In this paper we provide additional details about the earliest stages of damage in constructional materials, with special reference to cast metals and weldments.

EXPERIMENTAL PROCEDURES

Materials. The alloys selected for study were as follows: AISI 1020 steel, to correspond to 0.2% C-1.0% Mn steel, which is the most common material in hydraulic turbines; cast CA6-NM (13% Cr-4% Ni steel), another frequent choice for turbines and pumps; Type 304 stainless steel, occasionally chosen for small generating units; weld-deposited E308 stainless steel, the most common material for repair of cavitation-erosion damage; weld-deposited Stellite 21, an overlay option that is extremely resistant to cavitation damage; and weld couples of 1020/E308, CA6-NM/E308, and Type 304/E308 to simulate critical regions of repairs.

Cavitation-erosion samples 16mm in diameter and weighing about 10g were prepared according to ASTM-G32 [7]. Composite weldments were produced by milling half of a round (rolled AISI 1020) or a square (cast CA6-NM) bar and then filling the milled side with three passes of E308 from an electrode 2.4mm in diameter. Interpass temperatures were about 100°C. The Stellite 21 specimen was prepared by completely overlaying AISI 1020 with three layers, each about 1mm thick. Test pieces were polished with $1\mu m$ diamond grit before exposure to cavitation. Compositions of the specimens are given in Table 1 and pre-cavitation microstructures of the cast and welded exemplars in Figure 1.

Cavitation Erosion. Exposures were performed in a vibratory apparatus manufactured by Branson. The experiments were conducted with a vibration amplitude of $\pm 19\mu m$ at a frequency of 20kHz in distilled water at 20°C. Exposures were interrupted periodically for visual examinations of the specimens. Erosion attack was confirmed at low magnification with a portable microscope. Detailed characterizations of the early stages of erosion were performed by scanning-electron microscopy (SEM).

OBSERVATIONS

Wrought 1020 Steel. Erosion effects become visible almost immediately. After only five minutes exposure, grains and pearlite colonies are delineated, Figure 2(a), and small pits have formed. The pits are mostly at boundaries, but some damage can be seen within pearlite patches. Although the pearlite becomes obviously roughened and rumpled, damage develops more in the ferrite phase, starting at pearlite-ferrite boundaries. This is the same sequence reported by Wade and Preece [4] and by Akhtar and Stowell [8]. In Figure 2(b) damage is more widespread, but the

Table 1. Chemical Composition of Alloys								
	Composition, wt. pct.							
Alloy	C	Mn	Si	Ni	Cr	Co	Mo	Fe
1020 Steel	0.21	0.4	0.2	–	–	–	–	Bal.
CA6-NM	0.04	0.6	0.5	4.0	12.9	–	0.7	Bal.
Type 304 Stainless Steel	0.08	0.9	0.4	8.4	18.2	–	0.1	Bal.
E308 Stainless Steel	0.08	0.7	0.8	9.6	19.1	–	0.2	Bal.
Stellite 21	0.22	0.5	0.1	2.3	29.3	59.0	4.9	3.4

Figure 1 (a)

447

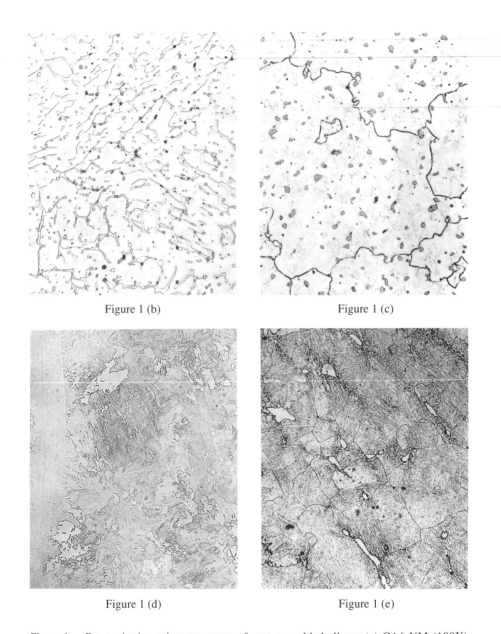

Figure 1 (b)

Figure 1 (c)

Figure 1 (d)

Figure 1 (e)

Figure 1.   Pre-cavitation microstructures of cast or welded alloys: (a) CA6-NM (100X);
(b) E308 weld metal (500X); (c) Stellite 21 weld metal (500X); (d) HAZ of 1020
steel (200X); (e) HAZ of CA6-NM (200X).

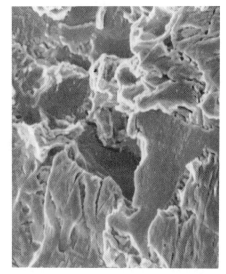

Figure 2 (a)   Figure 2 (b)

Figure 2.   Cavitation erosion of normalized 1020 steel: (a) Damage after 5 minutes (200X); (b) Material removal after 10 minutes (200X).

and Preece [4] and by Akhtar and Stowell [8]. In Figure 2(b) damage is more widespread, but the attack at boundaries and the removal of whole blocks of ferrite can still be distinguished.

Cast CA6-NM. This martensitic, precipitation-hardened steel is much more resistant to cavitation damage than 1020. Boundaries and microstructural features are only beginning to be visible after 15 minutes. At high magnification, Figure 3(a), surface rumpling by plastic deformation and cavitation attack at the martensite lath boundaries is clearly visible. As erosion proceeds, the lath structure becomes exaggerated, Figure 3(b). Material removal is much more uniform for CA6-NM than for 1020 steel, and the development of damage is very similar to that observed in wrought martensitic steels [5].

Wrought Type-304 Stainless Steel. The resistance of Type 304 to cavitation erosion is much greater than would be predicted from its hardness. It takes about 30 minutes before damage becomes unmistakable, at which time the features are quite dramatic. Figure 4(a) shows obvious evidence of deformation and upheavals at grain and twin boundaries. Material removal begins at these protrusions, primarily those at grain boundaries, Figure 4(b). Similar features were observed by Heathcock, *et al.* in Types 304 and 316 stainless steel and in Hadfield's steel [5], and by Preece and Brunton in copper-30% zinc [9]. Detailed examination of regions from which material was removed often reveals striated fracture surfaces, Figure 4(c).

Weld-Deposited E308 Stainless Steel. The response of E308 is in some ways similar to Type 304 in that linear deformation features appear after 30 minutes exposure, Figure 5(a). As the damage accumulates, however, the features become less linear and more reminiscent of an

449

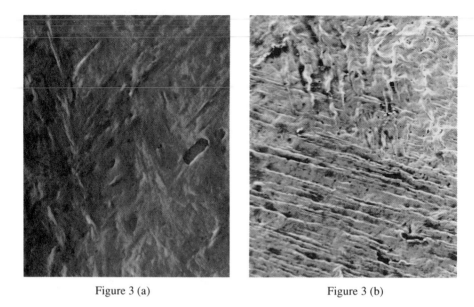

Figure 3 (a)   Figure 3 (b)

Figure 3.   Cavitation erosion of CA6-NM steel: (a) Deformation rumpling and pitting at lath boundaries (2000X); (b) More advanced stage of damage (500X).

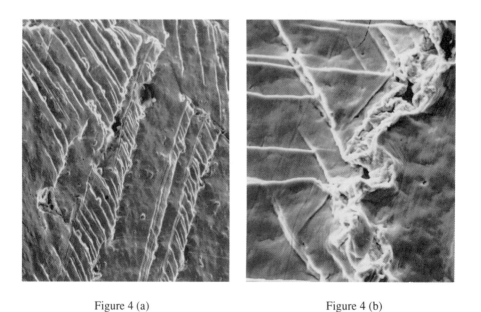

Figure 4 (a)   Figure 4 (b)

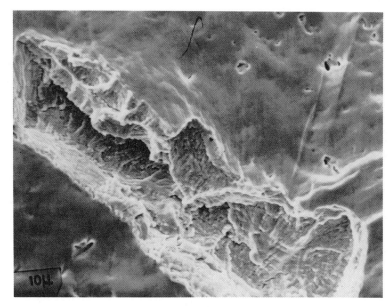

Figure 4 (c)

Figure 4.   Cavitation erosion of Type 304 stainless steel: (a) Linear deformation features and boundary definition (500X); (b) Material removal at upheaved grain boundary (2000X); (c) Striated fracture surface (2000X).

as-cast structure. Figure 5(b) is typical; dendrites are outlined by the surface upheavals and material loss has commenced within some of the dendrites.

Weld-Deposited Stellite 21. This material is extraordinarily resistant to cavitation erosion. Even after 75 minutes exposure, the only evidence of damage is the development of faint linear features and an occasional pit, Figure 6. Heathcock, *et al.* saw similar features in the matrix phase of Stellite 6 [6]. The main difference from the present work was that damage started at the massive carbides in Stellite 6, whereas in Stellite 21 there are very few hard-particle nucleation sites for material removal.

AISI 1020/E308 Weldment. The main features of this composite specimen are a columnar dilution zone on the E308 side of the fusion interface, the clearly-delineated interface itself (Figure 7(a)), and the heat-affected zone (HAZ) which shows erosion attack at the martensite lath boundaries (Figure 7(b)). Damage is accelerated in the HAZ, as is implied in Figure 7(a). After 135 minutes exposure, the deeply-ditched HAZ is visible to the unaided eye. Erosion in the HAZ progresses in much the same way as for wrought [5] and cast CA6-NM martensitic steels, and quite differently from the observations of Wade and Preece, who did not see any reflection of martensite crystallography in quenched and eroded 0.8% C steel. Accelerated attack in the HAZ nearest the fusion line is often observed in large, carbon-steel, hydro turbines

451

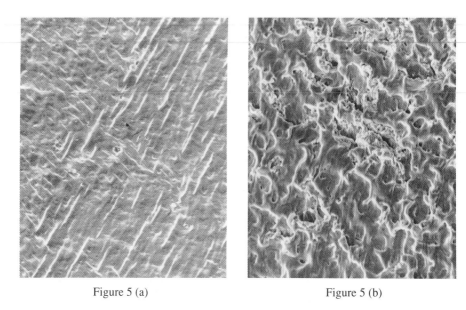

Figure 5 (a)   Figure 5 (b)

Figure 5. Cavitation erosion of weld-deposited E308 stainless steel: (a) Deformation features at early stage (500X); (b) Subsequent deformation of as-cast structure and initiation of material removal after 30 minutes (1000X).

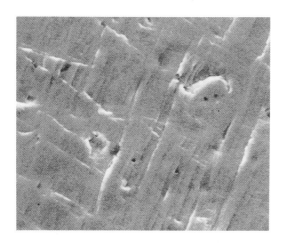

Figure 6. Linear deformation markings on Stellite 21 after 75 minutes exposure to vibratory cavitation (5000X).

and pumps that were weld-repaired with E308 or E309 [10]. We will comment on this phenomenon in the Discussion.

CA6-NM/E308 Weldment. Two interfaces are visible in this specimen: at the weld fusion line and at the transition from the HAZ to the CA6-NM base metal. Microstructures on each side of the HAZ-to-metal interface are shown in Figures 8(a) and 8(b), respectively; both reflect plastic deformation. The HAZ structure appears to be somewhat coarser, whereas the base metal shows more evidence of pitting at lath boundaries. Unlike the 1020/E308 weldment, however, prolonged

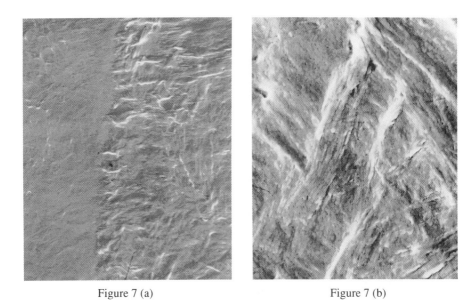

Figure 7 (a)   Figure 7 (b)

Figure 7.   Cavitation erosion of 1020/E308 weldment: (a) Damage transition from E308 (right
            side) to the HAZ of 1020 (left side) (500X); (b) Detail of deformation and pitting
            in the HAZ (2000X).

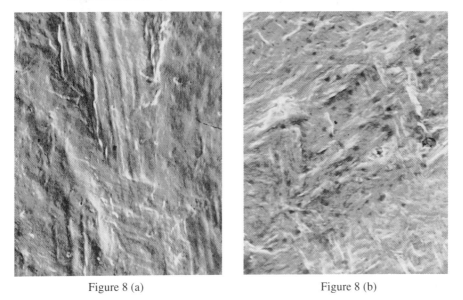

Figure 8 (a)   Figure 8 (b)

Figure 8.   Cavitation erosion of CA6-NM/E308 weldment: (a) HAZ adjacent to CA6-NM
            base metal (1000X); (b) Base metal adjacent to HAZ (1000X).

453

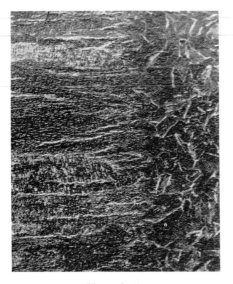 

Figure 9 (a)   Figure 9 (b)

Figure 9.   Cavitation erosion of Type 304/E308 weldment: (a) Overall view of weld metal (left
              side) and HAZ (50X); (b) Detail of deformation features on either side of fusion line
              (E308 on the left) (200X).

exposure does not develop great differences in the rates of attack in the various microstructures.
After 125 minutes exposure, only slightly more attack was seen in the CA6-NM than in the
E308. It can be deduced, therefore, that weld repair of CA6-NM with E308 is an appropriate repair
strategy.

   Type 304/E308 Weldment. An overall view of the fusion zone and interface with Type
304 base metal is shown in Figure 9(a), in which an equiaxed to columnar transition in the E308
weld metal can be seen, followed by another transition to equiaxed Type 304 base metal.
Details of the interface between E308 and Type 304 are contained in Figure 9(b). Here the initial stages of damage in the Type 304 HAZ are seen to be the same as in unwelded Type 304,
Figure 4. After 120 minutes of exposure, the fusion line was barely distinguishable, which
indicates that damage resistance is essentially unchanged in the HAZ.

DISCUSSION

   Micrographs of the early stages of cavitation erosion show that all of the alloys undergo
significant plastic deformation, induced by collapse of cavities in the liquid on (or near) the
metal surfaces. Although it is not intuitively obvious, McNaughton, *et al.* [11] show that
repeated random impulses or indentations normal to a metal surface is a form of cyclic defor-

mation and that damage accumulates with each reversal of plastic strain. In other words, we believe that cavitation erosion is primarily a fatigue process, as might be surmised from micrographs such as Figure 4(c).

Observation of fatigue-like failures from cavitation erosion is not new [12-14]. Correlations between erosion and fatigue were not successful, however, because correspondences were sought to performance measures such as endurance limit. If, instead, the parameters of strain-based fatigue analysis are invoked, excellent correlations can be obtained over the entire range of metallic materials from magnesium to Stellites [15]. The main determinant of damage susceptibility in cavitation erosion is the fatigue strength coefficient, σ_f', which is a measure of cyclic stress resistance. Correlations to material removal rates are further improved by considering the product of $\sigma_f' * n'$, in which n' (the cyclic strain-hardening exponent) can be thought of as an index of cyclic strain resistance. The inability of monotonic mechanical properties to correlate well with erosion behavior can be attributed to the fact that σ_f' is strongly mediated by cyclic strain hardening, and thus σ_f' is not simply related to any of the monotonic measures of stress and strain [15,16].

Reference to fatigue behavior guides understanding of several aspects of the present work. For instance, the excellent erosion resistance of Type 304 stainless steel arises in its low stacking-fault energy [16], which leads to planar deformation [17] and high cyclic strain-hardening (high σ_f' and high n'). Weld-deposited E308 also benefits from a stacking-fault energy nearly as low as that of Type 304 combined with a duplex microstructure (austenite plus ferrite), which is known to be associated with high cyclic strain-hardening and good resistance to low-cycle fatigue [18]. Stellite 21 is an extension of the argument: low stacking-fault energy and high strength [3] provide extraordinary resistance to cavitation erosion.

Although understanding the accelerated attack at the fusion side of the HAZ in 1020/E308 weldments is less straightforward, a reasonable explanation is possible. Strain-based fatigue analysis offers a conceptual scheme for identifying behavioral regimes. These regimes are defined primarily by the relative levels of plastic and elastic strain [19]. At high plastic strains, ductility is paramount for fatigue resistance. When cracks initiate early in life, propagation is the dominant failure mechanism. Conversely, at low plastic strains, strength governs behavior and crack initiation dominates. Since cavitation collapse imposes very localized, intense impulses on a surface, plastic strains are high. We can deduce, therefore, that cracks form early in the martensitic HAZ, which has little ductility and minimal resistance to crack propagation under cyclic strains.

CONCLUSIONS

The early stages of cavitation erosion in structural materials demonstrates that a mechanical process can develop all the usual metallographic manifestations of microstructure without the participation of chemical etching. This is testimony to the intimate relationship between damage resistance and microstructure. Furthermore, observations of the deformation modes, locations of initial damage, and the appearance of surfaces created by material removal, are all consistent with fatigue as the dominant mechanism of cavitation-erosion damage. Direct observation and implications of fatigue analysis both suggest a strategy of avoiding microstructures with significant proportions of constituents with low ductility and low strain-hardening capacity in situations that involve exposure to cavitation.

ACKNOWLEDGEMENTS

We thank the Electric Power Research Institute for support of this work under research contracts RP2866-1 and RP2426-13, managed by C.W. Sullivan and Dr. J. Stringer, respectively. Thanks are also due to J.A. Maasberg for the electron microscopy.

REFERENCES

1) C.M. Preece and I.L.H. Hansson, "A Metallurgical Approach to Cavitation Erosion", **Advances in the Mechanics and Physics of Surfaces**, Vol. 1, R.M. Latanision and R.J. Courtel, eds., New York, NY: Harwood Academic Publishers, p. 199, 1981.

2) C.M. Preece, S. Vaidya, and S. Dakshinamoorthy, "Influence of Crystal Structure on the Failure Mode of Metals by Cavitation Erosion", **Erosion: Prevention and Useful Application, ASTM STP 664**, W.F. Adler, ed., Philadelphia, PA: American Society for Testing and Materials, p. 409, 1977.

3) D.A. Woodford, "Cavitation-Erosion-Induced Phase Transformation in Alloys", **Metall. Trans.**, Vol. 3, p. 1137, 1972.

4) E.H.R. Wade and C.M. Preece, "Cavitation Erosion of Iron and Steel", **Metall. Trans. A**, Vol. 9A, p. 1299, 1978.

5) C.J. Heathcock, B.E. Protheroe, and A. Ball, "Cavitation Erosion of Iron and Steels", **Wear,** Vol. 81, p. 311, 1982.

6) C.J. Heathcock, A. Ball, and B.E. Protheroe, "Cavitation Erosion of Cobalt-Based Stellite Alloys, Cemented Carbides and Surface-Treated Low Alloy Steels", **Wear,** Vol. 74, p. 11, 1981-1982.

7) ASTM G32, "Standard Method of Vibratory Cavitation Erosion Test", **Annual Book of ASTM Standards**, Vol. 03.02, Philadelphia, PA: American Society for Testing and Materials, p. 187, 1984.

8) A. Akhtar and V.A. Stowell, "Cavitation Erosion of Hydraulic Turbine Steels Containing O.2 Pct. Carbon and 1 Pct Mn", **Journ. Mater. Energy Systems,** Vol. 4, p. 58, 1982.

9) C.M. Preece and J.H. Brunton, "A Comparison of Liquid Impact Erosion and Cavitation Erosion", **Wear,** Vol. 60, p. 269, 1980.

10) A.S. Rao, A. Akhtar, and D. Kung, "Interface Cavitation Erosion in Carbon Steel/Stainless Steel Weldments", **Proc. 14th Symp. Intl. Assoc. Hydraulic Res.**, p. 639, 1988.

11) W.P. McNaughton, R.H. Richman, and G.S. Beaupre, "Strain History and Hysteresis Effects for Elastic-Plastic Materials Subjected to Cyclic Contacts", accepted for publication in **Philos. Mag.**

12) B. Vyas and C.M. Preece, "Cavitation Erosion of Face Centered Cubic Metals", **Metall. Trans. A**, Vol. 8A, p. 915, 1977.

13) G.P. Thomas and J.H. Brunton, "Drop Impingement Erosion of Metals", **Proc. Roy. Soc. Lond. A**, Vol. 314, p. 549, 1970.

14) K.S. Zhou and H. Herman, "Cavitation Erosion of Titanium and Ti-6Al-4V: Effects of Nitriding", **Wear,** Vol. 80, p. 101, 1982.

15) R.H. Richman and W.P. McNaughton, "Correlation of Cavitation Erosion Behavior with Mechanical Properties of Metals", **Wear,** Vol. 140, p. 63, 1990.

16) R.H. Richman and W.P. McNaughton, "Fatigue Damage in Erosive Wear", **Morris E. Fine Symposium**, P.K. Liaw, J.R. Weertman, H.L. Marcus, and J.S. Santer, eds., Warrendale, PA: The Minerals, Metals, and Materials Soc., p. 383, 1991.

17) C.E. Feltner and P. Beardmore, "Strengthening Mechanisms in Fatigue", **Achievement of High Fatigue Resistance in Metals and Alloys, ASTM STP 467**, Philadelphia, PA: American Society for Testing and Materials, p. 77, 1970.

18) R.M. Ramage, K.V. Jata, G.J. Shiflet, and E.A. Starke, Jr., "The Effect of Phase Continuity on the Fatigue and Crack Closure Behavior of a Dual-Phase Steel", **Metall. Trans. A**, Vol. 18A, p. 1291, 1987.

19) R.W. Landgraf, "Control of Fatigue Resistance Through Microstructure — Ferrous Alloys", **Fatigue and Microstructure**, Metals Park, OH: American Society for Metals, p. 439, 1979

MICROSCOPY OF A 5W30 OIL IN THE 4-BALL TEST

J.S. Sheasby [1] and Z. Nisenholz [2]

ABSTRACT

The antiwear characteristics of a commercial 5W30 SF/SG oil were examined in a 4-ball wear machine. A profilometer was used to measure the growth of scars, and after subtraction of the thickness of the protective films, to estimate wear volumes. The development of the antiwear films was followed by optical and scanning electron microscopy, and compared to the films formed in the base oil with the ZDDP antiwear additive alone, or with one other additive.

[1] Department of Materials Engineering, The University of Western Ontario, London, Ontario, Canada. N6A5B9

[2] RAFAEL, Armament Development Authority, Ministry of Defence, Haifa, Israel.

INTRODUCTION

The 4-ball machine is the industry standard for measuring the effectiveness of antiwear additives in oils. Tests are usually performed in accord with the recommendations of ASTM D 4172 [1]. Typically the top ball is loaded to 147 N (15 kg) and rotated at 1200 rpm for 1 hour against the 3 bottom balls in the oil to be evaluated.

The only measure of performance described in the standard is the average size of the wear scar on the 3 lower balls as determined to an accuracy of 0.01 mm by optical microscopy. In many studies this diameter is measured over a range of loads and used to define transitions between lubrication regimes [2]. The actual rates of wear have been given lesser attention, and in the boundary lubrication mode are often considered negligible [3]. In practise however, the rate of boundary lubricated wear is reported to be the limiting factor in automobile valve train cam/finger follower surfaces [4].

The purpose of this paper is to demonstrate that the complex wear scars formed in 4-ball tests in fully formulated oils can be usefully examined to determine both the rate and a general appreciation of the mechanism of boundary lubricated wear. The oil used was a premium brand 5W30 SF/SG grade motor oil purchased from a local outlet. The antiwear additive in this oil is a primary/secondary zinc dialkyldithiophosphate (ZDDP) mixture and so is similar to essentially all engine oils in current use. The oil was tested new, and after service in one of the authors' car. To gain further perspective on the interaction between additives the oil was also partially built up from base stock .

EXPERIMENTAL PROCEDURES

Standard AISI E-52100 EP balls were worn in a Shell 4-ball lubricant testing machine. All the runs reported in this paper were performed with a load of 147 N (15kg), at 1200 rpm and 100°C. Times of the tests varied from a few seconds to 3 days, though 1 hour was used most often. The specimen cup was modified so that the oil could be filtered (Millipore 0.8 μm) during a run, and this was done for runs over 1 hour duration. An additional pipe into the cup gave control of the atmosphere over the oil when desired.

The resulting wear scars were examined and measured by optical microscopy, by SEM, and by a Dektak profilometer. Profiles were taken both parallel and perpendicular to the direction of sliding. Scanning Auger analysis was performed on selected scars.

RESULTS

1) Evolution of the scar in new oil.

SEM views of the three main stages of scar development are given in Figure 1. The dark patches in Figure 1 b, the scar after 1 hour of wear, are blue, otherwise the scars are basically brown. These relatively thick blue patches are present after 1 minute of wear and persist for about 6 hours. At higher magnifications the blue patches appear smooth in the SEM. Figure 2, whereas the brown films though initially smooth and homogeneous, progressively break up into small smooth pads on a rough background, Figure 2b.

2) Scar geometry

Profilometer traces parallel to the direction of sliding are given in Figure 3. In the first seconds the balls actually gained height due to the accumulation of the antiwear film. Thereafter the profiles of the scars

a

Figure 1: SEM micrographs of scars after a) 35, b) 60 min., c) 44 hours of wear at 100°C, 1200 rpm 147N load. a,b leading edge on right. c leading edge on left.

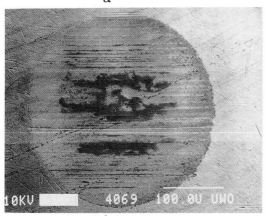

b

c

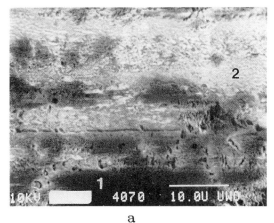

a

b

Figure 2: Higher magnification SEM micrographs of a) Fig. 1b and b) Fig. 1c. The dark patches in Fig. 2a appear blue by optical microscopy, and the lighter micro-wavy regions appear brown. The smooth film in Fig. 2b appears brown. The composition of points 1, 2 in Fig. 2a is given in Table 1.

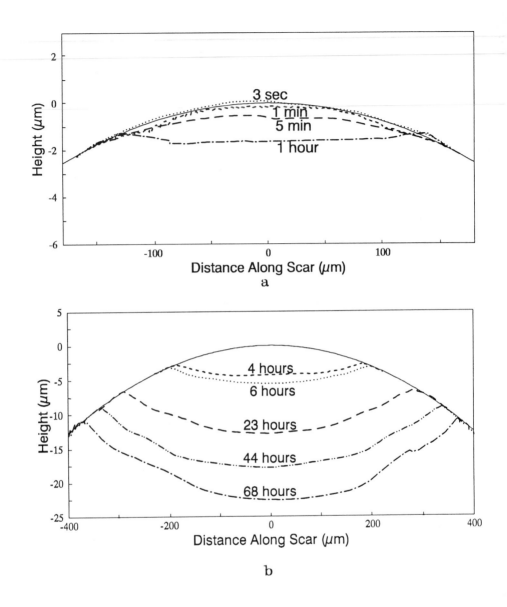

Figure 3: Profilometer traces across the wear scars parallel to the direction of sliding superimposed on the original ball geometry. The traces are labelled by time of sliding. All 147N load, 100°C, 1200 rpm. Leading edge on left.

developed logically as the top ball wore into the 3 lower balls.

In estimating the wear volume it was necessary to subtract the thickness of the antiwear films from the wear profiles, and then to mathematically rotate the modified profiles through 180°. The subtraction of the films was done with reference to the SEM picture for each scar. Both the rotation and subtraction procedures introduce uncertainties so the resulting volumes are not considered reliable below 10^3 μm^3. Wear volumes so calculated are given in Figure 4 together with the scar diameters as measured optically. After 27 m, or 1 min, of sliding the Log of the wear volume was proportional to the Log of the sliding distance: the slope of the line is 0.93.

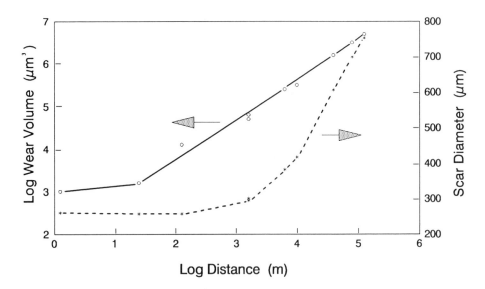

Figure 4. Wear scar volume per ball and optical scar diameter vs distance slid of balls in 5W30, 147N, 100°C, 1200rpm.

3) Composition of the surface films.

The Scanning Auger analyses of the blue and brown films, points 1 and 2 respectively in Figure 2a, are given in Table 1. The blue films were quite free of iron and though the S:P ratio was typical of the glassy films seen by other workers [5,6], the location of the P peak indicated that the P was more likely to be prsent as phosphide than phosphate. The brown films appear to be a mixture of iron and zinc oxides, sulphides and phosphides. The Ca came from the overbased detergent in the oil, and in part was present as carbonate.

4) Atmosphere.

N_2 was bubbled through the oil for runs lasting 60 mins and for 24 hours. In both cases the wear rate was reduced about 25% from the rate in air, Table 2. Scars were more heavily filmed than those formed in air, but otherwise were quite similar.

5) Oil pre-used in the 4-ball.

The oil that had been used in a 68 hour test was reused in a 1 hour test with new balls. The resulting scar appeared essentially the same as with new oil (Fig. 1b), though the wear volume was nearly twice as big, Table 2.

6) Oil pre-used in a car engine.

Standard 1 hour tests were done on the same brand of 5W30 oil after use in the authors car for 4, 1000 (twice), and 10 000 km. The latter included many short trips through a Canadian winter. Wear constants are given in Table 2. The scar from the 4 km oil was scratched but otherwise similar to Figure 1b. The scars from the 1000 km oils were significantly larger, and the patches of blue film were smaller and more uniformly distributed, Figure 5. The scar from the 10 000 km oil appeared bright and metallic by optical microscopy, though a thin patchy film could be detected by SEM, Figure 6.

TABLE 1:	Scanning Auger of the surface films on the sample shown in Figure 1b (at%) and 2a after 25s sputter

Blue film, location 1, Figure 2a

P	S	C	0	Fe	Zn	Ca
2.5	2.8	34.5	16.4	1.6	3.6	38.7

Brown film, location 2, Figure 2a

P	S	C	0	Fe	Zn	Ca
0.6	8.3	57.6	7.1	14	3.6	8.7

TABLE 2:	Wear Constants in 5W30 after various treatments 147N, 1200 rpm, 60 mins.

CONDITIONS	WEAR CONSTANT $m^3 .N^{-1}.m^{-1}$ $x10^{-20}$
New	27.4
New, N_2 atmosphere	18
Pre-used 68 hours in 4-ball	47.9
Used in car 4 km	26.2
Used in car 1,000 km	62.3
Used in car 10,000 km	117

a

b

Figure 5: SEM views of wear scar after one hour, 100°C, 1200 rpm test of 5W30 oil used in a car engine for 1000 km. Leading edge on left a) low magnification, the dark pads are blue. b) high magnification of centre of scar.

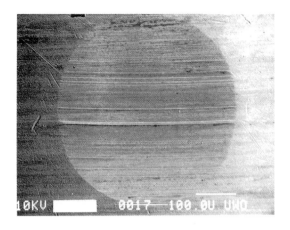

a

b

Figure 6: SEM views of wear scar after one hour, 100°C, 1200 rpm test of 5W30 oil used in a car engine for 1000 km. Leading edge on left a) low magnification, the scar is shiny and free of film by optical microscopy. b) higher magnification of the tope part of the scar.

TABLE 3: Effect of second additives on the antiwear capability of ZDDP 147N, 1200 rpm, 60 mins. Base stock solvent 150N	
ADDITIVES	WEAR CONSTANT $m^3 \cdot N^{-1} \cdot m^{-1}$ $x10^{-20}$
1.5 wt% Secondary ZDDP	1.4
1.5 wt% Secondary ZDDP + 0.1% Friction Modifier	23
1.5 wt% Secondary ZDDP + 6.0% Dispersant	12
1.5 wt% Secondary ZDDP + 2.0% Ca Detergent	27
5W30 Fully formulated	27.4

7) Oil with ZDDP and one other additive.

The authors have made an extensive evaluation of a commercial secondary ZDDP additive in the same base stock as the 5W30 of this study. The rate of wear in a standard 1 hour test in this formulation is less than 1/10 of that in the fully formulated oil, Table 3. The wear scars are covered in thick glassy pads that cannot be detected by optical microscopy until coated, Figure 7. The scars were also surrounded by a soft material, which has been termed friction polymer, that is easily lost on cleaning.

Second additives were put into this oil to investigate the origin of the differences between simple and fully formulated oils. Wear constants from standard tests in these oils are given in Table 3. The friction modifier and the detergent both degrade the wear performance to that of the 5W30 oil, and significantly change the appearance of the scars to that of the

a

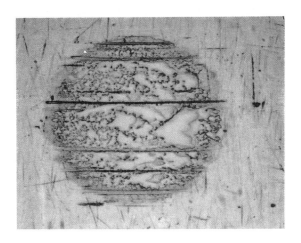

b

Figure 7: Optical micrographs of wear scar from one hour, 100°C, 1200 rpm test of 1.5wt% secondary ZDDP in base stock. Leading edge on left, the scar is 254μ diameter a) after washing, which removes the friction polymer from around the scar. b) after coating with gold to reveal the glassy pads.

471

formulated oil (except for the presence of friction polymer). The dispersant had a lesser effect on the appearance and wear rate, but as to be expected did prevent the deposition of friction polymer outside the scar.

DISCUSSION

This work used a 4-ball wear tester to follow the development of wear scars in a premium grade 5W30 oil for test times ranging from 3 s to 3 days. It differs from similar work in that every scar was examined by profilometry, and by optical and scanning electron microscopy. This represents a considerable increase in time and expense over the more usual measurement of scar diameter. One therefore has to consider critically whether the information gained was sufficient justification, as from a pragmatic view, much of the effectiveness of the oils could have been deduced from the scar diameter alone.

Models of the relationship between scar diameter and wear volume are available [7], and are useful for diameters greater than 300 μm. Wear volumes calculated for smaller scars are in considerable error due to the presence of surface films. These same films also restrict the accuracy of wear volumes calculated from profilometry, however with profilometry one can make informed corrections and so make reasonable estimates for scars down to 260 μm. The gain from 300 to 260 μm is significant when comparing the excellence, as opposed to the poorness, of oil formulations.

The films that formed on the wear scars varied in thickness, colour and composition with both position on the surface and the distance slid. In agreement with other workers [5,6] there appeared to be two basic types of film that were coloured blue and brown. The blue films were relatively thick (0.2 μm), and contained little iron. The thick films formed in the base stock with just ZDDP were clearly glassy, with the P Auger peak shifted to the phosphate location and the S:P ratio about 1:1. The nature of the blue films in the formulated oil is less clear, the S:P ratio was still 1:1, however the P peak was indicative of phosphide and the C (and Ca) content was very high. The change in the properties of the thick films with formulation should be persued as the fully glassy films were more than 10 times as wear resistant. The brown films were thinner and were probably mixed iron, zinc and calcium oxides, sulphides and phosphides. There was also a friction polymer like material present on and around the scars. The friction polymer was difficult to observe on scars that had been formed in the oils containing a dispersant, however subsidiary experiments in our Direct Observation Wear Machine [8], confirmed its presence whilst wearing.

The relative importance of friction polymer, and the blue and brown films for wear protection is ambiguous. The thick films formed by the secondary ZDDP alone in base stock gave exceptionally good wear protection. When the quantity and thickness of the film was reduced by friction modifier or detergent the wear rate increased by more than a factor of 10. Similarly the thinner films formed after using the oil in a car further increased the rate of wear. On the other hand, the absence of blue film on scars formed in tests of the 5W30 oil of longer than 6 hours duration did not result in an increased wear rate. Further, when the oil from these long tests was reused for 1 hour with new balls the blue film reformed, but the wear rate was twice that of new oil.

The correlations noted in this study between appearance and performance are therefore imperfect. Microscopy does however make one aware of the complexity of wear scars, and though the surface structure may not be the whole story, the whole story has to be compatible with the surface structure.

CONCLUSIONS

The 4-ball machine can be used to measure the low rates of wear given by effective antiwear additives. Care has to taken to account for the presence of protective films that can be thicker than the metal lost.

The origin of many of the features on wear scars can be accounted for, however their full significance in the antiwear action is not yet understood.

ACKNOWLEDGMENTS

This work was funded by a Natural Sciences and Engineering Research Council of Canada Strategic Research Grant, an Imperial Oil University Research Grant, and by an Ontario Ministry of Colleges and Universities URIF award. The authors would like to express their appreciation to T.A. Caughlin, W.A. Mackwood and M.P. Tangen who contributed to this study.

REFERENCES

1) Standard Test Method for Wear Preventive Characteristics of Lubricating Fluid (Four-Ball Method), ASTM Des.: D4172-82 1987.

2) C. M. Lossie, J. W. M. Mens and A. W. M. de Gee, "Practical Applications of the IRG Transition Diagram Technique", Wear, Vol. 129, 173, 1989.

3) R.C. Watkins, "The use of the hertzian dimension in wear scar analyses", Wear, Vol.91, 349, 1983.

4) J.C. Bell and T. Colgan, "Pivoted-follower Valve Train Wear: Criteria and Modelling",

Lub. Eng., Vol. 47, 114, 1991.

5) J.M. Georges, J.M. Martin, T.Mathia, Ph. Kapsa, G. Meille and H.Montes, "Mechanism of Boundary Lubrication With Zinc Dithio-phosphate", Wear, Vol. 53, 9, 1979.

6) M. Belin, J. M. Martin, J. L. Mansot, "Role of Iron in the Amorphization Process in Friction-Induced Phosphate Glasses", Trib. Trans., Vol. 32, 410, 1989.

7) I-Ming Feng, "A New Approach in Interpreting the Four-Ball Wear Results", Wear, Vol.5, 275, 1962.

8) J.S. Sheasby, T.A. Caughlin and J.J. Habeeb, "Observation of the antiwear activity of zinc dialkyldithiophosphate additives", accepted by Wear.

A STUDY ON MICROSTRUCTURE AND WEAR OF PLASMA (PTA)

DEPOSITED SUPER ALLOYS

* R. CHATTOPADHYAY

ABSTRACT

The superior heat resistant and creep properties of superalloys are utilised in high temperature applications like aeroengines, forging dies, hot shear blades, etc. One of the important developments in the area of superalloy application is the production of homogeneous atomised powders. These super-alloy powders are used to produce a wear-resistant overlay on cheaper substrate material. The powder materials selected for this study are Ni-Cr-Ti-Al, Ni-Cr-Mo, Co-Cr-W-C and Co-Cr-Mo-C type of alloys. The superalloy powders are deposited on mild steel plates through PTA (Plasma Trans-ferred Arc) process. The microstructure of plasma deposit is examined with both an optical and a scanning electron microscope with ECON detector. The high temperature hardness properties are studied with a hot hardness tester. The hardness and the wear properties of these Cobalt base and Ni-base alloys are compared. The difference in performance of Ni-Cr-Mo, Ni-Cr-Ti-Al, Co-Cr-W-C and Co-Cr-Mo-C in high temperature repeated impact wear appli-cation is correleted with the microstructure and the resultant mechanical properties at the elevated temperatures.

* Dy. General Manager (R&D), **EWAC ALLOYS LIMITED,**
A Subsidiary of Messers LARSEN & TOUBRO LIMITED, P.B.No.8933,
Saki Vihar Road,   Bombay - 400 072, India.

INTRODUCTION

Superalloys belong to a group of alloys which are mainly developed for aircraft industries. The excellent heat and creep resistance properties of these alloys are utilised in high temperature applications in industries other than aircraft e.g. glass, steel, power etc. One of the important developments in this area is the production of homogeneous atomised powders. Altogether new approach is to use the superalloy powders as wear resistant overlay on cheaper substrate material. An advanced technique for deposition of wear protective overlay is plasma transferred arc (PTA) process.

Superalloys derive their high temperature strength and hardness properties from their microconstituents in the microstructure. There is a definite relation between hardness of the material and wear even at elevated temperature. The objectives of this study is to find correlations between microstructure, hardness and wear. While wear can be related quantitatively with hardness, the relation with microstructure and wear rate is only qualitative.

EXPERIMENTAL PROCEDURES

Four different powder alloys are deposited on mild steel substrate through a Plasma Transferred Arc system of weld cladding. The chemical compositions of these alloys are determined by using a Vacuum Emission Spectrometer and of one with Energy Dispersive X-ray microanalysis. The microstructural studies are conducted in both Optical and Scanning Electron Microscope. The hardness at eleveted temperatures is determined by Nikon Hot Hardness Tester using 200 gm load, with 36° vickers indentor and indentation time of 10 seconds at respective temperatures. The wear properties at room temperature is evaluated by Dry Sand Rubber Wheel Abrasive Wear Test as per ASTM G65-85. The wear factor for each alloy is calculated from the wear volume loss for 6000 revolution at a load of 30 lbs. using 9" diameter wheel.

The impact test specimens for metallographic examination are prepared by repeated hammering at a temperature normally used for steel forging viz. 900°C.

RESULTS

The hot hardness values of the alloys are given in Table I. The room temperature wear coefficient of each alloy and high temperature hardness values are used to find the wear factor at different temperatures of the alloys (Table II). The comparitive number of cycles to initiate wear according to Zero Wear Limit equation for Alloy 2 with respect to Alloy 3 is indicated in Table VI.

478

The chemical composition of the alloys are shown in Table III to Table V.

DISCUSSION

The correlation between hardness and wear resistance in various wear situations are described by Archard (1).

The wear volume per unit sliding distance and unit load is given by (1)

$$VH = K \tag{1}$$

where K = wear coefficient which can be either for adhesive or abrasive wear. However, 'K' is constant for a particular material and microstructure. If there is no appreciable change in microstructure at eleveted temperature then the value of K determined at room temperature can be used for finding approximate Wear Factor (Table II).

In the case of impact wear the zero impact wear limit (2) was introduced to calculate the induction period or the number of cycles, No, after which the wear process starts. In case the induction period is not observed, then, No, indicates the number of cycles at which wear has reached half the surface finish.

Zero impact wear limit is expressed by an equation as follows (2)

$$No = 2000 \, (\Upsilon \cdot S_y / S_m)^9 \tag{3}$$

where Υ = Wear Factor of the material

S_m = Maximum peak pressure during contact

S_y = Uniaxial yield stress which is related to 'hardness' or shear yield stress and in case of ductile material as $\tau = 0.57 S_y$

The above equation is similar to relationship established by Bayer & Ku (3) for Zero sliding wear mode.

The wear process is strongly dependent on shear stress and may result in both surface and subsurface damage. The normal component of impact causes subsurface damage. The shear stress component of impact causes wear damage to the material to an extent depending on the shear yield strength of the material (τ) and thus the hardness of the material.

The microconstructual features contributing to yield stress (4) of an alloy can be shown by Hall-Petch equation (5):

$$S_y = S_o + Kyd^{-\frac{1}{2}} \tag{2}$$

where $\underline{\varsigma}_o$ = Friction stress = $\underline{\varsigma}_o{}^I + \underline{\varsigma}_o{}^{II}$

$\underline{\varsigma}_o{}^I$ = temperature independent path, arising from the resistance of random solute atoms (solid solution herdening), fine precipitates (precipitation hardening), and lattice defects (deformation hardening)

$\underline{\varsigma}_o{}^{II}$ = temperature dependent part arising out of Peirls - Nabarro stress (6)

Ky = measure of dislocation locking (strain hardening)

d = grain size

The importance of microstructure on yield stress (or hardness) and thus on wear is clearly shown in Hall - Petch equation.

The repeated impact wear at high temprature involves abrasive wear due to scale and adhesive wear as a result of metal to metal friction.

MICROSTRUCTURE , HARDNESS AND WEAR:

1. Ni-Cr-Mo Alloy (Alloy1)
The alloy belongs to type as indicated in UNS N10276 (Hastelloy type) and the composition is indicated in table III.

Normally , three types of carbides can be observed in the microstructure of alloy 1. viz. MC , $M_{23}C_6$ and M_6C : (7,8,9)

i) MC occurs as script or cubic morphology

ii) $M_{23}C_6$ are found at grain boundaries as irregular discontinuous blocking particles or as plate or as regular geometric shape.

iii) M_6C - precipitates as block form at grain boundaries

In the micrographs Ia & Ib the main bulk of the carbides appear to be $M_{23}C_6$ + M_6C at grain boundaries and MC inside the grains.

The composition of the alloy is in homogeneous gamma field, However, precipitation of TCP phases can occur due to segregation during solidification of the PTA weld. The chemical composition of ς, u, and P phases when analysed in the SEM-EDAX is found as containing weight per cent wise 33 Ni, 38 Mo, 6W, and 3 Fe (10). In PTAW deposits the formation of TCP phases may lead to hot cracking (11).

The hot hardness values of Alloy 1 are indicated in Fig. 1. There is an insignificant drop in hardness until a temeprature of 973°K is reached. Above 973°K, hardness drop is considerable at 1033°K. The exposure at 773°K results in the formation of A_2B type long - range order which may contribute to the strengthening at elevated temeprature at around 973°K, $M_{23}C_6$ carbides precipitate and thus contribute to hardening. The decrease in hardness above 973°K can be attributed as due to coarsening of precipitates and the softening of matrix. The inverse of hardness (Fig. 2) and wear properties (Fig. 3) also follow the same trend as observed for the variation of hardness with temepratures.

In the repeatedly impacted sample (900°C approx.) the surface scale remains adherent (Fig. 4). SEM photographs of the structure before and after impacting are shown in (Fig. 4).

Alloys of this type tested in hot upset tests in the temeperature range of 1000 - 1050°C exhibited good ductility. (12)

2. Ni-Cr-Ti-Al Alloy (Alloy 2)

Basically a Ni-Cr-Ti-Al precipitation hardening type of alloy with Mo and Co additions for further strengthening. This is a Nimonic type of alloy and the composition is indicated in Table IV. This alloy exhibits exceptionally good high temeperature properties due to gamma prime strengthening. The Ni based alloy 2 composition shows two kinds of phases in the microstructure as follows (7,8,9).

a) GCP type (Geometrically close packed) phases such as:

i) γ' - "gamma-prime" phases which is spherical (or as cubes) coherent with γ - matrix and primarily responsible for high temeperature strength of this alloy. The intermetallic compound approximates $Ni_3(Ti,Al)$ composition. The yield strength of γ' increases continuosly upto a temperature of 700°C several times the room temperature value, a tendency not observed in most of the alloys (13,14,15). The maximum strength of Ni_3Al occurs just above half the melting temperature (0.58 Tm) the melting point being 1396°C.

γ' is observed as spheres, cubes and also plates (16). The microstructure of alloy 2 is shown in Fig.5, which is similar to U 700 microstructure (7). The grain boundaries are marked by $M_{23}C_6$ precipitates (dark etched) and the white dots (spheres or cubes) are intermetallic precipitates.

ii) γ'' - Ordered BCT structure with composition Ni_3Cb

481

iii) n - Ordered BCT structure with composition Ni_3Ti

iv) ς - Orthorhombic structure

b) TCP (Teragonal Close-Packed) or A_2B types viz. ς, μ, x and laves. The TCP phases are generally detrimental and most commonly found are ς, μ and x. Computerised calculations (PHACOMP) (7) are used to control composition of Alloy 2,so as to eliminate TCP phase formation.

The elemental analysis of the PTA deposit carried out with SEM-ECON detector is shown in Fig.5 (also table IV). The microstructure observed in SEM are shown in Fig.6(17). The effect of structural change on the impacted specimen compared to non-impacted is marginal.

The repeated impacted specimen surface shows (Fig.6) adherent scale, presumably due to oxides of chromium and nickel. The consistantly high hardness (Fig.1) and excellent wear properties (Fig.3) even at the elevated temeperature is due to microstructural constituent like γ' and its unique property of increased strength with increase in temeperature upto 973°K. The net result of matrix softening and γ' hardening, is slow decrease in total hardness upto 973°K. Beyond this temperature the hardness decrease is faster and so also the wear rate. However,even at 1143°K this alloy has superb high temperature hardness and wear properties compared to all other alloys.

3. Co-Cr-W-C (Alloy 3) and Co-Cr-M-C (Alloy 4)

These are two cobalt based alloys with compositions indicated in Table V.

The PTA deposit of Alloy 3 (Fig.7) shows F.C.C. matrix and M_7C_3 type of carbides,the latter is present as eutectic (18). In Alloy 4 (Fig.8) predominant carbides are in the form of $M_{23}C_6$.

Subsequent heat treatment of PTA deposit can transform M_7C_3 into $M_{23}C_6$ (19,20). The general carbide transformation sequence is as follows:

$$M_7C_3 \text{ (or } M_3C_2) \rightarrow M_{23}C_6 \rightarrow M_6C \rightarrow MC$$

as the stronger carbides are added.

Precipitation and morphology (22) can be further related to the solubility of the carbide in cobalt which are as follows (at the temperature of 1260°C):

WC = 22%, Mo_2C = 13%, Cr_3C_2 = 12%, V_4C = 6%, CbC = 5%, TaC = 3%

Normal solution treatment temperature of alloy 3 is 1150°C (23).

Normally low residual ductility of cobalt based alloys can be improved by solution treatment.

The hardness values of both Co-Cr-W-C and Co-Cr-Mo-C show a drop at 923°K but still at a higher level compared to Ni-Cr-Mo type (Alloy 1) (Fig.1).

A comparison of impacted and non-impacted microstructure of Alloy 3 is quite revealing (Fig 7). After repeated impacts at 900°C, the microstructure of Alloy 3 shows spheroidised carbides and fine lamellar eutectic carbides. This microstructural change shall result in considerable improvement in toughness of Alloy 3. Inspite of poor thermal shock resistance and ductility at room temperature, this alloy shows excellent wear resistance at elevated temperatures due to refinement in microstructure with usage (Fig.7). The improved ductility resists early fatigue failures(24). However, coarsening of carbides at still higher temperature and softening of the matrix reduces considerably the hardness and thus the wear life. The surface of the alloy after repeated impact shows adherent oxide scale (Fig.7).

Alloy 4 behaviour is similar to Alloy 3. Alloy 4 possesses lower hardness and higher toughness compared to Alloy 3. Therefore wear rate of Alloy 3 is lower than Alloy 4 at temperatures above 800°K.

Following zero impact wear limit relationship (equation 3), the number of cycles for initation of wear in Alloy 2 compared to Alloy 3 is found as 512 times. Failure due to fatigue at a much lesser number of cycles than that obtained from above equation is normally encountered in practice. However, materials similar to Alloy 2 class are expected to survive almost 700 times the number of cycles to failure for die steel (composition 0.56C, 1.1Cr, 0.5Mo, 1.7Ni, 0.1V) (25). Garrison (24) observed that ductility can have a significant influence on wear life.

Hickl (26) made exhaustive study on the wear properties of PTA deposits of cobalt base and Nickel base alloys. For similar microstructure and wear, the cobalt based alloy is found to be harder than corresponding Ni based alloy at room temperature whereas the situation is just the reverse at elevated temperature (above 773°K).

According to Peng (27) wear on hot forging dies is due to abrasive action of oxides formed on the surface of the forged parts. The wear rate depends on the difference in hardness of this scale and the hardness of the die. The hardness of FeO scale at 800°C is 50 VPN (28) which is lower than Ni or Co-based alloys, (Alloy 1 to Alloy 4) and also nickel and cobalt oxides at 800°C. Hence loss of material due to abrasive wear of die overlayed with Alloy 1 to Alloy 4 is negligible. The wear is thus mainly due to repeated impact at elevated temperature followed by thermal fatigue failure.

CONCLUSIONS

1. The high temperature wear resistance properties under repeated impact of the four alloys are related to hardness and microstructure.

2. The microconstituents responsible for superior high temperature properties in Alloy 1, 3 and 4 are strong carbides. These are formed at elevated temperature and these are resistant to growth at temperature of use thus retaining the hardness and wear resistance properties.

3. Repeated impacts at high temperatures (900°C) transform coarse eutectic carbide in the PTA deposit to fine lamellar and sphereoidised carbides, for example in Alloy 3. This transformation results in improvement in ductility and therefore fatigue life.

4. In Alloy 2, the gamma-prime phase is responsible for good hot hardness and the wear properties.
5. The wear at high temperature normally follows the same pattern as that of hot hardness.

6. Zero Impact Wear Limit is a good indication of life under repeated impact than calculated from abrasive or adhesive wear equation.

This study indicates definite correlation between microstructure, high temperature hardness and the wear.

REFERENCES

1) Archard J.F Proc. Symp.on Interdisciplinary Approach to Friction & Wear, Ed. P.M. Ku, Antonio, Texas, Nov. 1967, NASA SP-1181, 267-

2) Engel Peter A, Impact Wear of Materials, Elsevier Scientific Publish ing Co., NY, 1978, p206.

3) Bayer, R.G. shalkey A.T., and Wayson, A.R. Machine Design, 41(1), 1969, 142.

4) Chattopadhyay, R.Ph.D. Thesis, University of London, 1975.

5) Petch, W.J. Proc. Fract. Conf. Swampscott, Mass, April, 1959 p54.

6) Chattopadhyay R., Tewari J.P., Bhatnagar S.S.. The structure and

properties of 9% Nickel Steel, Trans IIM, 1970, March, p8.

7) Earl W. Ross and Chester T. Sims: Nickel Base Alloys, Superalloy II, Ed. chester T. Sims & others, John Wiley and Sons, 1987, Chapter 4, p97

8) Superalloys - A Technical Guide, Ed, Elihu F. Bradley, ASM International, 1988, 42.

9) Brooks, C.R. Heat Treatment, Structure and properties of Nonferrous Alloys, ASM, 1982.

10) Link Thomas and Osterle Werner: X-ray Microanalysis in the Elec tron Microscope, pt IV: Superalloys, Practical Metallography, 28 (1991), 101

11) Ciestak, M.J. Hsdley, T.J., Romig. A.D. Met. Trans. 17A, 1986, 1891- 1906.

12) Waldherr Utrich, Pohl Michael, Noqueira R.A., Padilha A.F.,Pract. Met., 1989 (26), 174.

13) Westbrook, J.H., Trans AIME, 1957, 209, 898.

14) Davies R.G., and Stoloff, N.S., Trans AIME, 1965. 233, 714.

15) Fleischer, Robert L, Taub, Alan. I., "Selecting High Temperature Structural Intermetallic Compounds" The Material Science Ap proach, Journal of Met., 1989, September, 18.

16) Hagel W.C. and Battie, H.J., Iron & Steel Institute Special Rep. London, 64, 98.

17) White, C.H., "Metallography and Structure", The Nimonic Alloys, Ed.Betteridge and J. Heslop, 1974 (2nd Edition) p63, Edward Arnold (Publishers) Ltd. London.

18) Morral F.R. et. al. ASTM Tech. Rep. No. 08, 21.1.

19) McGinn P.J. et. al. , Met. Trans. A, Vol. 15A, June, 1984, 1099

20) Guyard G, et. al., J. Met. Sci., Vol. 16, 1981, 604.

21) Chester T. Sim, Cobalt base Alloys - Chapter 5. The Superalloys 1972, John Wiley and Sons, p150.

22) Edington J.W., et. al. Wear, 1976, Vol. 48, 131.

23) Desai, V.M. et. al. Wear, 1984, Vol. 94, p89.

24) Hickl A.J. Nickel base Alloys and alternative to Cobalt base Alloys for P/M, Wear and Environmental resistant components "Modern De velopments in Powder Met. Vol. 14, special Materials, Ed. Henry H. Hausners and others : Proc. Int. P/M. Conf. 1980 p455.

25) Thyssen Edelstahlwerke AG; Germany ; Literature on "High Tempra ture High Strength Nickel base Alloy for Wear Resistant Weld Overlay on Horizontal Forging Tools".

26) Garrison M. Warren jr. , Lechtenberg T.A., & Kin. J; "Ductility and the abrasive Wear Resistant of Hot Work die Steels" Wear, 1987, 116, p33

27) Peng, Q.F. "Improving Abrasive Wear by surface Treatment". Wear, 1989, 129, p194.

28) Ramalingam S. Stoichiometry of TIC and its significance to the performance of Hard Metal Components. Medical Science and En gineering, 1977, 29, p123.

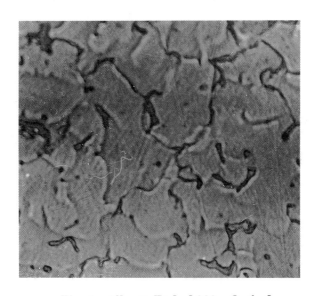

Fig. 4a. Alloy 1. Etched 800 x Optical

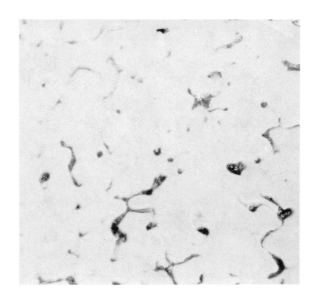

Fig. 4b. Alloy 1. Etched, Optical, hot impacted, 800 x

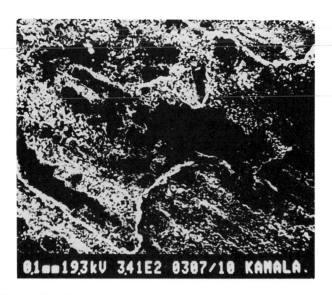

Fig. 4c. Surface after hot impacts SEM (unetched), Alloy 1

Fig. 4d. Alloy 1. Etched - SEM

Fig. 4e. Alloy 1. Etched. after hot impacts

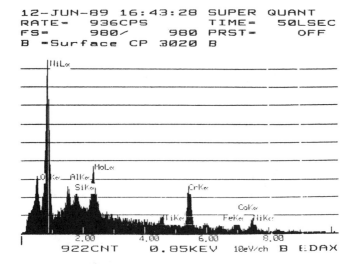

Fig.5 SEM-ECON Analysis of Alloy 2

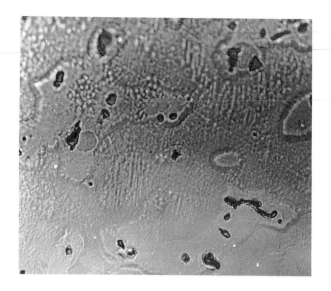

Fig. 6a. Etched Optical 800 x Alloy 2

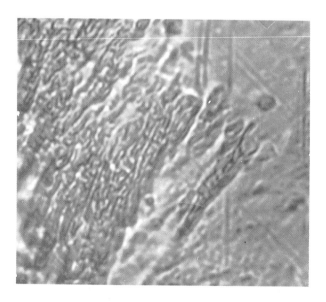

Fig. 6b. Etched Optical 800 x Hot impacted, Alloy 2

Fig. 6c. SEM Surface - Alloy 2, Hot Impacted

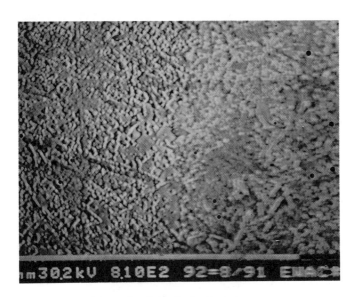

Fig. 6d. SEM Etched - Alloy 2

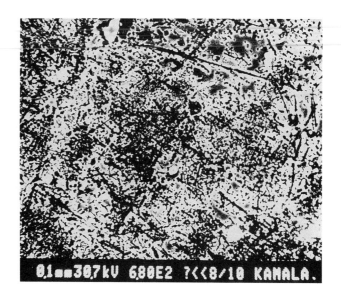

Fig. 6e. SEM Etched Hot impacted Alloy 2.

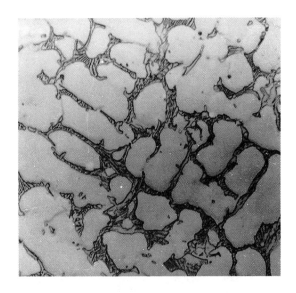

Fig. 7a. Etched Optical 800 x Alloy 3

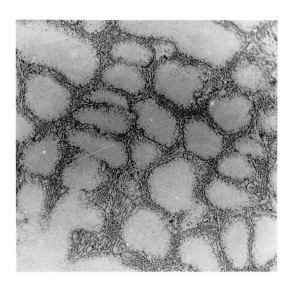

Fig. 7b. Etched Optical 800 x Hot impacted, Alloy 3

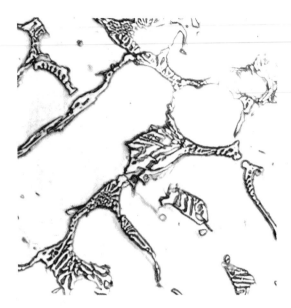

Fig. 7c. Etched Optical 1600 x Alloy 3

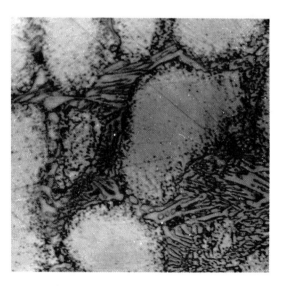

Fig. 7d. Etched Optical 1600 x Hot impacted, Alloy 3

Fig. 7e. Surface of Alloy 3 SEM

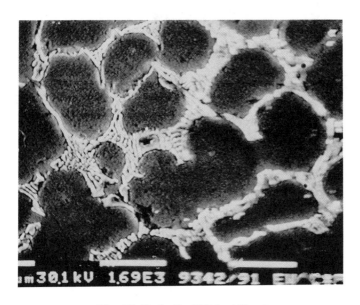

Fig. 7f. Etched , SEM   Alloy 3

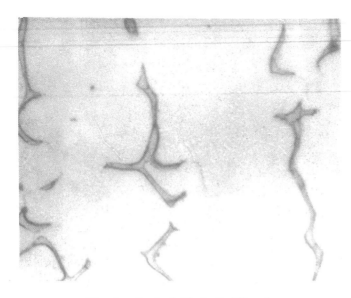

Fig. 8a. Optical, Etched, Alloy 4

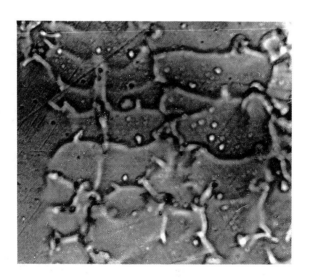

Fig. 8b. Optical, Etched, Alloy 4 Hot impacted

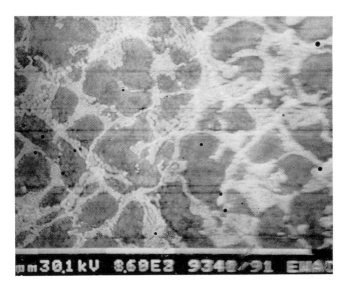

Fig. 8c. Etched, Alloy 4,SEM

TABLE I - HOT HARDNESS DATA IN HV SCALE

ALLOY	303^0K	698^0K	773^0K	813^0K	923^0K	973^0K	1033^0K	1143^0K
1	213	-	185	-	180	145	100	-
2	300	-	260	-	-	257	-	210
3	380	290	-	265	245	-	166	-
4	270	-	222	-	150	-	110	-

TABLE II WF AT DIFFERENT TEMPERATURES IN (^{o}K)

ALLOY	303	698	773	813	923	973	1033	1143	K VALUE @ RT
1	1.4	-	1.22	-	1.186	0.95	0.66	-	1.86×10^{-3}
2	1.9	-	1.6	-	-	1.58	-	1.29	1.99×10^{-3}
3	3.28	2.37	-	2.165	2.00	-	1.35	-	1.5×10^{-3}
4	2.84	-	2.32	-	1.57	-	1.152	-	1.17×10^{-3}

TABLE III – COMPOSITION OF ALLOY 1

ELEMENT	WT. %
C	0.03
Mn	0.8
Fe	1.0
Cr	15.0
Co	1.8
W	4.2
Mo	15.8
Ni	REST

TABLE IV – COMPOSITION OF ALLOY 2

ELEMENT	WT. %
Cr	18.73
Co	12.0
Mo	5.8
W	1.0
Al	1.7
Ti	2.86
Ni	REST

TABLE V - COMPOSITON OF ALLOY 3 & 4

ELEMENT	WT. %	
	ALLOY 3	ALLOY 4
C	1.18	0.28
SI	1.1	1.2
CR	28.0	27.2
W	4.0	-
MO	-	5.6
CO	REST	REST

TABLE VI - NO. OF CYCLES FOR INITIATION OF WEAR
IN ACCORDANCE WITH ZERO IMPACT WEAR LIMIT

TEMPERATURE	ALLOYS	$No_{(2)} / No_{(3)}$
$1043^{\circ}K$	ALLOY 2	512
	ALLOY 3	

NO. OF CYCLES FOR INITIATION OF WEAR IS 512 TIMES FOR ALLOY 2 COMPARED TO ALLOY 3.

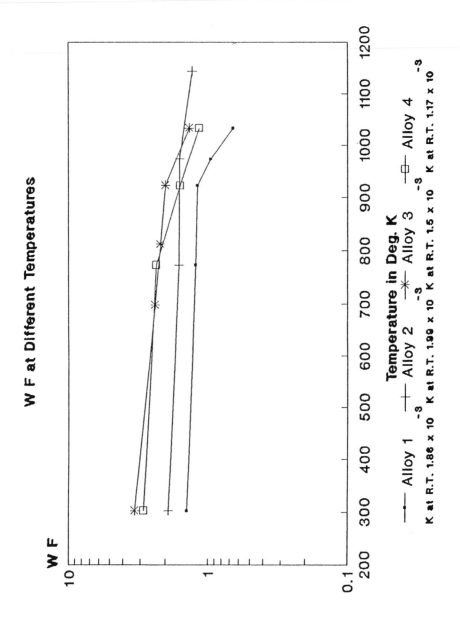

W F at Different Temperatures

1 / HV at Different Temperatures

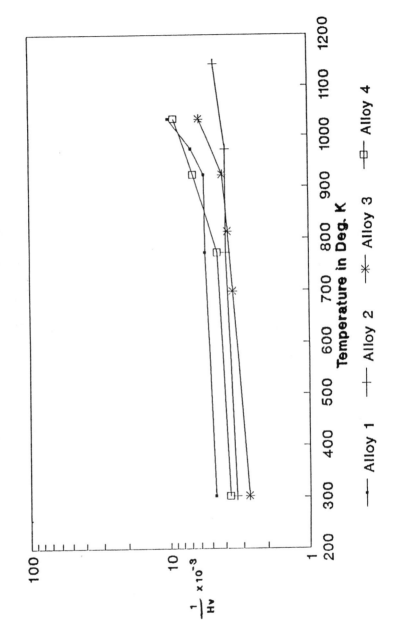

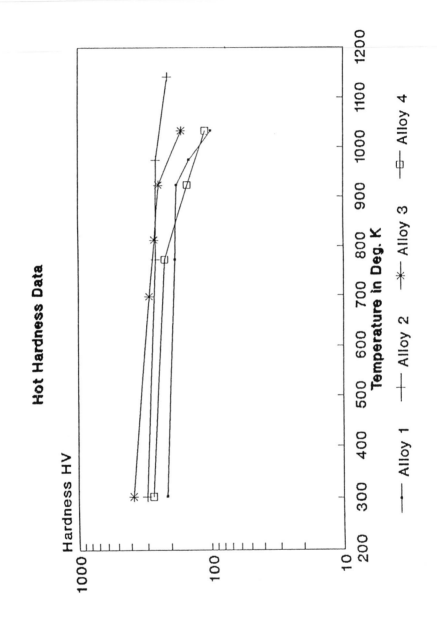

High Temperature Corrosion

LONG-TERM COMPATIBILITY OF A COBALT-BASE SUPERALLOY WITH

MOLTEN LiF-CaF$_2$ EUTECTIC COMPOSITION SALT

Robert P. Rubly[1]

ABSTRACT

An exposure test was performed to determine the long-term compatibility of the cobalt-base superalloy Haynes 188 with eutectic-composition LiF-CaF$_2$ salt. Sealed canisters filled with the salt were exposed to a temperature cycle from 811K to 1116K repeated over a period of 105 minutes for total exposure times of up to 20,000 hours. Following exposure, individual canisters were examined using optical and electron microscopy, energy-dispersive spectroscopy, and mechanical testing. Results indicate that overall compatibility of the salt-metal system is good. Mechanical properties were not affected as a result of exposure to the molten salt. There was evidence of mass transport within the canisters as Cr was found to be depleted from certain surfaces and deposited on others, apparently due to thermal gradients. Extrapolation of the rate of Cr depletion, based on diffusion kinetics, indicates that depletion should not result in significant degradation of the containment canisters over much longer exposure periods.

[1] Allied-Signal Aerospace Company, AiResearch Los Angeles Division, Torrance, CA 90509-2960.

INTRODUCTION

Solar dynamic power has been identified as an attractive power source for spacecraft, including Space Station Freedom, for which solar dynamic was selected as a candidate power system for the growth phase of that program. Solar dynamic power systems use heat energy from the sun to heat a working fluid which may then be used to drive a Rankine, Stirling, Brayton Cycle, or other type engine to produce electrical power. For orbiting spacecraft, solar dynamic power systems require a method for collecting and storing heat. Heat storage is required to provide a uniform power supply during the eclipse portion of the orbit. Various fluoride salts have been identified as ideal heat storage mediums due to their low melting ranges, chemical stability, and high latent heat of fusion [1].

Figure 1 shows a solar dynamic power system designed by Allied-Signal Aerospace Company for Space Station Freedom [2,3]. The system utilizes a parabolic concentrator to focus heat energy from the sun through an aperture and into a cylindrical container. The inner wall of this container is lined with a series of tubes which are heated by the solar flux and carry the heated working fluid to a Brayton Cycle engine. Each tube within this container is surrounded by a series of containment canisters as shown in Figure 2. These canisters are filled with a eutectic mixture of LiF and CaF_2 (LiF - 42.2 wt-% CaF_2)to serve as a heat storage medium. During the sunlight periods, the salt is melted and transfers heat to the working fluid. During eclipse periods, the salt gives up its heat to the working fluid while it cools and solidifies. The heat input includes a significant contribution realized from the latent heat of fusion of the salt.

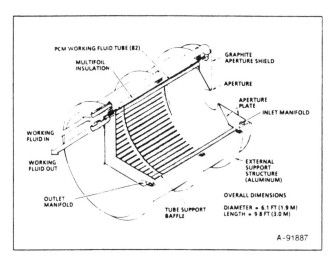

Figure 1. Space Station Freedom Solar Receiver.

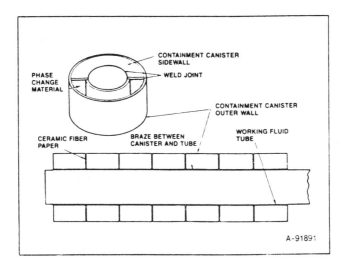

Figure 2. Receiver Tube and Containment Canister Configuration.

The material selected for the containment canisters was the cobalt-base superalloy Haynes 188© (HS188). This alloy was selected for its high-temperature strength and good fabricability. Compatibility with molten LiF-CaF$_2$ was also an important consideration since the system was to be designed for a 30 year lifetime. However, compatibility data for cobalt-base alloys in contact with molten fluoride salts are scarce. Due to the lack of data, a long-term test was initiated to evaluate the compatibility of Haynes 188 with molten LiF-CaF$_2$. A series of test canisters were fabricated, filled with LiF-CaF$_2$ and subjected to a repeated temperature cycle designed to expose the canisters to repeated melt-freeze cycles of the LiF-CaF$_2$ as well as long-time exposure to the molten salt mixture.

EXPERIMENTAL PROCEDURES

The test canisters were fabricated from HS188 sheet and tubing as shown in Figure 3. The outer wall was fabricated from formed and welded 1.27 mm sheet while the inner wall was fabricated from seamless tubing having a wall thickness of 1.12 mm. The sidewalls were fabricated from 1.27 mm and 0.81 mm sheet. The canisters were formed into a cup assembly as shown. The cups were filled with granular LiF-CaF$_2$ which was induction melted under a vacuum of 0.5 to 1.0 torr. After filling, the second sidewall was attached by electron-beam welding under vacuum to form the finished test canisters.

©Haynes 188 is a registered trademark of Haynes International, Inc.
Nominal alloy composition: Co-22Cr-22Ni-14W-0.1C.

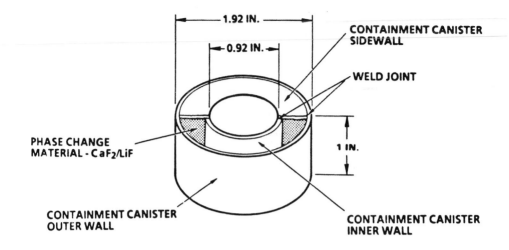

Figure 3. Fabrication of Test Canisters.

All canisters were exposed to a 10 cycle verification test to insure leak-tightness prior to commencement of the compatibility test. One canister was evaluated immediately after this 10 cycle test (18 hour total exposure time). Six test canisters were placed in an electrically heated, air-atmosphere furnace for the long-term exposures. The test exposure cycle is shown in Figure 4. Individual canisters were removed from the furnace at increasing time intervals up to a maximum total exposure time of just over 20,000 hours. Table I shows the specific exposure times for all seven test canisters.

After removal from the furnace, the external surfaces of the canisters were visually examined. The canisters were initially sectioned to allow examination of the internal surfaces. Additional sectioning was performed for metallurgical examination of various canister sections and mechanical testing. Specimens for metallurgical examination were mounted, polished, and examined in both the as-polished and etched conditions. Etching was performed by swabbing with a solution consisting of 30ml HCl, 7ml H_2O and 3ml H_2O_2. The specimens were examined using optical and electron microscopy. Chemical analysis was performed at several locations using energy dispersive spectroscopy (EDS). Reduced size tensile specimens were removed from the top sidewall for evaluation of mechanical properties.

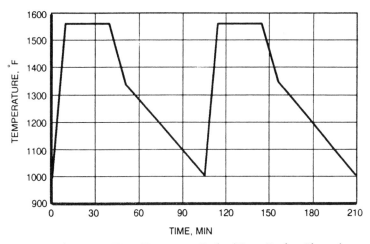

Figure 4. Test Exposure Cycle (Two Cycles Shown).

Table I
Exposure Conditions

Test Canister	Exposure Time (hours)
1	18
2	1,760
3	1,760
4	5,300
5	9,942
6	12,051
7	20,013

RESULTS

Visual Observations

The external surfaces of most canisters were uniform in appearance, having a dark green-black color. There was no indication of surface attack or deformation observed. Low magnification examination revealed an even, smooth surface texture except for one canister (Canister 5) which had a dimpled appearance. This canister was subsequently found to have suffered some form of accelerated oxidation on the external surface which led to wall thinning of the top sidewall. The cause of this condition was not determined, however it was shown to be an isolated case as none of the other canisters, even those with much longer exposure times, exhibited measurable wall thinning.

Preliminary sectioning was performed to allow examination of the solidified salt and internal canister surfaces. Figure 5 shows Canister 7 after initial sectioning. The solidified salt was relatively clean and colorless except a light green discoloration which was observed in localized areas of some canisters. The salt-exposed internal canister surfaces had a bright, shiny appearance, while the surfaces exposed to the void atmosphere (the area above the solidified salt) had a somewhat darker, dull appearance.

Figure 5. Canister 7 After Initial Sectioning.

Metallurgical Examination

Microscopic examination revealed slight surface attack on some internal surfaces. Figure 6 shows typical attack observed on the internal surface of the top sidewall of Canister 6. For the shorter exposure times, the attack resembled random pitting corrosion. With increasing exposure time, it became apparent that the attack was intergranular in nature. In all cases, the surface attack was confined to the upper internal canister surfaces, particularly on the internal surface of the top sidewall. Table II shows the maximum depth of attack observed on the test canisters. In general, the depth of attack increased with increasing exposure time, however, specific correlation with exposure time was inconsistent, especially for shorter exposure times.

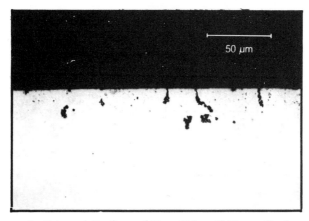

Figure 6. Internal Surface of Top Sidewall From Canister 6.

Table II
Maximum Depth of Surface Attack

Canister	Exposure (hours)	Pitting Depth (μ)
1	18	8
2	1,760	<3
3	1,760	<3
4	5,300	<3
5	9,942	43
6	12,051	38
7	20,013	76

The metallurgical specimens were also examined after etching to reveal grain structure. Figure 7 shows typical microstructures following various exposures and also for unexposed HS188 sheet. Exposure resulted in extensive carbide precipitation, both along grain boundaries and within grains. There was no apparent grain growth. A lighter-etching, carbide-free zone was observed along some internal surfaces, more noticeably with increasing exposure time. This layer was observed predominately in the top regions of the canisters.

For several canisters, chemical concentration profiles where measured on cross-sections using energy dispersive spectroscopy (EDS). Figure 8 shows Cr concentration profiles from Canister 6. The profiles indicated that Cr had been depleted from internal surfaces in the upper regions of the canisters. Depletion was observed to a maximum depth of approximately 30μ, with the most severe depletion observed on Canister 6. The depletion was found to coincide with the surfaces where the intergranular attack and carbide-free layers had been observed.

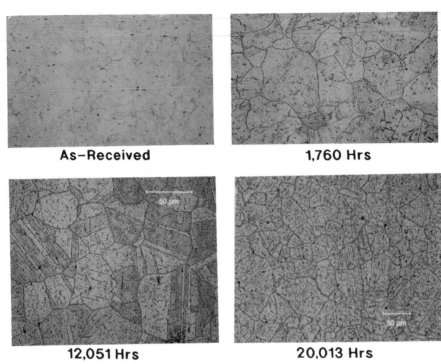

As-Received   1,760 Hrs

12,051 Hrs   20,013 Hrs

Figure 7. Microstructure of Haynes 188 Following Various Exposure Times.

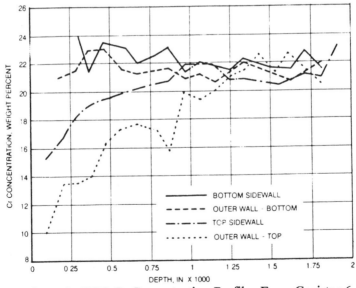

Figure 8. EDS Cr Concentration Profiles From Canister 6.

A continuous layer was observed along the bottom surfaces of Canisters 5, 6 and 7 as shown in Figure 9. The layer was observed on the bottom sidewall and on the inner and outer walls adjacent to the bottom sidewall, and appeared to increase in thickness with the increasing exposure. Figure 10 shows scanning electron microscope (SEM) photographs and EDS spectra taken on internal surfaces in the top and bottom areas of Canister 6. The surface in the top region was found to be covered with an array of particles, tentatively identified as tungsten carbides. EDS results from this area corresponded to the basic composition of HS188. In the lower region, the surface appearance was distinctly different. EDS indicated the composition of the lower surface to be rich in Cr.

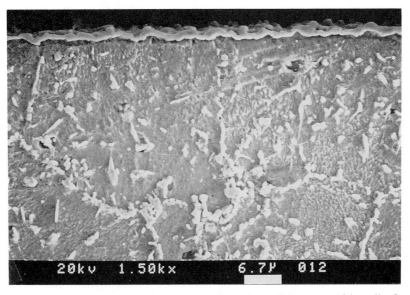

Figure 9. SEM Photograph of Cross-Section Through Bottom Sidewall of Canister 6 Showing Deposited Layer on Internal Surface.

Mechanical Testing

Figure 11 shows results of room temperature tensile testing of reduced size tensile specimens removed from the top sidewalls of the canisters. The results indicate only slight change in strength properties, but significant reduction in ductility following exposure. SEM examination of the tensile specimen fracture surfaces revealed a shift from ductile to intergranular fracture modes with increasing exposure time.

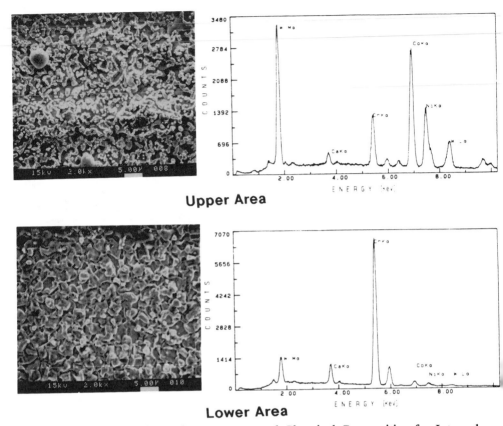

Upper Area

Lower Area

Figure 10. Comparison of Appearance and Chemical Composition for Internal Surfaces in Upper and Lower Areas of Canister 6.

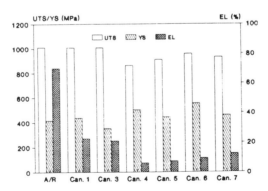

Figure 11. Tensile Test Results from Reduced Size Specimens Removed from Test Canisters After Exposure.

DISCUSSIONS

Chemical Compatibility

Thermodynamic analyses have shown that the LiF-CaF$_2$ salt mixture should not react with HS188 under the test conditions [1]. Preliminary observations indicated that the apparent pitting attack may have been due to contamination within the canisters. It is likely that some contamination was trapped within the canisters at the time of the final closure welds. The salt was melted under only a rough vacuum of 0.5 to 1.0 torr. Also, the canisters were not directly sealed after filling and thus were exposed to atmospheric conditions. The most detrimental contaminant is probably moisture since the presence of H$_2$O can result in the formation of HF.

While entrained contaminates may have been responsible for some surface attack, especially that observed at shorter exposure times, observations after longer exposure times indicate that the surface condition was not due to corrosive attack, but rather a result of the chromium depletion occurring during exposure. The observations of Cr depletion and deposition on different surfaces within the canisters indicate that mass transfer occurred during the test exposure. Koger [5,6] has reported mass transfer in dynamic molten salt systems due to temperature gradients within the system. While the current experiment was thought to be a closed-loop, static system, the positioning of the canisters in the test furnace with respect to the heating elements may have introduced temperature gradients during heating and cooling periods. To investigate the magnitude of any temperature gradients, thermocouples were attached to the top and bottom sidewalls of the Canister 7 during the test. The resulting temperature profiles indicated that the top sidewall was 45K to 55K hotter than the bottom sidewall during the heat-up portion of the cycle. The temperatures equalized after approximately 15 minutes at the maximum exposure temperature. During the cool-down portion of the cycle, there was a similar, but opposite effect. However, in this case, the bottom surface was only slightly hotter than the top surface.

The unintentional temperature gradient experienced by the canisters appears to have been sufficient to result in the observed Cr depletion and deposition. The mechanism for mass transport in fluoride salt systems has been attributed to the formation of CrF$_2$, and the variation in equilibrium concentration of CrF$_2$ with respect to temperature. In hot areas, the equilibrium concentration of CrF$_2$ is greater than in colder areas resulting in mass transfer as Cr is depleted from hot surfaces and deposited in colder areas. The thermodynamic stability of CrF$_2$ is higher than that of other fluorides formed from elements present in structural alloys (Fe, Ni, Co, Mo), and thus would be expected to form in preference of the other fluorides. While both LiF and CaF$_2$ are extremely stable and should not react with Cr, impurities, in the form of other active salts or moisture, can lead to

517

the formation of CrF_2.

In other experiments where Cr depletion has been observed, void formation and carbide dissolution have been reported in the depleted areas [6,7]. Void formation has been attributed to the Kirkendall effect. As Cr is depleted from a surface, outward diffusion of Cr occurs from the underlying region, resulting in an excess vacancy concentration. The vacancies tend to coalesce and grow forming voids, typically at grain boundaries. Carbide dissolution occurs due to the loss of Cr since Cr is a strong carbide-forming element. The similarity of these observations to those from the current investigation indicate that the apparent surface attack observed on the test canisters, especially for the longer exposure times, is probably a result of vacancy coalescence and not corrosive attack.

Although the magnitude of Cr depletion occurring during the compatibility test was relatively small, the effect of continued depletion over the proposed 30 year design lifetime of the system was a concern. If the rate of depletion is linearly extrapolated to a 30 year life, the total depletion expected would be significant. However, because of the dependency of the depletion process on the diffusion of Cr in the parent metal, it is not expected to continue at a linear rate.

The measured concentration profiles show that the surface Cr concentration has been reduced to low levels, and that a concentration gradient exists from the surface to the bulk alloy as a result of depletion. For depletion to continue, Cr must diffuse from the bulk alloy to the surface. The rate of depletion can not be faster than the rate of arrival of Cr at the surface. As depletion continues, the concentration gradient becomes less steep, and the diffusion distance becomes longer, both of which decrease the rate of arrival of Cr at the surface.

Solution to Fick's laws of diffusion show that the length of diffusion, i.e. the depth of penetration or depletion of a diffusing species, is proportional to the square root of time. For a homogeneous alloy held in an atmosphere which reduces the surface solute concentration to zero, the solute concentration in the alloy as a function time and depth from the surface is given by [8]:

$$C(x, t) = C_1 erf(\frac{x}{2\sqrt{Dt}}) \tag{1}$$

where: C = solute concentration
C_1 = initial solute concentration
erf = the error function
x = distance from surface
D = diffusion coefficient
t = time

Eq.1 describes a concentration gradient starting with a concentration of zero at the exposed surface and increasing to the bulk alloy concentration at a certain depth below the surface. Assuming that D does not change, Eq. 1 can be expressed as:

$$x = k\sqrt{t} \qquad\qquad (2)$$

where: k = constant for specific temperature, alloy and concentration ratio

Using results from the compatibility test, k can be calculated, and Eq. 2 can be used to predict depletion depths for longer exposure times. The greatest depletion was observed on Canister 6, having a depleted depth of 30μm after an exposure of 12,000 hours. Based on these data, and assuming the same time/temperature profile, the total depleted depth for an exposure of 30 years would be approximately 150μm. This depth represents a relatively small portion of the total wall thickness. It should be possible to design the actual canisters so that this level of depletion will not significantly affect the integrity of the canisters.

Mechanical Testing

The tensile test results have shown that exposure resulted in a large decrease in ductility. However, HS188 is known to experience a reduction in ductility following thermal exposure due to intergranular carbide precipitation. Data from Haynes International, the material supplier, show that room temperature tensile elongation decreases from 63% for unexposed material to 9% following exposure at 1033K for 8000 hours. Preliminary results from Wittenberger [4], from an extensive program to determine the effects of exposure to LiF-CaF$_2$ on the mechanical properties of HS188, indicate that mechanical properties after exposure to the molten salt or its vapor were the same as those following identical thermal exposure in a vacuum atmosphere. It appears, therefore, that the decrease in ductility measured on the test canister specimens is due to the effects of thermal exposure, and not a result of interaction with the molten salt.

CONCLUSION

The long-term compatibility test has shown that HS188 is compatible with LiF-CaF$_2$ in the temperature range of 811K to 1116K. The test canisters showed no evidence of significant degradation following exposure. Mechanical properties were not affected by exposure to the salt. Mass transport of Cr from hot to cold surfaces was observed due to slight thermal gradients experienced by the test canisters during the heat-up portion of the test cycle. Intergranular void formation and carbide dissolution were observed to very shallow depths on the Cr-depleted surfaces. Diffusion kinetics indicate that the rate of Cr depletion will decrease with continued exposure time and that Cr depletion should not be detrimental over much longer exposure periods.

ACKNOWLEDGEMENT

This investigation was conducted under contract from Rockwell International Corp., Rocketdyne Division and under the aegis of NASA Lewis Research Center.

REFERENCES

1) A. K. Misra and J. D. Wittenberger, "Fluoride Salts and Container Materials for Thermal Energy Storage Applications in the Temperature Range 973-1400 K", Proceedings of the 22nd Intersociety Energy Conversion Engineering Conference, Paper 879226, 1987.

2) H. J. Strumpf and M. G. Coombs, "Solar Receiver for the Space Station Brayton Engine", **J. Eng. Gas Turbines and Power**, Vol. 110, p 295, 1988.

3) H. J. Strumpf, R. P. Rubly and M. G. Coombs, "Material Compatibility and Simulation Testing for the Brayton Engine Solar Receiver for the NASA Space Station Freedom Solar Dynamic Option", Proceedings of the 24th Intersociety Energy Conversion Engineering Conference, Vol. 2, Paper 899076, p 895, 1989.

4) J. D. Wittenberger, "Tensile Properties of HA 230 and HA 188 After 400 and 2500 Hour Exposure to LiF-22CaF$_2$ and Vacuum at 1093 K", **J. Materials Engineering**, 1990.

5) J. W. Koger, "Fluoride Salt Corrosion and Mass Transfer in High Temperature Dynamic Systems", **Corrosion**, Vol. 29, p 115, 1973.

6) J. W. Koger, "Chromium Depletion and Void Formation in Fe-Ni-Cr Alloys During Molten Salt Corrosion and Related Processes", Advances in Corrosion Science and Technology, Vol. 4, M.G. Fontana and R.W. Staehle, Editors, p 245, 1974.

7) D. T. Bourgette and H. E. McCoy, "A Study of the Vaporization and Creep-Rupture Behavior of Type 316 Stainless Steel", **Trans. ASM**, Vol. 59, p 324, 1966.

8) P. G. Shewmon, Diffusion in Solids, J. Williams Book Co., p 14, 1983.

DEGRADATION OF 316L STAINLESS STEEL BY

MOLTEN ALUMINUM

James C. Marra[1]

ABSTRACT

When stainless steel contacts molten aluminum, dissolution of the steel occurs. A test program is ongoing to determine the influence of various conditions on the degradation of stainless steel as well as obtain mechanistic information regarding the dissolution process. Part 1 of the program examining the influence of temperature, time, and steel surface condition on the dissolution process is described in this paper. Apparent dissolution mechanisms at various temperatures are also discussed.

INTRODUCTION

In the unlikely event of a severe accident, involving complete loss of coolant, in the Savannah River Site reactors aluminum fuel and target elements could melt resulting in a pool of molten aluminum and debris contacting the stainless steel reactor vessel. In this scenario, temperatures of the melt as high as 1200°C have been hypothesized. This paper discusses the results of Part 1 of an ongoing study to examine the interaction of stainless steel with molten Al in severe reactor accident conditions.

---

[1] Westinghouse Savannah River Company, Savannah River Laboratory, P.O. Box 616, Aiken, SC 29802.

The attack of stainless steel, refractory metals and iron alloys by molten aluminum is a know phenomenon [1-5]. In interactions involving stainless steel and Al, growth of intermetallic layers occurs at the liquid aluminum-stainless steel interface. Dissolution of the stainless steel subsequently proceeds via a diffusion controlled process with iron being the rate controlling species [1]. Dybkov found that the dissolution was non-selective, i.e. the major constituent atoms of the stainless steel passed into the aluminum in the same ratios as were present in the steel [1].

A number of variables determine the rate and nature of the dissolution process. Three important factors are: temperature of the molten aluminum, time the stainless steel is exposed to the melt and stainless steel surface condition. Intuitively, an increase in temperature should lead to enhanced attack rates in a diffusion controlled process. It is also expected that the amount of attack would be enhanced due to the increased solubilities of the major stainless steel components in Al at higher temperatures as shown by the binary phase diagrams [6]. The rate of attack, however, is also influenced by the time the steel is exposed to the melt. As the components of the stainless steel are dissolved into the aluminum melt, a decrease in the dissolution rate occurs and dissolution ceases when the saturation concentration is reached. The surface condition also determines the nature of attack. Wetting of the steel surface is necessary for degradation to occur and the presence of oxide films on the steel surface is expected to impede the attack.

EXPERIMENTAL PROCEDURES

Immersion test coupons of 316L stainless steel approximately 1.9 cm (0.75 in) x 1.3 cm (0.50 in) x 0.16 cm (0.0625 in) were cut from larger stainless steel stock. To investigate the influence of surface oxide effects, three different coupon surface conditions were tested. One set of coupons was left "as-received" from the metals supply vendor (Metals Samples, Munford, AL). A second set was mechanically ground to minimize surface oxide film using 600 grit SiC. These coupons were additionally hand ground immediately prior to exposure to molten aluminum. A third set was treated to provide an extreme surface oxide condition on the coupons. This treatment consisted of annealing in air at 1100°C for 5 minutes followed by a water quench; acid cleaning in 2 volume percent HF + 10 volume percent HNO_3 in water for 20 minutes and thoroughly rinsing; and passivating in 25 volume percent HNO_3 for 30 minutes at room temperature followed by a water rinse.

Laboratory grade aluminum nuggets (99.95% purity - Noah Technologies Corporation, San Antonio, TX) were melted in an alumina crucible. Melt volumes were approximately 10 ml for each test. Figure 1 shows schematically the experimental configuration for the immersion tests.

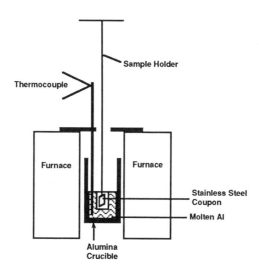

Figure 1. Schematic diagram of coupon immersion test apparatus.

Different samples representing each surface finish were exposed to each of the following temperature and time conditions: 700°C (5 minutes, 2 hours), 940°C (1 minute, 5 minutes, 30 minutes), and 1140°C (1 minute, 5 minutes). Following immersion, samples were mounted and polished to examine the cross-section of the coupon. An image analysis system facilitated dissolved area computations. A "thresholded image" was obtained for the remaining stainless steel coupon and an area dissolved was calculated. This technique was performed on four different sections of each tested coupon yielding an average area dissolved measurement. This value was then multiplied by the immersion depth to provide a volume dissolved value. The area technique was chosen over weight loss measurement for a few reasons.

The primary reason was that removal of the Al would limit microscopic examination of the interface and surrounding melt region. Additionally, difficulties in removing the intermetallic layers at the interface were expected to result in an underestimation of the dissolution. Finally, expanding the application base for the image analysis system was also desirable.

The dissolution of a solid metal into a liquid metal in a binary system can be described by the following expression [1]:

$$\ln\left(\frac{Cs - Co}{Cs - C}\right) = k\left(\frac{St}{V}\right) \qquad (1)$$

k = Dissolution Rate Constant (m/sec)
Cs = Saturation Concentration of Solid in Liquid
Co = Initial Concentration of Solid in Liquid
C = Measured Concentration of Solid in Liquid
S = Solid Surface Area
V = Volume of Melt
t = Time

This expression was used to calculate the dissolution rate constant (k) for the present experiments. Saturation concentrations for the stainless steel in Al were estimated from the Fe-Al binary phase diagram [6]. This is a reasonable value for the calculations as Fe is the major constituent in stainless steel and is proposed to be the diffusion controlling species [1,2].

Scanning electron microscopy (SEM) and electron dispersive spectroscopy (EDS) were performed on a number of the prepared sections to investigate the dissolution process and intermetallic formation at the interface.

RESULTS AND DISCUSSION

Variables Influence on Dissolution

The effect of the surface finish on the dissolution of stainless steel is shown by the micrographs in Figures 2 and 3. In Figure 2 it is evident that a ground surface enhances wetting of the stainless steel by the Al and leads to enhanced dissolution. Note the 'dimple-like" degradation of the ground sample in Figure 2a. Figure 3 shows attack on passivated samples where the protective oxide film was nonresistant

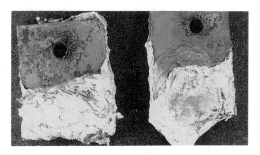

(a) Air Annealed/Passivated   600 Grit Ground   (Mag.=3X)

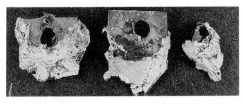

(b)   Air Annealed/Passivated   As-Received   600 Grit Ground   (Mag.=2X)

Figure 2.   Effects of surface oxide condition on dissolution of stainless steel in molten aluminum.

(a) Section 1 (Magnification=8X)

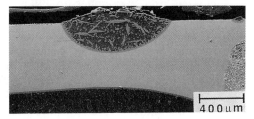

(b) Section 2

Figure 3.   Selective attack on annealed/passivated sample exposed to molten aluminum at 940°C for 30 minutes.

or easily breached in certain regions. It was quite evident that an unprotected (ground) stainless steel sample displays a general-type attack by molten Al, whereas an oxidized steel is attacked selectively much like "pitting" corrosion.

The dissolution rate constant calculations support these visual observations for the various sample surface conditions. At 940°C, the rate of dissolution of the ground sample was larger for all exposure times (Figure 4). This was especially evident for samples exposed for 30 minutes (Figure 2a). It is apparent that removing most of the protective oxide film led to "general-type" attack and perpetuated the degradation by allowing Al to contact a large amount of "fresh" stainless steel surface. The as-received samples and passivated samples showed comparably similar dissolution behavior for all test temperatures. In this case, the attack was selective in nature. The oxide film on the stainless steel inhibited contact between the Al and the steel, thus, leading to lower dissolution rates.

The calculated dissolution rate constants (Figures 4 and 5) generally decreased with increased immersion time. This was anticipated as an increase in concentration of stainless steel components in the Al melt occurs with increasing exposure times. Thus, for longer exposures the melt can "accommodate" progressively lower amounts of stainless steel components (namely Fe) and dissolution rate decreases. Eventually dissolution would cease when the saturation concentration of "stainless steel" in the aluminum is reached. Dybkov [1] determined elemental concentrations of stainless steel in Al when dissolution ceased. The concentrations of Fe from stainless steel generally coincided with the solubility of Fe in Al from standard phase diagrams. This led to the postulation that the dissolution is diffusion controlled and Fe is the rate controlling species. It is doubtful, however, that the dissolution rate constant for the ground samples remained essentially unchanged as the exposure time increased from 5 minutes to 30 minutes (Figure 4) due to the tendency towards melt saturation. This clearly points out the limits and potential errors associated with the "dissolved area" technique employed. For this reason, it was concluded that a weight measurement technique would provide overall more accurate results.

All three surface conditions showed similar dissolution trends with respect to increasing melt temperatures. A significant increase in dissolution rate is evident for an increase in temperature (Figure 5). This trend was consistent with the work of Niinomi, et. al [3] for iron alloy and Yeremenko, et. al for refractory metal dissolution in molten aluminum. However, this trend conflicts with the findings of Dybkov [1] for stainless steel dissolution in Al. It is postulated that the

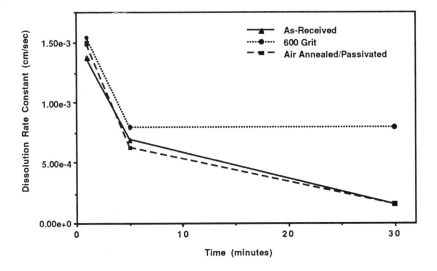

Figure 4. Dissolution rate constants for samples with various surface conditions exposed to molten aluminum at 940°C.

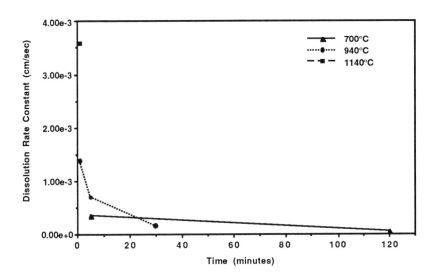

Figure 5. Dissolution rate constants for as-received samples exposed to various aluminum melt temperatures.

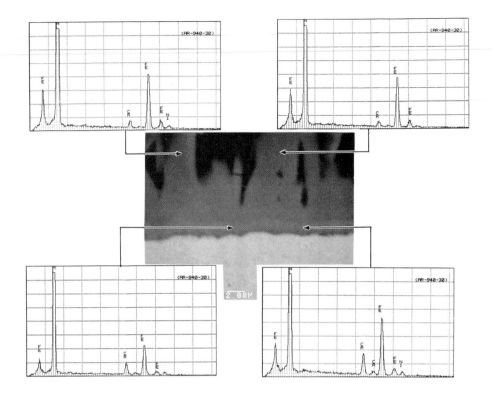

Figure 6. SEM micrograph of interfacial region of sample exposed at 940°C with corresponding EDS spectra.

relatively small temperature range (125°C) and rapid sample rotation (24 rad/sec on 1.1 cm diameter disks) led to no detectable gross changes in dissolution rate constant in Dybkov's experiments.

Dissolution   Mechanisms

A dual-phase interface region was observed for samples exposed at 700°C and 940°C (Figures 6 and 7). The intermetallic layer adjacent to

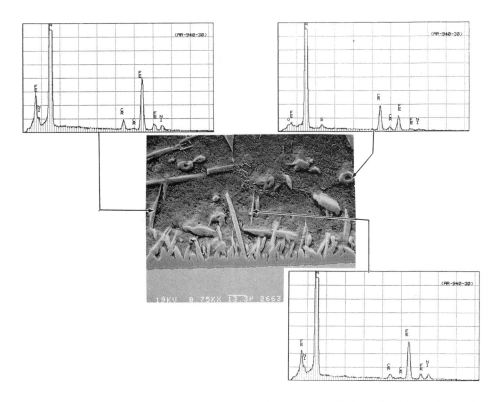

Figure 7. SEM micrograph and EDS spectra of interfacial region of sample exposed at 940°C.

the stainless steel is approximately 1 μm thick. The second intermetallic layer was observed to vary in thickness from about 5 μm to over 50 μm where plate-like regions were observed. The interfacial layers formed relatively rapidly as they were present in samples exposed for 1 minute and remained unchanged as little difference in interface morphology and dimensions was observed with varying exposure time. The two intermetallic layers varied significantly in Fe/Cr ratio as shown in the accompanying EDS spectra (Figure 6).

These observations correspond well with literature references involving this interaction [1,2]. It has been hypothesized that the layer adjacent to the stainless steel is based on Fe_2Al_5 while the layer adjacent to the melt corresponds to $FeAl_3$ intermetallic [1]. In both phases, Cr and Ni atoms are thought to partially substitute for Fe atoms. Dissolution is, thus, said to occur by passage of constituent atoms from the stainless steel through the interfacial region into the melt with the passage of Fe atoms being the rate controlling step. Significant amounts of intermetallic phases were observed in the molten matrix following solidification. These intermetallic compounds displayed two distinct morphologies: plate-like and tubular (Figure 7). The plate-like phases were very similar in chemical composition to the $FeAl_3$-base intermetallic interface layer. The unique tubular-shaped phases were chromium-rich and in some cases "filled" with pure Al phase. No plausible explanation for the formation of an intermetallic phase with this unique morphology is available at this time.

The dissolution behavior appeared significantly different at the elevated temperature of 1140°C. The diffusion of atoms appeared to occur in both directions (i.e. form the steel into the melt and the melt into the steel). The resulting interface for a sample exposed for 1 minute is shown in Figure 8. The interfacial layers were approximately the same thickness with some noticeable porosity primarily in the layer adjacent to the stainless steel. It is speculated that the porosity was due to a macroscopic "Kirkendahl-like" effect in which Fe (Ni and Cr) atoms diffused faster through the interface and into the aluminum melt than the Al atoms were diffusing through the interface region. It is interesting to note that the coupon has still maintained relatively the same dimensions and shape of the original coupon. For samples exposed for 5 minutes at 1140°C, the bar again had the approximate original dimensions, but now consisted of a porous "composite" of intermetallic phases surrounded by an aluminum matrix (Figure 9). This "composite" was similar to what was seen in the Al-rich (left hand) side in Figure 8b. From these micrographs the attack appeared much more rapid as the interface apparently moved across the stainless steel with increasing exposure time until the stainless steel was "consumed". The resulting intermetallic phases (again plate-like and tubular) provided some fascinating microstructures upon solidification.

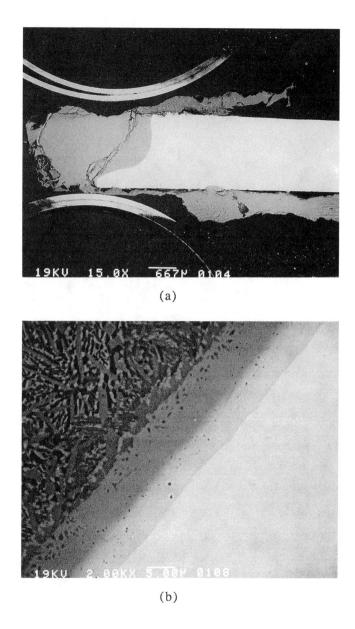

(a)

(b)

Figure 8. SEM micrographs of 600 grit ground sample exposed to molten aluminum at 1140°C for one minute.

(a)   Magnification = 8x

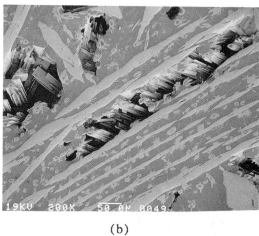

(b)

(c)

Figure 9.   SEM micrographs of air annealed/passivated sample exposed to molten aluminum at 1140°C for five minutes.

CONCLUSIONS

The following conclusions can be made from this work:

• A protective oxide film on stainless steel tends to inhibit the dissolution process and leads to a "pitting-like" attack.

• Dissolution rates increase with increasing temperatures.

• Dissolution rates decrease with increasing exposure time as the concentration of stainless steel components (namely Fe) increases in the melt.

• At the lower temperatures studied (700°C and 940°C) dissolution appeared diffusion controlled with the passage of atoms from the stainless steel into the Al melt.

• At 1140°C, diffusion of atoms appeared to occur from the stainless steel into the melt and from the Al into the steel causing rapid degradation of the steel.

FUTURE WORK

An experimental effort to investigate the effects of melt agitation on the dissolution process is planned. Additionally, the effects of uranium in the melt, which would be present from molten fuel elements, will be examined. These efforts will be concentrated at elevated temperatures (> 1100°C) in an attempt to better understand the apparent dual-diffusion mechanism and effects in reactor accident scenarios.

REFERENCES

1) V. I. Dybkov, "Interaction of 18Cr-10Ni Stainless Steel with Liquid Aluminum", **Journal of Materials Science**, Vol. 25, p 3615, 1990.

2) T. N. Nazarchuk, G. T. Kabannik and V. I. Dybkov, "Kinetics of Solution of 12Kh18N10T Steel in Molten Aluminum", **Soviet Materials Science**, Vol. 21, p 413, 1985.

3) M. Niinomi, Y. Ueda and M. Sano, "Dissolution of Ferrous Alloys in Molten Aluminum", **Transactions of the Japan Institute of Metals**, Vol. 23, p 780, 1982.

4) V. N. Yeremenko, Ya. V. Natanzon and V. I. Dybkov, "Interaction of the Refractory Metals with Liquid Aluminum", **Journal of Less Common Metals**, Vol. 50, p 29, 1976.

5) V. N. Yeremenko, Ya. V. Natanzon and V. I. Dybkov, "The Effect of Dissolution on the Growth of the Fe_2Al_5 Interlayer in the Solid Iron-Liquid Aluminum System", **Journal of Materials Science**, Vol. 16, p 1748, 1981.

6) T. B. Massalski, Editor-In-Chief, **Binary Alloy Phase Diagrams**, Vol. 1, ASM International, Materials Park, OH, p. 139, 1990.

CHARACTERIZATION OF SURFACE OXIDES ON Al-Li ALLOYS

Kamal K. Soni

ABSTRACT

The high temperature oxidation behavior of Al-Li alloys was studied with regard to the morphology and microchemistry of the oxide. Oxide morphology development is described in terms of nucleation and growth phenomena. Li distribution in the oxide and the underlying alloy was studied by secondary ion mass spectrometry and neutron depth profiling. Enhancement of oxidation by second-phase was also explored.

INTRODUCTION

The addition of Li to Al leads to reduced alloy density and increased elastic modulus but is accompanied by increased oxidation rate at high temperature [1]. Oxidation of Al-Li alloys occurs during high temperature processing of these materials such as solutionizing, diffusion bonding, rapid solidification and superplastic forming. Since Li oxidizes preferentially relative to Al, the alloy adjacent to the oxide is depleted of Li and the alloy properties are adversely affected. Therefore, it is important to study the oxidation mechanism, oxide characteristics and the associated effect on the near-surface alloy composition. Since Li cannot be detected by x-ray microanalysis, the techniques of secondary ion mass spectrometry (SIMS) and neutron

Enrico Fermi Institute and The Department of Physics
The University of Chicago
Chicago IL 60637

535

depth profiling (NDP) have been employed in the present study to investigate Li distribution in the oxide and in the underlying alloy.

EXPERIMENTAL PROCEDURES

Specimen Preparation: Polished Al-Li alloy samples were oxidized at 530 °C in air for times ranging from 5 min. to 1 hr. Subsequently, the specimens were cooled in air in order to retain the oxide layer on the surface; some specimens were water quenched after oxidation treatment in order to spall off the oxide layer and to examine the reacted alloy surface. Care was taken not to damage the surface layer and the specimens were not cleaned ultrasonically as organic solvents and water may dissolve Li-containing oxides.

Morphological Examination: The oxide morphologies were examined with an ETEC SEM at 20 kV or with a JEOL 840F SEM at 1 or 5 kV. The latter employs a field emission gun electron source and enables examination of the insulating surfaces at 1 kV without the need of a conductive surface coating.

Some oxidation experiments were performed in an Electroscan E3 environmental SEM (E-SEM) at 20 kV. The construction and operational principles of this instrument are described elsewhere [2]. For this experiment, the E-SEM was operated with an air pressure in the range 3-10 torr near the specimen surface. The specimens were heated and oxidized in the microscope and the development of the surface oxides was observed *in-situ* and recorded on a video tape. Polished specimens having dimensions of approximately 4 X 4 X 0.5 mm^3 were placed in an Al_2O_3 crucible for heating; thin specimens attain thermal equilibrium with the crucible rapidly.

SIMS Studies: Two configurations of SIMS were employed for the microchemical examination of the oxide layer and the near-surface alloy. In an ion microscope, a broad ion beam bombards the specimen to cause sputtering of the top atomic layers. The ejected secondary ions are analyzed according to their mass to charge ratio and their 3-D distribution is recorded in the form of ion images and depth profiles. A Cameca IMS 3F ion microscope was used in the present work and operated with an O_2^+ primary beam. In a scanning ion microprobe, a finely focused probe is rastered over the specimen surface and ion distribution maps are built up pixel-by-pixel. The lateral resolution with an ion probe, formed using a liquid metal ion source such as Ga, is nearly equal to the probe diameter (50-100 nm) and constitutes a major advantage of this technique compared to the ion microscope which has a lateral resolution of ~1 μm. However, the ion microprobe is unsuitable for obtaining deep depth profiles because of its slow erosion rate, whereas the ion microscope can be used to perform depth profiling up to a depth of ~10 μm. In this study, a scanning ion microprobe (SIM) developed at the University of Chicago (UC) was employed which is described by Levi-Setti *et al.* [3]. In the UC SIM, a finely focused Ga^+ probe is extracted from a liquid Ga source and accelerated to 40 keV. The probe is scanned on the specimen surface using a 512 X 512 raster spanning areas $(2.5 \ \mu m)^2$ to $(80 \ \mu m)^2$. The specimens were sputter-coated with a thin layer of Au-Pd in order to prevent charge build-up and also to preserve the surface during storage.

NDP Studies: NDP measurements were conducted at the National Institute of Standards and Technology (NIST), Gaithersburg, Maryland. The NDP principles and instrumentation are discussed in detail by Downing *et al.* [4,5]. A 20 MW reactor supplies an intense thermal neutron beam for the NDP experiment station which is collimated by passing it through steel, lead and cadmium successively. The ~1 cm diameter neutron beam impinges on the specimen at an angle to cause the following reaction with the ^6Li isotope:

$$^6Li + n = {}^3H \ (2055 \ keV) + {}^4He \ (2727 \ keV).$$

The tritium particles and the alpha particles are emitted isotropically with characteristic energies. The particles produced at the surface register full energy as indicated above, but those produced below the surface lose energy as they travel through the specimen on their way to a silicon surface barrier detector. The energy distributions of the detected signals are measured and translated into concentration-depth profiles [6]. The specimens were examined in the uncoated condition and had dimensions of ~25 mm X 25 mm X 3 mm.

RESULTS

Morphological Studies

Figs. 1(a,b) show morphology of the oxide formed on a polycrystalline Al-4.4 at.% Li alloy oxidized at 530 °C for 5 min. in air. Rapid oxidation can be identified at certain sites in the form of coarse oxide nodules (Fig. 1b). These nodules are surrounded by a thick oxide rim, while the rest of the oxide film is relatively uniform.

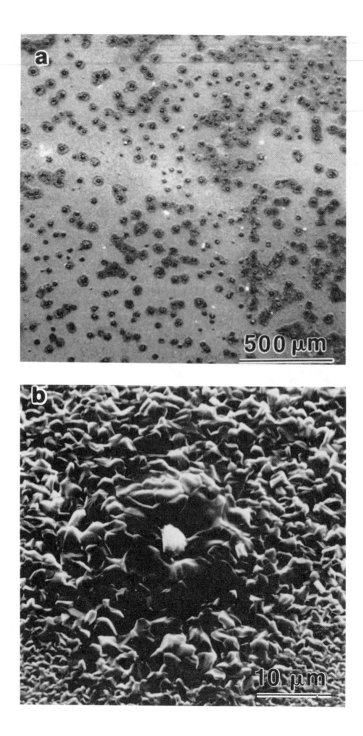

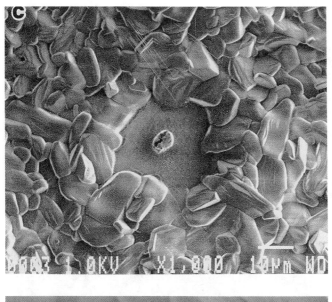

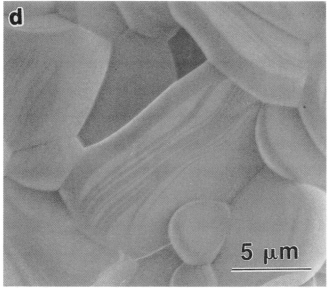

Fig. 1: SEM micrographs of a polycrystalline Al-4.4 at.% Li alloy oxidized at 530 °C and air-cooled. (a) 5 min., (b) 1 hr.

539

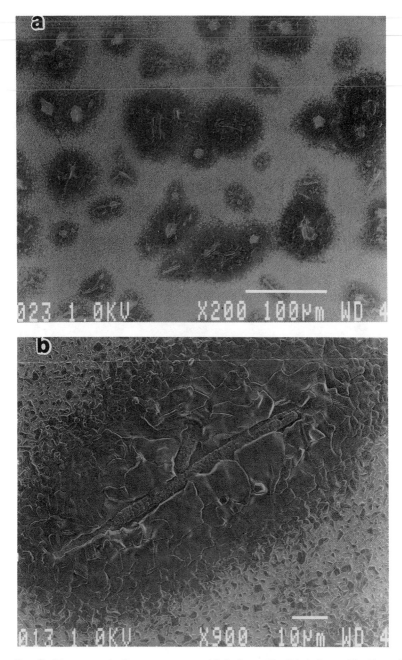

Fig. 2: Oxide morphology on an Al-3.6 at.% Li-1.0 at.% Fe alloy, oxidized at 530 °C for 5 min. and air-cooled.

This characteristic morphology was not present on a single crystal Al-Li alloy oxidized under identical conditions suggesting that the oxide nodules are associated with grain boundaries [7]. Partridge and Chadbourne [8] reported the presence of four characteristic oxide morphologies (types 1-4) on the surface of an Al-Li-Cu-Mg alloy oxidized at high temperature for short durations. The type 4 morphology described by these authors is similar to the nodules observed in the present study. The initial nodular oxide morphology observed after 5 min. transforms into that shown in Fig. 1(c) after 1 hr. of oxidation at 530 °C; the central nodule area becomes flat while coarse oxide grains develop in the rest of the film. This change in morphology is thought to be due to differences in growth rates at different locations. Fig. 1(d) displays facets on the surface of the oxide layer which are indicative of surface reconstruction at high temperature.

In order to examine the role of Fe-rich second-phase particles, an alloy containing a relatively large fraction of these particles was oxidized. Fe is a common impurity in Al-Li alloys and is present at ~0.06-0.3 wt.% level as Al_3Fe precipitates which are devoid of Li. Some previous studies attributed the high oxidation rate to the presence of these insoluble particles [8]. Figs. 2(a,b) show the surface of an Al-3.6 at.% Li-1.0 at.% Fe alloy oxidized at 530 °C for 5 min. [9]. The oxide is thicker in the vicinity of the second-phase; therefore the enhancement of oxidation can be attributed to these precipitates. This oxide also appears darker than the rest of the film in the secondary electron image; this contrast is probably due to the higher Li content of this oxide. The Fe-rich particles also oxidize, although to a much less extent.

Examination of the Oxide/Alloy Interface

The alloy surface was investigated with SEM and UC SIM after spalling the oxide by water quenching. Fig. 3(a) is an SEM micrograph of the alloy surface beneath an oxide nodule as described in Fig. 1(b). The bright circular halo is remnant of the oxide nodule on the surface. Similarly, Fig. 3(b) shows the surface of an Al-3.6 at.% Li-1.0 at.% Fe alloy, oxidized at 530 °C for 5 min. after oxide removal and corresponds to Fig. 2(b).

These reacted alloy surfaces were examined by the UC SIM in order to understand the Li distribution below the surfaces that are shown in Figs. 3(a,b). Figs. 3(c) and (d) show Li^+ maps of the specimens described in Figs. 3 (a) and (b) respectively after sputtering away the oxide residue so as to image the Li distribution in the underlying alloy caused by the observed morphological phenomena.

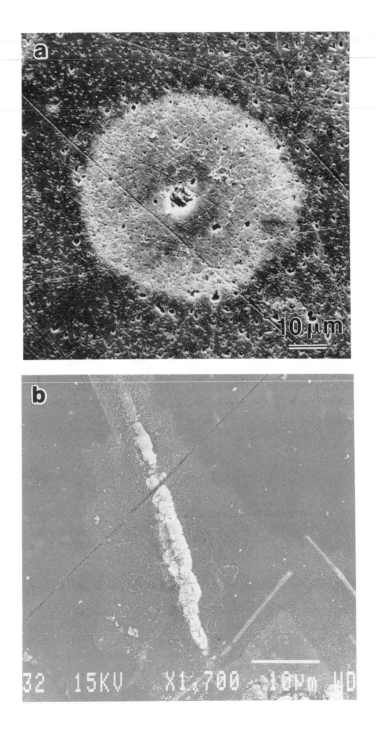

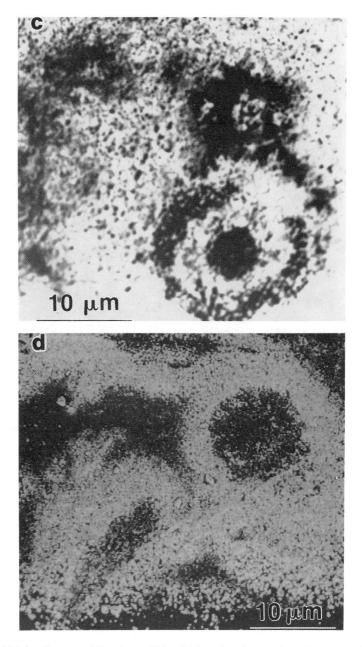

Fig. 3: Al-Li alloys oxidized at 530 °C for 5 min. and water-quenched to spall the oxide. (a,c) Al-4.4 at. % Li, (c,d) Al-3.6 at.% Li-1.0 at.% Fe. (a,b) SEM micrographs, (c,d) UC SIM ion maps of Li.

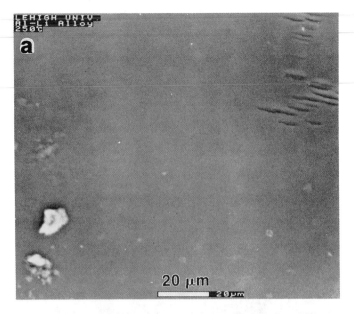

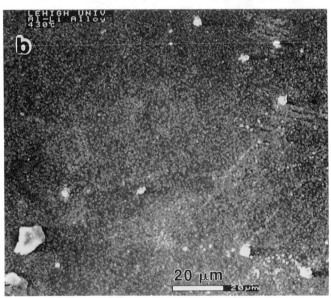

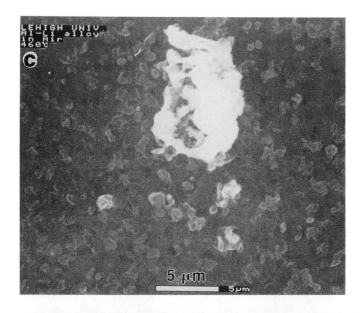

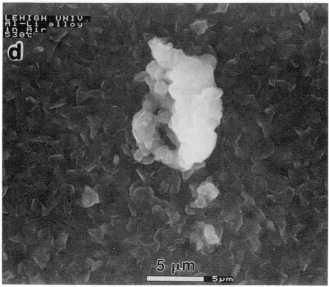

Fig. 4: Stills from *in-situ* oxidation studies of an Al-11.7 at.% Li alloy in an environmental SEM.   (a) 3.3 torr/250 °C, (b) 3.3 torr/430 °C, (c) 3 torr/460 °C, and (d) 5.7 torr/530 °C.

Fig. 3(c) reveals a circular Li depletion area surrounded by a Li depletion ring; the former corresponds to the oxide nodule while the latter is related to the thick oxide rim as displayed in Fig. 1(b) [7]. Fig. 3(d) shows Li in the alloy is segregated toward the alloy/second-phase interface [8].

In-situ *Oxidation Studies*

Oxidation of some Al-Li alloys was pursued *in-situ* in an E-SEM in order to achieve a better understanding of the evolution of oxide morphologies. Figs. 4(a-d) display a sequence of micrographs recorded from the surface of an Al-11.7 at.% Li alloy as the specimen was heated up to 530 °C at a pressure 3-10 torr. The originally flat, polished surface undergoes oxidation and develops coarse oxide eruptions at certain locations, unaided by any second-phase; the rest of the oxide film is composed of very fine oxide particles. One such oxide eruption is highlighted in Figs. 4(c,d): the growth of the central nodule is slow but the surrounding oxide grains grow coarser. Figs. 5(a,b) show development of the oxide morphology on an Al-3.6 at.% Li-1.0 at.% Fe alloy in the early stages of the reaction. In Fig. 5(a), a network of the needle-shaped Fe-rich second-phase precipitates is seen. Upon heating, oxide growth is seen originating in the vicinity of these particles in the lateral direction (Fig. 5b).

SIMS and NDP Analysis of the Oxidized Alloys

Fig. 6 shows a depth profile of the oxide film on an Al-10.4 at.% Li oxidized at 530 °C for 5 min. [7]. This depth profile was obtained by eroding $(2.5 \ \mu m)^2$ area with the Ga^+ beam in the UC SIM such that only upper layers of the oxide were sampled. This profile demonstrates that the surface is covered with a layer of Li_2O which contains negligible Al; beneath this layer, both Al-rich and Li-rich oxides are present which are probably Li_2O and $LiAlO_2$ [7]. SIMS analysis also revealed presence of trace amounts of C, Na, K, F, and Cl in the oxide film. Similar results were obtained for the Al-Li alloy containing Fe-rich second-phase, although some Fe was also detected in the oxide. A more detailed study of the oxide microchemistry is presented elsewhere [7,9,10].

In order to obtain depth profiles that extend from the oxide surface in to the underlying alloy, the Cameca ion microscope was employed which has a higher erosion rate than that of UC SIM. A depth profile of an Al-4.4 at.% Li alloy oxidized at 530 °C for 5 min. is given in Fig. 7 and demonstrates the relative change in Li concentration across the oxide/alloy interface [10]. The alloy adjoining the oxide is depleted

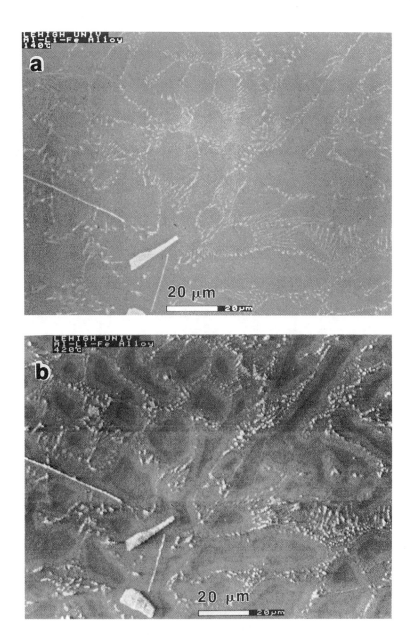

Fig. 5: *In-situ* oxidation of an Al-3.6 at.% Li-1.0 at.% Fe alloy at 3.3 torr. (a) 140 °C, (b) 420 °C.

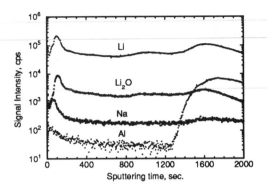

Fig. 6: UC SIM depth profile of the upper layers of the oxide film on an Al-10.4 at. % Li alloy oxidized at 530 °C for 5 min..

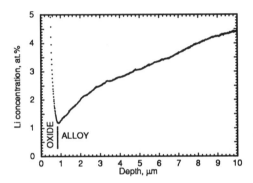

Fig. 7: SIMS depth profile of an Al-4.4 at. % Li alloy oxidized at 530 °C for 5 min..

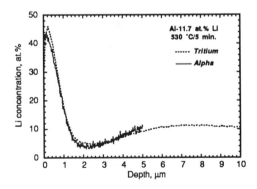

Fig. 8: Li concentration profiles of an Al-11.7 at. % Li alloy oxidized at 530 °C for 5 min. by neutron depth profiling.

because Li oxidizes preferentially relative to Al.

As discussed in experimental section, the basis of Li microanalysis by NDP is the reaction between a ^{6}Li atom and a neutron to produce a tritium particle and an alpha particle. Fig. 8 shows superimposed Li distributions obtained by analyzing both the tritium and the alpha particle distributions from an Al-11.7 at.% Li alloy oxidized at 530 °C for 5 min. [10]. The two profiles are nearly identical although the alpha signal is slightly noisy. The Li-rich peak belongs to the oxide film under which is the Li-depleted alloy surface. Because of interference from the alpha signal at lower energies, the tritium signal can be used to study a maximum depth of ~13 μm in Al-Li matrix. The useful depth range for the alpha signal is limited to ~5 μm. Because of this limitation, alloys oxidized for longer times could not be examined as the oxide thickness was greater than that accessible by NDP.

The above SIMS and NDP profiles also provide a good estimate of the oxide thickness which cannot be obtained from thermogravimetric measurements or Li-depletion studies. These profiles also enable a comparison of Li deficit in the alloy and amount of Li present in the oxide. Such comparison for the NDP profiles indicated that a dynamic equilibrium exists between the oxide and the alloy [10]. These profiles also demonstrate that the Li concentration at the oxide/alloy interface is non-zero. Since oxidation is a diffusion-controlled process, these depletion profiles were used to calculate diffusion data [10].

DISCUSSION

The nodular oxide morphology on a pure, polycrystalline Al-Li alloy (see Fig. 1b) can be rationalized in terms of oxide nucleation and growth phenomena. Oxide nucleation is favored at grain boundaries and other crystallographic defects. Most of the circular oxide patches were located on grain boundaries; the rest are probably associated with dislocations or represent random nucleation [7]. This conclusion is supported by the *in-situ* observations of oxide eruptions on polished surfaces (see Fig. 4a,b). In the initial stages, the oxide nuclei grow laterally along the specimen surface as evidenced by the circular residue present beneath the oxide nodules. Subsequently, oxide growth occurs normal to the specimen surface. These coarse oxide nodules may be responsible for the high oxidation rate of Al-Li alloys especially in the initial stages. Li for the enhanced oxide growth at these sites is drained from the underlying alloy. This conclusion is substantiated by the presence of circular Li depletion areas in the alloy as shown in Fig. 3(d). The growth of these nodules stops when the Li from the alloy is exhausted while the surrounding oxide grains continue to coarsen as demonstrated in Fig. 1(c) and also by the *in-situ* experiments (Figs. 4 a,b).

The incoherent second-phase particles were found to enhance the oxidation rate of Al-Li alloys [9]. Li segregates to the alloy/particle interface which acts a channel for fast outward diffusion of Li. This rapid diffusion of Li then results in accelerated oxidation in the vicinity of these particles. In the initial stages of the reaction, the oxide nucleates at the particle/alloy interface and spreads laterally as seen in Fig. 3(b) and more clearly in the *in-situ* studies (Fig. 5).

ACKNOWLEDGEMENTS

This work is funded by the National Science Foundation under grant DMR 9015868. Oxidation and SEM experiments were done at Lehigh University. The NDP and SIMS work was performed at NIST in collaboration with Drs. D. E. Newbury, R. G. Downing and G. Lamaze. The E-SEM experiments were conducted courtesy of Electroscan Corp. with the assistance of T. Hardt and T. Rice.

REFERENCES

1. P. G. Partridge, "Oxidation of Al-Li Alloys in Solid and Liquid States", **Inter. Mater. Rev.**, Vol. 35, p2765, 1990.
2. G. D. Danilatos, "Foundations of Environmental Scanning Electron Microscopy", in *Advances in Electronics and Electron Physics*, Vol. 71, p109, Academic Press, 1988.
3. R. Levi-Setti, J. M. Chabala, and Y. L. Wang, "Aspects of High Resolution Imaging with a Scanning Ion Microprobe", **Ultramicroscopy**, Vol. 24, p97, 1988.
4. R. G. Downing, R. F. Fleming, J. K. Langland, and D. H. Vincent, "Neutron Depth Profiling at the National Bureau of Standards", **Nucl. Instr. Methods Phys. Res.**, Vol. 218, p47, 1983.
5. R. G. Downing, J. T. Maki, and R. F. Fleming, "Applications of Neutron Depth Profiling", in *Microelectronics Processing: Inorganic Materials Characterization*, ACS Symp. Series 295, Ch. 9, Ed.: L. A. Casper, American Chemical Society, Washington, DC, 1986.
6. J. T. Maki, R. F. Fleming, and D. H. Vincent, "Deconvolution of Neutron Depth Profiling Spectra", **Nucl. Instr. Methods Phys. Res.**, Vol. B17, p147, 1986.
7. K. K. Soni, D. B. Williams, J. M. Chabala, R. Levi-Setti, and D. E. Newbury, "Morphological and Microchemical Phenomena in the High

Temperature Oxidation of Binary Al-Li Alloys", Accepted for publication in **Oxd. Met.**, 1991.

8. P. G. Partridge and N. C. Chadbourne, "High Temperature Oxidation of an Al-Li-Cu-Mg (8090) Alloy", **J. Mater. Sci.**, Vol. 24, p2765, 1989.

9. K. K. Soni, D. B. Williams, J. M. Chabala, R. Levi-Setti, and D. E. Newbury, "Role of Second-phase Particles in the Oxidation of Al-Li Alloys", Submitted for publication in **Oxd. Met.**, 1991.

10. K. K. Soni, Ph. D. Dissertation, Lehigh University, 1991.

THE REDUCTION OF CORROSION AND DEUTERIUM

PICKUP IN Zr-2.5Nb

R.A. Ploc[1], K.F. Amouzouvi[2] and C.W. Turner[1]

ABSTRACT

Control of the corrosion and deuterium ingress rates of Zr-2.5Nb can be regulated by small additions (<5 000 ppm by wt) of ternary elements. The slowest rate of corrosion of such ternary alloys does not necessarily produce the best long-term result. Small alloying additions significantly effect the specimen to specimen variability. Modifying the normal two-phase microstructure of Zr-2.5Nb CANDU pressure tube material by shot peening and aging decreases the corrosion rate by at least a factor of two. Similar results can be achieved by laser glazing and heat-treating the alloy. Early results suggest that the pickup rate of ^{2}H is also reduced. The corrosion rate of Zr-2.5Nb can be made less dependent on the alloy's microstructure, texture and micro-composition by coating with a thin, sintered layer of ZrO_2 obtained by dipping in a sol. Laboratory tests have shown a rate reduction for the coated material of at least an order of magnitude when compared with un-coated specimens.

---

[1]AECL Research
Chalk River Laboratories
Chalk River, Ontario K0J 1J0
Canada

[2]AECL Research
Whiteshell Laboratories
Whiteshell, Manitoba R0E 1L0
Canada

INTRODUCTION

Pressure tubes used in Canadian CANDU[1] nuclear reactors are made of an alloy of zirconium containing 2.5 wt% Nb. Because of different impurities and impurity concentrations in the starting materials, no two batches of the alloy have exactly the same composition. Beginning with chemical specifications originally developed for Zircaloy-2, the specifications for Zr-2.5Nb have been refined and modified, primarily based on the need to reduce the amount of material with a high capture cross-section for thermal neutrons (neutron efficiency is extremely important in the CANDU system, which uses natural UO_2 as a fuel), and maintain adequate strength and ductility in-reactor. None of the chemical specifications have been based upon the impurity's influence on the corrosion and deuterium pickup rates in-reactor (the CANDU system uses D_2O as both a coolant and moderator). Except for oxygen and niobium (2.5 to 2.8 wt%), the chemical specifications are set as low as practical with no regard to possible beneficial effects of deliberate additions. Current specifications are given in Table 1.

Table 1 Typical impurity limits for Zr-2.5Nb pressure tube material

Impurity	ppm(wt) maximum	Impurity	ppm(wt) maximum
Aluminum	75	Manganese	50
Boron	0.5	Molybdenum	50
Cadmium	0.5	Nickel	35
Carbon	150	Nitrogen	65
Chromium	100	Phosphorus	20
Cobalt	20	Silicon	100
Copper	50	Tantalum	100
Hafnium	50	Tin	100
Hydrogen	5	Titanium	50
Iron	650	Tungsten	50
Lead	50	Uranium	2
Magnesium	20	Vanadium	50

[1]CANDU: Canada Deuterium Uranium. Registered trademark

There have been several investigations into the effect of impurities such as oxygen, nitrogen and carbon [1-3] on corrosion, primarily on unalloyed zirconium or on Zircaloy-2. The effect of other elements, such as Fe, Ni, Cr and Sn have been examined, but much of the literature is contained in proprietary documents generated by corporations such as General Electric and Westinghouse, and government organizations such as the United Kingdom Atomic Energy Authority. Many of these studies are not directly applicable to Zr-2.5Nb because they have been performed on binary or quaternary alloys and often at temperatures or in environments unrelated to the CANDU system. Cox [4] has published limited data on impurity effects. The Zr/Nb alloys that Cox [4] tested were based on a 1 wt% Nb alloy rather than 2.5 wt%.

CANDU pressure tubes operate at about 570 K in pressurized D_2O at a pD (pH) of 10.5 (Room Temperature (RT), LiOH). Since impurity effects can change dramatically with temperature [5,6], studies performed at higher temperatures often lack relevance. Further, very little is known about the effect of impurity concentrations of less than 0.1 wt%. Because of synergistic effects of the elements, the determined performance of, say, Sn in unalloyed zirconium can not be extrapolated to deduce similar behaviour in other zirconium alloys. Binary alloys do not behave in the same manner as quaternary and ternary alloys.

A complicating factor in determining the role of impurities in Zr-2.5Nb is its two-phase structure. The microstructure of CANDU pressure tubes consists of elongated and flattened α-Zr grains embedded in a network of meta-stable β-Zr ($\approx$ 20% Nb). Limited information is available on impurity partitioning between the α- and β-Zr phases. In general, but certainly not in all cases, impurities segregate to the β-Zr phase or form ternary or quaternary precipitates. In the case of iron, nickel and chromium, it is known that their concentrations in solution in α-Zr are less than 100 ppm (parts per million). Irradiation usually smears impurities uniformly through the alloy, irrespective of its phase structure [7,8].

The corrosion rate of Zr-2.5Nb pressure tube material is slower in-reactor than out-reactor. The likely correlation for this has been given by Urbanic and Gilbert [9]. Out-reactor testing has demonstrated that the corrosion rate is dependent on the amount of cold-work, see figure 1 [10]. During fabrication of the pressure tube, α-Zr becomes supersaturated ($\approx$ 1.3%) with Nb, which does not readily precipitate with thermal aging alone. A combination of cold-working and aging is required or, as in the case of in-reactor exposures, irradiation alone. Precipitation of the niobium-enriched particles leaves the α-phase with approximately 0.6 wt% Nb, the equilibrium concentration at 883 K (610°C). The lowest corrosion rate is influenced not only by precipitation of the supersaturated Nb but also by the shape and distribution of the β-Zr network. Optimum resistance is achieved when the supersaturated Nb is precipitated from solution and the β-Zr phase is fully transformed into a uniform dispersion of small β-Nb precipitates [11].

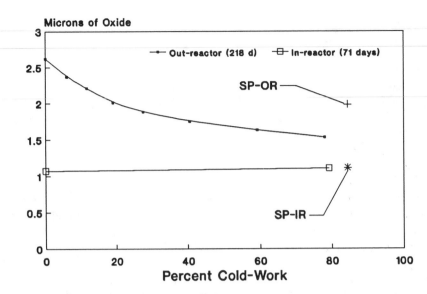

Figure 1 Corrosion rate as a function of percent cold-work [10]. IR, OR and SP are abbreviations for in-reactor, out-reactor and shot (glass) peened, respectively.

Body-centred cubic (bcc) β-Zr displays a faster diffusion rate for both oxygen and hydrogen, and corrodes more quickly than the hexagonal close packed (hcp) α-phase. Not only is it favourable from a corrosion point of view to break up the β-Zr network, but the discontinuous phase structure eliminates fast diffusion pipes for deuterium (hydrogen) into and in the alloy.

CANDU pressure tubes absorb deuterium at a rate of approximately 1 ppm/a, although a rate of three times this value has occurred in an abnormal situation. Coleman and Ambler published an equation for deuterium solubility in zirconium [12,13]; namely, $C = 2.4 \times 10^5 \times \exp(-35900/RT)$, where C is the concentration of deuterium and R and T have their usual meanings ($R = 8.314 \ J \cdot K^{-1} mol^{-1}$). At 573 K (300°C) the solubility of deuterium is about 130 ppm (wt). Considering that a pressure tube begins its in-reactor service already containing about 15 ppm hydrogen (introduced during fabrication), a tube could survive in a CANDU environment for approximately 100 years before being saturated with deuterium. Unfortunately, hydrides may form in localized areas well before uniform saturation occurs. Deuterium often migrates to localized cold spots and along tensile stress gradients to form brittle hydrides. Reduction of the ingress rate will reduce the amount of deuterium available to form hydrides and, disruption of fast diffusion pipes (breaking up the β-Zr network) will slow hydride formation in localized regions.

In general, Zr-2.5Nb is an excellent material for in-reactor use and we have chosen not to abandon the alloy but to look for techniques of making a good alloy even better. In particular, we have examined the possibility of reducing the corrosion and deuterium ingress rates by manipulating the parameters discussed above. The three techniques that are being examined are:

i) the systematic measurement of impurity effects, and whether small increases or decreases in concentration of specific impurities are beneficial. The author primarily responsible for this program is RAP.

ii) the disruption of the β-Zr network and precipitation of supersaturated Nb in the α-Zr phase by shot peening, laser glazing and aging. The author primarily responsible for this program is KFA.

iii) to reduce the dependence of the corrosion and hydriding rates on the alloy microstructure and composition by coating the pressure tube with a thin, dense layer of pure zirconia. The layer would also act as a stress-free barrier layer to cathodic and anodic oxidation currents. The author primarily responsible for this program is CWT.

RESULTS AND DISCUSSION

(a) Composition Effects

All Zr-2.5Nb-xX ternary alloys were made by adding measured amounts of the element X to the same Zr-2.5Nb stock material, analysis in Table 2. The stock and element X were arc-melted together six times (rotating each time) to form an alloy log approximately 50 to 55 grams in weight, 150 mm long and 15 mm in diameter. The log was then wrapped in zirconium foil, encapsulated under vacuum in a quartz tube and annealed for seven days at 973 K (700°C). The log was heated to 1023 K (750°C) in air and hot rolled into a strip which was machined into 20 to 30 parallel-sided corrosion coupons, each about 15 x 7 x 1 mm in size (about 1 to 1.5 grams in weight). The corrosion coupons were mechanically polished to 600 grit, then chemically polished in a pickling solution of 115/60/60/15 parts of $H_2O/HNO_3/H_2SO_4/HF$. Following the pickling procedure, the corrosion coupons were washed in distilled water at 353 K (80°C) for one hour and dried under tissue. Dimensioned and weighed coupons were corroded in 99.8% D_2O (0.2% H_2O) at 573 K (300°C), 1 245 psi and a pD of 10.5 (Li$_2$O) for various lengths of time. In every instance, the D_2O purity was checked since 2 ppb copper in the heavy water caused elemental copper to precipitate on the oxide surface. To ensure no free oxygen existed in the reaction chamber, a vacuum was

Table 2 Analysis of Zr-2.5Nb stock material in ppm (wt)

Element	ppm	Element	ppm	Element	ppm
B	0.8	Cl	7.2	Nb	2.55
C	161	Ca	<0.4	Mo	3
N	42	Ti	17	Cd	0.9
O	1047	V	1.7	Sn	6.8
Na	<0.02	Cr	61	Hf	61
Mg	0.02	Mn	8.1	Ta	54
Al	61	Fe	656	W	12
Si	43	Co	0.6	Pb	2
P	7	Ni	13	Th	0.03
S	2.4	Cu	4.2	U	0.6

pulled on the autoclaves when sealed. A small quantity of steam was bled off at 425 K (150°C) before being brought to the final temperature.

Following every autoclave exposure, each specimen was weighed four times, the highest and lowest readings were discarded and the remaining two were averaged. A specimen whose weight gain was close to the batch average was withdrawn for deuterium analysis. When there was significant variation in the average weight gains within a given batch, specimens with high and low values were also withdrawn for analysis. Since each alloy batch contained a limited number of specimens, care was exercised not to analyze too many specimens.

To date, the elements X added to Zr-2.5Nb are Al, Si, P, V, Cr, Mn, Fe, Ni, Mo and Sn, in ascending order of atomic number. Due to the early nature of the program and the large number of alloys that have to be prepared, the concentration range of X is limited. Normally, x varies from 100 to 5 000 or 10 000 ppm (wt). Control specimens of Zr-2.5Nb that have followed the same preparation route (i.e., melted, annealed, etc.) as the Zr-2.5Nb-xX ternary alloys were included in each autoclave for comparison.

The length of corrosion achieved to date varies because each alloy has entered into the program at a different stage. Figure 2 gives an indication of the relative susceptibilities to corrosion after 40 days' exposure. Some alloys have been

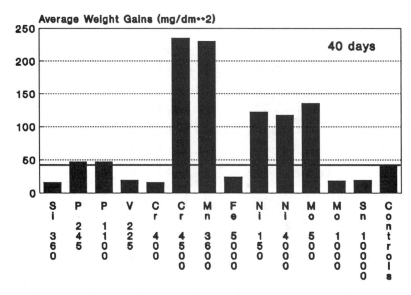

Figure 2 Relative corrosion rates after 40 days' exposure. Continuous, solid line represents the average weight gain achieved by the controls.

exposed for 365 days; for instance, the 360 ppm Si alloy, which continued to display good corrosion resistance. Every data bar in figure 2 represents an average of 15 to 30 specimens. Figure 3 displays the scatter, or variability within a given batch of material. For those batches with a large specimen to specimen variability, examination of the average weight gain (horizontal line through bar) gives an indication of the randomness of the variability. In general, small additions of Si, V and Cr (≤ 400 ppm) are beneficial. Additions of Fe up to at least 5 000 ppm and high concentrations of Mo (1 000 ppm) are beneficial. Phosphorous (≤ 1 000 ppm) slightly accelerates the corrosion rate. High concentrations (>1 000 ppm) of Cr and Mn are to be avoided. Tin suppresses the corrosion rate up to and including the tested maximum of 1 wt%.

Ni additions resulted in large specimen to specimen variability and were closely followed by Mo and high Cr. Additions of P or Mn apparently do not change the specimen to specimen variability. From the point of view of combined corrosion resistance and specimen variability low Si, V and Cr are advantageous. Tin and iron are also beneficial agents.

Only a limited amount of the data on deuterium pickup has been analyzed. There are a number of techniques that can be used to display these data. What was chosen for this report was to plot the percent theoretical deuterium pickup (where it is assumed that for every oxygen atom in the oxide, two atoms of

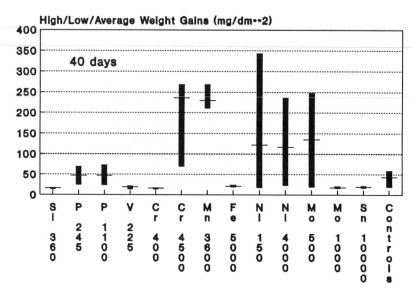

Figure 3 High, low and average (line through bar) weight gains as a
function of addition.

deuterium are produced) as a function of exposure in days, see figures 4a and 4b. However, it must be remembered that such plots do not give any indication of the corrosion rate. For instance, what appears to be a low deuterium pickup rate could be disastrous because of a high corrosion rate, and vice versa. In figure 4b, data for a 2 000 ppm Si alloy have been excluded for just this reason. After a few tens of mg/dm^2 pickup of oxygen, the specimens spalled badly. Figure 2 shows that Ni additions lead to high corrosion rates, but figure 4a shows that small amounts of Ni yield lower than usual deuterium pickup rates. Most of the alloys, when first immersed in the pH 10.5 lithiated D$_2$0 (573 K), experience a sharp peak in the deuterium pickup rate, but this decays quickly to about 5% theoretical. Generally, a low corrosion rate leads to a low, long-term deuterium pickup.

Care must be exercised in interpreting these data, as the corrosion results from the complete spectrum of concentrations have not been reported. In at least two instances, what seems to be sharp increases in the corrosion rate occur for very small changes in the amount of the ternary addition, notably for Ni and Cr.

(b) Surface Modifications

Shot peening and laser glazing [14] surface modification techniques were used to produce, within the specimen surface layers (≤ 100 μm), microstructures with

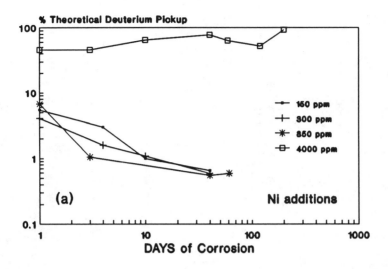

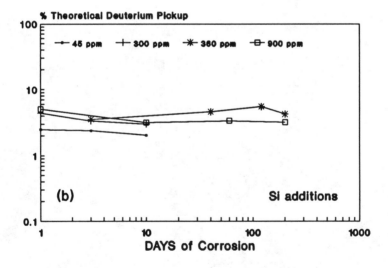

Figure 4 Percent theoretical pickup of deuterium as a function of exposure and the ternary (a) Ni concentration and (b) Si concentration.

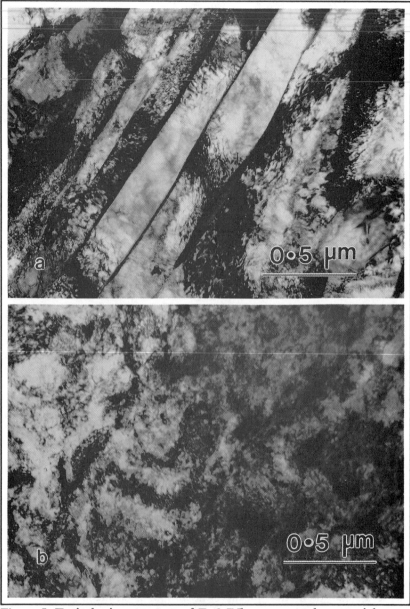

Figure 5 Typical microstructure of Zr-2.5Nb pressure tube material,
(a) as received, showing elongated α-Zr grains encased in β-Zr.
(b) after shot peening.

improved corrosion resistance. Thermal aging was performed at 773 K (500°C) for 24 hours or at 673 K (400°C) for 72 hours. Autoclave corrosion tests were performed in pressurized water at 573 K with a pH of 10.5 (RT) on specimens machined (25 x 12.5 x 4 mm) from Zr-2.5Nb pressure tubes. Specimens received either a chemical or electro-polish, followed by washing in distilled water prior to autoclaving.

Figure 5a is an example of the typical, as-received microstructure of Zr-2.5Nb pressure tube material. Figure 5b reveals the microstructure after shot peening and figure 6, after aging at 773 K for 24 hours. Similar microstructures can be achieved by laser glazing and heat-treating. Figure 7a is the microstructure after glazing and, figure 7b after heat-treating at 723 K for 24 hours. The laser-glazed specimens show an α' martensitic structure.

Cold-working enhances precipitation from supersaturated solid solutions before aging [15]. As shown here, heat-treating shot-peened material results not only in the precipitation of the β-phase from the supersaturated α-Zr, but also in

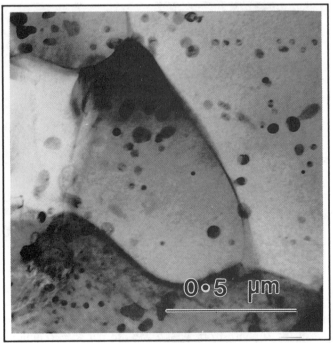

Figure 6 Zr-2.5Nb pressure tube material shot peened and aged 24 hours at 773 K, showing a recrystallized structure with a fine and uniform dispersion of Nb-enriched β-particles.

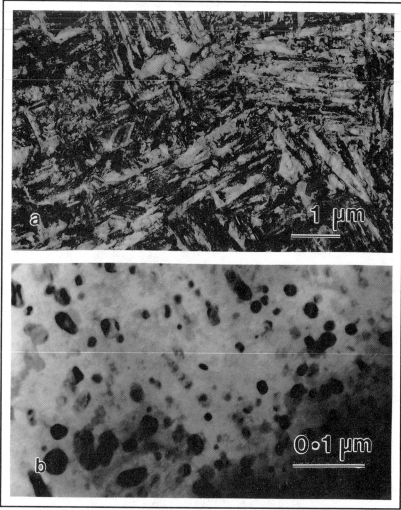

Figure 7 Microstructure of pressure tube material (a) after laser glazing,
(b) after heat-treating at 723 K for 24 hours.

the break-up of the β-Zr network. Heat-treating at 723 K reduced the Nb concentration of the α-phase from 1.3-1.5 wt% to 0.6 wt% (equilibrium value). The same can be said of the laser-glazed specimens.

Table 3 gives some early corrosion results taken from shot-peened material. Similar results were obtained from laser-glazed specimens. Oxide film thicknesses were measured using an FTIR (Fourier Transform InfraRed) spectrometer.

Table 3 Oxide thicknesses (μm) grown in 573 K water (pH 10.5)

MATERIAL	17 days	41 days
As received (cold drawn)	1.2	1.5
Ground and Electro-polished (EP)	0.9	1.2
Peened and EP	0.7	0.9
Peened, heat-treated (673 K, 72 hours) and EP	0.6	0.8
Peened, heat-treated (773 K, 24 hours) and EP	0.5	0.7

The combination of shot peening and heat-treating Zr-2.5Nb pressure tube material resulted in at least a factor of two reduction in the corrosion rate, out-reactor. Out-reactor rates are approximately twice those for in-reactor, so whether the surface treatment described here will confer any additional benefits is unknown until in-reactor studies are completed. Electro-polishing, as opposed to chemical polishing also reduces the corrosion rate, likely the result of eliminating the surface fluorine contamination that arises from chemical polishing. Figure 1 suggests that shot peening (with no-heat-treatment) reduces the out-reactor (SP-OR) corrosion rate but has little effect on in-reactor rates (SP-IR). However, this observation is not conclusive since the study that gave rise to figure 1 used glass shot rather than steel, which produces a greater depth of surface modification and, the specimens were not heat-treated.

Most of the corrosion studies on the shot-peened and laser-glazed materials have been performed in light water (H_2O) for short periods of time. Therefore, data on hydrogen ingress is limited and cannot be easily analyzed. Studies are now underway using D_2O, which will allow an easy distinction to be made between the starting hydrogen content of the pressure tube and the deuterium which is absorbed during the corrosion process. After 15 months of corrosion in light water, the amount of hydrogen absorbed is what was expected from a constant pickup rate of approximately 1 or 2% theoretical and a reduced corrosion rate.

(c) Zirconia coatings

A sol-gel process was developed based on the hydrolysis and condensation of a zirconium alkoxide solution in its parent alcohol according to the reactions:

hydrolysis $\quad Zr(OR)_4 + nH_2O \rightarrow Zr(OR)_{(4-n)}(OH)_n + nHOR \quad$ (1)

condensation $\quad 2Zr(OR)_{(4-n)}(OH)_n \rightarrow \{Zr(OR)_{(n)}(OH)_{(n-1)}\}_2O + H_2O \quad$ (2)

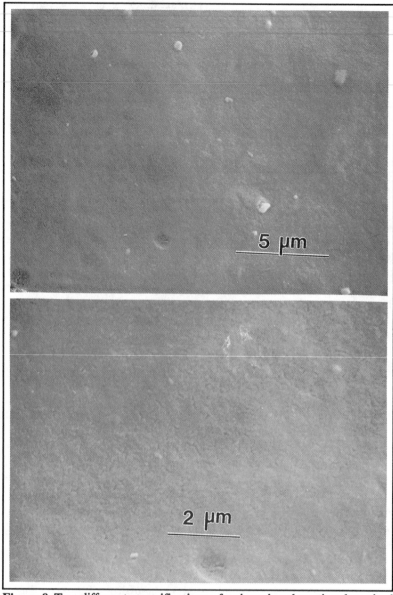

Figure 8 Two different magnifications of a zirconia sol coating deposited by dipping and withdrawal at 20 mm/s and sintering at 873 K for one hour.

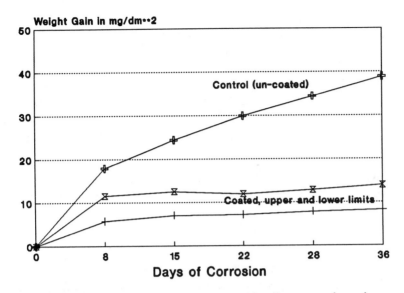

Figure 9 Results of corrosion tests on coated and un-coated specimens. The two lower curves define the data envelope for the different specimens (different sols).

with the overall reaction given by:

$$Zr(OR)_4 + 2H_2O \rightarrow ZrO_2 + 4HOR \quad (3)$$

The hydrolysis of zirconium is rapid compared with condensation and unless the zirconium is complexed with a complexing agent [16], such as acetylacetone, the reactions described by equations 1 and 2 lead to the formation of a precipitate rather than a gel. By suitable control of the reaction conditions, however, the precipitation leads to the formation of a stable zirconia sol that can be used as a starting material for a sol-gel ceramic coating.

The coatings are applied to the substrate by dipping. The coating thickness increases with pull-rate, sol viscosity, and sol concentration, in accordance with the theory of dip coating [17]. Coating thicknesses up to 150 nm per application after sintering were achieved by this method. Figure 8 shows two scanning electron microscope micrographs (different magnifications) of a typical coatings sintered for one hour at 873 K (600°C) under an atmosphere of dry argon.

Figure 9 contains data showing the effect of the sol-gel ceramic coatings on the corrosion rate of some thin-walled, Zr-2.5Nb tubing in 573 K (300°C) water at a pH of 10.5 (at RT). After an initial rapid weight gain, the coated specimens corroded at a rate eight times lower than the un-coated controls. Tests currently in progress (i.e. beyond the 36 day data given in figure 9) show a 20-fold reduction in the corrosion rate of the coated specimens compared to the un-coated controls during the first 75 days of exposure. The 20-fold reduction in corrosion rate was apparently for specimens which were inadvertently heated into the α- plus β-Zr two-phase region (>883 K). Zr-2.5Nb quenched from the two-phase region corrodes at a rate of approximately twice that for material held in the single-phase α-Zr region. The deposited and densified synthetic zirconia coating appears to render the substrate relatively insensitive to its microstructure.

Because of technical difficulties, only limited data are available on the deuterium pickup performance of the coated materials. As in the case of the shot-peened specimens, deuterium ingress apparently parallels the corrosion rate; i.e., the rate is likely to be about 5% of the theoretical limit.

CONCLUSIONS

The corrosion resistance of Zr-2.5 wt% Nb, at least on a short-term basis, can be significantly improved by increasing (within tight limits) the concentration of specific impurities such as Si, V or Cr. Additions of Fe and Sn up to much higher concentrations (~ 1 wt%) are also beneficial. There are indications that some ternary alloys with low corrosion rates (like those containing V) may develop early spalling. The elements which lead to improved corrosion resistance also decrease specimen to specimen variability. Concentrations of Ni up to 850 ppm (wt) lead to the lowest deuterium pickup rates, but cause severe specimen variability and high corrosion rates. In general, most of the ternary alloys absorb deuterium at about 5% of the theoretical limit.

Shot peening and laser glazing followed by heat-treating significantly modifies the surface of Zr-2.5Nb pressure tube material to depths approaching 100 μm. The modified microstructure results in a factor of two reduction in the out-reactor corrosion rate. Early indications are that the hydrogen pickup rate parallels the reduction in the corrosion rate.

A zirconia sol-gel and coating technique has been developed that reduces the corrosion rate of Zr-2.5Nb by an order of magnitude. Some modifications of the sol may be reducing the rate by a factor of twenty over a period of three months' exposure. Studies are not sufficiently mature to analyze the deuterium pickup rates, but early indications are that these rates are also low.

Combining the three techniques discussed above, namely, modifying the alloy chemistry, shot peening and coating with a thin synthetic zirconia layer are likely to significantly improve the long-term corrosion and deuterium ingress rates of CANDU pressure tubes.

ACKNOWLEDGEMENTS

The authors would like to thank the technical staff who have and are so fully contributing to the program. These are M.C. Jacklin, J.A. Roy, D.W. Smith and P.A. Lavoie at the Chalk River Laboratories and L.J. Clegg and R.C. Styles at the Whiteshell Laboratories. Funding for this project was provided by the CANDU Owners Group (COG) through working party 33, work package 3111.

REFERENCES

1) D.E. Thomas, "The Metallurgy of Zirconium", chapter 11 (II), editors B. Lustman and F. Kerze, Jr., McGraw Hill, N.Y., 1st edition, (1955)pp 608-640.

2) S. Kass, J.D. Grozier and F.L. Shubert, "Effects of Silicon, Nitrogen and Oxygen on the Corrosion and Hydrogen Absorption of Zircaloy-2", Corrosion - National Association of Corrosion Engineers, Vol. 20 (1964)pp 350t-360t.

3) S. Kass and J.D. Grozier, "Effect of Carbon Content on Corrosion and Tensile Properties of Zircaloy-2", Corrosion - National Association of Corrosion Engineers, Vol. 20 (1964)pp 158t-165t.

4) B. Cox, "Oxidation of Zirconium and its Alloys", Advances in Corrosion Science and Technology, Volume 5, edited by M.G Fontana and R.W. Staehle, Plenum Press, N.Y., chapter 5 (1976)pp 173-391.

5) B. Cox, P.G. Chadd, J.F. Short, "Further Studies of Zirconium Niobium Alloys", part 15 of series "Oxidation and Corrosion of Zirconium and its Alloys", United Kingdom Atomic Energy Authority, Research Group Report, AERE R-4134, (1962).

6) B. Cox, "The Effect of Some Alloying Additions on the Oxidation of Zirconium in Steam", United Kingdom Atomic Energy Authority, Research Group Report, AERE R-4458, (1963).

7) M. Griffiths, R.W. Gilbert and G.J.C. Carpenter, "Phase Stability, Decomposition and Redistribution of Intermetallic Precipitates in Zircaloy-2 and -4 During Neutron Irradiation", J. Nucl. Mater., 150(1987)pp 53-66.

8) M. Griffiths, "A Review of Microstructure Evolution in Zr Alloys During Irradiation", J. Nucl. Mater., 159(1988)pp 190-218.

9) V.F. Urbanic and R.W. Gilbert, "Effect of Microstructure on the Corrosion of Zr-2.5Nb Alloy",presented at the IAEA Technical Committee Meeting on Fundamental Aspects of Corrosion of Zirconium-Base Alloys for Water Reactor Environments, Portland, Oregon, 11-15 September 1989.

10) V.F. Urbanic, Unpublished results.

11) F. Garzorolli, H. Ruhman and S. Trapp-Pritsching, "Status of Siemens Examination of Zr-2.5Nb",Technical Report U6 431/89/e001 , Siemens Ag. Unternehmenscherich, (1989).

12) C.E. Coleman and J.F.R. Ambler, "Measurement of Effective Solvus Temperature of Hydrogen in Zr-2.5 wt% Nb using Acoustic Emission", The Metallurgical Society of C.I.M. Annual Volume, (1978)pp 1-4.

13) J.J. Kearnes, "Terminal Solubility and Partitioning of Hydrogen in the Alpha Phase of Zirconium, Zircaloy-2 and Zircaloy-4",J. Nucl. Mater., 22(1967)pp 292-303.

14) K.F. Amouzouvi, L.J. Clegg and R.C. Styles, "Surface Modifications of Zirconium Alloys by Laser Glazing", Surface Engineering, edited by S.A. Meguid, Elsevier Applied Science, New York, (1990)pp 270-279.

15) J.E. LeSurf, "The Corrosion Behaviour of Zr-2.5 wt% Nb Alloy",Applications - Related Phenomena for Zirconium and its Alloys, ASTM STP 458, American Society for Testing and Materials, Philadelphia, (1969)pp 286-300.

16) M. Shane and M.L. Mecartney, "Sol-Gel Synthesis of Zirconia Barrier Coatings", J. Mat. Sci., 25(1990)pp 1537-1544.

17) I. Strawbridge and P.F. James, "Thin Silica Films Prepared by Dip Coating", J. Non-Cryst. Solids, 82(1986)pp 366-372.

APPLICATION OF A FAST X-RAY DIFFRACTION METHOD TO STUDIES OF HIGH TEMPERATURE CORROSION OF IRON.

V. Kolarik, M. Juez-Lorenzo, N. Eisenreich and W. Engel

Fraunhofer-Institut ICT, Joseph-von-Fraunhofer Straße 7, 7507 Pfinztal 1, Germany

ABSTRACT

New experimental possibilities using X-ray diffraction were applied for studying the high temperature oxidation of iron. The measuring system consists of an X-ray diffractometer combined with a high temperature device with a programmable temperature controller. Using fast detectors series of more than 300 patterns per day were measured in isothermal and non-isothermal experiments in air at ambient pressure. Energy dispersive X-ray diffraction was applied to studies of the whole oxide layer, whereas angle dispersive measurements were used for Fe_3O_4 layers below 840 K. X-ray diffraction with grazing incidence was used for thin Fe_2O_3 films. The lattice plane distances were calculated as a function of temperature or time allowing an in situ identification of the oxidation products. A numerical treatment of the diffraction data, the difference method, yields curves $Y(T)$ representing the relative changes during the series. After calibration to the end thickness of the oxide layer, which was measured microscopically, the $Y(T)$ curves allow isothermal or non-isothermal kinetic evaluation.

INTRODUCTION

The development of high temperature resistant alloys requires detailed knowledge of the corrosion reactions [1]. Conventionally high temperature corrosion is studied under isothermal conditions in various experiments at different temperatures in a defined atmosphere. The corrosion products are identified after cooling and the temperature dependence of the oxidation rate constant is usually determined from these series [2,3,4,5].

However, as phase transitions or solid phase reactions may occur on cooling, the in situ identification of the corrosion products at high temperatures is an important issue [1]. Thermogravimetry (TG) is often applied to obtain non-isothermal experiments in situ. This method, however, monitors only weight changes, but does not identify the structure of the oxide layer.

The combination of an X-ray diffractometer and a high tempeature device allows the required in situ investigation of the corrosion processes [6]. Series of diffraction diagrams can be recorded with fast detectors during a freely selectable temperature program. A numerical procedure applied to the isothermal or non-isothermal series of diffraction patterns allows determination of the oxidation rate constants.

EXPERIMENTAL

Measuring System

The system consists of an X-ray diffractometer with a high temperature device. The sample temperature is controlled by a freely selectable program changing the temperature stepwise or continuously between 290 and 1970 K. On each temperature step or after defined time intervals, diffraction patterns are recorded. Series of up to 300 diffraction patterns are measured during one day.

Energy dispersive measurements were performed with the continuous spectrum of a tungsten tube at a constant diffractometer angle using a Si(Li)-detector. The method allows exposure times of 60 to 120 s per diffraction spectrum. This kind of measurements was preferably used for the investigation of thick oxide layers, where a relatively high penetration depth is required.

The angle dispersive patterns are recorded with a position sensitive proportional counter, which permits a fast scanning of an angular range yielding short exposure times (100 to 120 s), or with a scintillation counter. The penetration depth of the X-ray beam is taken into account selecting the appropiate X-ray tube.

Finally, angle dispersive X-ray diffraction with grazing incidence [7] was applied to the investigation of thin layers below 1000 Å using a scintillation counter.

Measurements

The experiments were carried out with Fe-ARMCO in air at ambient pressure. The relative moisture of the air in the laboratory was 30% at a temperature of 297 K.

Isothermal experiments were performed at 973, 1073 and 1193 K with energy dispersive X-ray diffraction. In this case penetration depths of more than 200 μm are required and realized with a tungsten tube. The energy dispersive diffraction spectra were recorded from 12 to 32 KeV with $2\theta = 20°$.

The growth of an Fe_2O_3 film on Fe_3O_4 was observed at 723 K by means of X-ray diffraction with grazing incidence. The patterns were recorded at incidence angles of $\Phi = 0.3; 0.4; 0.5; 0.6; 0.7$ and $0.8°$. The critical angle for total reflection on Fe_2O_3 using a Cu-tube is $\Phi_c = 0.33°$.

Non-isothermal kinetics of the oxidation of iron below 840 K was studied heating samples from 298 to 823 K in steps of 20 K with an average heating rate of 4.7 K/min. Angle dispersive X-ray diffraction with Cu-tube was used.

Evaluation

The evaluation of the diffraction peaks was carried out by means of a least squares fit procedure using a Gaussian curve yielding peak position, intensities and peak widths.

The kinetic parameters were obtained by the difference method [8], where two diffraction patterns are subtracted. The sum $Y(t_j)$ of the absolute differences between corresponding channels correlates with the changes in the diffraction diagrams.

$$Y(t_j) = \sum_{i=1}^{n} |Y_i(t_1) - Y_i(t_j)|$$

j = number of the current diagram
i = number of the current channel
n = total number of channels
t = time (in non-isothermal series temperature)

573

The difference method can be applied to the whole patterns or to selected intervals containing definite peaks, yielding separate Y(t) curves for the base material and the formed oxides.

$$Y(t)_{total} = Y(t)_{metal} + Y(t)_{oxide} + Y(t)_{background}$$

The method applied on intervals containing the peaks of the oxides yields a Y(t) curve showing the relative increase of the oxide layer. The curve was normalized to 1 and the $Y(t_j)$ values were multiplied by the final thickness of the oxide layer that was measured microscopically. The resulting curve x(t) represents the growth of the oxide layer.

Isothermal Kinetics

In the isothermal case the calculated curve that is fitted to the measured curve x(t), describes the parabolic time law, which is expected at high temperatures [1,2]:

$$x(t) = \sqrt{2 k t} \qquad (1)$$

The parabolic oxidation rate constant k is determined directly by the least squares fit.

Non-Isothermal Kinetics

In non-isothermal experiments the growth of the oxide layer depends on two parameters: temperature and time. For obtaining temperature dependence of the parabolic rate constant, the x(T) curve must be described by a function f = f(T,t) [9].

Starting with the differential equation for parabolic time dependence

$$\frac{dx}{dt} = \frac{k}{x} \qquad (2)$$

and introducing the heating rate $\alpha = dT/dt$ we obtain

$$x \, dx = \frac{k}{\alpha} \, dT \tag{3}$$

The temperature dependence of reaction rates is described by the Arrhenius equation:

$$k = z \, e^{-\frac{E}{RT}} \tag{4}$$

where R is the gas constant, E activation energy and z the preexponential factor. In our case z is a normalization factor. Inserting (4) in (3) we obtain after integration the following relation for x(T):

$$x(T) = \sqrt{\frac{2z}{\alpha} \int e^{-\frac{E}{RT}} \, dT} \tag{5}$$

The equation (5) is fitted to the measured curve x(T) by means of a least squares fit procedure obtaining values for the fit parameters z and E. After the determination of these parameters the temperature dependence of the oxidation rate constant can be described by the equation (4).

RESULTS AND DISCUSSION

Isothermal Series

The difference curves Y(t) of the isothermal series at 973, 1073 and 1193 K show the relative changes in the oxide layer (Fig. 1). At these temperatures a parabolic time law is expected [1]. The described evaluation procedure calculates the time dependence of the oxide layer thickness x(t) from the difference curves Y(t). The parabolic time law (1) was fitted to the x(t) curves (Fig. 2). The values obtained for the parabolic oxidation rate constant are in agreement with the literature [1] (Fig. 3).

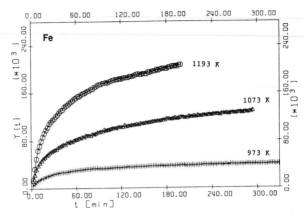

Fig. 1 Difference curves Y(t) of the isothermal oxidation of Fe.

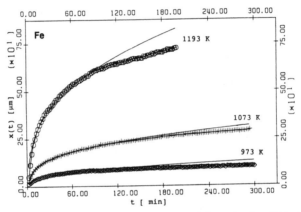

Fig. 2 Growth of the total oxide layer x(t) with the fitted curves.

Series with Grazing Incidence

The isothermal formation of an Fe_2O_3 film on Fe_3O_4 at 723 K was observed with grazing incidence. At this temperature very thin layers of Fe_2O_3 are expected.

The difference method was applied to the diffraction patterns that were recorded at equal incidence angles Φ selecting only the intervals containing Fe_2O_3 peaks. The Y(t) curve grows parabolically as long as the X-ray beam penetrates completely the Fe_2O_3 film. The Y(t) curve, calculated from the diffraction diagrams with an incidence angle $\Phi = 0.3°$, ceases from increasing after 430 min. While the Fe_3O_4 peaks disappear in the diagram with $\Phi = 0.3°$, they are still observed with $\Phi = 0.4°$. As the penetration depth of the X-ray beam as a function of the incidence angle Φ is known, the thickness of the Fe_2O_3 film after 430 min at 723 K can be estimated to be 50 Å.

The Y(t) curve was normalized to 1 at 430 min and calibrated to 50 Å yielding the growth of the Fe_2O_3 film x(t). A parabolic curve (1) was fitted to the x(t) curve (Fig. 4). The parabolic rate constant of Fe_2O_3 at 723 K was estimated in the order of magnitude 10^{-15} g^2 cm^{-4} s^{-1}.

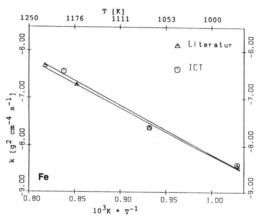

Fig. 3   Parabolic oxide rate constant of Fe in comparison with literature [1].

Non-Isothermal Kinetics

Below 843 K Fe_3O_4 and small amounts of Fe_2O_3 are formed in air [1]. The formation of Fe_3O_4 on heating was observed in situ by angle dispersive X-ray diffraction.

The difference method was applied to the intervals of the diffraction patterns containing peaks of Fe_3O_4. The resulting Y(T) curve was calibrated to the microscopically measured end thickness of the oxide layer D = 4 μm yielding the growth of the layer with temperature x(T). A calculated curve (5) was fitted to the measured x(T) curve yielding an activation energy of E = 120 KJ. Fig. 5 shows the measured and the calculated x(T) curve. Inserting the obtained parameters z and

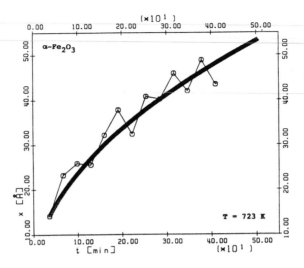

Fig. 4 Growth of the Fe$_2$O$_3$ layer x(t) with the fitted curve.

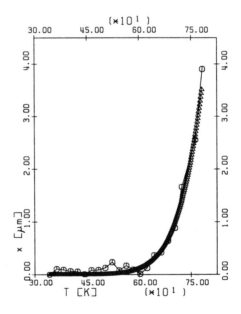

Fig. 5 Measured growth of the oxide layer x(T) and the fitted curve.

E in equation (4) the temperature dependence of the parabolic oxidation rate constant was determined as shown in Fig. 6. The results are in agreement with literature [1].

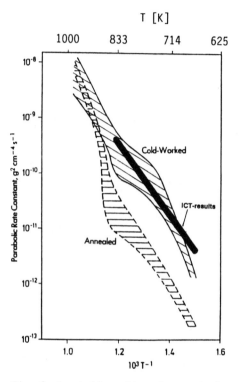

Fig. 6 Parabolic oxide rate constant
of Fe as function of temperature
in comparison with literature [1].

CONCLUSIONS

X-ray diffraction is a suitable method for the investigation of high temperature corrosion and its kinetics. The formed oxides can be identified in situ and the temperature dependence of the oxidation rate constant can be determined by means of the difference method in series of isothermal experiments or in a single non-isothermal experiment.

Very thin layers that are not observed by microscopy can be studied by means of X-ray diffraction with grazing incidence. The penetration depth can be easily calculated allowing kinetic evaluation.

REFERENCES

[1] A. Rahmel und W. Schwenk; Korrosion und Korrosionsschutz von Stählen.; Verlag Chemie (1977)

[2] P. Kofstad; High Temperature Corrosion; Elsevier Applied Science LTD (1988)

[3] R.W.J. Morssinkhof, T. Fransen, M.M.D. Heusinkveld, P.J. Gellings; High Temperature Corrosion Properties of Thin Alumina Films, Deposited by MOCVD.; Proc. 9th European Congress on Corrosion, Royal Netherlands Industries Fair, FU-108, (1989)

[4] A.S. Khanna, P. Kofstad; Oxidation of 304 Stainless Steel in Oxygen and Oxygen Containing 2% Water Vapour.; Proc. 11th Int. Corrosion Congress, Assoc. Italiana di Metallurgia, 4.45, (1990)

[5] E. Otero, A.J. Criado, P. Hierro, A. Pardo, M. Baladía; Corrosion Behaviour of a stainless steel 12cr 1Mo 0.3V 0.2C at temperatures higher than 973 K.; Proc. 22nd Int. Annual Conf. of ICT, "Combustion and Reaction Kinetics", 114-1, (1991)

[6] M. Juez-Lorenzo, V. Kolarik, N. Eisenreich, W. Engel, A. J. Criado, E. Otero; Surface Oxidation of Steel Studied by Fast X-ray Diffraction.; Proc. 9th European Congress on Corrosion, Royal Netherlands Industries Fair, FU-273, (1989)

[7] Michael F. Toney, Ting C. Huang; X-Ray Depth Profiling of Iron Oxide Thin Films; J. Mater. Res., 3(2), 351, (1988)

[8] N. Eisenreich, W. Engel; Difference Thermal Analysis of Crystalline Solids by the Use of Energy-Dispersive X-ray Diffraction.; J. Appl. Cryst. 16, 259, (1983)

[9] N. Eisenreich; Direct Least Squares Fit of Chemical Reaction Curves and its Relation to the Kinetic Compensation Effect; J. Thermal Anal. 19, 289, (1980)

Coatings

ALUMINIZED STEELS FOR MOLTEN CARBONATE FUEL CELLS

Thomas E. Swarr[*]

ABSTRACT

Aluminide coatings on 316SS and INCO 825 separators showed that corrosion protection was provided by a β- aluminide phase that formed a $LiAlO_2$ scale during exposure to carbonates. Diffusion studies on aluminized 310SS demonstrated coatings were formed by the inward diffusion of Al. Coatings similar to those on 316SS and INCO 825 tested in stacks were produced at diffusion temperatures 250°C lower than the standard process, simplifying manufacturing and reducing costs.

INTRODUCTION

Molten carbonate fuel cell (MCFC) power plants offer clean and efficient electric generation using coal-derived fuels and other hydrocarbon feedstocks. The molten Li_2CO_3- K_2CO_3 electrolyte is an aggressive environment that severely limits the materials of construction. One area subject to particularly severe corrosion is the "wet seal", shown schematically in Figure 1. A ceramic matrix, typically $LiAlO_2$, is flooded with electrolyte and compressed between metal flanges to form a gas seal isolating the fuel and oxidant flow fields from the external environment. The flux of reactant gases through the matrix can create localized corrosion cells[1].

Cells and stacks fabricated using uncoated 316SS experienced wet seal corrosion. The attack contributed to performance decay by consumption of electrolyte, development of gas leaks, and deposition of metallic corrosion products causing electrical shorts. In some cases, the buildup of corrosion products actually pried the cell apart.

*International Fuel Cells, South Windsor, CT 06074

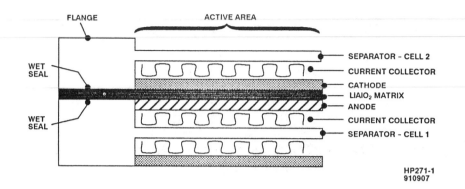

Figure 1. Schematic of MCFC Wet Seal

Early research using potentiostatic immersion test methods identified aluminum-containing ferritic stainless steels as potential candidates for MCFC structural applications[2,3]. In out-of-cell laboratory tests, alumina scales formed on uncoated ferritic stainless steels and aluminized 316 showed excellent stability[4]. Protection was attributed to the formation of an adherent scale of $LiAlO_2$[5].

Although a variety of coating methods borrowed from the gas turbine industry have been used to aluminize steels for MCFC applications, there has been limited effort reported in the literature to optimize the coatings for fuel cell applications[6]. The development of MCFC systems has reached the stage where cost and scale-up concerns are receiving greater priority. The "standard" high temperature (1050°C) diffusion cycle used to form the aluminide coating imposes cost and manufacturing constraints. Development of lower temperature diffusion cycles would simplify manufacturing by reducing distortion of the separator plates during fabrication. An improved understanding of the formation and degradation mechanisms is needed to evaluate low cost alternative coating processes.

Materials selection for the aluminized wet seal is constrained by design requirements imposed by the interior current carrying, or active, area of the cell. The separator flange comprising the wet seal is an integral part of the separator plate and provides the field depth for the electrodes and gas flow. Individual cells are stacked vertically to sum voltage. Cost constraints dictate that the separator plate should be as thin as possible, typically 0.3mm. Because of high current flows through the stack, the separator plate must not form an electrically resistive scale that contributes high ohmic losses. Thus, selection of a separator material involves a trade-off between high Ni alloys for corrosion resistance and mechanical properties and high Fe alloys for low cost. Aluminide coatings on a range of alloys were characterized to assess the effect of alloy content and diffusion temperatures.

EXPERIMENTAL PROCEDURES

Coated AISI 316SS and INCO 825 flanges were sectioned from stacks tested previously to provide a baseline description of the wet seal coating. An aluminum overcoat ≈25μ thick was applied to the flanges using ion vapor deposition (IVD) or slurry methods. All coatings were diffusion treated at 1050°C in vacuum. Test duration was about 2000 hours. Samples of IVD coated and slurry coated AISI 310SS were prepared with diffusion cycles at 800 and 1050°C. In addition, a second series of IVD coated 310SS samples were wrapped in alumina paper, and sealed in evacuated Vycor tubes. The samples were heat treated in a box furnace, and samples were pulled at temperatures of 700, 800, and 900°C and air quenched to characterize the formation of the diffusion coatings.

Samples were characterized using standard metallographic procedures. Metallographic sections were examined in the as-polished condition or after etching electrolytically with 10 percent oxalic acid or after immersion in a solution of 30 ml lactic acid, 30 ml acetic acid, 20 ml hydrochloric acid, and 10 mil nitric acid[7]. Electron microscopy was performed using a Cameca MBX microprobe. Samples were characterized by secondary and backscatter electron images, x-ray images, and elemental traces. Semi-quantitative elemental profiles were generated using a 100μ window in 2μ steps across the coatings. XRD analyses were performed using graphite monochromated Cu K-α radiation on a Bragg-Brentano horizontal diffractometer. Microhardness profiles were measured on a Leco M-400 hardness tester using a 50 gram load and 15 second hold time.

RESULTS

Flanges fabricated from 316SS and INCO 825 were characterized in the as-coated condition and after exposure to carbonates in subscale stack tests before and after testing to assess the effect of base metal composition on the corrosion resistance of aluminized flanges. A comparison of as-coated 316SS and INCO 825 is provided in Figure 2. The 316SS flanges showed a duplex coating comprising an outer, essentially single phase layer ≈35μ thick covered with a thin oxide scale and an inner multiphase layer ≈50μ thick. The single phase layer was fairly uniform in Al, Cr, and Fe. The Cr content was 8 to 10 w/o, and the Al content was ≈25 w/o. There was a decreasing Ni gradient in the outer layer toward the surface. The inner layer had a heavy concentration of precipitates at the interface that decreased gradually toward the base metal. There was an enrichment in Ni at the interface between the two layers of the coating. The Al concentration profile showed a step decrease to ≈7 w/o at the inner layer, gradually decreasing to ≈4 w/o across the layer, and dropping sharply to zero at the well defined interface with the base alloy. The coating-substrate interface was extremely planar. This inner layer is likely a transition to the ferritic phase as the aluminized samples were magnetic.

INCO 825 showed a more complicated three layer coating which was just over 75μ thick. An outer multiphase zone measuring ≈25μ thick showed a bright second phase in the outer half which appeared concentrated at grain boundaries. In addition to the thin oxide scale on the external surface, there was also some internal oxidation penetrating along the grain boundary phase. The inner half of this layer showed a band of fine, dark precipitates. The outer zone was backed by an inner aluminide zone comprising a 25μ thick band of Cr-rich particles and a precipitate free denuded zone 10 to 15μ thick. The interface between the base metal and diffusion coating was characterized by a lamellar precipitate that extended out into the denuded zone and consolidated into a dense continuous phase at the coating-substrate interface. Elemental analysis of the coating showed significantly more scatter than the 316SS coating because of the numerous second phase particles. The Al trace indicated a continuous gradient through the coating, with a concentration of 20 to 25 w/o in the outer 25μ of the coating.

As-coated samples of 316SS analyzed by x-ray diffraction (XRD) showed that the major phase was β- (Fe,Ni)Al. Samples were covered with a thin scale of α-alumina and contained Cr- carbides, Cr_3C_2 and Cr_7C_3. The samples were electropolished to remove ∼1 mil of material and reexamined. Despite the band of second phase precipitates observed in the inner layer, the XRD pattern showed essentially single phase β- aluminide or one-to-one intermetallic. There was some distortion of relative intensities and broadening of the peaks, so it is possible that the two phases are different chemistries of the β- aluminide phase.

Results for the as-coated INCO 825 sample were similar. The pattern showed primarily β- aluminides with some Cr- carbides and a trace of alumina for both the external and polished surfaces. This sample was considerably more difficult to polish. Metallography showed an uneven polish with some of the original surface still present. Thus, the carbides and alumina observed on the polished sample could be a result of remnants of the original surface that survived the electropolishing.

Neither 316SS nor INCO 825 coatings showed any significant morphological change or broadening of the diffusion zone in 2000 hours of cell or stack testing. The outer scale on both samples converted to $LiAlO_2$ and the major phase in all samples was β- (Fe,Ni)Al. Thickness of the oxide scale varied from 5 to 20μ. No carbides were evident in the 316SS pattern. A trace of Li and K carbonate was detected, probably due to entrapped electrolyte in the porous oxide scale. The outer surface of the INCO 825 showed Cr_3C_2 and Cr_7C_3. The inner layer, after electropolishing, showed a trace of $Cr_{23}C_6$.

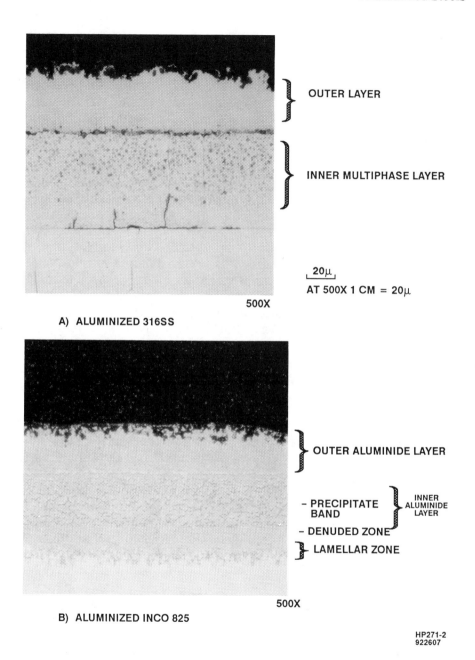

A) ALUMINIZED 316SS

B) ALUMINIZED INCO 825

HP271-2
922607

Figure 2. Effect of Base Alloy on Diffusion Coating Morphology

587

Samples of aluminized 310SS were prepared for comparison with 316SS and INCO 825 using IVD and a proprietary slurry method. A representative photomicrograph of slurry coated 310SS is shown in Figure 3. Morphology of the IVD coating was similar. The overall coating thickness was 85μ, and the general structure was similar to INCO 825 coatings. The outer 20μ showed discrete second phase particles with a higher concentration near the external surface that was preferentially oxidized. Distribution of Fe, Cr, and Al in the matrix was uniform across the layer. This layer was backed by a band ≈40μ thick characterized by discrete Cr-rich particles showing a bright contrast and a blotchy zone of dark precipitates, probably carbides. The inner 10μ of this band was relatively precipitate-free and showed Ni enrichment. Adjacent to the base alloy was a 25μ thick multiphase zone with a lamellar structure. There was no band of Cr enrichment at the coating-substrate interface as observed in INCO 825.

A series of 310SS samples coated by IVD were heat treated in sealed Vycor tubes to simulate a vacuum cycle. Samples were heated at 60°C per hour, removed at 700, 800, and 900°C, and air quenched to characterize the formation of the diffusion coating. Photomicrographs are shown in Figure 4. The sample pulled at 700°C showed a single phase coating backed by a narrow band, < 2μ thick, of a darker phase. The outer surface of the coating was oxidized and showed numerous defects, e.g., pores, cracks. The samples were wrapped in alumina paper during heat treatment, and the paper apparently generated sufficient oxygen partial pressures due to dissociation to oxidize the coating.

At 800°C, the dark band had broadened into a complex multiphase layer adjacent to the base alloy. The dark phase had grown to ≈6μ thick with fingers extending into the bright, outer coating. A thin band of the lamellar phase had developed at the coating-base metal interface. At 900°C, the 310SS developed a coating microstructure similar in appearance to those observed on INCO 825 heat treated at 1050°C, although somewhat thinner, 50μ. The darker phase had grown outward to the surface, leaving a nearly continuous horizontal band of a brighter, second phase ≈15μ from the surface. Finally, there was a zone of lamellar structure backed by a continuous band of a bright phase at the coating-base alloy interface.

A second series was heat treated in sealed Vycor tubes for 10 hours at 800°C. The coating morphology was very similar to the 900°C sample shown in Figure 4. Overall coating thickness was 35μ. The band of brighter , second phase particles was less developed, suggesting more complete inter-diffusion. There were fingers of the brighter phase extending inward from the outer surface. The lamelar zone was ~7μ thick and was backed by a continuous band of a brighter phase shown to be Cr- enrichment in the INCO 825 samples.

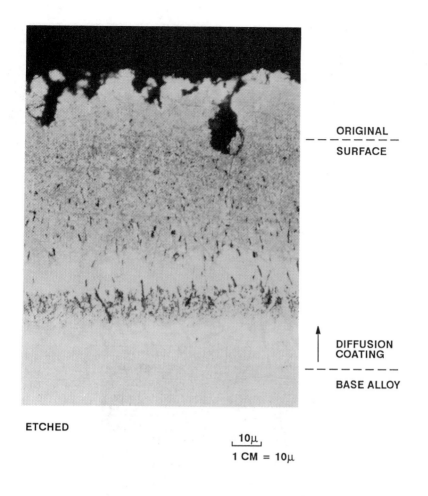

ORIGINAL
SURFACE

DIFFUSION
COATING

BASE ALLOY

ETCHED

10μ
1 CM = 10μ

HP271-3
922607

Figure 3. Slurry Coated 310SS Diffusion Treated at 1050°C

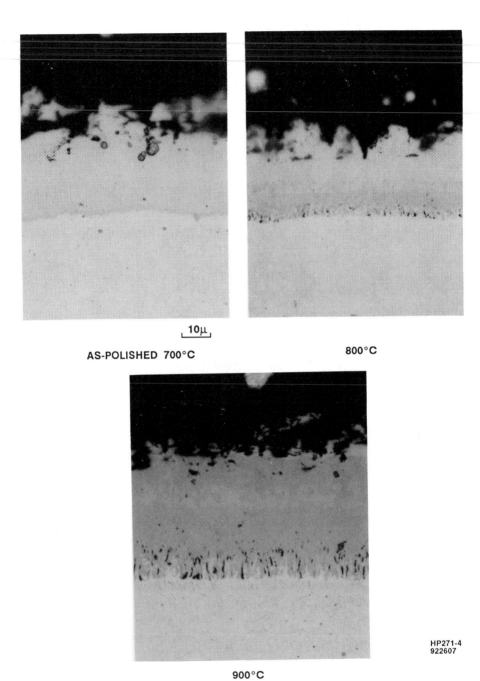

AS-POLISHED 700°C

800°C

900°C

HP271-4
922607

Figure 4. Diffusion Treatment of IVD Coated 310SS

A major concern with the low temperature diffusion cycles is brittleness of the coating because of limited diffusion of the Al into the base metal and higher resultant Al contents in the aluminide phase. Microhardness profiles were taken to compare the various coatings. Results for the coatings treated at 1050°C are shown in Figure 5. Schematics of the coating morphology are also shown so that the hardness data can be correlated with structural features. The 316SS coating showed a relatively flat hardness profile, ranging from 400 to 450 VHN across the coating. All samples yielded low hardness readings near the surface. However, the indentation was distorted, and these data are suspect. 310SS showed very high values, 700 to 900 VHN, in the band of fine precipitates. The hardness then dropped off steeply to a value of $\approx$190 VHN in the base metal. INCO 825 showed somewhat more scatter, with readings of 600 to 700 VHN in the two phase band, dropping to 450 to 500 VHN in the denuded zone and then peaking at 700 VHN again in the lamellar zone. All samples showed similar hardness in the base metal, 180 to 200 VHN. Because of the thin coatings developed at 800°C, readings were taken only in the mid-plane of the coating. Readings were in the range of 550 to 600 VHN. The width of the indentation exceeded the width of the lamellar zone and the indentation was distorted, with cracks initiating at the interface between the band of Cr enrichment and the base metal. Readings in the base metal indicated a hardness of 210 VHN.

DISCUSSION

These coatings can be classified as high activity or inward diffusion type coatings [8]. The initial Al overlay melted during the diffusion cycle. Substrate elements were dissolved in the melt, followed by resolidification in an Al- rich phase such as ϵ- (Fe, Ni) Al_3 or δ- (Fe, Ni)$_2Al_3$ with simultaneous inward diffusion of Al through the solid phases. The morphology of the INCO 825 and 310 SS coatings closely matched the three layer structures observed on nickel-base superalloys [7,8]. 316SS coatings yielded a duplex structure which was attributed to solubility effects.

591

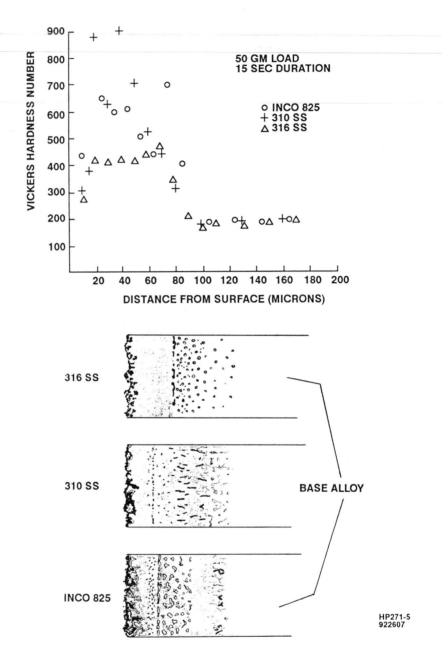

Figure 5. Microhardness Profiles of Aluminized Coatings

The outer aluminide layer on INCO 825 and 310SS was relatively free of second phase precipitates because the composition is controlled by the solubility of the substrate elements in the Al- rich melt. Because of the limited solubility of Cr in the Al- rich phases, Cr was concentrated in a nearly continuous line of precipitates at the interface. The band of heavy second phase particles was a zone formed by the inward flux of Al through the solid phases reacting with the base alloy. As Al diffused inward, the reaction eventually formed a hypostoichiometric β- phase where Ni diffusion dominates mass transport. The formation of additional coating was then controlled by the outward diffusion of Fe and Ni, resulting in the precipitate free zone. The formation of single phase β continued until the decreasing flux from the outer layers could no longer maintain the proper phase state for predominate Al diffusion. Subsequent conversion of the outer layers proceeded by the outward diffusion of Fe and Ni, and the β- phase formed during this process contained the second phase precipitates formed during the initial stages of coating. The complex lamellar band resulted from the supersaturation of the substrate because of the outward diffusion of Fe and Ni, with Ni appearing to play a dominant role. This layer has been characterized as containing σ and carbides [9].

The 316SS coating showed a two layer structure similar in appearance to that observed for outward diffusing or low activity coatings [8]. However, the formation mechanism was similar to that of the INCO 825, with the structural differences attributed to the higher solubility of Fe- rich aluminides combined with the lower alloy content of 316SS compared to INCO 825. The lack of a denuded zone and a band of Cr- enrichment behind the coating suggests that inward diffusion of Al predominated during the entire formation cycle. The limited amount of second phase precipitation in the coating contributed to the flatter concentration and microhardness profiles.

Metallography of post-test samples showed that the corrosion protection of these coatings was provided by the formation of an alumina scale that converted to $LiAlO_2$. The porous structure of this scale suggested a classical hot corrosion mechanism [10]. Despite the differences in coating structure and composition, the performance during testing in carbonates was similar. This indicates that the key parameter was the ability of the outer aluminide layer to form the alumina scale and to reheal when the scale was fluxed by a hot corrosion mechanism. The similar performance of these coatings suggests a relatively wide composition range where these conditions were met. Experience with gas turbines has shown that the β- aluminide phase has the best oxidation resistance of the aluminides [9].

Examination of the 310SS coating after diffusion at 1050°C showed a structure similar to that observed in INCO 825. The key difference was the lack of a band of Cr- enrichment at the coating-substrate interface. This observation suggests that diffusion in the base metal was sufficient to replace Fe and Ni as these elements react with the inward flux of Al to form the denuded zone. The distribution of second phase particles results in an Al concentration gradient and microhardness profiles very similar to that of INCO 825.

The series of microphotographs presented in Figure 4 were consistent with the formation mechanism described above. Successive photomicrographs illustrate the initial formation of a bright outer layer, which is probably the Al- rich intermetallics, either ϵ- or δ- aluminides. A narrow band of the darker phase indicates the reaction of the inward flux of Al to form the β- aluminide, which then grew outward, eventually converting the entire coating. A bright layer at the coating-substrate interface is indicative of Cr- enrichment and suggests that at these temperatures, diffusion rates in the substrate were not sufficient to match the flux to the coating. A very similar coating was obtained by holding 10 hours at 800°C. Characterization of the coating indicated that the outer aluminide layer was similar to that obtained at higher diffusion temperatures. Although the coating was thinner, higher Al content of the outer aluminide layer may actually improve corrosion resistance.

CONCLUSION

Characterization of aluminide coatings on 316SS and INCO 825 separators has shown that protection is provided by the ability of the β- aluminide phase to form and reheal alumina scales that convert to $LiAlO_2$ during exposure to carbonates. Diffusion studies on 310SS showed a similar mechanism of formation with a resulting coating morphology between the extremes shown by 316SS and INCO 825. These results demonstrate that by adjusting diffusion heat treatment temperatures to compensate for Ni content of the base alloy, the desired coating structure can be maintained. In addition, the initial melting of the aluminum overcoat and dissolution of base metal constituents plays a critical role in controlling the chemistry of the outer layer of these aluminide coatings. Additional work is required to better define the point where depletion of Al to oxidation reactions or continued diffusion into the base alloy inhibits the healing process.

ACKNOWLEDGEMENTS

Portions of this work were sponsored by the Department of Energy under contracts DE-AC21-87MC23270 and DE-AC21-91MC27393. The electron microprobe analyses were performed by C. Burila at the United Technologies Research Center.

REFERENCES

(1)   R.A. Donado, L.G. Marianowski, H.C. Maru, and J.R. Selman, "Corrosion of the Wet-Seal Area in Molten Carbonate Fuel Cells I. Analysis," J. Electrochem. Soc., Vol. 131, p 2535, 1984.

(2)   G.J. Janz and A. Conte, "Potentiostatic Polarization Studies in Fused Carbonates- II. Stainless Steel," Electrochimica Acta, Vol. 9, p 1279, 1964.

(3)   H.J. Davis and D.R. Kinnibrugh, "Passivation Phenomena and Potentiostatic Corrosion in Molten Alkali Metal Carbonates," J. Electrochem. Soc., Vol. 117, p 392, 1970.

(4)   R.A. Donado, L.G. Marianowski, H.C. Maru, and J.R. Selman, "Corrosion of the Wet-Seal Area in Molten Carbonate Fuel Cells II. Experimental Results," J. Electrochem. Soc., Vol. 131, p 2541, 1984.

(5)   R.B. Swaroop, J.W. Sim, and K. Kinoshita, "Corrosion Protection of Molten Carbonate Fuel Cell Gas Seals," J. Electrochem. Soc., Vol. 125, p 1799, 1978.

(6)   C.Y. Yuh, P. Singh, L. Paetsch, and H.C. Maru, "Development of Aluminized Coatings for MCFC Wet Seal Applications," Corrosion '87, San Francisco, CA, Paper 276, 1987.

(7)   G.W. Goward, D.H. Boone and C.S. Giggins, "Formation and Degradation Mechanisms of Aluminide Coatings on Nickel-Base Superalloys", Trans ASM, Vol. 60, p 228, 1967.

(8)   G.W. Goward and D.H. Boone, "Mechanisms of Formation of Diffusion Aluminide Coatings on Nickel-Base Superalloys", Oxidation of Metals, Vol. 3, p 475, 1971.

(9)   S.P. Cooper and A. Strang, "High Temperature Stability of Pack Aluminide Coatings on IN738LC", in High Temperature Alloys for Gas Turbines, ed. R. Brunetaud, Dordrecht, Reidel, 1982.

(10)   R.A. Rapp, "The Hot Corrosion of Metals – Theory and Experiment", Fourth Conf. on Gas Turbine Materials in a Marine Environment, Annapolis, MD, 1979.

METALLURGICAL EVALUATION OF CHROMIZED

COATINGS FOR BOILER TUBING APPLICATIONS

F. V. Ellis*

ABSTRACT

A technique for diffusing chromium into the surface of components fabricated from carbon or low alloy steels has found widespread application in the power generation, pulp and paper industries. "Chromized" components exhibit superior resistance to a variety of aggressive corrosive mechanisms, including high temperature oxidation, sulfidation, and chloride-related attack. The unique properties of a chromized product are reflected in the structure of the chromized layer, which consists of a chromium-rich ferritic matrix interspersed with varying amounts of chromium carbides. Processing variables affect the amount of chromium carbide in the chromized layer, and thereby determine the extent of the decarburized zone that develops in the original material structure. The purpose of this study of chromized boiler tubing was twofold; first, metallographic examinations were performed to characterize relevant structural features, and second, measurements were performed to determine the tensile and creep properties.

*   Tordonato Energy Consultants, Inc.
4156 S. Creek Road, Chattanooga, TN 37406

Microstructural Science, Volume 19

INTRODUCTION

Chromized components exhibit superior resistance to a variety of aggressive corrosive mechanisms, including high temperature oxidation, sulfidation, and chloride-related attack. Because of this resistance, chromized coatings have found increased usage in hostile furnace environments characteristic of utility and chemical recovery boilers.

The chromizing process is similar to the well known pack carburization process in several respects. First, the material to be chromized is placed in a retort and surrounded by the chromizing powder. Second, the chromized coating is applied via a diffusion mechanism. Due to the lower diffusivity of chromium as compared to carbon, the process temperature for chromizing is sufficiently high causing grain growth to occur in the base material during the process hold time. In order to refine the grain size of the as-chromized material, a renormalizing and tempering heat treatment is usually performed prior to service for commercial applications involving boiler tubing. The third and final analogous feature of the two processes is that only thin surface coatings of the materials are applied. These thin surface coatings have enhanced properties when compared to the underlying base material. Typical chromized coating thicknesses developed for corrosion resistance are in the range of 75 to 300 micrometers.

This paper discusses the metallographic examinations performed on chromized low alloy boiler tubing as part of a routine quality control program. The microstructural features of interest include the thickness of the chromized layer, the degree and morphology of the porosity in the chromized layer, the extent of the decarburized zone in the base material, and base material grain size and hardness. Because the thermal processing history and microstructure for chromized components differs from that for typical materials of construction, a limited tensile and creep rupture testing program was conducted on chromized tubing material.

METALLURGICAL EVALUATION

All samples of chromized tubing examined were removed from components such as waterwall panels, corner panels and burner panels that were destined for service in a fossil fuel boiler. In all cases, the chromizing was performed at a commercial shop facility with extensive experience and capability to chromized large boiler components, such as full-size waterwall panels and similar types of furnace parts. The evaluations of the finished chromized product were based on the metallurgical condition of the tubing samples and was to be consistent with the known capabilities of the

598

commercial shop. In total, thirty two tubes were received for examination. All but two of the tubes had a specified minimum wall thickness of 6.4 mm and outside diameters in the range of 32 mm to 38 mm.

For the purposes of the evaluation, a ring section was cut from each tube sample. These sections were then mounted in bakelite, polished, and etched using a 2% nital etchant. The microstructure of the tubing and the chromized layer were then examined to determine the effects of the chromizing operation on the final condition of the processed tubing. In particular, the thickness of the chromized layer, the depth of the decarburized zones at the O.D. and I.D. surfaces of the tubing substrate, and the relative amount of porosity observed in the chromized layer were measured. In addition, the effectiveness of the final normalize and temper heat treatment was verified, and the hardness of the tubing material was determined in an area that had not been decarburized.

Figure 1 shows a through-wall photomacrograph of a typical tube section with a chromized coating applied to the outside surface of the tube. The microstructure of the chromized layer consists of a chromium-rich ferritic matrix interspersed with varying amounts of chromium carbides. Microprobe measurements for the chromized layer indicate a gradient in the chromium content with values typically above 40% in the region of the parallel band of pores and about 12% to 18% at the base material-coating interface. Other readily identified features shown in Figure 1 include the decarburized zone in the tube base material below the chromized coating, the portion of the tube wall within a normal base material structure, and the shallow decarburized zone located at the inside diameter of the tube.

The chromized layer thickness depends on the position of the tube surface relative to the bottom and top of the retort. For the purposes of this paper, the bottom of the retort position is termed as the "casing side" of the tube while the top of the retort position is termed as the "furnace side". In general, the average chromized coating thickness was relatively uniform around the circumference of the tube with slightly thicker coatings on the furnace side of the tube as compared to the coating on the casing side of the tube. The average furnace side chromized layer thicknesses ranged from 225 to 575 micrometers with the majority of the values measuring less than 350 micrometers. However, on the casing side of the tube the range of chromized coating thicknesses had a higher degree of variability. In addition, there were anomalous cases that were occasionally observed in localized areas on the casing side of the panels where the chromized layer was not fully developed as shown in Figure 2. This condition has been observed occasionally in localized areas on the sides of the panels facing the bottom of the retort. Even though the chromium is transported in the gaseous phase from the powder to the part, the explanation for this condition is related to the fact that apparently within the retort the casing sides (down-sides) of the panels do not maintain as intimate a contact with the chromizing powder as do the furnace sides (up-sides).

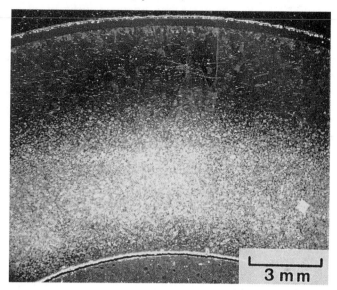

Figure 1. Through-wall photomacrograph of chromized boiler tube.

The porosity rating system adopted for these evaluations is based on an arbitrary classification scheme developed by Henry [1] following examination of a number of chromized samples in which a broad range of porosity was evident in the chromized layer. These classifications are given in Table 1. The photomicrographs in Figure 3 compares the porosity of two separate chromized layers that were rated as Class 2 and Class 5 by the definitions of Table 1. The porosity in one particular chromized layer may vary slightly from position to position such that, for example, one region would be rated as Class 2 while an adjacent region would be Class 3. When this condition occurred the chromized layer was denoted as Class 2/3 in the evaluation procedure. For chromized layers with average furnace side thicknesses measuring less than 350 micrometers, the porosity class ratings vary from Class 1/2 to Class 3/4 with a Class 2 rating being typical. Thicker chromized coatings usually had Class 4 and Class 5 porosity ratings. At present, there is no valid correlation between the degree of porosity in the chromized layer and the service performance of the chromized tube. With continued service, it may be possible and desirable to determine if such a correlation exists.

TABLE 1. Porosity Rating System for Chromized Coatings

Class	Description
1	Minimal porosity, with all pores located in a band running parallel to, and usually within 25 to 50 micrometers of the O.D. surface.
2	Moderate porosity, but with all pores located within the band running parallel and close to the O.D. surface.
3	Moderate to heavy porosity, with the majority of the pores located in the band running parallel to the O.D. surface; a few isolated pores, often elongated, extending down into the chromized layer.
4	Moderate to heavy porosity, with a large number of the pores located within the band adjoining the O.D. surface, but with a significant number of elongated pores extending down into the chromized layer.
5	Heavy porosity, with a large number of elongated pores extending down in to the chromized layer.

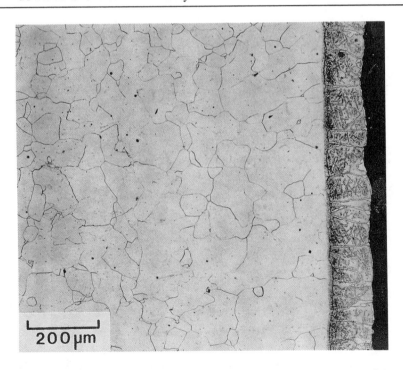

200 μm

Figure 2. Anomalous chromized layer on casing side of tubing which was not fully developed.

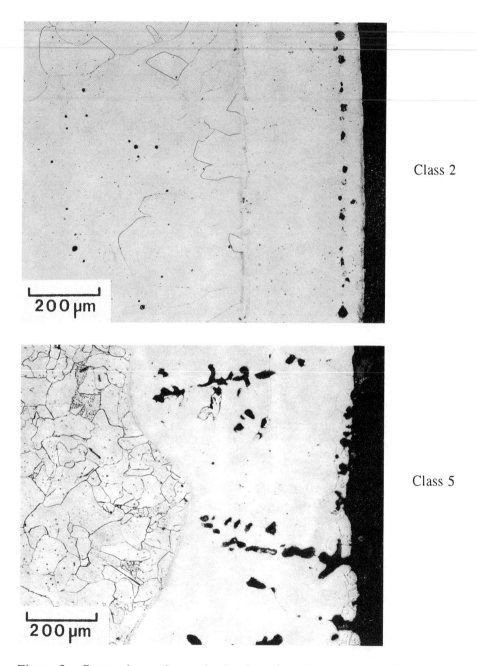

Class 2

Class 5

Figure 3. Comparison of porosity in chromized layer rated as Class 2 and Class 5.

The depth of decarburization depends on the amount of carbide precipitation in the chromized layer which is dependent on the cooling rate in certain critical temperature ranges for the chromized tubing. The depths measured in this study are considered typical for the process thermal history used at the time. In general the furnace side of the tube had a larger decarburized zone than the casing side of the tube. The maximum measured depth of the decarburized zone divided by the specified minimum wall thickness of 6.4 mm is plotted as a function of the average furnace side chromized layer thickness in Figure 4. For chromized layers measuring less than 350 micrometers in thickness, the decarburization depth is between 25 to 50 percent of the total wall thickness. The thicker coatings tend to have decarburized zone depths greater than 50 percent, while the decarburized zone located at the inside diameter of the tube is considerably less, generally less than 250 micrometers. It is the extent, and presumed strength of the decarburized zone material [2,3] that raises concerns about the tensile and creep properties of chromized components when subjected to in-service loading.

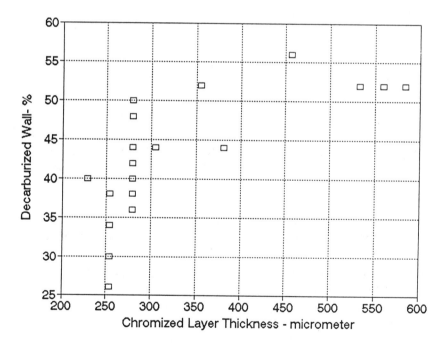

Figure 4. Depth of decarburized zone as a percent of minimum wall thickness versus chromized layer thickness for 6.4 mm wall thickness tubing.

In all cases, the normalize and temper heat treatment was successful in reducing the as-chromized grain size of the tubing base material in regions that were not decarburized. Hardness values (Vickers, 30 Kg load) for the base material ranged from 121 HV to 285 HV with an average hardness value of 197 HV.

TENSILE AND CREEP PROPERTIES

A single heat of 2.25Cr-1Mo steel tubing was used for the tensile and creep test program. The tube had a 63.5 mm outside diameter and a 4.7 mm wall thickness. Uniaxial tensile and creep-rupture tests were performed on the chromizing tubing material in the as-chromized, chromized plus normalized, and chromized plus normalize and tempered heat treatment condition. The tensile and creep specimens are commonly termed "dog bone" or strip specimens because of their characteristic shape and orientation. For these specimens, the tensile axis is oriented in the longitudinal direction of the tube. The specimens had a gage section of full wall thicknesses by 6.4 mm with a 50 mm gage length. For tubing in service applications, the thickness of the applied chromized coating is not used in satisfying the minimum wall thickness requirements. However, for these tests the coating thickness was included for the purpose of calculating the specimen's cross sectional area.

Figure 5 compares the ultimate tensile strength of the chromized 2.25Cr-1Mo tubing material and the minimum value obtained from the National Research Institute for Metals compilation for bare 2.25Cr-1Mo tubing [4]. At room temperature, the tensile strengths for the chromized tubing is approximately equal to or slightly below minimum strength material from the NRIM data compilation [4]. At an elevated temperature, the chromized tubing given the normalized only, or the normalized and tempered heat treatment is equal to or greater than the observed minimum for virgin bare tubing.

The uniaxial stress rupture properties measured at 650°C are compared in Figure 6 for the chromized tubing, bare tubing, and minimum rupture time from the NRIM data compilation [4] for virgin bare tubing material. The rupture strength of the bare tubing used for the chromized base material is above that for the observed bare tubing minimum. The rupture time for the chromized tubing does not depend strongly on heat treatment conditions and in all instances was approximately half of the minimum rupture time for the bare tubing material. The reduced creep strength for the chromized tubing is associated with the decarburized zone in the base material and not any inherent property of the chromized layer material.

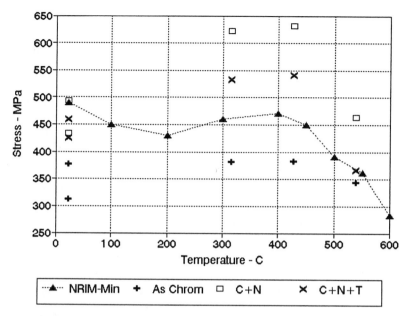

Figure 5. Comparison of ultimate tensile strength as a function of temperature for chromized tubing and minimum strength bare tubing material.

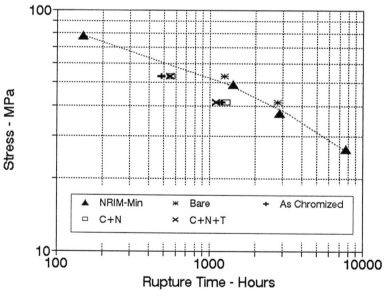

Figure 6. Comparison of stress rupture properties at 650°C for chromized tubing, bare tubing, and bare tubing minimum strength material.

SUMMARY

The metallographic examination of commercially chromized tubing has revealed that generally uniform chromized coatings of greater than 250 micrometers are obtained. The porosity in the chromized layer was classified using an arbitrary rating system. The usual porosity rating on this scale was Class 2 to Class 3, indicating moderate to heavy porosity with the majority of the pores located in a band parallel to the O.D. surface. For tubes of 6.4 mm specified minimum wall thickness, the decarburized zone depth varies from about 25 to 50 percent of minimum wall thickness. The renormalization and tempering heat treatment employed following the tube chromizing process was effective in reducing the grain size of the base material in the areas that were not decarburized.

The tensile strengths of chromized 2.25Cr-1Mo tubing given standard post chromizing heat treatment are at or above minimum strengths compared to bare tubing material over a temperature range from room temperature to 540°C. The measured creep strength of chromized tubing for thin wall thickness tubes (4.7 mm) was below minimum strengths observed for bare tubing strengths due to the low creep strength of the decarburized zone material.

REFERENCES

1. J. F. Henry, Unpublished Research, Chattanooga, TN.
2. R. L. Klueh, "The Effect of Carbon on 2.25Cr-1Mo Steel; (1) Microstructure and Tensile Properties," J. of Nuclear Materials, Vol. 54, 1974, p. 41.
3. R. L. Klueh, "The Effect of Carbon on 2.25Cr-1Mo Steel; (2) Creep-Rupture Properties," J. of Nuclear Materials, Vol. 54, 1974, p. 55.
4. National Research Institute for Metals, "Data Sheets on the Elevated-Temperature Properties of 2.25Cr-1Mo Steel for Boiler and Heat Exchanger Seamless Tubes (STBA24)," NRIM Creep Data Sheet No. 3B, 1986, Tokyo, Japan.

THE ROLE OF PROTECTIVE COATINGS

IN METAL JOINING

John W. Yardy[1]

ABSTRACT

Properties of protective coatings are discussed in relation to surface engineering where surface technology, materials selection, and processing methods are combined to match conditions operating in manufacturing and service. The role that protective coatings play in the joining process is described with the help of examples from successful industrial applications of surface engineering where metallography is used to evaluate the suitability of materials and processes.

INTRODUCTION

For many years, protective coatings have been used almost exclusively to protect the underlying basis material against corrosion or wear without exploiting simultaneously any other property that the coating might have, apart from the aesthetic. Within the last decade or so, more importance has been attached to surface treatment/material combinations in an effort to meet the progressively-increasing demand for better engineering performance. Thus the concept of surface engineering has evolved around the design of surfaces to match the operating conditions that prevail in both service and manufacture of parts, particularly where more expensive bulk materials can be replaced with cheaper materials. Surface engineering covers a wide range of industrial activities influencing not only surface technology

---

[1] Danfoss A/S, DK-6430 Nordborg, Denmark

but also materials selection and manufacturing methods, where joining technology, in particular, has benefitted from this innovative concept.

A sound knowlegde of the individual technologies involved is required before attempting to select new untried combinations of materials and processes. Thus manufacturing companies like Danfoss, with a broad experience in materials selection, well established plating shops, and a multitude of metal joining facilities, have been successful in combining technologies to create new products, reduce costs, and improve performance and reliability.

The following description of the role of protective coatings in metal joining is confined to industrial experience with metallic coatings, and joining methods such as soldering, brazing and welding. Other combinations employing adhesives or painted surfaces are beyond the scope of this paper.

PROPERTIES

Protective coatings.

When discussing the role of protective coatings in metal joining it is necessary to consider the properties of these coatings and their interaction with factors such as materials selection and joining techniques. The properties of protective coating are covered by 4 main areas of interest

> Appearance
> Corrosion resistance
> Tribology
> Joining

The aesthetic or cosmetic value of a coating is often to be found in its colour, lustre or lack of such. The main object being to give the component a uniform and attractive finish. These coatings can be so thin that they only offer protection against tarnish, e.g. brass plated with 2 μm nickel. However, much thicker coatings, usually >30 μm, are needed to effectively protect the substrate against corrosion, this applies equally for coatings which are either anodic (zinc on iron) or cathodic (nickel on iron) to the substrate.

Apart from being hard, >500 HV, a number of coatings that are used in tribology applications to reduce friction and/or wear, also offer a degree of protecton against corrosion e.g. electroless nickel or physical vapour deposition (PVD) coatings.

Coatings with low melting points like Sn and Sn/Pb alloys not only increase wetting and adhesion during the soldering process but also afford the substrate, be it copper tracks on printed circuit boards or component lead wires, protection against oxidation prior to soldering. Other coatings, such as nickel, act as a barrier on lead wires of copper to prevent diffusion and the formation of brittle intermetallic compounds of copper and tin after soldering.

The useful properties of protective coatings are not restricted to these examples alone, equally important for some applications are coatings that increase or even restrict electrical and/or thermal conductivity, reduce electrical/thermal losses or enhance the reflection of light and heat.

Substrate material.

The interaction between protective coatings and the selection of substrate or basis material becomes obvious when materials such as brass can be replaced by a cheaper material such as coated steel. However, the shaping process is just as economically important as composition, especially with the growing public emphasis on materials conservation. This has caused the demise of conventional machining in favour of near-net-shape forming processes like drawing, pressing, cold forging, hot stamping and powder metallurgy.

Joining methods.

The joining process must meet several requirements in order to be accepted for joining surface coated parts:

- The protective coating must not be affected in any way that will cause its essential properties to be impaired or lost
- The basis material must not be damaged or lose its properties, especially when strength is at a premium.

In the case of thermal joining processes, the heating cycle must be kept short to restrict excessive heating of the parts. The rapid heat transfer during resistance heating (< 1 sec.) usually meets these requirements. Rapid heating is however, not always advantageous if a molten weld pool forms, as in the case of laser heated copper or electroless nickel coatings on steel. The weld pool becomes contaminated with copper and phosphorous respectively, and the resulting embritlement causes cracking during the rapid solidification and cooling that follows. Typical heating times of 15-20 seconds for torch and induction brazing, and also wave soldering are rapid enough for these methods to be used succesfully for joining coated parts. Furnace brazing involves much longer heating and cooling times and is therefore unsuitable.

SELECTION AND APPLICATION

Coating classification.

The metallic coatings used in metal joining applications are usually electrolytic, electroless or hot dipped coatings. The harder, refractory like coatings, produced by PVD are difficult to combine with thermal joining processes.

There are three main types of metallic coatings, active, passive and barrier. All three types of coating protect the substrate against oxidation and corrosion prior to joining.

- Active coatings melt and then wet surfaces and join components together on solidification.
- On melting, passive coatings are pressed out of the joint gap and in doing so expose and clean the surface allowing solid state bonding reactions to occur.
- Barrier coatings neither melt nor are they pressed out of the joint gap. They form a physical barrier to protect the substrate from combining with the molten solder or brazing alloy.

Active/passive coatings.

During resistance joining, where the parts are heated on an electrical resistance welding machine, coatings will behave either actively, passively or both, depending upon the temperature achieved, the melting point of the coating, and the relative movement of the parts. The resulting resistance joint is usually a combination of brazing, diffusion bonding/welding and possibly fusion welding. Typical active/passive coatings include tin, zinc, copper and electroless nickel. In comparison electrodeposited nickel has too high a melting point (1728K) for it to be used to any advantage in resistance joining applications.

Electroless nickel deposits are amongst the most versatile active coatings used in resistance joining. They offer uniform coverage of complex shapes as well as corrosion - and wear resistance [1]. These deposits are amorphous containing typically 8-10% P and have a maximum melting range between

1155-1313K, which is below that of most substrate materials. However, electroless nickel coatings are three times more expensive than electroplated nickel and are notoriously brittle. Accordingly coatings, prior to resistance joining, are only 2-10 μm thick. This limits cost and provides protection against mild forms of corrosion. To ensure a strong joint, the thickness of coating remaining as filler metal after joining should be $\leq 0,5$ μm.

The hardness of electroless nickel deposited on low carbon steel increases from c. 600 $HV_{0,06}$ to c. 1100 $HV_{0,06}$ after heat treatment at 673K for 1h [1]. This change in hardness is accompanied by the formation a crystalline mixture of nickel and nickel phosphide (Ni_3P), which has no adverse effect on joining.

Electrodeposited copper is another metallic coating used extensively in metal joining. Even though it has a melting point 200K above electroless nickel, copper is easily squeezed out of the joint during resistance joining and constitutes a passive coating just like tin and zinc coatings. The thickness of copper coatings is, as in the case of electroless nickel, 2-10 μm, which makes it an attractive altenative to electroless nickel, as copper coatings are at least six times cheaper.

Copper and silver are used as active coatings in furnace brazing where the coating, on melting, takes on the role of filler metal between closely fitting parts. In transient liquid phase brazing, tightly fitting silver plated brass parts are heated until the silver coating diffuses rapidly into the brass substrate. At a given concentration of silver the melting point of the substrate is below the furnace temperature and the substrate melts. However, due to the concentration gradient between the molten and solid brass, silver continues to diffuse into the solid brass, raising the melting point of the molten brass which then solidifies to form a strong sound joint between the two parts.

Barrier coatings.

The improved throwing power of modern plating baths ensures that copper can be deposited more uniformly than previously achieved. Thus it is now possible to plate the internal surface of small tubes with a suitable thickness of copper, without excessive deposits building up on the external surface. This increased throwing power increases the suitability of copper as a barrier coating during brazing. Consequently, low cost Cu-P-(Ag) brazing alloys can be used to braze copper plated steel parts. The molten Cu-P-(Ag) brazing alloy wets and flows easily on the copper surface but cannot come in contact with the steel substrate, thus avoiding the formation of brittle iron phosphide films. However, as copper coatings are eventually dissolved by

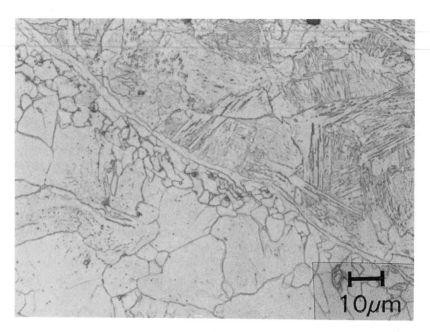

Figure 1. Resistance joint with active electroless nickel coating.

Figure 2. Brazed joint with electroless nickel barrier coating

Figure 3. Resistance joint with traces of passive copper coating.

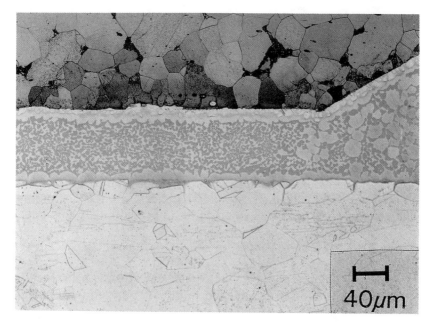

Figure 4. Brazed joint with copper barrier coating on steel.

the molten brazing alloy, only methods with short brazing times such as torch and induction brazing are suitable. It is also interesting to note that copper plated steel parts heat much quicker to brazing temperature than the copper parts they often replace. These factors, together with the fact that Cu-P-(Ag) brazing alloys do not require the application and subsequent removal of additional flux, account for a rapid increase in the use of copper as a barrier coating on steel.

Both electrodeposited and electroless nickel coatings are used as barrier coatings in soldering and brazing applications, i.e. on copper wire and zink-aluminium alloys [2] respectively. However, only electrodeposited nickel has a high enough melting point for the coating to remain intact when furnace brazing with copper or nickel based filler metals. Free machining brass and steel are coated with electrodeposited nickel prior to funace brazing, to prohibit dezincification and to avoid contamination of the filler metal with the free machining constituents, lead and/or manganese sulphide. Similar effects can also be achieved with active coatings and resistance heating [3]. During furnace brazing or heat treatment at temperatures > 1100K, nickel diffuses into the steel substrate thereby increasing the adhesion and corrosion resistance of the coating [4]. However, contra diffusion of iron into the nickel coating may cause a decrease in corrosion resistance depending upon coating thickness, and heating temperature and time.

ASSESSMENT

Although it is a destructive test method, metallography is a convenient and widely used method for obtaining accurate assessments of coating characteristics and joint configurations, be it process control or development work.

Optical microscopy, is used to measure, for instance, coating thickness and uniformity, filler metal flow, and joint porosity. Scanning electron microscopy, together with energy dispersive microanalysis is especially useful when investigating the structure and chemical composition of coatings, filler metals and interfaces. There are also standardised tests for coating adhesion, corrosion resistance and hardness, and it is customary to test joints for pressure tightness when part of a pressure tight component.

ADVANTAGES AND CONCLUSIONS

Apart from the obvious cost benefits when using cheaper basis materials and brazing alloys, there are additional benifits to be gained from taking a surface engineering approach to the joining of coated parts. These include the possibility to join parts where:

- sections are of unequal thickness.
- Melting points and or electrical/heat conductivity of the basis materials are very dissimilar.

Less obvious, but equally rewarding, are improvements in production flow and stock keeping. By plating before assembly, which is the reverse of normal practice, the following advantages are frequently gained:

- Parts difficult to handle are joined during the final stage of assembly.
- Assemblies of two or more basis materials, that are invariably difficult to plate, can be plated under optimum conditions as individual parts.
- Instead of keeping finished products in stock in a number of versions, parts or sub-assemblies can be assembled from stock on receipt of the customers' individual requirements, and dispatched without delay.

However, it must be emphasized that these advantages, can only be achieved through a closely integrated multidisiplanary approach.

ACKNOWLEDGMENTS

The auther would like to thank Elli Grau and Hanne Søllingvraa for their assistance in preparing the manuscript.

REFERENCES

1) E. Schemeling, Eigenschaften und beurteilung von chemisch reduktiv abgeschiedenem Nickel, **Metaloberfläche** Vol. 41, p. 353, 1987.
2) R.L. Peaslee, **Welding Journal** Vol. 70, p. 118, 1991.
3) S. Skytte Jensen, **Modstandsvejsning af automatstål**, Technical University of Denmark, Lyngby, Denmark (Nov. 1990).
4) M.I. Nogin, A.V. Ryabchenkov, V.V. Ovsyankin and Yu.S. Slotin, The protective and electrochemical properties of nickel-phophous coatings, **Zashechita Metallov** Vol. 13, p. 466, 1977.

STRESSES IN AND BENEATH A SURFACE COATING

DUE TO ROUGH SURFACE CONTACT

S. J. COLE[1] and R. S. SAYLES

ABSTRACT

A method for calculating the subsurface stresses arising from the two-dimensional, dry, frictionless contact of two elastic bodies with real rough surfaces, where one body has a rigidly bonded surface layer is presented. The model uses surface profile data directly recorded with a profilometer. Verification of the model by comparison with test case results is presented. The effect of layer thickness, material properties and surface finish on the subsurface stress distribution in the layer and substrate is examined.

INTRODUCTION

Surface engineering processes are frequently applied to components for use in tribological applications, and often give rise to layered structures where one or more layers of a material with certain elastic properties are bound to a substrate with different properties. In use, these structures are often exposed to contact stresses. Integral

---

[1]Tribology Section, Mechanical Engineering Department,
Imperial College, London SW7 2BX, United Kingdom.

transforms are the usual mathematical tool employed in the stress analysis of such structures and many papers exist in the literature dealing with two- and three-dimensional layered elastic bodies subject to normal and transverse surface pressure loading [1..6].

Surface roughness plays an important role in both dry and lubricated contact, and has been the subject of much attention , with the underlying aim of explaining some of the more fundamental observations made in the study of tribological processes, such as friction, wear and lubrication. However, the effect of surface roughness on subsurface stresses in layered bodies arising from contact has received little attention to date, although Sainsot et al.[7] followed an approach broadly similar to that used here. The research presented here takes the surface pressure distribution from a contact model developed by the authors [8], which simulates rough layered body contact, and calculates the stresses in the layer and substrate due to this surface loading.

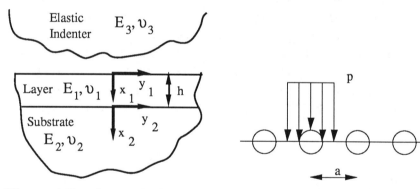

Figure 1 Notation and Coordinates for Layered Elastic Solid

Figure 2 Surface and Pressure Representation

THE CONTACT MODEL

The contact model is described in detail in reference [8]. In summary, the model simulates the two-dimensional dry, frictionless, elastic contact of an upper body of arbitrary shape against a half-space which has a rigidly bonded surface layer of thickness h, as in figure 1. The surfaces are represented as arrays of points a fixed distance, the sampling interval a, apart as in figure 2, where each point has associated with it a height value, relative to some datum. The model calculates the surface contact pressure distribution that arises when the bodies are in contact under a given normal load, in terms of constant pressure elements centered on the surface profile points as in figure 2.

EXPRESSIONS FOR THE STRESSES

Using the coordinates and notation of figure 1, the following expressions are derived in the appendix to this paper for the stresses in the layer due to a constant unit normal surface pressure on the interval $-a/2 \leq y_1 \leq a/2$

$$\sigma_{x_1} = \frac{-1}{\pi h} \left[\int_0^{s_0} s^2 \left(\frac{\alpha_1}{\psi} + \beta_1 \psi \right) \cos s\zeta \, ds + 2h \int_{s_0}^{\infty} \frac{1 + s\xi}{s\psi} \sin \frac{sa}{2h} \cos s\zeta \, ds \right] \quad (1)$$

$$\sigma_{y_1} = \frac{-1}{\pi h} \left[\begin{array}{l} \int_0^{s_0} s \left\{ \frac{2B_1' - s\alpha_1}{\psi} - \psi \left(2D_1' + s\beta_1 \right) \right\} \cos s\zeta \, ds \\[2mm] + 2h \int_{s_0}^{\infty} \frac{1 - s\xi}{s\psi} \sin \frac{sa}{2h} \cos s\zeta \, ds \end{array} \right] \quad (2)$$

$$\tau_{xy_1} = \frac{-1}{\pi h} \left[\begin{array}{l} \int_0^{s_0} s \left\{ \frac{B_1' - s\alpha_1}{\psi} + \psi \left[D_1' + s\beta_1 \right] \right\} \sin s\zeta \, ds \\[2mm] -2h\xi \int_{s_0}^{\infty} \frac{1}{\psi} \sin \frac{sa}{2h} \sin s\zeta \, ds \end{array} \right] \quad (3)$$

and the stresses in the substrate are given by

$$\sigma_{x_2} = \frac{-1}{\pi h} \int_0^{s_0} \frac{s^2 \alpha_2}{\psi} \cos s\zeta \, ds \quad (4)$$

$$\sigma_{y_2} = \frac{-1}{\pi h} \int_0^{s_0} \frac{s}{\psi}\left(2B'_2 - s\alpha_2\right)\cos s\zeta \; ds \quad (5)$$

$$\tau_{xy_2} = \frac{-1}{\pi h} \int_0^{s_0} \frac{s}{\psi}\left(B'_2 - s\alpha_2\right)\sin s\zeta \; ds \quad (6)$$

where $\psi = e^{s\xi}$, $\alpha = A' + B'\xi$, $\beta = C' + D'\xi$, $\xi = y/h$, $\zeta = x/h$, s is the product of h and the Fourier transform integrating variable, and s_o is a value of s after which simplifying assumptions may be made about A'_1, B'_1, C'_1, D'_1, A'_2, B'_2, the derived stress function coefficients, which are found by solving the set of simultaneous equations in table 1.

$$\begin{bmatrix} s^2 & 0 & s^2 & 0 & 0 & 0 \\ -s & 1 & s & 1 & 0 & 0 \\ 1 & 1 & t^2 & t^2 & -t & 0 \\ -s & (1-s) & st^2 & t^2(1+s) & st & -t \\ -sf_1 & j_1 - sf_1 & st^2f_1 & t^2(j_1 + sf_1) & st\!f_2 & -\gamma\!j_2 \\ k_1 & k_1 - \frac{2}{s} & k_1t^2 & t^2\left(k_1 + \frac{2}{s}\right) & -\gamma k_2 & \frac{2\gamma}{s} \end{bmatrix} \begin{bmatrix} A'_1 \\ B'_1 \\ C'_1 \\ D'_1 \\ A'_2 \\ B'_2 \end{bmatrix} = \begin{bmatrix} P \\ 0 \\ 0 \\ 0 \\ 0 \\ 0 \end{bmatrix}$$

$$t = e^s, \quad v' = \frac{2-v}{1-v}, \quad v'' = \frac{v}{1-v}, \quad \gamma = \frac{(1-v_2^2)E_1}{(1-v_1^2)E_2}, \quad f = 1 - v',$$

$$j = 3 - v', \quad k = 1 + v'', \quad P = \frac{2h}{s}\sin\frac{sa}{2h}$$

Table 1. Matrix Equation for the Derived Stress Function Coefficients

INTEGRAL EVALUATION

The integrals in the range $[0, s_o]$ must be evaluated numerically, and the work here uses the Numerical Algorithm Group (NAG) library routine D01ANF, an adaptive routine designed to evaluate integrals of the form

$$\int_A^B g(x)w(x)\,dx, \quad w(x) = \sin(\omega x), \cos(\omega x) \quad (7)$$

The integrals in the range $[s_0, \infty]$ may be expressed in terms of integrals of the form

$$I_a = \int_{s_0}^{\infty} \frac{\sin bs}{e^{as}}\,ds, \quad I_b = \int_{s_0}^{\infty} \frac{\cos bs}{e^{as}}\,ds, \quad I_c = \int_{s_0}^{\infty} \frac{\sin bs}{s\,e^{as}}\,ds \quad (8)$$

I_a and I_b can be analytically integrated by parts but I_c does not have a closed form solution and must therefore be evaluated numerically; NAG routines D01AN/SF are suitable.

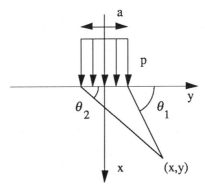

Figure 3 Elastic Half-Space with Uniform Pressure Strip Loading.

VERIFICATION

When the layer and substrate have identical material properties, ie.$\gamma=1$, table 1 gives

$$A'_1 = \frac{P}{s^2}, \quad B'_1 = \frac{P}{s}, \quad C'_1 = D'_1 = 0 \quad (11)$$

With regard to the layer, this effectively makes $s_o=0$ (P is the Fourier transform of the applied load). By application of the closed form solutions for the following integrals (Dwight[9], Gradshtein [10])

$$\int_0^\infty \frac{\sin mx \cos nx}{e^{ax}}dx \ , \int_0^\infty \frac{\sin mx \cos nx}{xe^{ax}}dx \ , \int_0^\infty \frac{\sin mx \sin nx}{e^{ax}}dx \quad (12)$$

it can be shown that the above expressions for the stresses in the layer reduce to those for the stress in a half-space subject to constant normal pressure strip loading, which, using the notation of figure 3, are given by the following (Johnson [11])

$$\sigma_x = \frac{-P}{2\pi}\left[2(\theta_1 - \theta_2) - (\sin 2\theta_1 - \sin 2\theta_2)\right] \quad (13)$$

$$\sigma_y = \frac{-P}{2\pi}\left[2(\theta_1 - \theta_2) + (\sin 2\theta_1 - \sin 2\theta_2)\right] \quad (14)$$

$$\tau_{xy} = \frac{P}{2\pi}(\cos 2\theta_1 - \cos 2\theta_2) \quad (15)$$

Although this gives evidence of the correctness of the mathematical approach, it gives no information about the accuracy with one evaluates the integrals. To this end, the expressions for the stresses in the layer and substrate were evaluated for a wide range of ξ, ζ and h for $\gamma=1$, and the results compared with the half-space values as calculated from equations (13) to (15); excellent agreement was obtained.

The expressions for the stresses were also evaluated for a wide range of γ and h at the surface, interface and at large distances from the load; excellent agreement with the boundary conditions was achieved.

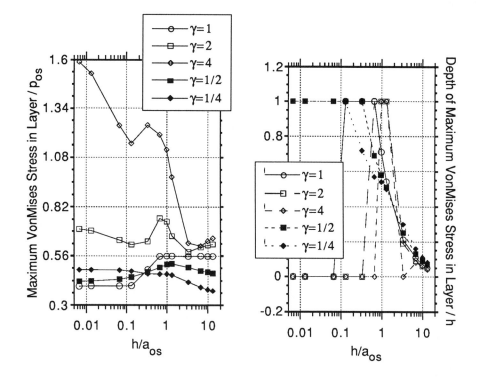

Figure 4 <u>Variation of Maximum VonMises Stress in Layer with</u> <u>γ and h.</u> $\gamma = E_{lay}/E_{sub}$.

Figure 5 <u>Variation of Depth of Maximum VonMises Stress in</u> <u>Layer with γ and h.</u>

RESULTS

A series of simulations were carried out, using the model in [8], where a smooth cylinder was loaded against a smooth surfaced layer, with a particular elastic modulus and thickness, bonded to a substrate. From the calculated surface contact pressure distributions, the stresses in the layer and substrate were calculated. Figures 4 to 7 show how the magnitude and depth of the maximum VonMises stress in the layer and substrate vary with layer thickness and elastic modulus. The results are normalised with respect to the cylinder-substrate contact under the same conditions, which has a maximum pressure p_{o_s} and a contact half-width a_{o_s}.

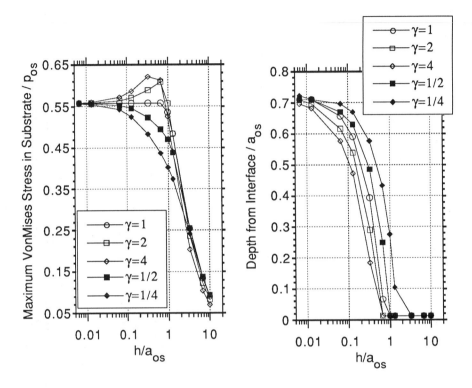

Figure 6 <u>Variation of Maximum VonMises Stress in Substrate with γ and h. γ=E$_{lay}$/E$_{sub}$.</u>

Figure 7 <u>Variation of Depth of Maximum VonMises Stress in Substrate with γ and h.</u>

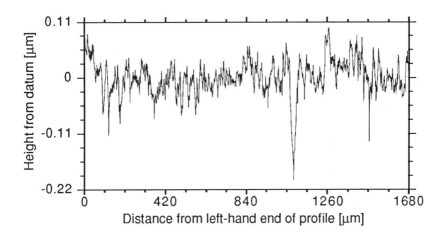

Figure 8 <u>Digitised Surface Profile.</u> 1176 points, 1.434μm sampling interval . RMS roughness 0.034μm, CLA roughness 0.025μm.

Contact simulations, where a cylinder was loaded against a layered body, were carried out for a stiff and a compliant layer for three layer thicknesses, and where the layer surface was first assumed to perfectly smooth and then to have the surface profile shown in figure 8. The cylinder diameter was 20mm, the applied normal line loading was 0.1N/μm, $E_{2,3}$=200GPa, υ=0.3, thus p_{o_s}=0.59GPa and a_{o_s}=107.64μm. The stress distributions in the layer and substrate arising from the surface contact pressure distributions as calculated by the contact simulations are shown in figures 9 to 14.

DISCUSSION

Much of the behaviour in figures 6 and 7, which show the effect of layer thickness and stiffness on the maximum VonMises stress in the substrate for smooth contact, can be explained by consideration of the change in contact pressure which the layer causes; a stiff layer causes the contact pressure to become narrower but taller, the effect increasing with increasing layer thickness, and this increased maximum contact pressure leads to an increase in the subsurface stress, whilst the reduced contact width brings the maximum stress

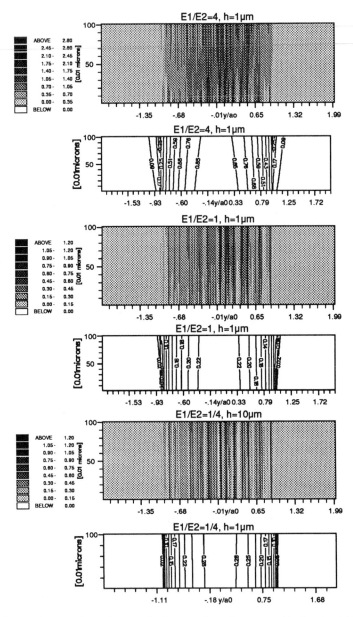

Figure 9 VonMises Stress in Layer, Rough & Smooth Contact [GPa]

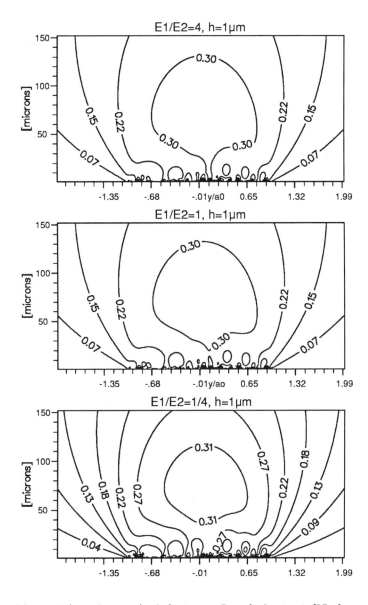

Figure 10 VonMises Stress in Substrate, Rough Contact [GPa]

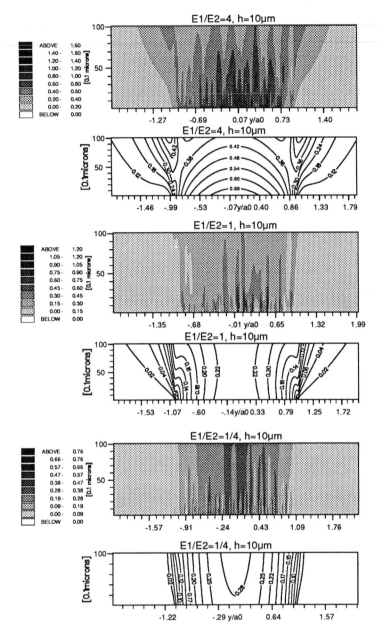

Figure 11 VonMises Stress in Layer, Rough & Smooth Contact [GPa]

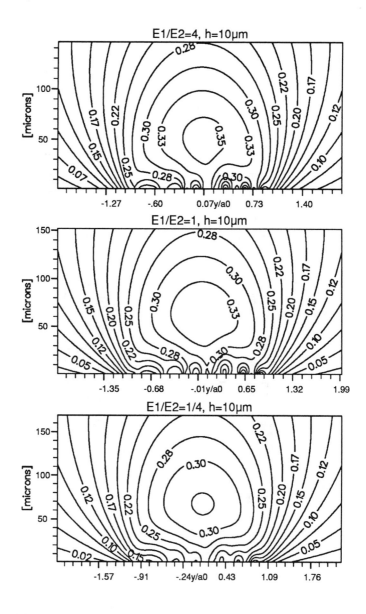

Figure 12 VonMises Stress in Substrate, Rough Contact [GPa]

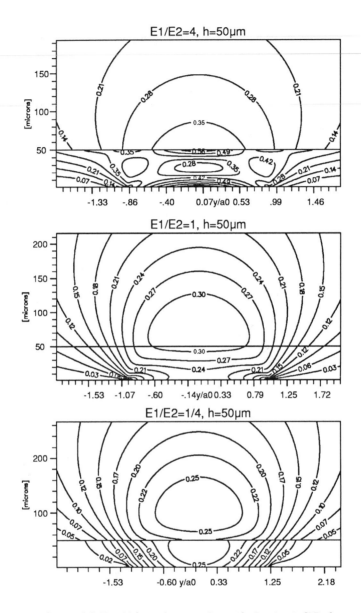

Figure 13 VonMises Stress, Smooth Contact [GPa]

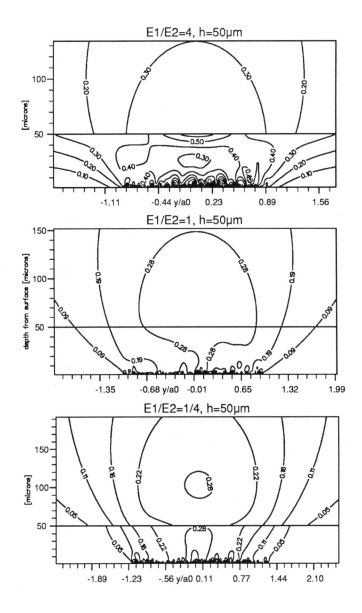

Figure 14 VonMises Stress, Rough Contact [GPa]

closer to the surface. Similarly, a compliant layer yields a wider surface pressure distribution with a lower peak pressure, and this reduces the magnitude of the subsurface stress, whilst moving the maximum stress deeper into the substrate. Thus, the presence of the layer has little effect on the stress in the substrate other than that caused by the change in contact pressure.

Figures 4 and 5 show that when an elastically dissimilar layer bonded to a substrate is in smooth contact, the induced stresses are very different to those which would arise if the layer had the same elastic properties as the substrate. When the layer is thick, the presence of the substrate has little effect on the stress distribution in the layer, and the difference in the stresses arises from the change in contact pressure, as discussed above. However, as the layer becomes increasingly thin its deformation is increasingly governed by the substrate and the elastic dissimilarity raises the stress level.

When surface roughness is considered, the idealised smooth body single contact is replaced by a series of discontinuous contacts between the asperities. The interaction of the stress fields due to each contact yields a complex resultant stress distribution, particularly close to the surface. The load bearing capacity of the asperities, which is higher for a stiff material and lower for a compliant material, means that for a stiff coating the surface pressure distribution is composed of fewer, taller, narrower pressure peaks than the uncoated case, whilst a compliant layer produces more, wider, lower peaks, behaviour that increases with layer thickness [8]. These changes in the surface pressure distribution are reflected in the stress fields presented in figures 9 to 14. These figures demonstrate that when surface roughness is taken into consideration the situation is very much changed from the smooth case. Figure 9 shows that for a 1μm layer 4 times stiffer than the substrate, the maximum VonMises stress in smooth contact is roughly 4 times greater than the uncoated case at the same depth, but when the surface roughness is allowed for, although the stress levels are higher than the smooth case, the difference between the stiff layer and uncoated case is roughly 2.5. Again, when the stiff layer is 10μm thick, for smooth contact figure 10 suggests that the maximum stress should be about 3 times greater but with surface roughness the difference is about 1.3. Similarly, for the 1 and 10μm compliant layers, smooth contact leads to the expectation of increased stress levels but surface roughness keeps the stresses at or beneath the uncoated case.

Figures 13 and 14 show that apart from a disturbed zone close to the surface, the stress levels in a thick layer are not greatly affected by the surface roughness.

CONCLUSIONS

A method for calculating the subsurface stress distributions arising from the two-dimensional dry frictionless contact of layered elastic solids with real rough surfaces is described. The sample results presented show how the subsurface stress distributions for a given surface roughness, which can be significantly different from the smooth case, change with layer thickness and elastic modulus. The surface roughness results in a series of contacts with larger h/a_{o_s} ratios than the single smooth case contact patch, and thus although the roughness increases the stress levels in the layer above those in the smooth case, the difference in stress levels between the case of a layer elastically equal to the substrate and one dissimilar to it are not as great as smooth contact theory would suggest, particularly for thin layers. The results show that apart from the changes induced by the changed surface pressure distribution, the stress distribution in the substrate is not greatly affected by the presence of the layer.

The results emphasize that to obtain accurate knowledge of contact stress levels in layered bodies, the effect of surface roughness must be included.

APPENDIX

From [4], the stresses in the layer and substrate are given by

$$\sigma_{x_{1,2}} = -\frac{1}{2\pi}\int_{-\infty}^{\infty} \omega^2 G_{1,2} e^{-i\omega y}\, d\omega \qquad (A4)$$

$$\sigma_{y_{1,2}} = \frac{1}{2\pi}\int_{-\infty}^{\infty} \frac{d^2 G_{1,2}}{d x_{1,2}^2} e^{-i\omega y}\, d\omega \qquad (A5)$$

$$\tau_{xy_{1,2}} = \frac{1}{2\pi} \int_{-\infty}^{\infty} i\omega \frac{d\,G_{1,2}}{d\,x_{1,2}} e^{-i\omega y}\, d\omega \quad (A6)$$

where G is the Fourier transform of the Airy stress function , and is given by

$$G(x,\omega) = (A + Bx)e^{-|\omega|x} + (C + Dx)e^{|\omega|x} \quad (A1)$$

A, B, C, D are the 'stress function coefficients' and are in general functions of x. By applying the boundary conditions in (A2) for the layered elastic solid in figure 1 subject to normal surface pressure loading $p(y)$, it is found that $C_2 = D_2 = 0$ and the six simultaneous equations in table A1 for the remaining stress function coefficients are obtained.

$$\sigma_{x_2, x_2 = \infty} = 0, \; \sigma_{x_1, x_1 = 0} = -p(y),\; \tau_{x_1 y_1, x_1 = 0} = 0,\; \sigma_{x_1, x_1 = h} = \sigma_{x_2 x_2 = 0},$$

$$\tau_{x_1 y_1, x_1 = h} = \tau_{x_2 y_2, x_2 = 0},\; u_{1, x_1 = h} = u_{2, x_2 = 0},\; v_{1, x_1 = h} = v_{2, x_2 = 0} \quad (A2)$$

$$
\begin{bmatrix}
\omega^2 & 0 & \omega^2 & 0 & 0 & 0 \\
-|\omega| & 1 & |\omega| & 1 & 0 & 0 \\
\dfrac{1}{q} & \dfrac{h}{q} & q & hq & -1 & 0 \\
\dfrac{-|\omega|}{q} & \dfrac{1-|\omega|h}{q} & |\omega|q & q(1+|\omega|h) & |\omega| & -1 \\
\dfrac{-|\omega|f_1}{q} & \dfrac{j_1 - |\omega|hf_1}{q} & |\omega|qf_1 & q(j_1 + |\omega|hf) & |\omega|f_2 & -f_2 \\
\dfrac{|\omega|k_1}{q} & \dfrac{|\omega|hk_1 - 2}{q} & |\omega|qk_1 & q(|\omega|hk_1 + 2) & -\gamma|\omega|k_2 & 2\gamma
\end{bmatrix}
\begin{bmatrix}
A_1 \\ B_1 \\ C_1 \\ D_1 \\ A_2 \\ B_2
\end{bmatrix}
=
\begin{bmatrix}
P \\ 0 \\ 0 \\ 0 \\ 0 \\ 0
\end{bmatrix}
$$

$$q = e^{|\omega|h},\; v' = \frac{2-v}{1-v},\; v'' = \frac{v}{1-v},\; k = 1 + v'',\; f = 1 - v',\; j = 3 - v'$$

Table A1 Matrix Equation for the Stress Function Coefficients

The term P in table A1 is the Fourier transform of the applied normal pressure. In the work in this paper constant elemental pressures are used. For a constant unit pressure applied over the interval $-a/2 \leq y \leq a/2$

$$P = \int_{-\infty}^{\infty} p(y) e^{i\omega y}\, d\,y = \int_{-a/2}^{a/2} \cos \omega y\; d\,y = \frac{2}{\omega} \sin \frac{\omega a}{2} \quad (A3)$$

As G is an even function, the range of the integrals may be reduced from $[-\infty, \infty]$ to $[0, \infty]$. By making the changes of variables $\zeta = y/h$, $\xi = x/h$

and $s = \omega h$, substituting for the derivatives of $G(\xi, s)$, and making the following substitutions,

$$A'_1 = \frac{A_1}{h^2}, \ B'_1 = \frac{B_1}{h}, \ C'_1 = \frac{C_1}{h^2}, \ D'_1 = \frac{D_1}{h}, \ A'_2 = \frac{A_2}{h^2}, \ B'_2 = \frac{B_2}{h} \quad \text{(A 7)}$$

the expressions for the stresses may be written as follows

$$\sigma_{x_{1,2}} = -\frac{1}{\pi} \int_0^\infty \frac{s^2}{h} \left[\alpha e^{-s\xi} + \beta e^{s\xi} \right] \cos s\zeta \ ds \quad \text{(A8)}$$

$$\sigma_{y_{1,2}} = \frac{-1}{\pi h} \int_0^\infty s \left\{ e^{-s\xi} \left(2B'_{1,2} - s\alpha \right) - e^{+s\xi} \left(2D'_{1,2} + s\beta \right) \right\} \cos s\zeta \ ds \quad \text{(A9)}$$

$$\tau_{xy_{1,2}} = \frac{-1}{\pi h} \int_0^\infty s \left\{ e^{-s\xi} \left[B'_{1,2} - s\alpha \right] + e^{s\xi} \left[D'_{1,2} + s\beta \right] \right\} \sin s\zeta \ ds \quad \text{(A10)}$$

where

$$\alpha = A' + B'\xi, \qquad \beta = C' + D'\xi \quad \text{(A11)}$$

Noting that the integral is in the positive region, the simultaneous equations of table A1 may be rewritten as in table 1.

For large values of s, it is clear from table 1 that

$$A'_1 = \frac{2h}{s^3} \sin \frac{sa}{2h}, \qquad B'_1 = \frac{2h}{s^2} \sin \frac{sa}{2h}, \qquad C'_1 = D'_1 = 0 \quad \text{(A12)}$$

Denoting by s_0 the value of s after which these assumptions can be made, equations (1) to (6) in the main body of the paper are obtained.

NOMENCLATURE

a = profile sampling interval

A,B,C,D = stress function coefficients

A',B',C',D' = combination of A,B,C,D and layer thickness

E = Young's modulus of elasticity

h = layer thickness

p = maximum contact pressure

P = Fourier transform of $p(y)$

s = variable derived from Fourier transform integrating variable $(= \omega h)$

s_0 = value of s after which may make assumptions about s.f.c.s

v = transverse deflection

a_{o_s} = Hertzian contact half width of indenter on half-space of substrate material

G = Fourier transform of Airy stress function

$p(y)$ = applied surface pressure

p_{o_s} = Hertzian max. pressure of indenter contacting half-space of substrate material

$t = es$

u = normal deflection

x = normal coordinate

y = transverse coordinate

$\beta = C' + D'\xi$

γ = ratio of material properties of layer to those of substrate

ξ = normalised normal coordinate $(=x/h)$

$\alpha = A' + B'\xi$

$\psi = e^{s\xi}$

v = Poisson's ratio

ω = Fourier transform integrating variable

ζ = normalised transverse coordinate $(=y/h)$

REFERENCES

[1] Burmister,D.M., Journal of Applied Physics, 16,89-96,126-127,296- 302,1945.

[2] Chen,W.T., "Computation of Stresses and Displacements in a Layered Elastic Medium", International Journal of Engineering Science, Vol. 9, 1971, pp.775-800.

[3] Pao,Y.C., Wu,T.S., Chiu,Y.P., "Bounds on the Maximum Contact Stress of an Indented Elastic Layer",ASME Journal of Applied Mechanics, Vol. 38, Sept. 1971, pp.608-614.

[4] Gupta,P.K., Walowit,J.A., "Contact Stresses Between an Elastic Cylinder and a Layered Elastic Solid", ASME Journal of Lubrication Technology, April 1974, pp. 250-257.

[5] Chiu,Y.R., Hartnett,M.J., "A Numerical Solution for Layered Solid Contact Problems with Application to Bearings", ASME Journal of Lub. Tech., Oct. 1983, Vol.105, pp. 585-590.

[6] King,R.B., O'Sullivan,T.C., "Sliding Contact Stresses in a Two-Dimensional Layered Elastic Half-Space", International Journal of Solids and Structures, Vol.23, No.5, 1987, pp.581-597.

[7] Sainsot,P., Leroy,J.M., Villechaise,B., "Effect of Surface Coatings in a Rough Normally Loaded Contact", 'Mechanics of Coatings', Proc. 16th Leeds-Lyon Symposium on Tribology, Lyon, France, Sept. 1989, Tribology Series 17, Elsevier.

[8] Cole,S.J., Sayles,R.S.,'A Numerical Model for the Contact of Layered Elastic Bodies with Real Rough Elastic Surfaces', to be presented at STLE-ASME Tribology Conference, Oct. 1991, to be published in ASME Journal of Tribology.

[9] Dwight,H.B.,"Tables of Integrals and Other Mathematical Data", 4th Ed., Macmillan, 1961.

[10] Gradshtein,I.S., Ryzhik,I.M.,"Table of Integrals, Series and Products", Academic Press, 1980.

[11] Johnson,K.L., "Contact Mechanics",Cambridge Uni. Press, 1985.

METALLOGRAPHIC EVALUATION OF THERMAL SPRAYED COATINGS

A. R. Geary
Pratt & Whitney
East Hartford, Connecticut

ABSTRACT

Today's advanced thermal sprayed coatings, used for the protection of various alloys in gas turbine engines, must be metallographically prepared in a highly controlled manner if the real microstructure is to be revealed. A vacuum impregnation technique using a mixture of a fluorescent dye (Rhodamine "B") and epoxy, which has been used for two decades, allowed investigators to study the presence of both preparation and service induced defects in ceramic and metallic coatings, when those defects were surface connected. The procedures used to prepare and examine thermally sprayed ceramic/metallic coatings by optical and electron microscopy is described. Additionally, a methodology for using this approach to resolve coating related preparation problems is presented.

INTRODUCTION:

Metallographic preparation of thermal sprayed materials have shown varying levels of porosity. Investigators question the porosity levels indicated and whether they are in fact present or an artifact of the preparation techniques used. This paper attempts to resolve this issue by comparing metallographic preparation techniques with SEM evaluations of a cryogenically fractured surface. A careful review of these changes has resulted in the development of a unique approach to improving coated sample quality. While many laboratories use both hand and automated polishing equipment, it is felt that many problems result from inconsistencies prior to sample polishing procedures. Metallic and ceramic coated turbine components are most susceptible to cracking, chipping or pullout during the initial preparation stages when conventional techniques are employed. Therefore, a low viscosity epoxy mounting media such as Stycast [1] or its equivalent, can be used to totally infiltrate the porous materials and preserve the metallic/ceramic structure. Due to variations in sample integrity resulting in various porosity levels, measured amounts of Rhodamine B, which is a biological stain can be stirred into the encapsulating media. Infiltrating a porous coating with a dyed media traditionally produces a clear distinction between interconnected porosity to the gas surface and isolated porosity regions. However, further enhancement of coating microstructure can be studied after cryogenic fracture and subsequent evaluations on the SEM.

PREPARATION: MOUNTING

The principal purpose for metallographically preparing thermal sprayed materials for optical and SEM evaluations is to determine relative levels of porosity and/or crack development. To preserve the inherent porosity and/or cracking commonly found in most metallic and ceramic materials vacuum impregnation is required to ensure that the investigator has a sound metallographic sample. This is most important when image analysis [2,3] data is required of a suspected coating problem. However, it should be noted that not all thermal sprayed metallic materials exhibit characteristic porosity or defects. For many years researchers have pursued an improved method of preparing coated materials, thus establishing a reliable finished micro-crossection. The following parameters are recommended to assist investigators while preventing grinding or polishing damage to the sample surface:

o Should samples require sectioning, it is <u>important</u> to thoroughly dry the sample prior to mounting. Drying in an oven for 5/10 minutes, set at 150°F will improve moisture evaporation and allow a uniform infiltration of the epoxy media.

o By weight measure equal amounts of Stycast part A and B and place in an oven, set at 150 to 200°F; for approximately 15 minutes prior to mixing. This will allow the Stycast to reach a low viscosity, which should improve mixing capabilities. For equivalent potting medias it is suggested that the recommended manufactured mixing procedures are followed.

o Spray metal cups with a mold release agent and in the case of silicone rubber cups it is suggested that grease be used, by applying a thin film on the inner cup wall. This approach will allow for a easy separation of container and finished sample, after curing.

o After mixing Stycast part A & B together with the approximate amount of Rhodamine "B" (fluorescent dye-see Exhibit I), pour a small amount of the warm Stycast over the sample surface; if large samples are mounted it is suggested that investigators lay sample flat and pour just enough potting media to cover the sample, DO NOT FILL CUP TO THE BRIM. A small amount of potting media in the cup will allow the investigator to evacuate in a vacuum oven for extended periods of time without spillage. Extended periods of evacuation times may be required for total infiltration to produce a supported sealed metallic and ceramic structure. This type of early preparation technique will be required when investigators are anticipating subsequent image analysis of polished crossections.

o Place the sample in a vacuum oven or a vacuum impregnation apparatus and infiltrate the dyed Stycast into the coating (when applicable) for approximately 15 to 60 minutes at 225°F. These times will vary depending on the inherent porosity levels which are connected to the gas surface. Stycast is used because it has a longer curing time, thus allowing extended periods of evacuation time. The sample should be completely impregnated with the fluorescent dyed Stycast when air bubbles no longer appear on the surface. It usually is necessary, to break the vacuum and re-evacuate for this to occur. Remove sample and completely fill cup to the brim. Place cup back in the oven for curing. No additional evacuation is required at this time. Minimum curing times are estimated to be 2 hours for Stycast, however, longer periods of curing times are more desirable. In the case of equivalent potting medias investigators are encouraged to use manufacturer mixing procedures, which may require a change in the infiltration time and procedure.

EXHIBIT I

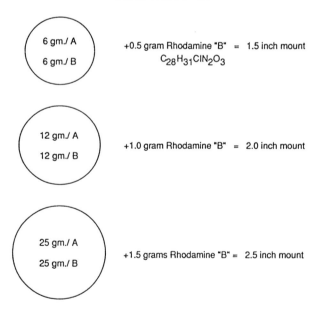

6 gm./ A 6 gm./ B	+0.5 gram Rhodamine "B" = 1.5 inch mount $C_{28}H_{31}ClN_2O_3$
12 gm./ A 12 gm./ B	+1.0 gram Rhodamine "B" = 2.0 inch mount
25 gm./ A 25 gm./ B	+1.5 grams Rhodamine "B" = 2.5 inch mount

PREPARATION: POLISHING

The grinding and polishing procedures for both metallic and ceramic materials were performed on the Struers abra-system, Table 1 and with limited polishing on a Struers Prepamatic, which is a fully automatic microprocessor system, Table 2. Comparable results were achieved on both systems, however, the fully automated system is more desirable because of its simplicity and repeatability.

TABLE 1.   SEMI-AUTOMATIC PREPARATION FOR METALLIC AND CERAMIC MATERIALS.

Grinding

	1	2	3	4	5	6	7
Equipment	Abraplan	Abrapol	Abrapol	Abrapol	Abrapol	Abrapol	Abrapol
Grain (Size No.)	150 Mesh	240 SiC	320 SiC	400 SiC	600 SiC	2400 SiC	4000 SiC *
Speed (RPM)	1400	300	300	300	300	300	300
Force (Newtons)	200	200	200	200	200	200	200
Time (Sec/Min)	Until Flat	15 Sec.	15 Sec.	15 Sec.	15 Sec.	15 Sec.	15 Sec.

Polishing

	1	2	3	4
Equipment	Abrapol	Abrapol	Abrapol	Syntron Δ
Grain (Size No.)	6 μm Diamond	1 μm Diamond *	Alumina Powder, 0.3 μm Linde "A"	Alumina Powder, .05 μm Linde "B"
Polishing Cloth	Canvas Cloth	Nylon Cloth	Microcloth	Microcloth
Polishing Speed (RPM)	150	150	150	-
Force (Newtons)	200	200	150	-
Lubricant Type	DP Green §	DP Green	Alumina Powder, 0.3 μm Linde "A"	-
Time (Sec/Min)	3 Min.	3 Min.	30 Sec.	30 Min.

* - ultrasonically clean
Δ - microstructural inhancement
§ - ethylene glycol and alcohol mixture

641

TABLE 2. FULLY-AUTOMATIC PREPARATION FOR METALLIC AND CERAMIC MATERIALS.

Disc/Pad	Grit	Lubricant	Removal/Time
Stone	150 mesh stone	Water	50 sec.
P-M	6 μm	DP Blue *	100 sec.
DP Mol (Wool)	6 μm	DP Blue	100 sec.
DP Dur (Silk)	1 μm	DP Blue	180 sec
DP Chem	OP-S	-	60 sec.

* ethylene glycol and alcohol mixture.

RESULTS AND DISCUSSION:

The difficulty in validating porosity/defect levels [4] in metallic and ceramic materials [5] has resulted in the use of both optical and SEM examinations. The thermal sprayed samples prepared for this paper are examples of a select group of plasma coated microstructures utilized to emphasize the fluorescent dye techniques.

This approach allows investigators to distinguish void level/crack behavior throughout a metallic sample and supports interpretation efforts on materials where Rhodamine B was not employed. The penetration of the dyed epoxy throughout a candidate sample can be detected on a microscope equipped with a polarized light system. Generally, this effort works well on metallic materials; whereas, translucent materials (ceramic) have a history of poor behavior when evaluated using polarized light due to the fluorescent nature of the material.

A typical phenomenon which occurs during preparation in many laboratories is illustrated in Figure 1a. Here investigators have "tin canned" metallic material during preparation. "Tin-canning" is defined as the act of camouflaging a void (in this case interconnected porosity) by filling it with smeared residue coating and alloy material during metallographic preparation. Presumably, this occurs during the grinding sequence. Subsequent diamond polishing will produce an acceptable surface, regardless of whether the material is characteristic of the part or "tin-canned" during

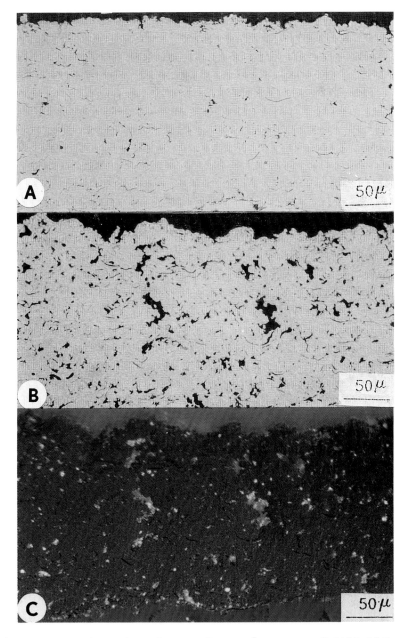

Figure 1. Microstructure of an Air Sprayed M/CrAlY Coating: (a) Poor Preparation, (b) Proper Preparation, (c) Same as Above Area Vacuum Infiltrated with Dyed Epoxy.

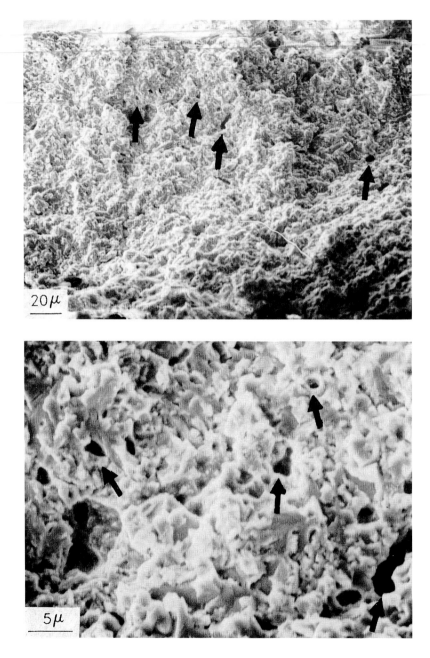

Figure 2. Cryogenic Fracture of a M/CrAlY Coating Showing Voids (Arrows).

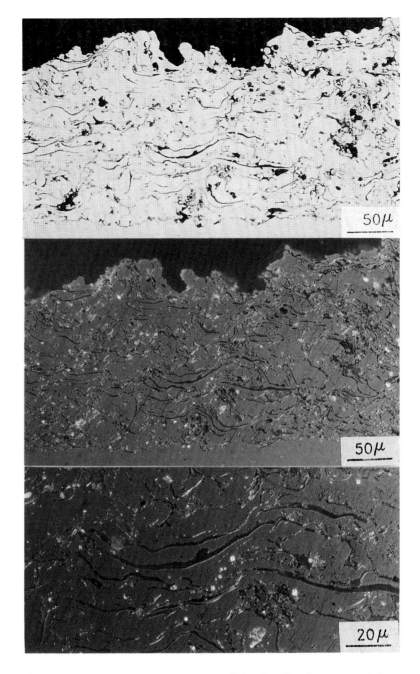

Figure 3.   Plasma Sprayed Nickel Aluminum Coating.

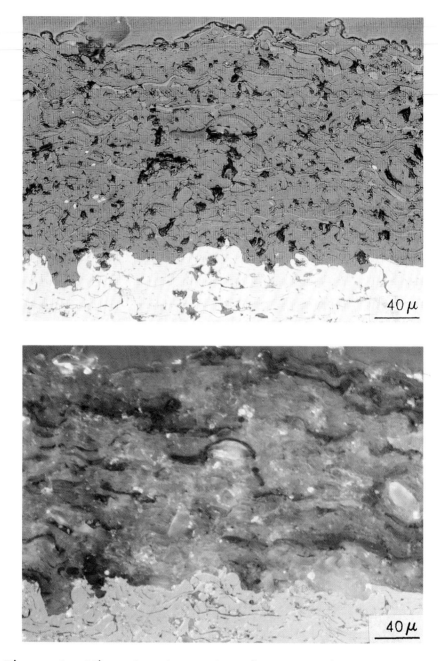

Figure 4.   Microstructure of a Nickel Aluminum Bond Coat
with a Top Coat of Aluminum Oxide.

grinding. In Figures 1b and 1c the same air sprayed coating was vacuum infiltrated prior to preparation. The final results clearly illustrate the porous nature of the coatings.

To further enhance investigations of the porous nature of the M/CrAlY material a cryogenic fracture of the coating was studied, Figure 2. Here we see a noticeable distribution of voids across the fractured surface. This then supports investigator efforts along with the impregnation of the interconnected porosity [6] of air sprayed materials.

In the case of plasma sprayed nickel aluminum bond-coat, Figure 3 a clear distinction of what is commonly called stratified oxides by some investigators can be identified as linear porosity. In this case interconnected porosity exists as a result of splat cooled coating material striking a surface and solidifying as it flattens out. Vacuum infiltration of the part prior to the preparation sequence clearly depicts the microstructural character in real life. In this case very little oxide exists.

Dyed epoxy when heated will lower its viscosity and reduce surface tension. This will ensure penetration of dyed epoxy into inherent defects. The use of Stycast allows the investigators sufficient time for evacuation before curing.

The in-situ characterization of an aluminum oxide top coat over a bond coat of nickel aluminum is presented in Figure 4. Here we see typical void distribution throughout both coatings (note the presence of the red dye) along with areas of noticeably dark regions in the ceramic material. The somewhat fluorescent nature of the aluminum oxide coating makes for some discussion as to the authenticity of its make-up. However, the black and white to color comparisons (with the use of polarizing light) using proper cross polarization settings add considerably to the study of fluorescent materials. This is an advantage in light optical microscopic examinations in that it increases the contrast of the polished surface under polarized light.

Finally, during a turbine blade tip experimental coating development program investigators recognized that unusual microstructures were common to a routinely

647

prepared sample; and the usual mounting and preparation techniques may not apply to this type of sample. The removal of SiC grits during grinding and subsequent polishing from a vacuum sprayed NiCoCrAlY matrix was subject to further review. Bakelite versus epoxy mounting media was the issue in this case. Vacuum infiltration methods were employed [7] and resulted in microstructures similar to those presented in Figure 5. Good adherence of the SiC grits along with infiltration of obvious voids to the alloy coating interface, with the dyed epoxy is necessary to maintain sample integrity.

CONCLUSIONS:

The following conclusions were made from this study with regard to simplicity, quality and repeatability:
1. The preparation sequence prior to actual sample polishing is a major element in maintaining true sample integrity.
2. Observations of polished crossections evaluated using Rhodamine "B" and Stycast mixtures is a procedure recommended for a select group of samples to validate coating quality.
3. The realization of a unique method for improved characterization of candidate porous materials with a discernible dye mixture (utilizing polarized light) has benefited investigators in determining coating excellence.
4. A strong emphasis for todays metallographic laboratories is toward automated polishing equipment simply because of its reproducibility.

ACKNOWLEDGEMENT:

The author would like to recognize Mr. D. Lathrop for his metallographic assistance in automated sample development.

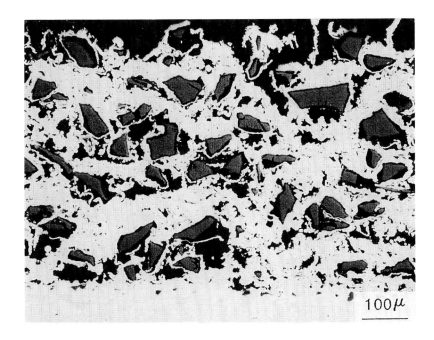

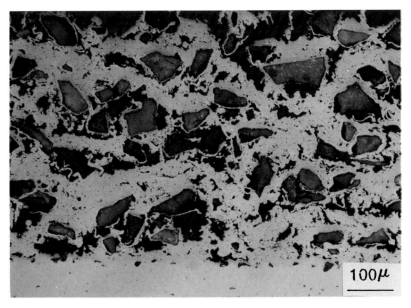

Figure 5. Plasma Sprayed Metallic with Ceramic Grits
Vacuum Infiltrated with Rhodamine "B".

649

REFERENCES:

1) Stycast^R 1269 A & B Crystal Clear Epoxy; Emerson & Cuming, Inc.
2) J. C. Oppenheim, "Application of Quantitative Image Analysis in the Advanced Materials Laboratory," Microstructural Science, Vol. 17, 1989.
3) D. B. Fowler, "A Method for Evaluating Plasma Spray Coating Porosity Content Using Stereological Data Collected by Automatic Image Analysis." Microstructural Science, Vol. 18, 1990.
4) G. A. Blann, D. J. Diaz & J. A. Nelson, "Raising the Standards for Coating Analysis", Advanced Materials & Processes, Vol. 136, issue 6, December 1989.
5) W. J. Brindley & T. A. Leonhardt, "Metallographic Techniques for Evaluation of Thermal Barrier Coatings", Materials Characterization 24:93-101, 1990.
6) H. Herman, "Plasma Sprayed Coatings", Scientific American, September 1988.
7) G. F. Vander Voort, "Metallography Principles and Practice", McGraw-Hill, Inc., 1984.

Advanced Materials

OPTICAL TECHNIQUES

FOR MICROSTRUCTURAL CHARACTERIZATION

OF FIBER-REINFORCED POLYMERS

Luther M. Gammon
Boeing Commercial Airplane Company
Seattle, WA 98124-2207

ABSTRACT

Microscopic analysis of the morphology of fiber reinforced polymers (FRP) requires techniques different from those commonly employed in metallography. Since most polymeric materials are semi-transparent to light, conventional reflected light microscopy techniques are not suited to reveal details of the microstructure. There are also problems associated with the preparations of polished cross sections of fiber reinforced polymers owing to the extreme differences in hardness and mechanical behavior of the fiber and matrix. This paper discusses some of the new approaches to the preparation and optical microscopic examination of fiber reinforced polymer composites. Techniques using dye penetrants to identify matrix microcracks will be outlined. The use of a) etchants, b) epi (epidermal) -fluorescence, c) transmitted phase contrast , and d) transmitted differential interference contrast can be used to bring out the polarized light microscopy for the study of semi-crystalline thermoplastics such as PEEK will be discussed.

INTRODUCTION

Composites are a heterogeneous laminated system, composed of thermoplastic or thermoset polymers, and typically reinforced with nylon, Kevlar, glass or carbon fibers. Understanding the morphology of these heterogeneous materials which results from processing or environmental exposure is key. Microscopy provides the tools to examine and understand composites. An historical, step-by-step overview of the

653

research and development techniques conducted by Boeing Materials Technology, over the last twelve years to observe polymer morphology is discussed in this paper.

DISCUSSION

The first step in conducting any microstructural characterization is proper sample preparation. Good microscopy is ninety-eight percent sample preparation. Producing a flat, scratch-free, strain-free cross-section without introducing artifacts requires both a science and an art. Polymers are usually softer than the fibers, and without special care, one will round off the fibers and erode away the resin. Figure 1 shows a carbon fiber thermoset properly cross-sectioned. On a properly cross-sectioned specimen, thinning of the polymer adjacent to the fiber results in the prism

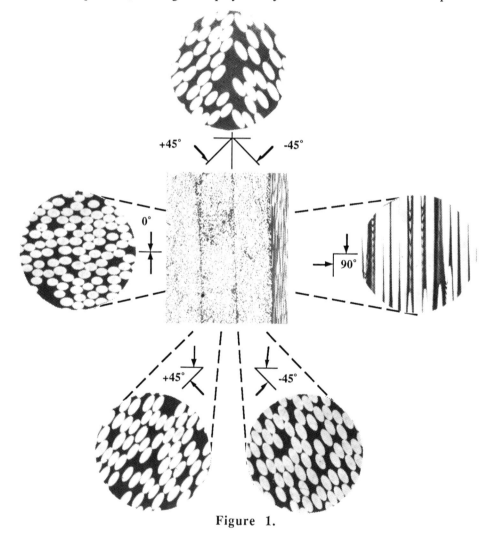

Figure 1.

effect. Note this optical wedge adjacent to one side of the 90° fiber. No relief is seen around the fibers, nor is there rounding of the fibers.

Most composite microscopy involves looking for porosity. Many composite polymers, thermosets and thermoplastics, appear light gray under epi-bright field illumination and contrast readily with the black-appearing porosity. Some composites, such as fiberglass/epoxy, exhibit such low reflectivity that little or no contrast between porosity and resin exists. This low contrast can be enhanced with translucent dyes. The dye is applied to the cross-section and back-polished off the surface. The translucent dyed porosity can then be observed with epi-dark field illumination.

Examples of ply-counts and fiber orientations can be seen in Figure 1. The sample was prepared by cutting a 45° angle to the plan view surface. With this technique, one can measure the ply rotation accurately. On standard cross-sections, checking fiber orientations requires multiple sections and does not have a high degree of accuracy. However, with the 45° plan view cross-section technique, only one cross-section is needed. The fiber orientations can be observed on this single cross-section within ±2°. On a photomicrograph or TV monitor, one can overlay a transparent protractor and measure the major axis of the fiber ellipse. For example: a zero orientation will read zero degrees, a +45° orientation will read 45°, a -45° orientation will read 135°, and a 90° orientation will read 90°.

Composites are susceptible to damage during cutting, grinding and polishing, making microcracks extremely difficult to observe. It is important to proceed carefully when working with microcracks, so that cracks are not altered or created. Microcracking is the result of stresses which surpass the ultimate strength of the polymer or polymer/fiber interface bond. By a careful study of microcracking, one can determine the durability of composites in service and investigate the relationship of process parameters and environmental exposure to microcracks.

Microcracks have high aspect ratios; therefore, observing them can be elusive. As with porosity under epi-bright field illumination, most resins appear light gray,while larger cracks and some microcracks appear black. However, microcracks in composites with translucent fibers such as fiberglass or Kevlar can be invisible with bright field. This type of microcrack is best observed by dyeing, back-polishing, and illuminating with EPI dark field. Microcracks in some composites such as carbon fiber reinforced thermoplastics remain invisible with either of the above techniques. For these composites, one can use a fluorescent penetrant or a laser dye with a compatible solvent and observe with epi-fluorescents.

Before using a pigmented solvent on a resin, the effects of solvent on that particular resin should be determined. For example, methylchloroform, used in many penetrants, can cause microcracks in thermoplastics. It is important to first observe composites with epi-bright field for subtle signs of microcracking before applying a dye. By following this procedure, one can enhance the contrast without causing or worsening the microcracking.

Morphology on some multiphase systems is enhanced by etching with one of two techniques. First, solvent can selectively swell or attack one or more of the phases. Some solvents cause preferential swelling, which can be observed with differential interference contrast. The phases can also be saturated with a fluorescent pigment and observed with epi-fluorescents. Secondly, etching with oxidizing chemicals can be used, and the effects can be observed with epi-bright field, epi-differential interference contrast. However, epi-phase contrast is usually preferred when observing polymers.

Often detailed multiphase morphology is only visible with transmitted light using thin section microscopy. This technique requires polishing two parallel surfaces, one to four microns a part, while keeping each surface flat. The preparation must not create residual stresses that will distort the microstructure. Microtoming is not an acceptable thin section preparation technique because severe distortions result from residual stresses during cutting. To observe the sample with transmitted light, more than one contrast mode is useful.

Thermoplastic (PEEK, PPS) crystallinity can be observed with transmitted polarized light. Amorphous thermosets can be observed with transmitted Hoffman, phase, and differential interference contrast. The advantage of transmitted light along with the full complement of contrast modes is evident. Contrast can vary over a wide range, depending upon optical characteristics of the sample: polarized light, circular polarized light, phase contrast, Hoffman contrast, darkfield, and differential interference contrast. For example, phase intermingling in honeycomb sandwich material is seen using transmitted polarized light in Figure 2 and transmitted phase contrast in Figure 3.

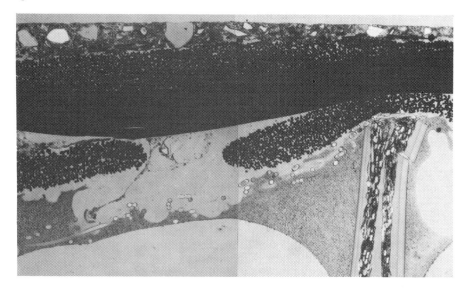

Figure 2.   Transmitted Polarized Light (100X magnification)

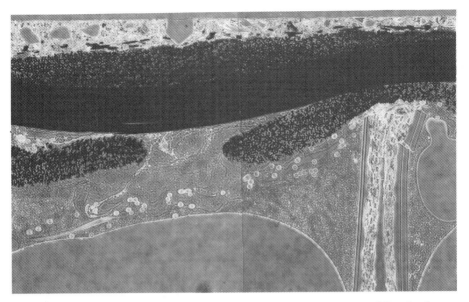

Figure 3. Transmitted Phase Contrast (100X magnification)

CONCLUSION

Polymer composites are optically semitransparent. Each system has its own optical properties and refractive index. No single optical technique works for every system. Therefore, it is important to have the full complement of epi- and transmitted light contrast modes. Optical microscopy of polymer composites is often overlooked in favor of electron microscopy. However, while optical microscopy is complementary to electron microscopy, in many cases it is preferred and can reveal information unobtainable with any other tool.

657

MICROSTRUCTURE OF MECHANICALLY PROCESSED POLYMERIC MATERIALS

J. Pan[1] and W.J.D. Shaw[1]

ABSTRACT

Mechanically alloyed polymers are a new class of materials currently under development. They are made using a special processing technique of high energy grinding resulting in a fine powder which is subsequently consolidated well below the materials melting point temperature. Two polymeric materials, polyamide and polyethylene were processed by means of mechanical alloying. This has resulted in material structures not previously thought possible. These two materials are compared with the same materials processed using regular melting practices. The mechanical property test shows that the strength of mechanically processed polyamide has increased considerably. The results from x-ray diffraction, optical and electron microscopy indicate that the mechanically processed polymeric materials results in definite alternations of the material crystal structure and atomic bonding. All indications to date suggest that it should be possible to create mechanically alloyed polymeric materials.

[1] Department of Mechanical Engineering, The University of Calgary, Calgary, Alberta, Canada, T2N 1N4

659

INTRODUCTION

In 1968 The International Nickel Company developed the mechanical alloying process[1] during a program of oxide dispersion strengthening, ODS, applied to high temperature nickel. The intent was to strengthen the matrix with both gamma prime precipitation and oxides alloyed directly into the material on a very uniform and fine scale. The intended use for this special alloy was in the gas turbine industry. Today the process of mechanical alloying methods has evolved to a position of allowing the engineering of composite metallic powders of many different alloy combinations. The advantages are the extremely fine grain size produced and the ability to combine thermally incompatible materials in an interactive manner. Thus material alloys that were previously impossible to produce using conventional techniques can now be made using this new technique.

The term mechanical alloy describes a very unique type of material, where its properties are obtained from specialized mechanical processing. Materials that are made by mechanical alloying are first introduced into a high energy ball mill and are ground over a long period of time producing an extremely fine powder by the mechanism of fracture and cold welding. The powder is then consolidated below the melting point of the material by using pressure, time and temperature combinations.[2]

Alloyed polymeric materials have been continually developed over the last few decades. The technique of producing polymeric alloys to date are based upon using chemical reactions. Recently, a novel idea of creating polymeric materials by means of mechanical alloying has arisen. If this concept works, then a new class of materials will come into existence.

The main objective of this work is to lay down some ground work to determine whether it is possible to create mechanically alloyed polymeric materials. The preliminary study looks specifically at whether polymeric powder particles can be ground small enough without degrading the material and then consolidated below their melting point. The microstructure and x-ray diffraction behavior of mechanically processed polymeric materials are characterized in this paper.

EXPERIMENTAL PROCEDURES

Mechanical Milling

The materials used in this study are polyamide (nylon 6,6, Du Pount Product) and low density polyethylene (commercial trade mark XXKS, Du Pount Product). Reactor grade polyamide and polyethylene pellets were processed in a specially designed shaker ball mill (see Figure 1). The ball mill has an overall acceleration of 12.3 g's while operating at frequency of 1750 Hz. The atmosphere inside the process chamber was not controlled in this initial study but rather consisted of an air environment. The ball mill chamber was cooled using liquid nitrogen to a temperature below -150°C. The material was processed for 24 hrs at the end of which the material was removed and the powder size was measured.

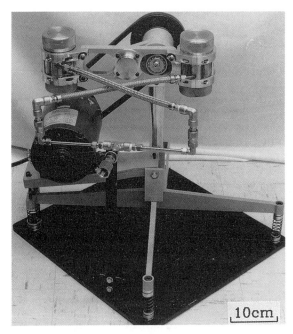

Figure 1. Shaker ball mill with insulation removed.

Consolidation

The processed material was transferred to a consolidation press where it was heated to 80°C under vacuum conditions for 20 hrs to de-gas the material. Following this, the temperature was raised to 233°C for polyamide powders and 90°C for polyethylene powders under a pressure of 68.95 MPa for a period of 48 hrs. At the end of this time period, a solid billet had formed and was removed from the consolidation equipment. The melting point temperature as measured for regular processed polyamide and low density polyethylene were found to be 256°C and 115°C respectively.

Polishing And Etching

Solid billets of polyamide and polyethylene were cut and mounted in thermoset plastic with the longitudinal and transverse surface exposed. The exposed surfaces were polished mechanically from 240 grit, 600 grit, to 6μm, 1μm, and finally 0.05μm. As polyethylene and polyamide are soft materials, in order to get rid of all scratches, specimens were cooled by immersion in liquid nitrogen during the polishing process.

A number of etching techniques were developed and evaluated in order to provide good differentiation of the various microstructural areas. It was found that the best etchant for polyamide and polyethylene was concentrated xylene regent and the etching condition for polyamide material was 3 to 4 minutes at 75°C[3] and 20 to 60 seconds at 70°C for polyethylene.

661

RESULTS AND DISCUSSION

Powder Characteristics

Typical powder morphology after 24 hrs of milling for polyamide and polyethylene are shown in Figures 2 and 3. Both figures show conglomerations of larger particles being made up of many fine small sized particles; as well as a few very fine small sized particles are apparent by themselves. Thus, the overall average particle size is approximately 3 microns for polyamide powders and 5 microns for polyethylene powders. It appears that the fracture and cold welding has occurred in the mechanical alloying process since many of the larger particles are conglomerations of fine individual particles.

An overall relation of the polyamide particle size with grinding time is shown in Figure 4. This figure indicates that even after processing for a period as short as 6 hrs, the powder particle size has become quite small.

X-ray Diffraction Study

X-ray analysis was conducted on materials that were consolidated into a solid form from the mechanically processed powders and compared to the regular thermal melt materials. The results for polyamide and polyethylene are shown in Figure 5 and 6 respectively. It was found that for the most part, many of the reflected peaks on the left and right areas of the x-ray diffraction curves are identical between mechanically processed polyamide, MPPA, and regular thermal melt polyamide, PA. The same comparative result was obtained between mechanically processed polyethylene, MPPE, and thermal melt polyethylene, PE. However, for both cases the major peaks in the middle area are considerably different. Generally, a lower intensity, a shifting and missing one of the major peaks were found in both figures. The x-ray analysis data definitely indicates that alterations have occurred in the atomic bonding of polymeric materials due to the mechanical alloying technique.

Microstructure

Figure 7 is a optical micrograph of regular thermal melt polyamide. It is in general agreement with the features of similar material published by others.[4] However, the microstructure of mechanically processed polyamide is quite different (see Figure 8). The small grain size is a principle characteristic of MPPA material. Two distinct regions are seen in this figure. It is possible that one area is crystalline with the other region being amorphous. X-ray diffraction study has shown that mechanically processed polyamide has a lower crystallinity compared to regular thermal melt polyamide. Thus it is reasonable that there is a distinct amorphous area in the microstructure of MPPA material.

The microstructure of thermal melt and mechanically processed polyethylene are shown in Figures 9 and 10. Again, the mechanically processed polyethylene shows a much finer grain size compared with the regular thermal melt material.

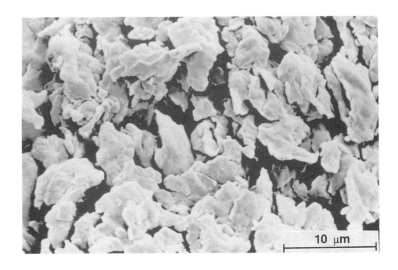

Figure 2. Polyamide powder particle size after 24 hours of processing.

Figure 3. Polyethylene powder particle size after 24 hours of processing.

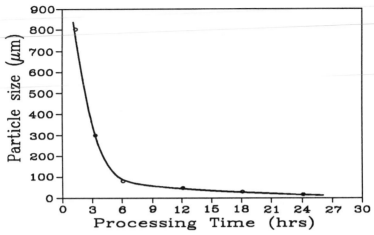

Figure 4. Polyamide particle size changes with processing time.

Mechanical Property Evaluation

The compressive stress-strain test shows that the mechanically processed polyamide is stronger by 15% as compared to the regular processed polyamide material (see Figure 11). However, the ductility has been reduced by approximately 25%. Additionally, the regular processed polyamide material failed by buckling while mechanically alloyed material failed by cleavage.

Measurements of hardness of the polyamide material resulted in an increase of approximately 7.5% for the mechanically processed material over that of the regular processed polyamide (see Figure 12).

Since the mechanically processed material has a lower crystallinity compared to the regular processing materials, it is probable that the increase of strength results from the fine grain size and the change of atomic bonding between the polymeric molecular chain in the mechanically processed material.

CONCLUSIONS

This preliminary study has shown that polymeric materials can be processed using the mechanically alloying method without suffering major degradation. Initial processing of polyamide and polyethylene by the mechanical alloying technique has successfully resulted in the creation of unusual types of polymeric materials. The mechanically processed polyamide and polyethylene powders are quite small (approximately 3 and 5 microns respectively), and can be consolidated relatively easily at temperatures well below their melting points which is not possible to do for

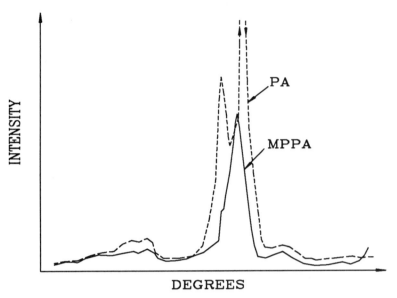

Figure 5. X-ray diffraction behavior of polyamide material.

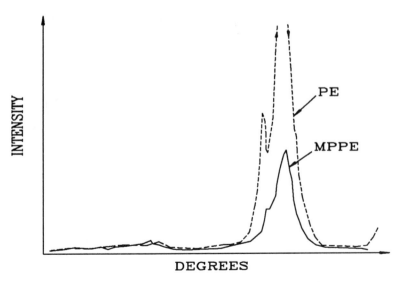

Figure 6. X-ray diffraction behavior of polyethylene material.

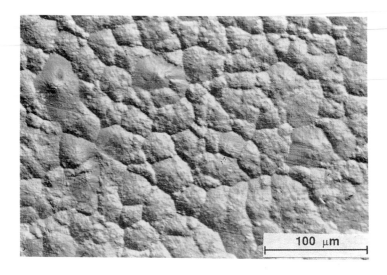

Figure 7. Optical micrograph of regular thermal melt polyamide.

Figure 8. Optical micrograph of mechanically processed polyamide.

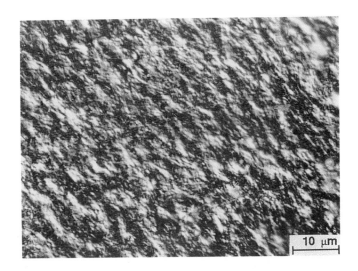

Figure 9. Optical micrograph of regular thermal melt polythylene.

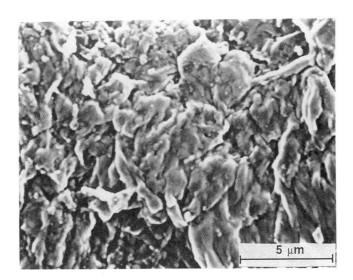

Figure 10. SEM micrograph of mechanically processed polythylene.

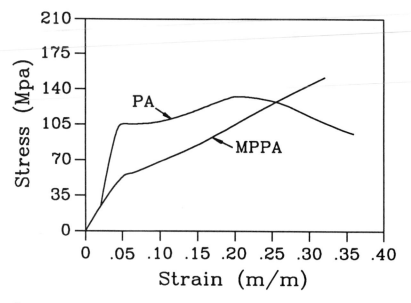

Figure 11. Compressive stress/strain behavior of polyamide material.

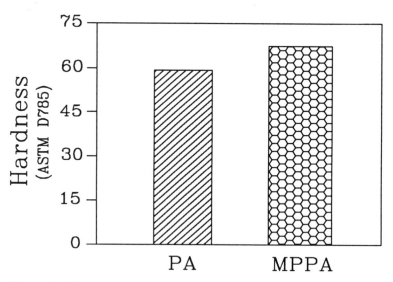

Figure 12. Hardness comparison of thermal melt processed polyamide (PA) *vs* mechanically processed polyamide (MPPA).

any regular polymeric materials. The physical properties of mechanically processed polyamide have not degraded but rather have improved somewhat over that of the regular polyamide. The results from x-ray diffraction, optical and electron microscopy show that the mechanical processing of polyamide and polyethylene results in definite alterations of the material crystal structure and atomic bonding. All information to date strongly suggests that it should be possible to create mechanically alloyed polymeric materials with enhanced properties.

ACKNOWLEDGMENTS

The authors are grateful to The Institute For Chemical Science and Technology and The Natural Science and Engineering Research Council Of Canada for financial support of this work. Thanks are also given to Clay Elston of Du Pont Company for his continued input and support of this program. Additionally sincere thanks to the design engineer, Mike Sutherland who has provided continued dedication towards overcoming the equipment problems.

REFERENCE

1) J. S. Benjamin, "Dispersion Strengthened Superalloys By Mechanical Alloying", **Metall. Trans.**, Vol. 1, p.2934, 1970

2) P. S. Gilman and J. S. Benjamin, "Mechanical Alloying", **Ann. Rev. Mater. Sci.,** Vol. 13, p.279, 1983

3) S. Y. Hobbs, "Polymer Microscopy", **J. Macromol., Sci.-Rev. Micromol. Chem.**, C19 (2), p.221, 1980

4) L. Bartoxiewicz and Z. Mencik, "An Etching Technique To Reveal The Supermolecular Structure Of Crystalline Polymers", **J. Polym. Sci., Polym. Physics Edit.,** Vol. 12, p.1163, 1974

PLASMA ETCHING A CERAMIC COMPOSITE

David R. Hull[1], Todd A. Leonhardt[2], and William A. Sanders[3]

ABSTRACT

Plasma etching is found to be a superior metallographic technique for evaluating the microstructure of a ceramic matrix composite. The ceramic composite studied is composed of silicon carbide whiskers (SiC_w) in a matrix of silicon nitride (Si_3N_4), glass, and pores. All four constituents are important in evaluating the microstructure of the composite. Conventionally prepared samples, both as-polished or polished and etched with molten salt, do not allow all four constituents to be observed in one specimen. As-polished specimens allow examination of the glass phase and porosity, while molten salt etching reveals the Si_3N_4 grain size by removing the glass phase. However, the latter obscures the original porosity. Neither technique allows the SiC_w to be distinguished from the Si_3N_4. Plasma etching with $CF_4+4\%O_2$ selectively attacks the Si_3N_4 grains, leaving the SiC_w and glass in relief, while not disturbing the pores. An artifact of the plasma etching reaction is the deposition of a thin layer of carbon on Si_3N_4, allowing Si_3N_4 grains to be distinguished from SiC_w by back scattered electron imaging.

INTRODUCTION

The need for more efficient propulsion systems requires materials to operate at increasingly higher temperatures in oxidizing environments. Ceramics are a leading

---

[1]NASA Lewis Research Center, Cleveland, OH 44135.

[2]Sverdrup Technology Inc., Lewis Research Group, Brook Park, OH 44142.

[3]Analex Corporation, Brook Park, OH 44142.

671

candidate to provide high temperature strength and oxidation resistance. A major problem with ceramics is their low fracture toughness, which leads to brittle catastrophic failures. A method of increasing the toughness has been to introduce interfaces to deflect a propagating crack. One material system being studied is a ceramic composite composed of SiC_w in a matrix of Si_3N_4. The SiC_w provide the interfaces to deflect the crack tip [1].

The fabrication of SiC_w/Si_3N_4 matrix composites involves several variables; ie. whisker morphology and volume fraction, glass composition and volume fraction, consolidation temperature and pressure. Metallography plays a critical role in the evaluation of the effects of these variables on the composite. Metallography provides the ability to observe porosity, Si_3N_4 grain and SiC_w size, percent glass and distribution of the phases. Correlation of microstructural information with mechanical property data can be used to modify fabrication processes to further improve mechanical properties.

In metallography, etching is the critical step in revealing a material's microstructure. Etching is performed by selective removal of material from a specimen by a chemical reaction. Different features in the microstructure react at different rates, providing topographic relief observable by optical and electron microscopy [2]. An etchant commonly used on Si_3N_4 has been molten salts (eg. potassium hydroxide, KOH, used at 673°K). The molten salt dissolves the intergranular glass phase that surrounds each Si_3N_4 grain, allowing the grains to be observed.

Plasma etching is a dry etching technique developed for the fabrication of microelectronics in the late 1960's [4-7]. The technique uses an electric discharge excited by a radio frequency source to produce chemically reactive fragments from an appropriate gas in a vacuum of 0.001 to 2 torr. The fragments react with the specimen creating volatile products which are pumped away. For silicon based materials the gas used is CF_4, which breaks down to CF_x and reacts with Si to form SiF_y (gas) [4].

The use of plasma etching for revealing the microstructure of a sialon was first reported in 1983 [8]. Since then, others have used plasma etching to reveal the microstructure of Si_3N_4 [9-11]. In these materials, the β-sialon or Si_3N_4 are selectively removed, with the glass phase remaining in relief, thus revealing the microstructure. Only one reference was found on plasma etching a ceramic composite (SiC/Si_3N_4) [13]. Plasma etching was used to remove Si_3N_4, thereby isolating the glass phase for chemical analysis in the transmission electron microscope (TEM). The purpose of this paper is to evaluate the use of plasma etching in revealing the microstructure of a SiC_w/Si_3N_4 matrix composite by examination in the scanning electron microscope (SEM).

EXPERIMENTAL

The SiC_w/Si_3N_4 composite used for the plasma etching studies was formulated to yield a fully dense composite containing 10 percent SiC_w and 14 percent glass by volume. Pre-milled Si_3N_4 powder was blended in hexane with SiC_w, and Ce_2O_3/SiO_2 sintering aid. The homogeneous blended slurry was then pressed into a 5.1 cm diameter by 0.7 cm thick disc which was dried, and then sintered at 2413°K for 4 hours under 2.5 MPa nitrogen overpressure. The bulk density of the disc was 3.31 g/cm^3 (immersion method). The disc was then sectioned to provide specimens for the plasma etching studies.

Metallographic preparation consisted of mounting the specimen in a two part epoxy, followed by vacuum degassing, and curing for 12 hours. The subsequent grinding / polishing steps that were used are listed in Table I.

TABLE I : Metallographic Preparation

Step	Diamond Abrasive, μm	Lubricant	Time, min.	Comments
Grinding	115	water	till planar	
(fixed abrasive)	65	water	1	
	20	water	1	
Lapping	6	alcohol	6	Iron/copper disc
(non-fixed abrasive)	6	alcohol	9	Plastic disc
Polishing (slurry)	6	alcohol	3	Hard synthetic
	3	alcohol	2	cloth
	1	alcohol	2	
Final polishing (vibratory)	0.5	water soluble oil	16 hr.	High nap synthetic cloth

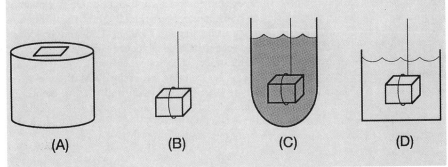

Figure 1. - Schematic diagram of steps involved in molten KOH etching. (a) Mounted and polished specimen. (b) Specimen removed from mount and wrapped in Pt wire. (c) Specimen immersed in molten KOH at 673°K for 10 to 20 seconds. (d) Specimen ultrasonically cleaned in water.

Etching of the specimens was performed by two methods; molten salt or plasma etching. Etching with molten KOH was performed as outlined in Figure 1. The specimen was removed from the epoxy mount, wrapped in platinum wire and immersed in molten KOH at 673°K for 10 to 15 seconds. Plasma etching was performed using the system shown in Figure 2. The mounted sample is placed in the vacuum chamber and evacuated to 0.1 torr. The sample is preheated using nitrogen at 0.2 torr and plasma RF power of 100 watts for 10 minutes. Next, etching is performed using $CF_4+4\%O_2$ at 0.3 torr and plasma RF power of 100 watts for 2 to 5 minutes.

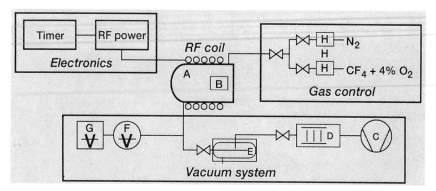

Figure 2. - Schematic diagram of plasma etching system. (a) Vacuum chamber. (b) Specimen. (c) Mechanical vacuum pump (fomblin oil). (d) Sorption trap. (e) Liquid nitrogen trap. (f) Thermocouple gauge. (g) Vacuum read out. (h) Flow meter.

Specimens were sputter coated with palladium and examined in a SEM using back scattered electron imaging and an accelerating voltage of 20 kV. TEM specimens were prepared by conventional techniques. A 3 mm disk was ultrasonically drilled from a 0.5 mm thick section, ground on both surfaces with 15 and 3 μm diamond to a thickness of 140 μm, and dimpled from each side to a thickness of less than 20 μm. Final thinning was performed by ion milling to perforation using argon at 5 kV and 12 degree incidence angle. Electrical conductivity of the specimen was achieved by evaporation of approximately 10 nm of carbon onto one surface. The specimen was examined in the TEM at a accelerating voltage of 120 kV. After examination in the TEM, the specimen was affixed to a glass slide with small drop of carbon paint and plasma etched as described above, with a preheat of 5 minutes and an etch of 2 minutes. The specimen was removed from the glass slide and re-examined in the TEM. Electron energy loss spectroscopy (EELS) was performed to obtain chemical information.

RESULTS

Back scattered electron imaging of the as-polished composite allows the identification of three microstructural features (Figure 3). The white phase is the CeO_2 rich glass phase as determined by x-ray energy dispersive spectroscopy (XEDS), the black areas are regions of porosity, and the gray phase is SiC_w or Si_3N_4. Measurement of the size and shape of the gray phase is possible, but SiC_w and Si_3N_4 can not be differentiated due to their identical average atomic number (SiC:(14+6)/2=10; Si_3N_4:(3*14+4*7)/7=10).

Molten salt etching removes the intergranular glass leaving the SiC_w and Si_3N_4 grains in relief (Figure 4). Back scattered electron imaging again can not differentiate SiC_w from Si_3N_4. The white particles were identified by XEDS as contamination from the platinum crucible during etching. Obtaining reproducible etching results is difficult as shown by the influence an additional 5 seconds had on the depth of etching on a second specimen (Figure 4b).

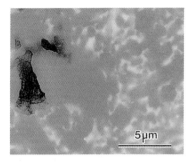

Figure 3. - Back scattered electron image of as-polished SiC$_w$/Si$_3$N$_4$ matrix composite.

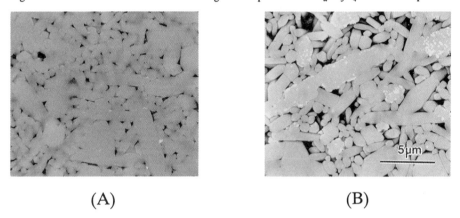

(A)   (B)

Figure 4. - Back scattered electron images of SiC$_w$/Si$_3$N$_4$ composite following molten KOH etching at 673°K. (a) Etched 10 seconds. (b) Etched 15 seconds.

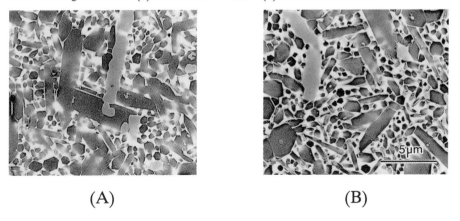

(A)   (B)

Figure 5. - Back scattered electron images of plasma etched SiC$_w$/Si$_3$N$_4$ composite (10 minutes N$_2$ preheat and CF$_4$+4%O$_2$ etch for times shown). (a) 3 minutes. (b) 5 minutes.

Plasma etching selectively attacks the Si_3N_4 leaving the SiC_w and glass in relief (Figure 5). Back scattered electron imaging reveals the CeO_2 rich glass as the white phase. An atomic number difference between Si_3N_4 and SiC_w is observed, with the SiC_w being lighter, and thus, apparently higher in average atomic number. As etching time is increased to 5 minutes, the contrast between SiC_w and Si_3N_4 increases (Figure 5b). Examination of the specimen surfaces at higher magnification, using secondary electrons, reveals a mottled structure on the Si_3N_4, with the SiC_w and glass appearing unaffected (Figure 6). The longer etching time (5 minutes) removed more of the Si_3N_4.

TEM of an as prepared specimen shows the typical faulted structure of the SiC_w, an intergranular glass phase, and Si_3N_4 grains containing dislocations (Figure 7a). After plasma etching for two minutes, examination of the exact same region shows the SiC_w and glass are intact, but the Si_3N_4 is removed as evidenced by the absence of the dislocations (Figure 7b). The evaporated conductive carbon film is intact with a mottled deposit on it. Electron diffraction of the deposit showed an amorphous pattern and EELS shows carbon as the only element present (Figure 7c).

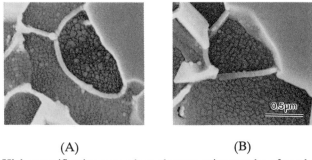

(A)   (B)

Figure 6. - High magnification secondary electron micrographs after plasma etching. (a) Etched 3 minutes. (b) Etched 5 minutes.

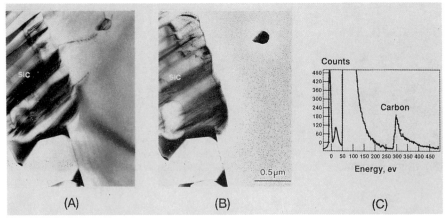

(A)   (B)   (C)

Figure 7. - TEM bright field of SiC_w/Si_3N_4 matrix composite. (a) As prepared. (b) After plasma etching 2 minutes. (c) Electron energy loss spectrum of deposits.

DISCUSSION

Many problems exist with the conventional microstructural characterization techniques applied to SiC_w/Si_3N_4 composites. Examination of both as-polished and molten salt-etched surfaces with SEM does not reveal the complete microstructure of the SiC_w/Si_3N_4 composite. Neither technique allow distinction between SiC_w and Si_3N_4 grains in the SEM. Molten salt etching problems also include the requirement for the sample to be removed from the epoxy mount. This is necessary because the epoxy decomposes at 473°K and entrapped moisture can cause a small gas explosion, splattering molten salt. Etching of small delicate specimens cannot be performed because of destruction of the sample upon removal from the mount. Determining etching times is difficult because etching rate varies with the size of the sample. Larger samples have a higher heat capacity and require a longer etching time. Examination and handling of the sample is complicated when it is not in a uniform size mount, especially during observation with an inverted metallograph, where glass cover slips must be used to support the sample. Finally, if the sample is over-etched, it must be remounted for repolishing.

Plasma etching overcomes all of the problems discussed above. The temperature of the sample during plasma etching remains less than 358°K, using the etching conditions discussed. The epoxy mount can withstand temperatures to approximately 423°K, therefore the sample can remain in the mount. Another benefit of the lower etching temperature is the ability to partially etch a sample by simply covering a portion with adhesive tape (Figure 8). This is very useful for samples that might require unetched areas for electron microprobe studies.

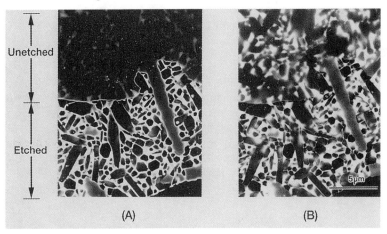

Figure 8. - Partial etching of SiC_w/Si_3N_4 matrix composite. (a) Secondary electrons. (b) Back scattered electrons.

The apparent atomic number difference between SiC_w and Si_3N_4 is explained by the deposition of carbon during removal of Si_3N_4, as identified by TEM (Figure 7). A schematic drawing of a cross section of the composite surface is shown in Figure 9.

Although the chemical reactions that occur in the plasma etching process are not fully understood [6,7], the chemical reaction shown in Equation 1 provides a basic understanding of the source for the carbon. Fragments of CF_x, generated by the $CF_4+4\%O_2$ plasma, react with Si_3N_4 to form volatile products of SiF_y and N_2 which are removed by the vacuum pump, while carbon is deposited on the Si_3N_4. For the etching to proceed the carbon must be removed by reaction with the $4\%O_2$ to form CO and/or CO_2 (Equation 2). The selective carbon deposition provided by the chemical reaction at the Si_3N_4 surface provides a decrease in the apparent average atomic number of the Si_3N_4, while the SiC_w surface remains unetched (Figure 6).

$$CF_x(gas) + Si_3N_4(solid) \rightarrow SiF_y(gas) + N_2(gas) + C(solid) \quad (1)$$

$$C(solid) + O_2(gas) \rightarrow CO, CO_2(gas) \quad (2)$$

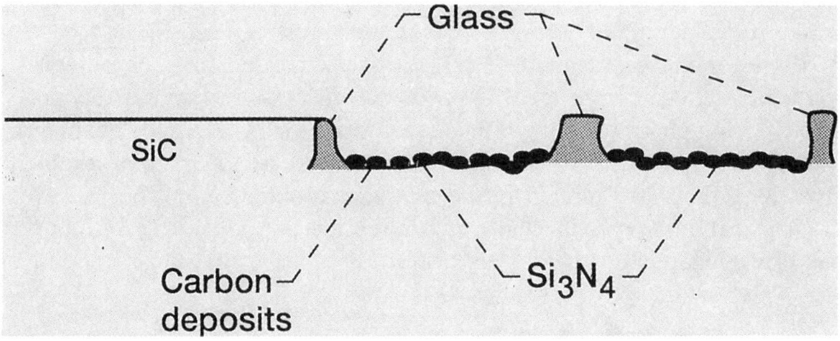

Figure 9. - Schematic cross-section of composite after plasma etching.

CONCLUSIONS

The microstructure of a SiC_w/Si_3N_4 matrix composite is revealed by plasma etching with $CF_4+4\%O_2$. All four microstructural constituents: Si_3N_4 grains, SiC_w, pores, and intergranular glass are distinguishable in one specimen. The selective deposition of carbon on Si_3N_4 allows the SiC_w to be imaged by back scattered electrons. Ability to keep the specimen in the mount allows examination of small, delicate specimens and facilitates imaging on an inverted metallograph. Partial etching saves time when both unetched and etched specimens are required for analysis.

REFERENCES

1. Evans, A.G., "Perspective on the Development of High-Toughness Ceramics", **Journal of American Ceramic Society,** Vol. 73, No. 2, p 187, 1990.

2. Kehl, G.A., **The Principles of Metallographic Laboratory Practice,** McGraw-Hill, 1949.

3. Lange, F.F., "Relation Between Strength, Fracture Energy, and Microstructure of Hot-Pressed Si_3N_4", **Journal of American Ceramic Society,** Vol. 56, No. 10, p 518, 1973.

4. Bersin, R., "A Survey of Plasma Etching Processes", **Solid State Technology**, p 31, 1976.
5. Wolf, E.D., Adesida, I., and Chinn, J.D., "Dry Etching for Submicron Structures", **Vacuum Science and Technology A**, Vol. 2, p 464, 1984.
6. Cox, T.I., and Dshmukh, V.G.I., "Fundamental Studies of Ion Surface Interaction in Dry Etching", Royal Signals and Radar Establishment, Malvern (England), 1985.
7. Singer, P.H., "Dry Etching of SiO_2 and Si_3N_4", **Semiconductor International**, p 98, 1986.
8. Chatfield, C., and Norstrom, H., "Plasma Etching of Sialon", **Communications of the American Ceramic Society**, p C-168, 1983.
9. O'meara, C., Nilsson, P., and Dunlop, G.L., "The Evaluation of Beta-Si_3N_4 Microstructures Using Plasma-Etching as a Preparative Technique", **Journal De Physique**, Colloque C1, supplement au n.2, Tome 47, p. c1, 1986.
10. Sieben, K.N., and Lovington, W.M., "Plasma Etching of Si_3N_4", **Microstructural Science**, Vol. 16, p 319, 1988.
11. Mitomo, M., Sato, Y., Yashima, I., and Tsutsumi, M., "Plasma Etching of Non-oxide Ceramics", **Journal of Materials Science Letters**, Vol. 10, p 83, 1990.
12. Mitomo, M., Yoh-ichiro, S., Ayuzawa, N., and Yashima, I., "Plasma Etching of α-Sialon Ceramics, **Journal of American Ceramic Society**, Vol. 74, No. 4, p 856, 1991.
13. Ostereicher, K.J., and Pink, F.X., "Isolation of Intergranular Glassy Phase in SiC / Si_3N_4 Composites by Plasma Etching", **Journal of American Ceramic Society / Communications**, Poster at 89th Annual Meeting Ceramographic Exhibit, Vol. 70, No. 10, October, 1987.

MICROSTRUCTURE OF ULTRA-HIGH NITROGEN STEELS

C N McCowan[*], A Tomer[**], and J W Drexler[***]

ABSTRACT

A series of ultra-high-nitrogen steels were made for this study to evaluate the effects of compositional variations on their microstructure. The alloys are based on an Fe-20Cr-N composition, they differ primarily in Ni and Mn contents: One binary Fe-N alloy was also included in the study. Nitrogen contents for the alloys are higher than those typical of conventionally melted alloy steels because they were melted in a hot-isostatic-pressure furnace at nitrogen pressures of approximately 200 MPa (2,000 Atm). The microstructure of the alloys are always a mixture of nitride and austenitic (and/or ferritic) phases, but the morphology of the phase mixture varies greatly with alloy composition. For example, the binary Fe-N alloy contains iron nitrides (Fe_4N) in an alpha-iron Widmanstatten structure: the 73Fe-20Cr-1Ni-6Mn alloy has a eutectoid-like structure (86%) with some nitride (10% CrN) and austenite (4%) present. Other alloys have as many as three nitride (Cr_xN) morphologies in their structures: large classical dendrites, smaller bush-like dendrites, and very fine nitrites in eutectoid structures. Because the balance of phase morphologies is likely to govern the properties of these alloys, this study focuses on characterizing the structural features of these experimental alloys by morphology, composition (WDS analysis), and hardness.

[*] NIST, Materials Reliability Division, Boulder, Co

[**] Guest researcher at NIST, from the NRCN, Israel.

[***] University of Colorado, Boulder, Co

INTRODUCTION

Nitrogen has been used as an alloy addition to stainless steel for strengthening, creep resistance, and corrosion resistance.[1234] Nitrogen gas is readily added to molten steel during processing or welding, and it is an inexpensive, non-strategic material that helps to stabilize austenite in the steel against martinsitic transformation during deformation or cooling.[5678] Because nitrogen is 20 to 30 times more effective than nickel at stabilizing austenite, small additions of nitrogen (0.1 wt.%) can replace several percent of nickel or manganese additions without loss in austenite stability.[91011]

The solubility of nitrogen in Fe-20Cr-Ni stainless steel is approximately 0.2 wt.% at 1600 C, 1 atmosphere, for Ni contents of 0 to 40 wt.%.[12] For pure Fe at 1600 C, 1 atmosphere, the solubility limit of N is much lower (0.004 wt.%) Special processing techniques are needed to attain alloys with nitrogen contents that exceed these liquid solubility limits. The alloys investigated in this study were processed in a hot-isostatic-pressure furnace (HIP) at high nitrogen pressures so nitrogen contents could be increased. The alloys compositions were formulated to yield general information regarding the affect of Cr, Ni, and Mn additions on the structure of ultra-high nitrogen stainless steels.

MATERIAL AND PROCEDURES

The alloys were melted in a hot-isostatic-pressure furnace at nitrogen pressures of approximately 2,000 Atm (30,000 psi).[13] The alloys were slowly cooled from the 1750 C processing temperature while the pressure was held constant. The samples are small castings that have cylindrical shapes and weigh approximately 100g each.

The compositional matrix used for the alloys is shown in Table 1. One binary Fe-N alloy was included in the matrix, but the remaining samples have 20 wt.% Cr. To the Fe-Cr base alloy composition, 3 levels of Ni and Mn were made: The Fe-20Cr-12Ni composition is similar to several type 300 series austenitic stainless steel compositions, and the matrix revolves around this type of stainless steel composition.

The samples were prepared for evaluation in the light microscope and the scanning electron microscope (SEM) using conventional grinding techniques (to 5 micron), followed by diamond polishing. Much of the work was possible without etching, due to the high contrast between the nitrides and the matrix. When etches were used, the composition is noted in the figure cation.

Image analysis was used to measure the area fraction of phases and phase morphologies in the alloys. Samples were evaluated at 200X on polished surfaces: 20 fields per sample were counted.

	Fe	Cr	Ni	Mn
A	100			
B	80	20		
C	78	20	1	1
D	73	20	1	6
E	67	20	1	12
F	73	20	6	1
G	67	20	12	1
H	68	20	6	6
I	62	20	6	12
J	62	20	12	6
K	58	20	12	12

Table 1: The compositions of the alloys (wt.%).

Microhardness measurements (Vickers) were made on polished surfaces of the samples. The load used for the tests was 100g. Measurements were made in "single phase" regions of the matrix at 50X.

RESULTS AND DISCUSSION

The microstructures for six of the alloys are shown in Figure 1. These

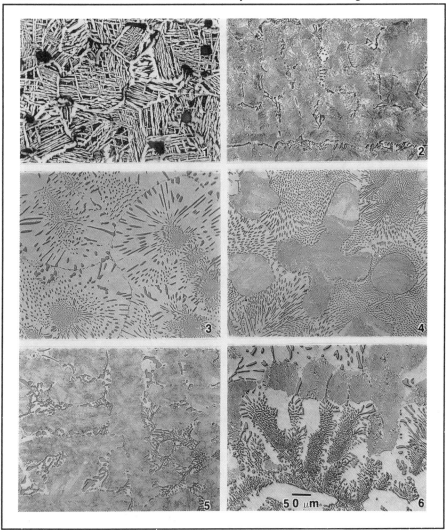

Figure 1: The microstructures of alloys A, D, G, B, I, and K are shown in order from 1 to 6.

examples represent the variations in structures observed for the alloys evaluated in this study. The binary Fe-N alloy, Figure 1-1, has a widmanstatten structure, similar to Fe-C alloys: here, the phases are Fe_4N and α-Fe (Table 2). The plate-like features in the microstructure have the lowest N content in the structure. The matrix is a fine 2-phase structure, smaller plates, with a N content near 3 wt.% (Table 3).

We presume this fine structure is a mixture of Fe_4N and α-Fe, for which the Fe-N binary diagram has a eutectoid reaction at 2.35 wt.%:

$$\gamma\text{-Fe} = \gamma'_{\centerdot} + \alpha\text{-Fe.} \quad (1)$$

where α-Fe and γ-Fe are very similar to the α and γ phases in the Fe-C alloys, and $\gamma'_{\centerdot}$ is an ordered fcc phase (Fe_4N). Although the bcc and bct phases cannot be distinguished with xray diffraction analysis, the hardness value would indicate this is not a martensitic structure.

The ternary Fe-20Cr-N alloy, Figure 1-4, has several levels of eutectoid-like structure. It appears that primary dendrites transformed to produce a fine lamellar structure, and a coarser 2-phase structure surrounds these regions. Possibly the coarse 2-phase structure is made up of primary nitrides that formed from the liquid.

The addition of 12 wt.% Ni to the ternary, Figure 1-3, resulted in a larger grain (cell) size and a predominately coarse 2-phase structure. The matrix phase in this alloy is a more continuous matrix phase (66% in this alloy).

The addition of Mn to the ternary alloy, Figure 1-2, promoted the formation of fine 2-phase structures (86% in this alloy). In Figure 2, the lamellar structure typical of these alloys is shown. Although the two Mn alloys shown in Figures 1-2 and 1-5 have similar phase morphologies and phases identified, alloy I (higher Mn alloy) has the finer structure of the two alloys. In Figure 3, the microstructure of alloy E and its N map are shown. Even in

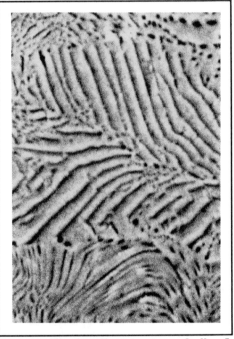

Figure 2: The lamellar structure of alloy I: 5mm = 1 micron.

areas surrounding the large primary nitrides in the sample, fine 2-phase structure is visible. In figure 1-6, the high Ni, even with high Mn, promoted a coarse 2-phase structure. Two chromium nitride phases are present in this alloy (CrN and Cr_2N).

In Table 3, the compositions of the phases measured in 7 of the alloys are given. In these alloys, the N content of the single phase matrix regions was generally near 0.5 wt.%. Although this would be considered supersaturated for the binary Fe-N and Fe-20Cr-Ni alloys, it maybe near equilibrium concentration for higher Mn alloys.[14] Why the N content of these various matrix phases in the alloy are so similar is not clear. Two phase regions in the alloys typically had N contents of 1 to 2 wt.%: Mn and Ni contents in these regions are similar to that in the single-phase-matrix regions. Primary nitrides had between 17 and 24 wt.% N, normally about 2 wt.% Fe, and with the exception of alloys E and K, low Mn and Ni. The stoichiometric amounts of Cr and N in the CrN and Cr_2N nitrides are approximately 79Cr-21N and 82Cr-12N respectively. High Mn alloys had appreciable amounts of Mn in the primary nitride phase.

Table 2: Summary of phases identified in the alloys by x-ray diffraction analysis, area fractions of phase morphologies present in the alloys, and Vickers microhardness results of the matrix phase.

ID	Phases (X-ray)	Phase Morphology %			Hardness H_v
		MF	NF	EF	
A	$bcc+Fe_4N$	--	--	--	203
B	$fcc+CrN+bcc$	21	26	53	380
C	$bcc+CrN$	19	16	65	470
D	$bcc+fcc+CrN$	4	10	86	400
E	$bcc+fcc+CrN+Fe_4N$	18	28	54	340
F	$fcc+CrN+bcc$	63	23	15	440
G	$fcc+CrN+bcc$	66	19	15	240
H	$fcc+CrN+bcc$	23	7	71	300
I	$fcc+CrN$	6	7	88	310
J	$fcc+CrN$	28	20	52	250
K	$fcc+CrN+Cr_2N$	51	27	23	245

* MF=matrix phase, NF=nitride phase, EF=eutectiod

SUMMARY

Alloy composition consistantly changed the nitride morphology of the samples. Both Ni and Mn promote an fcc matrix phase, but Ni additions tend to result in a coarse nitride morphology.

Table 3:Results of the microprobe analysis (WDS).

		Fe	Cr	Ni	Mn	N
A	P1	98.7	0.0	0.0	0.1	2.9
	P2	98.5	0.0	0.0	0.0	0.5
B	P1	83.2	16.5	0.0	0.0	0.0
	P2	82.1	16.0	0.0	0.0	0.5
	P3	--.-	--.-	-.-	-.-	-.-
C	P1	82.4	14.7	1.1	1.0	0.5
	P2	81.1	15.2	1.0	0.7	2.0
	P3	2.4	72.7	0.0	0.5	24.3
E	P1	79.2	10.6	1.0	7.8	0.5
	P2	73.8	16.2	1.0	7.3	1.9
	P3	2.1	75.8	0.0	3.6	18.5
F	P1	82.8	6.9	7.5	1.1	0.5
	P2	74.3	16.3	5.8	0.6	2.0
	P3	2.2	79.6	0.1	0.7	18.1
G	P1	71.6	13.8	13.5	1.1	0.5
	P2	67.8	18.7	11.6	0.6	1.3
	P3	1.6	80.4	0.2	0.5	17.3
K	P1	63.1	14.5	13.3	8.2	0.5
	P2	64.2	15.2	12.6	7.5	1.0
	P3	2.5	73.2	0.19	3.5	20.6

* P1=matrix, P2=eutectoid, P3=nitride

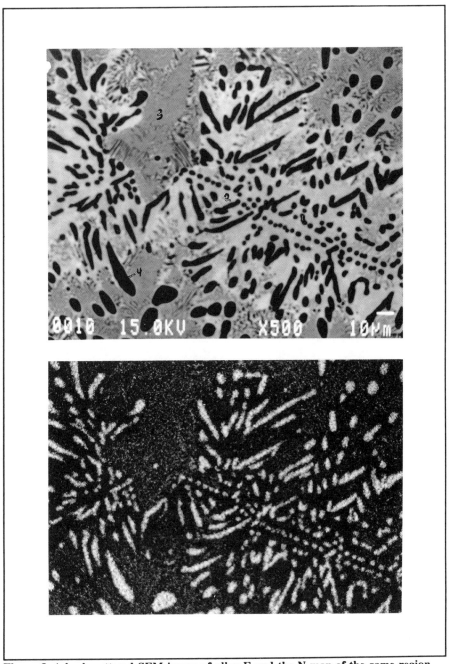

Figure 3: A backscattered SEM image of alloy E and the N map of the same region.

MICROSTRUCTURAL STUDIES OF INTERNALLY

OXIDIZED COPPER ALLOYS

J. Groza and S. Farrens[1]

ABSTRACT

Current theoretical theories on thermal conductivity and high temperature strength are interpreted in terms of microstructural requirements for dispersion strengthened (DS) copper alloys. These requirements involve a pure and annealed matrix and fine, chemically stable and physically insoluble hard particles. Experimental assessment of the microstructural characteristics of internally oxidized $Cu-Al_2O_3$ such as matrix purity, and particle-matrix interfacial properties may be used as guidelines for an improved design of DS copper alloys.

INTRODUCTION

The interest in dispersion strengthened (DS) copper alloys has increased recently due to a renewed need for high-conductivity, high temperature resistant materials. These materials are of critical importance to the development of new applications such as plasma interactive components and high heat flux aerospace structures. Consequently, a significant research effort was mobilized into the development of new DS copper alloys. Much of that effort on alloy development has been reviewed in an earlier paper [1]. It has been recognized that the main requirements for copper alloys to be used as actively cooled structures are a high conductivity copper matrix and a dispersion of fine, hard, and stable particles for high temperature, long term stability. There are numerous innovative

---

[1]University of California at Davis, Davis, California 95616

processing methods aimed at producing DS copper alloys that combine the above requirements to lead to high strength and good thermal conductivities. To name just a few, these alloys comprise the GLIDCOP series that are internally oxidized, MXT and XD alloys obtained by liquid state reactions, and rapidly solidified alloys. In spite of this considerable alloy development effort, there are only a few attempts at microstructural and property characterization of the existing DS copper alloys. Certainly, a better theoretical understanding of their structures is needed in order to properly tailor them for high temperature strength and conductivity. We will first highlight our own efforts in the alloy design concept and point out the relevant microstructural requirements for DS copper alloys for elevated temperature applications. We will then describe our microstructural studies aimed at characterizing one of the existing DS copper alloy. These studies will lead to further refinements in alloy design, including matrix chemistry and structure, and particle phase selection.

ALLOY DESIGN CONCEPT

Dispersion strengthened (DS) alloys consist of metal matrices that are reinforced by fine, hard and stable particles that are added in small proportions, typically below 5 vol %. While at room temperature any precipitates are efficient dislocation obstacles, at high temperatures, retention of the fine particle size is a major condition for good mechanical strength. In addition, for actively cooled structures, good thermal conductivity is essential to maximize the efficiency of active cooling and minimize thermal strains in the material. As expected, a pure copper matrix is the first requirement for high thermal conductivity. The particles that reinforce this matrix must therefore be completely insoluble, stable, and not react with the copper matrix at processing and service temperatures. By considering these alloys as particulate composites, their thermal conductivity may be calculated using the following formula given by Schroeder [2]:

$$\sigma = \sigma_m \frac{1 + 2f \dfrac{1 - \sigma_m/\sigma_p}{2\sigma_m/\sigma_p + 1}}{1 - f \dfrac{1 - \sigma_m/\sigma_p}{2\sigma_m/\sigma_p + 1}} \tag{1}$$

where σ_m and σ_p are matrix and particle conductivities, respectively, and f is the dispersoid volume fraction. This formula was initially derived for electrical conductivity. Although the thermal conductivity in metals is determined by both electron flow and lattice heat transfer, only electrons are responsible for heat flow at high temperatures. Therefore the linear Wiedemann-Franz law which relates thermal to electrical conductivity may be safely used. The above formula was verified to give accurate results for $f < 5\%$ in Al - Al_2O_3 alloys for which experimental measurements of electrical resistivity are available [3]. For instance, the error between the experimental and calculated conductivity of Al - 2.06 vol% Al_2O_3 is 2.97%. In using formula (1), it is interesting to note that the intrinsic conductivity of second-phase plays only a minor role. For example, an increase in particle conductivity by a factor of 5 results in only 1% change of overall thermal conductivity at $f = 0.05$. Although the alloy conductivity is relatively insensitive to the intrinsic conductivity of the particle, impurity content has a far greater effect. Thermal conductivity degrades quickly by the introduction of elements from the dispersoids into the copper matrix. Elements in solid solution are of particular concern since even small amounts are especially harmful. Similarly, a cold-worked matrix will yield an overall lower conductivity compared to an annealed condition. In conclusion, the major requirements for high conductivity DS alloys are high temperature stable and insoluble particles in an annealed, pure matrix.

For long time applications of heat resistant materials, the effect of dispersoids on creep strength is of prime importance. To rationalize the high stress exponents found in dispersion strengthened materials, a threshold stress is considered so that the theoretical description of dislocation motion past incoherent dispersoids may be applied. A more complete treatment of creep theories applied to DS alloys may be found in a previous paper [1]. A new model developed by Arzt and coworkers [4-6] describes creep in dispersion strengthened alloys by considering that dislocations are pinned on the departure side of the particles. An attractive interaction between the dislocation and the particle/matrix interface that was experimentally observed leads to a detachment stress given by:

$$\sigma_D = \sqrt{1-k^2}\ \sigma_{OR} \tag{2}$$

where k is an interaction parameter and σ_{OR} is the Orowan stress. No relaxation takes place or there is no particle/dislocation interaction for k=1. This is the case when matrix and particle lattices match perfectly or particles are coherent. At the other end of the spectrum, a strong attractive interaction occurs when k=0 and dislocation completely relaxes its energy. The detachment process is thermally activated and the time for this detachment is significant in comparison to the time required for a dislocation to move from one particle to another and to climb over the new particle. Based on this assumption, Arzt and coworkers developed a new equation for creep rate:

691

$$\dot{\varepsilon} = \dot{\varepsilon}_o \exp\left[-\frac{Gb^2 r}{k_B T}(1-k)^{3/2}\left(1-\frac{\sigma}{\sigma_D}\right)^{3/2}\right] \qquad (3)$$

in which $\dot{\varepsilon}_o$ is a reference strain rate equal to $3D_v \rho \lambda/b$, where D_V is the volume diffusivity, ρ is the density of mobile dislocations, λ is the mean particle spacing, G is the shear modulus, b is the Burgers vector, r the mean particle radius, k_B is the Boltzmann's constant, T is temperature, and σ is the applied stress.

The new equation for creep has valuable predictive capabilities for the design of high temperature strength materials. Two of these theorized issues - the threshold stress and optimum particle size - were addressed in details elsewhere [1], so we concentrate only on the interaction parameter k. It should be as low as possible. Its value depends on the particle/matrix interfacial properties and material. For instance, Rosler and Arzt found that Al_4C_3 obtained by mechanical alloying has a stronger interaction $(k=0.75)$ with the aluminum matrix than Al_xFe_yCe (k=0.95) obtained by rapid solidification [6]. They argue that carbides are poorly bonded with the aluminum matrix and thereby strongly attract dislocations at high temperatures. The explanation for less efficient dispersoids in rapidly solidified alloys evolves from the strong interfacial bonding as required by the nucleation process. We conclude that highly attractive obstacles will be the poorly bonded dispersoids, i.e. incoherent precipitates with probably high interfacial energy. Further studies on the relationship between interfacial bonding and k parameter values will strongly enhance our chances of selecting the best dispersoids and processing routes to ensure high temperature strength. To date only creep tests have been used to evaluate k. These k values for different DS copper alloys are currently sought and will be reported in a later paper.

EXPERIMENTAL PROCEDURE

The dispersion strengthened copper alloy utilized in this work is Glidcop Al-15 (C15715) low-oxygen grade, that has 0.7 vol % Al_2O_3. It was produced by SCM Metals, Cleveland, Ohio using an internal oxidation process [7]. In this process, Al in the atomized Cu-Al powders is preferentially oxidized by oxygen supplied by copper oxide. The residual copper oxide is reduced by subsequent heat treating in a reducing atmosphere. The final material is consolidated and extruded from a bar preheated at approximately 1113 K. This Al-15 low-oxygen grade material has boron additions in order to reduce excess oxygen and prevent blistering during subsequent high temperature exposure. The boron content was measured at Sandia National Laboratories as 0.0153 wt %.

Samples for thin foil TEM were sliced parallel and normal to the extrusion direction and mechanically thinned to about 0.12 mm. Disks 3 mm in diameter were then mechanically punched from the thinned sheets and electrolytically thinned by jet polishing with a solution of 33% nitric acid in methyl alcohol at -30 to -40°C. Thin foils were examined using 120 kV Philips CM 12 transmission electron microscope equipped with EDS detector and 1500 kV Kratos HVEM.

RESULTS and DISCUSSION

The thermal conductivity of annealed GLIDCOP Al-15 is 365 W/mK (93 % IACS) [7]. For comparison, the thermal conductivity of GLIDCOP Al-60 with 2.7 vol % Al_2O_3 is 322 W/mK (78 % IACS). Our calculated values using formula (1) are 387 and 377 W/mK, respectively for the two alloys. These discrepancies in the thermal conductivity values may be accounted for by considering the main factors that decrease alloy conductivity: cold work and impurities. Since the conductivity is a very sensitive function of the amount of cold work and the dependence of conductivity with temperature demonstrates no recovery phenomena up to 673 K, as reported by SCM for annealed alloys, we exclude the possibility of a cold worked matrix. It is worth mentioning that the above variation of conductivity with temperature is also indicative of Al_2O_3 particle stability. This stability was further proved by Stephens et al., who found that the alumina dispersoid particles remain small (about 50 nm) even when recrystallization of matrix takes place [8]. However, they also showed that annealing above 1273 K leads to a transformation in dispersoid type from γ-Al_2O_3 to $9Al_2O_3$-$2B_2O_3$ line compound. Therefore, we assume that residual elements in the copper matrix after internal oxidation may be responsible for lower than calculated conductivities in Cu-Al_2O_3 alloys. EDS analysis of the Cu matrix in the TEM revealed residual aluminum and oxygen amounts but no boron. However, since Stephens et al. reported that B_2O_3 compound was formed after long high temperature exposure, there may be some boron in the copper matrix that we were unable to detect. At these low levels all elements are expected to be in solid solution and thus lower the copper conductivity. We note that copper high conductivity may be restored after long annealing when B_2O_3 is completely formed, if perfect stoichiometric boron additions can be made.

Typical microstructures of as-extruded GLIDCOP Al-15 alloy are shown in Figures 1 and 2. A common feature revealed by TEM studies was the existence of a well developed dislocation structure that includes both dislocation tangles and cells (Fig. 1). Dislocation tangles indicate that the deformation structure

693

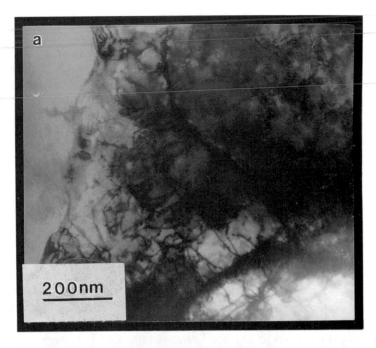

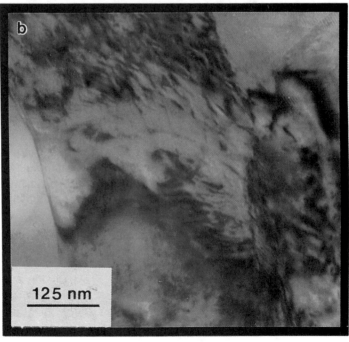

Figure 1. Typical dislocation configurations in as-extruded GLIDCOP Al-15 alloy.

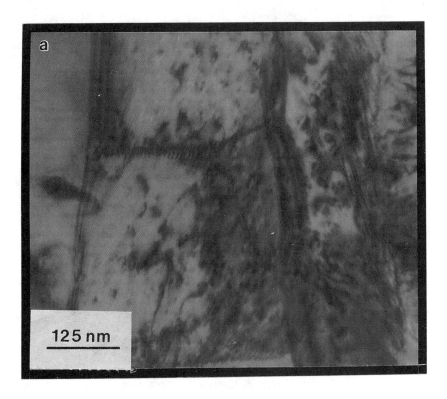

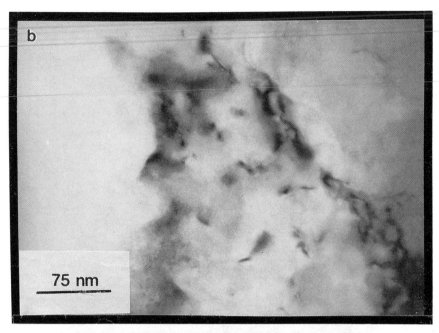

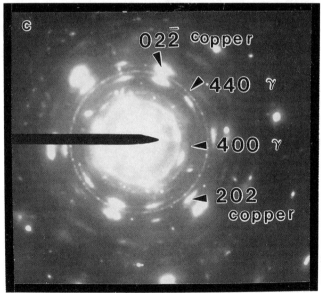

Figure 2. a) and b). Particle microstructure and distribution in GLIDCOP Al-15 alloy. c) Selected area diffraction pattern from an isolated copper grain. Ring pattern is indexed as γ Al$_2$O$_3$.

has been retained. The formation of dislocation networks that is observed in Figure 1b is a characteristic of high strain rate deformation structure and dispersoid-free matrix. Also visible in Figures 1a and 2a are planar defects which are consistent with low SFE of copper. At the same time, the cell presence suggests that the copper matrix is undergoing dynamic recovery during the extrusion process (Fig. 1c). The dislocation density present in this material is substantial and the strain contrast effects from dislocations hindered observation of the particles. Nevertheless, the material is characterized by a uniform distribution of very fine particles, about 20.4 ± 5.2 nm in diameter and a mean spacing of 34 ± 14 nm (Fig. 2). The diffraction ring pattern of the fine particles corresponds to the tetragonal γ Al_2O_3, in agreement with previous investigations [8]. Occasional particles of size in excess of 30 nm were also seen in the thin foil samples and they are suspected to be α-Al_2O_3 structure. Strain effects were observed at very small particles that are indicative of a possible coherent bonding between precipitates and the matrix. However, examination of most of the particles indicates they are incoherent with the copper matrix (Fig. 2b). Since Al_2O_3 is formed by a solid state reaction, lattice matching is expected between precipitate and the matrix, at least at small particle sizes. Coherent Al_2O_3 in internally oxidized Cu-Al_2O_3 has been previously reported by Mott and Nabarro [9]. Daneliya et al., also found that internal oxidation of Cu-Al alloys produces γ Al_2O_3 plates distributed in the {111} matrix planes [10]. This agrees with recent theoretical studies of Cu-Al_2O_3 interface that showed a preferred orientation in which close packed planes in both lattices are parallel [11]. At later stages, Al_2O_3 has a tendency to coarsen and therefore to become incoherent with the matrix. In addition, deformation of GLIDCOP Al-15 provides an extra driving force for structural change including coherency loss due to additional defects introduced by extrusion. Therefore, the possibility exists that the coherency strains to be relaxed yielding mostly incoherent particles, as observed in this investigation.

The strong Al_2O_3 tendency to coarsen noticed by several investigators may point out a high value for the interfacial energy [12]. This agrees well with Daneliya et al., where adsorption of Ti atoms on the surface of growing Al_2O_3 is reported to lower (the initially high) Al_2O_3 interfacial energy [10]. The interest on the degree of coherency stems from its possible connection to the k parameter. The magnitude of k value depends on the interfacial properties but no relationship between k and other interface parameters has been yet established. Optimum high temperature strength is provided by fully incoherent particles with the lowest k value. A relationship between the interfacial energy value and the k parameter value may give a better quantitative insight on the design parameters, such as selection of particle and processing route for high temperature strength.

CONCLUSIONS

1. For good elevated temperature thermal conductivity, insoluble, stable particles in a pure, annealed copper matrix are required. The intrinsic particle thermal conductivity plays an insignificant role in the overall conductivity of DS alloy at volume fraction < 5%. GLIDCOP Al-15 has a lower than theoretical conductivity due to impurities in the matrix such as aluminum, oxygen and, possibly boron from internal oxidation process.

2. TEM studies of the as-extruded alloy revealed a complex dislocation structure with deformation as well as recovery dislocation configurations. Planar defects were observed as a characteristic of low SFE materials. Dislocation networks were also observed.

3. The particles obtained by internal oxidation processing are generally fine and uniformly distributed. Most of them are incoherent with the matrix. Deformation during extrusion appears to contribute to the loss of coherency.

4. The results of this study suggest that the interfacial energy values may be used to evaluate the k parameter to be used for improved design of creep resistance DS alloys.

ACKNOWLEDGEMENTS

Special thanks are due to Prof. J. Gibeling at UCD for stimulating discussions and help in Glidcop material procurement. We are grateful to Ken Anderson and Lilian Davila for TEM specimen preparation.

REFERENCES

1. J. Groza, "Heat Resistant Dispersion Strengthened Copper Alloys" submitted to **J. Mater. Eng.,** 1991

2. K. Schroeder, "Experimental Aspects of Electrical and Thermal Properties of Metals", in **High Conductivity Copper and Aluminum Alloys,** edited by E. Ling and P. W. Taubenblat, p. 1, TMS-AIME, Warrendale, PA, 1984.

3. J. S. Benjamin and M. J. Bomford, "Dispersion Strengthened Aluminum Made by Mechanical Alloying", **Met. Trans.,** vol. 8A p. 1301, 1977.

4. E. Artz, "High Temperature Properties of Dispersion Strengthened Materials Produced by Mechanical Alloying: Current Theoretical Understanding and Some Practical Implications" in **New Materials by Mechanical Alloying Techniques**, edited by E. Arzt and L. Schultz, p. 185, DGM Informationsgesellschaft, Oberursel, 1989.

5. J. Rossler and E. Arzt, "Creep Equation for Dispersion Strengthened Materials", **Acta Met.**, vol. 38 p. 671, 1990.

6. J. Rossler and E. Arzt, "Creep in Dispersion Strengthened Aluminum Alloys at High Temperatures - A Model Based Approach" in **New Materials by Mechanical Alloying Techniques**, edited by E. Arzt and L. Schultz, p. 279, DGM Informationgesellschaft, Oberusel, 1989.

7. A. V. Nadkarni, "Dispersion Strengthened Copper Properties and Applications" in **High Conductivity Copper and Aluminum Alloys**, edited by E. Ling and P. W. Tautenblat, p. 77, TMS-AIME, Warrendale, PA., 1984.

8. J. J. Stephens, J. A. Romero and C. R. Hills "Grain Growth in a Commercial Dispersion Strengthened Copper Alloy" in **Microstructural Science**, vol. 16, edited by H. J. Cialoni, et.al., p. 245, 1988.

9. D. M. Williams and G. C. Smith, "A Study of Oxide Particles and Oxide-Matrix Interfaces in Copper", in **Oxide Dispersion Strengthening**, p. 509, Gordon and Breach, New York, 1968.

10. Y. P. Daneliya, M. D. Teplitskiy, and V. I. Solopov, "Precipitate Morphology and Dispersion Hardening in Internally Oxidized Cu-Al-Ti-Zr Alloys", **Fiz. Metal. Metalloved**, 47, p. 595, 1979.

11. C. A. M. Mulder and J. T. Klomp, "On the Internal Structure of Cu- and Pt-Sapphire Interface", **J. Phys. Colloc.** C4, 46, p. 111, 1985.

12. T. Takahashi, Y. Hashimoto, K. Koyama and K. Suzuki, "Preparation of Particle-Dispersion-Strengthened Copper by the Application of Mechanical Alloying" in **Proc. 4th Intern. Symp.on Science and Technology of Sintering**, edited by S. Somiya et. al., p. 659, Elsevier Applied Science, London, 1987.

MICRO-INDENTATION HARDNESS TESTING

OF PLASMA-COATED TITANIUM ALLOYS

A.T. Barbuto and D.W. McKee[1]

Abstract

Ductile plasma-sprayed coatings, such as MCr and MCrAlY types, are effective oxygen barriers on alpha+beta titanium alloys. It is important to know what happens to the alloys beneath these coatings as they are exposed to high temperatures over long periods of time, including the depth of embrittlement caused by oxygen penetration into the metal. It is also known that some coating elements such as Fe, Co, Ni and Cr diffuse a short distance into the alloy substrate, forming a beta-stabilized zone, and the thickness of this zone is important. A micro-indentation hardness testing profile is a quick way of obtaining information about these effects. Light load hardness profiles were made starting at the coating-substrate interface and traversing inward with spacings of as small as 5 mils. Micro-indentation hardness testing profiles detected the presence, if any, of sub-surface embrittlement ("alpha case") resulting from oxygen diffusion. Typical hardness data from uncoated titanium pins showed a large increase in hardness values, indicative of surface embrittlement resulting from oxidation, extending over 100 mils in from the surface. The coatings that were evaluated effectively prevented sub-surface embrittlement in Ti6-4 and Ti6-2-4-2 during high temperature oxidation exposures. The zone of embrittlement extended for only a few microns below the surface.

[1]General Electric Corporate Research & Development P.O. Box 8, Schenectady, New York 12301

INTRODUCTION

Presently, a significant amount of research is being performed on very high temperature alpha+beta titanium alloys because of their desirable properties such as light weight and very high strength to weight ratios. However, these alloys are prone to sub-surface embrittlement caused by in-service oxidation at elevated temperatures. Plasma-sprayed coatings, such as MCr and MCrAlY types, appear to be effective oxygen barriers on two of these alloys (Ti-6Al-4V and Ti-6Al-2Sn-4Zr-2Mo). A metallographic technique has been developed which uses micro-indentation hardness testing profiles to evaluate the effectiveness of these coatings as oxygen barriers. A description of sample preparation and micro-hardness testing methods will be given along with photomicrographs and graphs to show the usefulness of this technique.

EXPERIMENTAL

The titanium alloy pins used in this study were EDM machined from commercial heat treated Ti 6-4 and Ti 6-2-4-2 bar stock. The final machined size of the test pins was 1/8 inch diameter by 1 1/4 inch length. The pins were buffed, polished and then grit blasted to slightly roughen the surface prior to plasma spraying. The specimens were preheated to approximately 800-900°C, and the coating applied with low pressure spray in an inert atmosphere of argon and helium. The coating was applied by first spraying one side of the pin and then placing it in a fixture and spraying the ends and other side to insure a uniform coating around the entire pin. Various plasma coatings were applied and then tested at a wide range of temperatures in flowing air. The coatings were metallographically examined for adherence quality followed by the sub-surface embrittlement evaluation by micro-indentation hardness testing profiles.

METALLOGRAPHIC SAMPLE PREPARATION

The titanium alloy pin samples were embedded in Lecoset 7000— a hard metallographic mounting media — which is a three part cold mounting material with excellent specimen edge retention properties. It air cures in 6-10 minutes with a peak cure temperature of 190-210°F. Sectioning was carried out with a low speed cut-off machine equipped with a diamond blade. This procedure helps to keep the coatings intact and relatively flat during the sectioning. All of the subsequent metallographic sample preparation was performed on a fully automatic polishing machine (Struers Prepamatic) that can be programmed for easy reproducibly. The samples were first rough ground on a 20 µm diamond pad at high wheel revolution (1000 rpm) programmed to remove 200 µm of material. Fine grinding was done on a 6 µm diamond disc designed to lap off material quickly while maintaining sample flatness; 100 µm of material was removed with this disc. The entire polishing procedure was carried out on hard, short nap nylon polishing cloths to help maintain sample flatness. Coarse polishing was done with 6 µm diamond abrasive for 5 minutes followed by 3 µm diamond abrasive for 3 minutes, both using ethanol as a lubricant. The final polish was performed with 1 µm diamond abrasive for 3 minutes. The contact force on each specimen was maintained at 30-35 newtons. The use of this procedure results in a sample which was ready for micro-indentation hardness testing and is scratch free, flat, and with negligible or no residual surface damage (Fig.1 & 2).

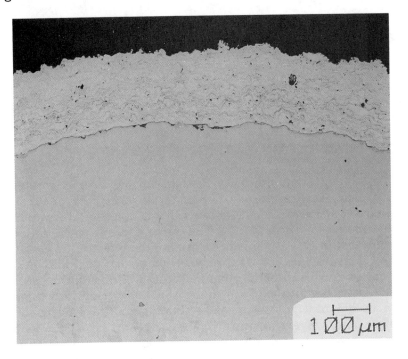

Figure 1. Ti-6Al-2Sn-4Zr-2Mo alloy plasma coated. Metallographically prepared by automatic polishing machine. Brightfield illumination. 100X

Figure 2. Higher magnification of figure 1. Brightfield illumination.
200X

MICRO-INDENTATION HARDNESS TESTING

Micro-indentation hardness testing can be directly related with
quality control of materials. In general terms, hardness is "the
resistance to permanent indentations by an indenter of fixed size and
shape under a known load" [1]. The purpose of the micro-indentation
hardness test in this study is to determine the effectiveness with which
various plasma-sprayed coatings serve as oxygen barriers. This
information can be directly compared to the surface hardness zone of
the uncoated samples oxidized under the same conditions. Light load
hardness indentations were made starting at the coating-substrate
interface and traversing inward with spacings as small as 5 mils
using a Knoop diamond indenter under a load of 5 grams. The Knoop
indentations ranged in diagonal length from 10 μm to 15 μm. Three sets
of micro-indentation hardness profiles were made on each specimen.

For the coated samples, a total of 10 to 15 impressions were taken in
each profile with a total distance of 100 to 125 mils traversed (Figure 3).

Hardness values were also obtained near the center of the specimens to insure the absence of surface effects. Ten to fifteen impressions were also made on each uncoated specimen (Figure 4). However, a larger distance—400 to 500 mils—from the edge was traversed due to more extensive oxygen penetration. Decreases in the length of the hardness impressions reveal the presence and depth of embrittlement near the surface caused by oxidation in the uncoated specimen.

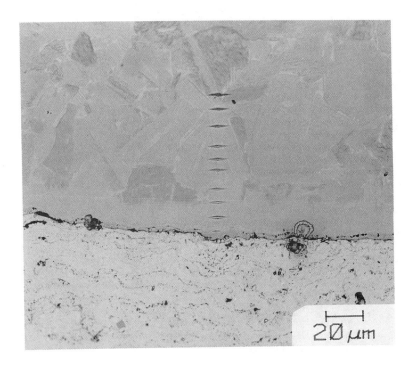

Figure 3. Plasma spray-coated Ti-6Al-2Sn-4Zr-2Mo alloy oxidized for 10008 hrs at 825°C. Metallographically prepared by automatic polishing machine. Brightfield illumination. 500X

The hardness tests were done on a Leco M-400-G2 Digital Micro-indentation Hardness Testing Machine equipped with a television measuring system. The indentations are conveniently and accurately measured on the screen of the monitor. The diagonal length measurement information is sent to an IBM PC AT microcomputer which is linked to the hardness tester by way of a standard parallel interface. The data is converted to hardness numbers and tabulated.

Finally, micro-indentation hardness profiles were plotted from these data utilizing a Macintosh computer and Kaleidograph plotting program software.

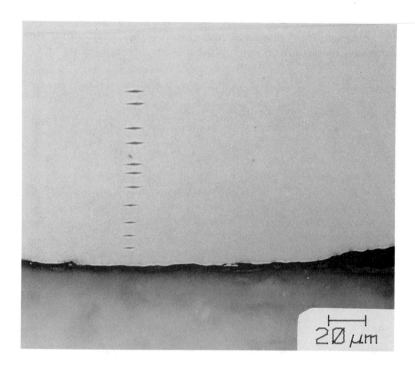

Figure4. Uncoated Ti-6Al-2Sn-4Zr-2Mo alloy metallographically prepared by automatic polishing machine. Brightfield illumination. 500X

RESULTS

Figure 5 illustrates a typical specimen indentation profile for a coated sample which shows very little difference in hardness as a function of depth below the surface. Thus, the coating has effectively prevented the diffusion of oxygen into the base metal and associated embrittlement of the alloy. Typical hardness data from uncoated titanium pins (Figures 6 and 7) showed a large increase in hardness values near the surface which is indicative of surface embrittlement resulting from oxidation. The zone of embrittlement extended over 100 mils in from the surface.

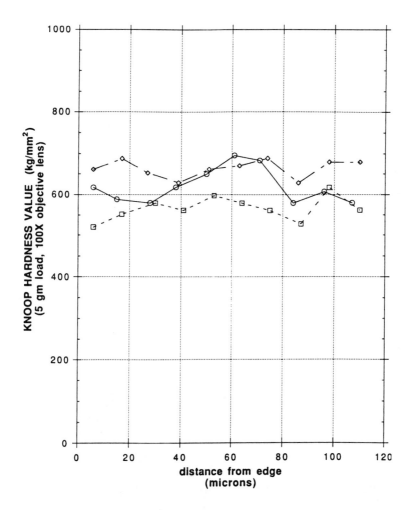

Figure 5 Hardness profile for Ti-6Al-2Sn-4Zr-2Mo alloy specimen that was plasma coated and heat treated 1000hrs at 730°C. Specimen shows very little sub-surface embrittlement.

707

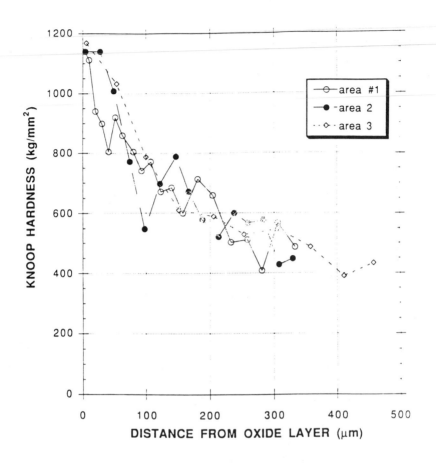

Figure 6 Typical hardness profile plot of uncoated specimen.

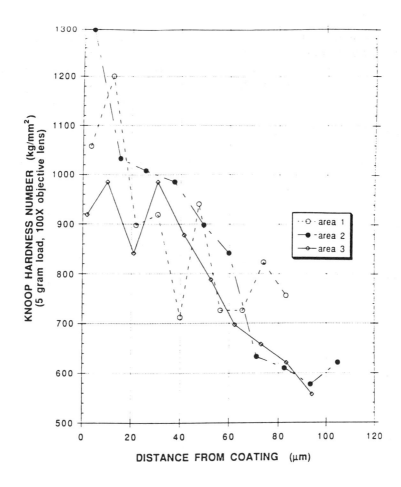

Figure 7 Hardness profile of uncoated Ti-6Al-2Sn-4Zr-2Mo alloy exposed to same oxidation conditions as specimen shown in Figure 5. Specimen shows sub-surface embrittlement.

CONCLUSIONS

Light load hardness testing is a useful tool for evaluating the effectiveness of plasma-sprayed coatings as oxygen barriers. In this study, the technique showed that the MCr and MCrAlY type coatings that were evaluated effectively prevented sub-surface embrittlement in Ti6-4 and Ti6-2-4-2 during high temperature oxidation exposures.

Micro-indentation hardness testing in conjunction with automatic sample preparation is a quick and simple procedure for extracting information which can be used to predict coated part life and service response. The total test time from the mounting of the specimen to the generation of the final hardness profile plots is less than two hours.

ACKNOWLEDGEMENT

The author wishes to thank A.S. Holik for technical assistance in the preparation of this manuscript.

REFERENCES

1) V.E.Lysaght,"The How and Why of Microhardness Testing", Metals Progress,August 1960.

MICROSTRUCTURE CONTROL OF GAMMA

TITANIUM ALUMINIDES THROUGH POWDER METALLURGY

D. Eylon, C. M. Cooke, and Y-W. Kim

ABSTRACT

Powder metallurgy (PM) approach for gamma titanium aluminide alloys was studied. Ti-48Al-3Nb-2.5Cr-1Mn (at%) and smaller quantities of Ti-Al binaries were studied. Powders produced by the plasma rotating electrode process (PREP) and by the gas atomization (GA) process were filled into evacuated titanium cans and subsequently hot isostatically pressed (HIP'd) at a range of temperatures and pressures. Rod and plate shape compacts were used for microstructure study, while larger cylindrical consolidates were used for a closed-die isothermal forging study. Powders HIP'd at 1800°F (980°C)/38 ksi (260MPa)/6 hr had the finest gamma structure (5-10 μm), while coarse lamellar $\alpha_2+\gamma$ structure was obtained in $\gamma+\alpha$ solution treated compacts. The best closed-die forgability was for the Ti-47Al (at%) binary alloy. The finer microstructures of the GA powder particles resulted in compacts with finer structure when HIP'd below the alpha transus temperature.

Metcut-Materials Research Group, P.O. Box 33511, Wright-Patterson AFB, OH 45433-0511

711

INTRODUCTION

Powder metallurgy (PM) of titanium alloys was the
subject of a thorough investigation and development in the
last twenty years [1, 2]. This effort was aimed at lower
cost products through net-shape technology, and improved
properties through microstructure refinement and
uniformity. The low cost net-shape approach was achieved
by using a blend of titanium sponge fines and powder of
Al-V master alloy [3, 4] and is known as the blended
elemental (BE) method. The relatively inexpensive
products were the result of the low cost of the titanium
sponge fines, and the low cost of the cold pressing and
the sintering processes. On the other hand, the inherent
chlorides in NaCl or MgCl reduced sponge allowed a maximum
of only 99.8% density to be achieved, even after
post-compaction hot work densification [5, 6]. As a
result, such parts are produced today in large quantities
only for aerospace applications not requiring good dynamic
properties [2]. For applications requiring good dynamic
properties, such as high cycle fatigue, prealloyed (PA)
powder is the only PM option. Such powders are produced
by atomizing an ingot or a molten stream of the alloy. To
date, only two types of clean prealloyed titanium alloy
powders are available in the United States in commercial
quantities. The plasma rotating electrode process (PREP)
powder [7, 8] is made by spin-atomizing an alloy rod. The
tip of the rod is melted by a plasma heat source and the
ejected droplets are cooled into an argon filled chamber.
The gas atomized (GA) powder [9] is made by high pressure
argon gas atomization of a stream of molten alloy poured
through a refractory metal orifice. The powder produced
has a wider range of particle sizes (compare Figures 1a
and 1b), higher solidification cooling rate, and as a
result much finer particle microstructure (compare
Figures 1c and 1d). On the other hand, the GA powder has
a higher portion of irregular shaped particles when
compared to the PREP powder (Figures 1a and 1b). By
applying the proper HIP conditions, it is easy to produce
100% dense compacts free of contaminants and with
isotropic and uniform microstructure and properties [2].
As a result, thorough studies of aerospace titanium
alloys, such as Ti-6Al-4V [10] and Ti-829 [11, 12],

produced powder compacts with properties equal to and even slightly exceeding wrought alloys. However, in spite of the high levels of properties, titanium prealloyed PM products are not used yet on a commercial basis. This is because the cost of the powder is not compensated by the slight advantage in mechanical properties. In addition, the part producers are still not confident that on a large scale production basis, occasional contamination may not get into the compacts. Previous works have shown that even an occasional 50 μm ceramic contaminant may cause an order of magnitude degradation in high cycle fatigue life [13].

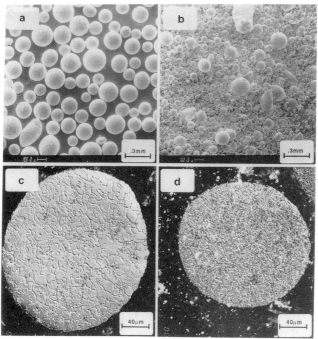

Figure 1. The morphology and microstructure of PREP powder (a and c), and GA powder (b and d). Original magnifications are 50X and 400X, respectively.

Recent advances in titanium alloys based on the ordered phases α_2, $\alpha_2+\gamma$, and γ have renewed the interest in PM. The relatively brittle ordered titanium aluminide alloys require forging temperatures higher than ordinary

titanium alloys. While a typical forging temperature of
Ti-6Al-4V is 1600-1850°F (850-1000°C), alpha-2 alloys are
forged above 1900°F (1050°C), and gamma alloys above
2000°F (1100°C). In addition, machining reduces the
surface integrity of ordered alloy components, leading to
frequent surface related and greatly premature fatigue
failures. True net-shape PM may eliminate the need for
heavy machining and may improve fatigue strength. In
addition, the fine PM structures when compacted below the
beta or alpha transus temperature of an ordered alloy, may
be more suitable for forging and may present higher room
temperature (RT) ductility. A PM study of the alpha-2
alloy Ti-25Al-10Nb-3V-1Mo (wt%), showed that by
sub-transus compaction it is possible to produce fully
dense complex shape products (such as an impeller), with
RT elongation as high as 6% [14, 15]. This was the result
of the highly equiaxed fine structure, and the unusually
high level of retained beta phase. Previous work already
demonstrated that gamma alloy powders can be consolidated
into fully dense products with a range of microstructures
[16, 17]. The objectives of the present work were to
study in gamma PM alloys: (a) the properties of HIP
compacts; (b) the range of microstructures available
through various HIP and solution treat and age (STA)
treatments; (c) the forgability of powder compacts; and
(d) the comparison between the PREP and the GA prealloyed
powders.

EXPERIMENTAL PROCEDURES

Material

The materials selected for this work were the PREP
(Figure 1a) and the GA [9] (Figure 1b) prealloyed powders.
These are the only two sources available for prealloyed
spherical titanium powders in commercial quantities at a
level of cleanliness suitable for aerospace quality
products. The main body of work was done on PREP and GA
Ti-48Al-3Nb-2.5Cr-1Mn (at%) (Ti-33Al-7Nb-3.3Cr-1.4Mn(wt%))
powders produced by Nuclear Metals, Inc. (NMI) and
Crucible, respectively. The chemical compositions in at%
and wt% as reported by the producers are provided in
Table 1. In addition, small amounts of binary and ternary

composition GA powders were made by Crucible in the form of HIP'd plates and cylindrical compacts. A picture of such compacts HIP'd at 1850°F (1000°C)/29 ksi (200MPa)/4 hr is shown in Figure 2.

TABLE 1
CHEMICAL COMPOSITION OF ALLOY POWDER

Alloy/Source	Al (at%) wt%	Nb (at%) wt%	Cr (at%) wt%	Mn (at%) wt%	O wt%	N wt%	C wt%	H wt%	Fe wt%
Ti-48Al-3Nb-2.5Cr-1Mn									
Target, at%	(48.0)	(3.0)	(2.5)	(1.0)					
Target, wt%	32.9	7.1	3.3	1.4					
GA Powder, wt%	33.4	7.0	3.3	1.4	0.090	0.0070	0.012	0.0019	0.06
PREP Bar Stock, wt%	32.95	6.93	3.58	2.17	0.043	0.004	0.020	0.0034	0.06
PREP Powder, wt%	33.00	6.94	3.31	2.09	0.042	0.005	0.005	0.0007	NA
Binary GA Powders									
Ti-44Al	(43.7)				0.041	0.002	NA	NA	0.02
Ti-46Al	(46.5)				0.060	0.002	NA	NA	0.04
Ti-47Al	(47.0)				0.054	0.004	NA	NA	0.04
Ti-48Al	(48.2)				0.055	0.002	NA	NA	0.03
Ti-53Al	(53.0)				0.068	0.004	NA	NA	0.01
Ternary GA Powder									
Ti-50Al-2Cr	(49.7)		(1.8)		0.075	0.002	NA	NA	0.03

Figure 2. Cylindrical and plate compacts HIP'd at 1850°F (1000°C)/29 ksi (200MPa)/4 hr.

The sieve analysis of both powders is in Table 2. While the median of the PREP powder is at 250 µm, the median of the GA powder is at 177 µm. By comparing Figures 1a and 1b, it is clear that the GA powder is finer, however, it has twice as high oxygen level (900 ppm vs. 420 ppm, Table 1).

TABLE 2

SIEVE ANALYSIS OF Ti-48Al-3Nb-2.5Cr-1Mn POWDERS

(Percent of powder under)

US Standard No.	-35	-45	-60	-80	-100	-120	-170	-200	-230	-325
Sieve Opening, µm	500	354	250	177	149	125	88	74	63	44
GA Powder	100	91	72	53	44	NA	NA	17	NA	6
PREP Powder	100	89.3	52.0	20.4	NA	8.5	2.1	NA	0.7	NA

Powder Compaction

Preliminary HIP compaction experiments were conducted on powders filled into 0.5 in. diameter x 2 in. long CPTi tubes. The HIP conditions, the best densities, and the resulting microstructures are listed in Table 3. The plate listed in Table 3 was used for tensile test evaluations of the as-compacted material reported in Table 6. The compact densities were evaluated by optical microscopy.

TABLE 3

BEST RESULTS OF PREP AND GA Ti-48A1-3Nb-2.5Cr-1Mn
POWDERS COMPACTED AT RANGE OF HIP TEMPERATURES

Type of Powder	Compact Shape	HIP Condition, °F/ksi/hr	Optical Density, %	Average Gamma GS, μm	Comments on Microstructure
GA	Plate	2150/15/4	99.99	60	Highly twinned
GA	Bar	2100/30/6	99.9	50	Highly twinned
PREP	Bar	2100/30/6	99.99	50	Highly twinned
GA	Bar	2000/30/6	99.9	20	Twinned
PREP	Bar	2000/30/6	99	20	Twinned
GA	Bar	1900/35/6	92[a]	10	
PREP	Bar	1900/35/6	99	15	
GA	Bar	1800/40/6	99.99	5	
PREP	Bar	1800/40/6	99.99	15	

[a]Leak in can.

Additional larger plate compacts (3 in. x 4 in. x
0.5 in. canned dimensions) and cylinders (3 in. diameter x
4 in. tall canned dimensions) were HIP'd at 1800°F/40 ksi/
6 hr (at Wright Laboratory) or at 1850°F/20 ksi/4 hr (at
Crucible). The lower HIP temperature was selected to
produce the finest compact structure (Table 3). The plate
compacts were aimed at mechanical testing of net-shape HIP
products while the cylindrical compacts were all
isothermally forged. The composition as well as the HIP
and forging conditions of the larger compacts are listed
in Table 4.

717

TABLE 4

PROCESSING CONDITIONS OF LARGER
COMPACTS OF GAMMA ALLOY POWDERS

Alloy	Powder	Compact Shape	HIP Temp., °F	Compact Optical Density, %	Forge Temp. and Dwell, °F/min	Final Forged Shape	Forg-ability
Quadrinary[a]	GA	Cylinder	1800	100	2150/5	Disk	Poor
Quadrinary	GA	Cylinder	1800	100	2150/5	Pancake	Fair
Quadrinary	PREP	Cylinder	1800	100	2150/5	Pancake	Good
Ti-48Al-2Cr	GA	Cylinder	1850	100	2150/5	Disk	Fair
Ti-48Al-2Cr	GA	Plate[b]	1850	--	--	--	--
Ti-44Al	GA	Cylinder	1850	100	2150/1	Pancake	V. Good
Ti-44Al	GA	Plate[b]	1850	100	--	--	--
Ti-46Al	GA	Cylinder	1850	97[c]	2150/2	Pancake	Poor[d]
Ti-46Al	GA	Plate[b]	1850	100	--	--	--
Ti-47Al	GA	Cylinder	1850	100	2150/2	Pancake	V. Good
Ti-47Al	GA	Cylinder	1850	100	2150/1	Disk	Good
Ti-47Al	GA	Plate[b]	1800	100	--	--	--
Ti-48Al	GA	Cylinder	1850	100	2150/2	Pancake	V. Good
Ti-48Al	GA	Plate[b]	1850	97[c]	--	--	--
Ti-53Al	GA	Cylinder	1850	100	2150/2	Pancake	Good
Ti-53Al	GA	Plate[b]	1850	100	--	--	--

[a] Quadrinary alloy: Ti-48Al-3Nb-2.5Cr-1Mn.

[b] Plates were not forged.

[c] Possible leak in the can.

[d] Due to porosity.

The alpha transus temperature of the quadrinary alloy was determined by DTA at 30°C/min heating rate to be:

GA loose powder: 1365°C; GA HIP compact: 1360°C
PREP loose powder: 1345°C; PREP HIP compact: 1345°C

Post-Compaction Treatments

To study the range of possible PM microstructures, the heat treatments listed in Table 5 were performed on coupons taken from the 2150°F HIP'd plate listed in Table 3.

TABLE 5
SOLUTION TREATMENT AND MICROSTRUCTURE OF
GA Ti-48Al-3Nb-2.5Cr-1Mn HIP'd COMPACTS

ST Temp.[a] and Time, °C (°F)/hr	HIP Temp., °F	Microstructure Characterization
None	1850	Fine (5 μm) equiaxed gamma
None	2150	Coarse (60 μm) equiaxed gamma
1000 (1830)/6	2150	Same as as-HIP'd
1200 (2190)/6	2150	Same as as-HIP'd
1290 (2290)/3	1850	Fine (15 μm) equiaxed gamma
1350 (2460)/1	2150	Coarse gamma + 20% lamellar
1400 (2550)/0.5	2150	Coarse lamellar

[a]Cooling rate 100°C/min.

Light metallography and scanning electron microscopy were performed on selected conditions. The best photomicrograph results were obtained from samples polished but left unetched when observed under polarized light.

Isothermal Forging

Isothermal forging was performed by the Ladish Company. All forgings were done at 2150°F, 0.1 min⁻¹ strain rate, and 150 tons peak load with a dwell at peak load. The forging temperature and the dwell times are listed in Table 4. Most alloys were forged into 6 in. diameter disks using an outside confinement die which simulated closed-die forging. Three alloys were also forged into 5 in. diameter subscale simulated disks to study the die fill capability of the alloy powder

719

compacts. The forged shapes and the quality of the forgings (forgability) are listed in Table 4.

Tensile Testing

Tensile tests were performed on 0.13 in. diameter x 0.68 in. long nominal gage section specimens at 0.005 in./in./min strain rate to 0.2% yield followed by 0.03 in./min crosshead rate to fracture. Test coupons were taken from PREP and GA as-HIP'd Ti-48Al-3Nb-2.5Cr-1Mn bars and from 1250°C and 1400°C solution treated GA bar and plate material (Table 3). The tensile tests were performed at a temperature range of RT to 1000°C as listed in Table 6. The plate was HIP'd at 2150°F and the bars were HIP'd at 1800°F.

RESULTS AND DISCUSSION

Powder Compacts and Microstructures

The compacts listed in Table 3 were used to evaluate the range of HIP conditions needed to fully compact the gamma alloys, and the resulting microstructures. Because of a slight leak in some cans, not all resulted in full density. However, from the most successful compacts it is clear that full density could be obtained under either 2150°F/15 ksi, or 2000°F/30 ksi, or 1800°F/45 ksi HIP. While all the HIP conditions experimented resulted in full density, they had a different effect on the as-HIP'd microstructure. The 2150°F cycle produced a coarse 60 µm gamma grain size, while the 1800°F cycle produced a fine 10 µm and 5 µm grain sizes of the PREP and the GA compacts, respectively (Table 3 and Figure 3). A comparison of the microstructure of the GA compacts at a HIP temperature range of 1800°F to 2150°F is shown in Figure 4. The microstructure of the PREP compacts HIP'd in the same conditions were similar.

TABLE 6
TENSILE TEST RESULTS OF
Ti-48Al-3Nb-2.5Cr-1Mn POWDER COMPACTS

Test Temp., °C (°F)	0.2% YS, MPa (ksi)	UTS, MPa (ksi)	El., %	RA, %
PREP As-HIP Bar				
24 (75)	-[a]	620 (90)	-[b]	0.5
GA As-HIP Bar				
24 (75)	-[a]	772 (112)	-[b]	0.3
GA 1250°C ST Bar				
24 (75)	-[a]	579 (84)	0.4	0.3
GA 1400°C ST Plate				
24 (75)	-[a]	448 (65)	0.6	0.3
24 (75)	-[a]	455 (66)	0.5	0.8
700 (1290)	-[a]	400 (58)	1.1	0.5
700 (1290)	413 (60)	420 (61)	1.9	0.5
850 (1560)	372 (54)	413 (60)	-[b]	0.9
850 (1560)	-[a]	393 (57)	-[b]	0.3
1000 (1830)	241 (35)	296 (40)	2.1	1.2
1000 (1830)	186 (27)	255 (37)	11.3	12.6

[a]Specimen failed before 0.2% yield was reached.

[b]Multiple fracture; elongation could not be measured.

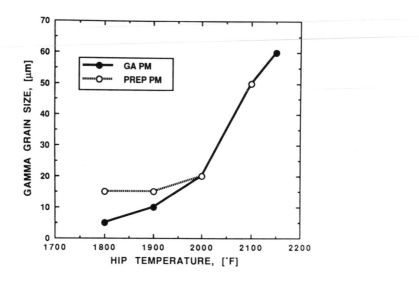

Figure 3. Effect of HIP temperature on the gamma grain size of the Ti-48-3-2.5-1 alloy of PREP and GA powder compacts.

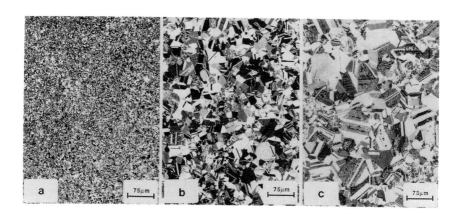

Figure 4. Microstructure of GA Ti-48-3-2.5-1 powder compacted at (a) 1800°F/40 ksi, (b) 2000°F/30 ksi, and (c) 2150°F/15 ksi. Original magnification is 200X.

It is interesting to note that both powders resulted in the same compact gamma grain size when HIP'd in the 2000°F to 2150°F range (Figure 3). However, the finer structure GA powder resulted in a substantially finer structure then the PREP powder when compacted in the 1800°F to 1900°F range. Because of the microstructure refinement at the lower HIP temperatures, the subsequent HIP study was carried out at 1800°F to 1850°F range.

Microstructure modification treatments (Table 5) were performed on the 2150°F/15 ksi HIP'd GA plate listed in Table 3. The alpha transus temperature of this compact is 1360°C. The as-HIP'd microstructure is shown in Figure 4c. The heat treatments performed in the $\alpha_2+\gamma$ region (1000°C) and in the lower end of the $\alpha_2+\gamma$ region (1200°C) did not result in any noticeable change in microstructure from the as-HIP'd condition on the optical microscopy level. The ST performed close to the alpha transus temperature at 1350°C, produced a coarse equiaxed gamma structure with 20% lamellar α_2/γ structure (Figure 5a). The above-transus treatment at 1400°C produced α_2/γ lamellar structure with a prior alpha grain size on the order of 250 μm (Figure 5b). The coarse as-HIP'd structure of this plate, the result of the high temperature HIP cycle, resulted in a coarse structure in all subsequently solution treated samples. However, the 1850°F GA powder compact, produced after 1290°C ST an equiaxed structure with only 15 μm gamma grain size (Figure 5c). These process conditions were selected for pre-heat treatment of some of the forging PM preforms, since it was expected to produce good ductility at forging temperatures.

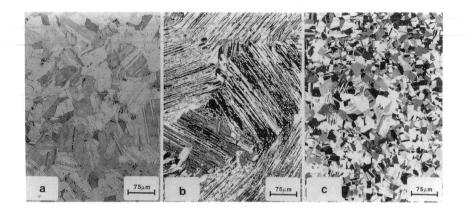

Figure 5. Microstructures of GA powder compacts HIP'd and solution treated at (a) 2150°F HIP+1350°C ST, (b) 2150°F HIP+1400°C ST, and (c) 1850°F HIP+1290°C ST. Original magnification is 200X.

Tensile Properties of HIP'd Compacts

The tensile results of the as-HIP'd and HIP+ST compacts are listed in Table 6. These results include PREP and GA powders tested in as-HIP'd conditions. In addition, the GA compacts were tested at room temperature after 1250°C and 1400°C ST and in the 700°C to 1000°C test temperature range after 1400°C ST. All process conditions displayed very low tensile elongation at room temperature. The compact with the finest gamma grain structure (1800°F HIP'd GA powder, Figure 4a), produced the highest tensile strength -772MPa (112 ksi). The somewhat coarser structure PREP compact, produced only 620MPa (90 ksi) tensile strength at room temperature. The 1400°C ST GA compacts with the very coarse prior alpha structures (Figure 5b) produced the lowest strength of 448MPa (65 ksi). The effect of the gamma or prior alpha grain size on the tensile strength is illustrated in Figure 6a. The effect of test temperature on tensile strength is illustrated in Figure 6b for the 1400°C ST GA compact. Since PM compacts occasionally contain porosity that affect the tensile elongation, the elongation plot in Figure 6b is based on the best tensile test results in

724

Table 6. It should be noted that the tensile elongation (EL) is plotted on a X10 scale for graphical clarity. As seen, above 800°C, there is a substantial increase in elongation and decrease in strength.

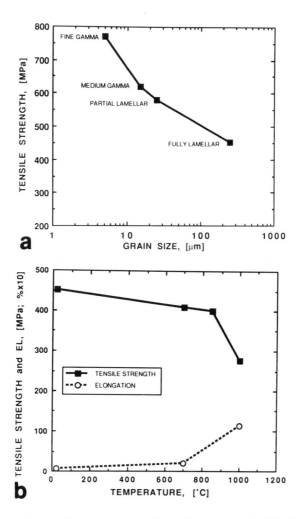

Figure 6. Tensile strength of GA Ti-48-3-2.5-1 PM compacts (a) grain size effect and (b) test temperature effect.

Forgability of PM Compacts

At this point, only partial results are available. As-HIP'd compacts of the compositions listed in Table 4 were subjected to isothermal forging at 2150°F. As seen from this table, the best forging results were obtained from the Ti-47Al and the Ti-48Al binary alloys. Of the quadrinary alloy, the PREP compact produced a better quality forged pancake than the GA compact. However, additional PM compacts after various ST conditions are now being evaluated for closed die forgability. An illustration of low and high quality forged pancakes and disks is shown in Figure 7.

Figure 7. PM low and high quality compacts forged into disks and pancakes.

SUMMARY AND CONCLUSIONS

The use of powder metallurgy for making fully dense products with a range of microstructures was experimented. The processed compacts were evaluated as net-shapes and also as isothermal forging preforms. The commercially available PREP and GA prealloyed Ti-48Al-3Nb-2.5Cr-1Mn (at%) powders were evaluated and compared.

Hot isostatic pressing (HIP'ing) was found to be a very effective method to produce good compacts with full density achieved under 40 ksi pressure at temperatures as low as 1800°F.

Lower temperature compaction produced fine gamma grain structures. Powders with ultrafine grain structure, such as the GA process powder, produced compacts with finer grain structure.

At a higher HIP temperature (above 2000°F), the finer grain structure powders had no effect on the microstructure refinement of the compacts.

Solution treatment of powder compacts at temperatures below the alpha transus temperature produced grain structures which were related to the as-HIP'd structure. Coarse grain compacts (produced by high temperature HIP) yielded coarse ST structures. Fine grain compacts (produced by lower temperature HIP), yielded finer ST microstructures.

The tensile strength of the PM compacts is related to their gamma or prior alpha grain size. The fine grain conditions produced the highest strength while the coarse structures produced the lowest strength.

All samples tested at room temperature to 700°C range displayed low tensile elongation. Only over 800°C the alloy developed tensile elongation, but at the expense of tensile strength.

The PREP Ti-48-3-2.5-1 compacts and the GA Ti-44Al, Ti-47Al, and Ti-48Al compacts produced very good forging pancakes. The Ti-47Al alloy also showed a good closed-die forgability.

ACKNOWLEDGMENTS

The authors wish to acknowledge William Nachtrab and John Nicholson of NMI, Fred Yolton and John Moll of Crucible for their assistance in providing the powders. The assistance of Fred Yolton in producing some of the HIP cans is highly appreciated. The authors also wish to thank Miss Karen A. Sitzman for manuscript preparation.

REFERENCES

1. F. H. Froes and D. Eylon, "Powder Metallurgy of Titanium Alloys," International Materials Reviews, Vol. 35, No. 3 (1990), pp. 162-182.

2. D. Eylon and F. H. Froes, "Titanium P/M Products," A chapter in Metals Handbook, Tenth Edition, Volume 2, Properties and Selection: Nonferrous Alloys and Special-Purpose Materials, ASM International, Materials Park, OH (1990), pp. 647-660.

3. D. Eylon, F. H. Froes, L. Parsons, and E. J. Kosinski, "Titanium PM Parts for Aerospace Applications," Materials Engineering (February 1985), pp. 35-37.

4. F. H. Froes, D. Eylon, and R. G. Rowe, "Titanium Powder Metallurgy -- Products and Applications," 1986 International Conference on Titanium Products and Applications, Vol. II, Titanium Development Association, Dayton, OH (1987), pp. 758-781; also in Powder Metallurgy in Defense Technology, Vol. 7, ed. by W. J. Ullrich, MPIF, Princeton, NJ (1987), pp. 121-144.

5. Y. Mahajan, D. Eylon and F. H. Froes, "Microstructure Property Correlation in Cold Pressed and Sintered Elemental Ti-6Al-4V Powder Compacts," Powder Metallurgy of Titanium Alloys, ed. by F. H. Froes and John E. Smugeresky, TMS-AIME, Warrendale, PA (1980), pp. 189-202.

6. J. Weiss, D. Eylon, M. W. Toaz, and F. H. Froes, "Effect of Isothermal Forging on Microstructure and Fatigue Behavior of Blended Elemental Ti-6Al-4V Powder Compacts," Metallurgical Transactions A, Vol. 17A, No. 3 (1986), pp. 549-559.

7. E. J. Kosinski, "The Mechanical Properties of Titanium P/M Parts Produced from Superclean Powders," Progress in Powder Metallurgy, Vol. 38, ed. by J. G. Bewley and S. W. McGee, MPIF, Princeton, NJ (1983), pp. 491-501.

8. P. R. Roberts and P. Loewenstein, "Titanium Alloy Powders Made by the Rotating Electrode Process," Powder Metallurgy of Titanium Alloys, ed. by F. H. Froes and J. E. Smugeresky, TMS-AIME, Warrendale, PA (1980), pp. 21-35.

9. C. F. Yolton, "Gas Atomized Titanium and Titanium Aluminide Alloys," P/M in Aerospace and Defense Technologies, Vol. 1, compiled by F. H. Froes, MPIF, Princeton, NJ (1989), pp. 123-132.

10. A. A. Sheinker, C. R. Chanani, and J. B. Bohlen, "Evaluation and Application of Prealloyed Titanium P/M Parts for Airframe Structures," International Journal of Powder Metallurgy, Vol. 23, No. 3 (1987), pp. 171-176.

11. N. Osborne, D. Eylon, and J. P. Clifford, "Evaluation of Titanium Alloy Powder Compacted to Near Net Shape," Final Technical Report, AFWAL-TR-88-4250, Wright-Patterson AFB, OH (February 1989).

12. N. R. Osborne, D. Eylon, F. H. Froes, and J. P. Clifford, "Compaction and Net-Shape Forming of Ti-829 Alloy by Powder Metallurgy Rapid Omnidirectional Processing," Advances in Powder Metallurgy 1989, Vol. 3, compiled by T. G. Gasbarre and W. F. Jandeska, MPIF, Princeton, NJ (1989), pp. 373-386.

13. S. W. Schwenker, D. Eylon, and F. H. Froes, "Influence of Foreign Particles on Fatigue Behavior of Ti-6Al-4V Prealloyed Powder Compacts," Metallurgical Transactions A, Vol. 17A, No. 2 (1986), pp. 271-280.

14. W. J. Porter, N. R. Osborne, D. Eylon, and J. P. Clifford, "Rapid Omnidirectional Compaction (ROC) of Titanium Aluminide Prealloyed Powders," Advances in Powder Metallurgy 1990, Vol. 2, compiled by E. R. Andreotti and P. J. McGeehan, MPIF, Princeton, NJ (1990), pp. 243-257.

15. N. R. Osborne, W. J. Porter, and D. Eylon, "Mechanical Properties of Powder Metallurgy Ti-829 and Ti-25Al-10Nb-3V-1Mo Produced by Rapid Omnidirectional Compaction," SAMPE Quarterly, Vol. 22, No. 4 (July 1991), pp. 21-28.

16. K. R. Teal, D. Eylon, and F. H. Froes, "Compaction of Titanium Aluminide Powders by Hot Isostatic Pressing (HIP)," Sixth World Conference on Titanium, Part II, ed. by P. Lacombe, R. Tricot, and G. Beranger, Les Editions de Physique, Paris, France (1989), pp. 1121-1125.

17. Y-W. Kim and J. J. Kleek, "Microstructure and Tensile Properties of a Powder Metallurgy Gamma Titanium Aluminide Alloy," PM'90 - World Conference on Powder Metallurgy, Vol. 1, The Institute of Metals, London, England (1990), pp. 272-288.

IDENTIFICATION OF INTERMETALLIC COMPOUNDS IN TANTALUM

T. K. Chatterjee[1] & P. Kumar[2]

ABSTRACT

Tantalum exhibits attractive properties under high strain-rate loading conditions. These properties are developed through the judicious application of microalloying and thermo-mechanical processing. Powdered tantalum, microalloyed with silicon and molybdenum, exhibits a unique subgrain structure, depending on the micro alloying elements added. Some of the precipitates observed within the subgrains and low angle grain boundaries were identified as the intermetallic compounds, $MoSi_2$ and $TaSi_2$, using Selected Area Electron Diffraction (SAED) methods. A model to correlate these features with strengthening mechanisms is proposed on the basis of physical and thermodynamic properties for the "tantalum system". A ribbon-like subgrain structure was observed when samples were heat treated at 1300 deg C. Grain boundary ledges believed to be a source of dislocations were observed in powdered tantalum microalloyed with silicon. The subgrain boundaries were extensively broadened when silicon and molybdenum were used in combination. This broadening was not observed when molybdenum was not present in the alloy.

----

1. US Army Research Development & Engineering Center Picatinny Arsenal, NJ 07806-5000 United States of America
2. Cabot Corporation Boyertown, PA 19512 United States of America

INTRODUCTION

Applications for tantalum include the use in high-temperature environments such as hardware for furnaces and components for electrical capacitors. Although the high-temperature exposure for tantalum usually occurs in vacuum, it should be noted that tantalum can be exposed to significant concentrations of oxygen for very short periods of time.

The high temperature stability in an oxidizing atmosphere is usually imparted by alloying additions, such as silicon and tungsten. In principle, molybdenum can also be used; however this material is not commercially available.

The improvement in properties is due to the following mechanisms:
a. Improvement in the oxidation resistance by adding silicon [1-3].
b. Dispersion strengthening [4-5] by adding ThO_2 and Y_2O_3.
c. Increase in the recrystallization temperature by microalloying.

It is interesting to note that microalloying (i.e. additions less than 1000 ppm) is highly effective in altering the mechanical and physical properties of tantalum. For example, when silicon and yttrium are added at less than 50 ppm the recrystallization temperature of tantalum is significantly raised [6].

While the mechanical property improvement in tantalum is well established, the effect on microstructure is not well documented; nor is the effect of microstructure on the high strain-rate deformation properties of tantalum well understood.

The ultimate objective of this on-going research is to improve the high strain-rate deformation behavior of tantalum by means of microstructural modifications. As a first step, the microstructure of the microalloyed material was characterized using transmission electron microscopy methods. The data are presented in this paper.

EXPERIMENTAL PROCEDURES

Powder metallurgy methods were used to produce tantalum sheet approximately 0.38 mm thick. Three types of sheet materials were produced; one with silicon added, another with molybdenum added, and another with silicon and molybdenum combined. These materials contained 25 ppm carbon, 150 ppm oxygen and 15 ppm nitrogen. Other impurities were below the detection limits. Table 1 lists the mechanical properties, grain size and microhardness of the annealed sheet samples.

Table 1: Properties of Annealed Tantalum Sheets

Property	Sample A	Sample B	Sample C
Mo (ppm)	200	- -	200
Si (ppm)	- -	400	400
Grain Size (ASTM No.)	10	9	8
Micro-Hardness (DPH No.)	140	137	126
Yield Strength (ksi)	66.7	66.1	55
Tensile Strength (ksi)	74.4	75.7	61.0
Elongation (%)	22	28	34

RESULTS

The powdered metal strip, 0.38 mm thick, annealed at 1300 deg C, contained an abundance of broadened subgrain boundaries, dislocations, and ribbon-like structure.

Sample "A" Results:

These microalloyed P/M samples contained 200 ppm Mo. A bent extinction contour C and ribbon-like polygonized structure R inside the contour are shown in figure 1. Low angle grain boundaries L and numerous dislocation nodes are seen in figure 2. Figure 3 shows the formation of a polygonized ribbon-like structure P by low angle grain boundaries indicated by arrows. All these features indicated above have already been observed previously [6] in tantalum P/M specimens containing 45 ppm Si.

Sample "B" Results:

These specimens contained 400 ppm Si. The electron micrograph shown in figure 4 reveals large grain boundaries which intersect at the point T. The subgrain S and segregated precipitates P near the grain boundary and within the grain interior are seen in this electron micrograph. Selected area electron diffraction analyses confirmed the presence of TaSi2 precipitates. A similar picture taken from another area of the same specimen is shown in figure 5 at a much higher magnification The interfacial dislocation fringes D, generated by interactions between the precipitate and matrix area are clearly seen. The arrow indicates the location of dense dislocations adjacent to the grain boundary.

Figure 6 shows dislocation networks forming a bent ribbon-like structure. These structures tend to form polygonized regions P when proper heat treatment is given. These ribbon-like structures were also observed in P/M tantalum samples containing 45 ppm Si as well as in pure tantalum ingot material [6].

Figure 7 shows the presence of low angle grain boundaries L and a precipitate P. This microstructure was also observed in P/M tantalum sample with 45 ppm Si [6].

Figure 8 shows ledges within the grain boundary and dislocations D emitting from the ledges. The "zig-zag" structure and dotted contrast are believed to arise from dislocation segments near the foil surface [7,8].

Figure 9 shows that the precipitates prefer to locate themselves near the grain boundary junction T. Extinction contours and bend contours can be seen in the area A,B and C [9].

Sample "C" Results:

These samples contained 400 ppm Si, 200 ppm Mo. The electron micrograph in figure 10 for a specimen from this sample reveals subgrain structure, 2 to 10 microns in size. Grain boundary broadening can be seen at the location indicated by the arrow. The precipitates P inside the grain are also seen distinctly. The electron micrograph in figure 11 was taken from another area of the same specimen and showed grain broadening effects also.

Dark field electron microscopy techniques were used to identify the precipitates lying within the broadened grain boundary and within the grains themselves. The bright field and the corresponding dark field electron micrographs are shown in figures 11 and 12 respectively. The precipitates lying within and adjacent to the grain boundaries were identified as MoSi2 and "virgin" Ta. The grain boundary broadening

734

indicated by the area D in figure 3 shows the mobilization of MoSi2 precipitates to another subgrain.

DISCUSSIONS

The results obtained from an examination of specimens A through C indicate the following. The broadening of the grain boundary in figures 10 and 11 is due to the presence of dislocation lines within the grain boundary of the annealed P/M material. It is recognized that the familiar dislocation-like structure is not readily apparent in figures 10 and 11, however their presence is readily confirmed for the P/M sample shown in figure 6. These dislocations are usually called dislocation ledges.

This suggests that substantial deformation of the grain and subgrain boundaries takes place when dislocation lines are trapped within the grain boundaries. It has been confirmed [10] that a high dislocation density within the grain boundary causes extensive strain effects.

For P/M tantalum samples microalloyed with Mo and Si,diffusion of the intermetallic compounds, MoSi2 and/or TaSi2 through the grain boundaries produces broadening because of the strain caused by these precipitates. The dark field electron micrograph shown in figure 12 confirms the presence of MoSi2 within the grain boundary and the grain interior.

It was observed for tantalum containing only silicon that the average subgrain size was on the order of 0.4 microns. For tantalum containing only molybdenum, the average grain size was on the order of 1 to 3 microns. Thus the presence of silicon helps to form a fine subgrain structure exhibiting good ductility.

The presence of silicon induces dislocations emanating from the grain boundary ledges and form a number of three-fold dislocation nodes believed responsible for the high ductility found in this material10. Patches of TaSi2 and other intermetallic compounds near the grain boundary indicate a higher annealing temperature is necessary to promote improved solid solution strengthening in this material. However when molybdenum is present in combination with silicon, these patches are not observed. The sample containing no silicon exhibits no such subgrain structure.

735

The ribbon-like structure is believed to be responsible for the higher ductility observed in this material. Similar beneficial effects were also observed by the presence of oxides as dispersoid [5,11]. However the polygonized ribbon-like structure has been observed previously in rolled and annealed tantalum [12].

The detection of molybdenum disilicide inside the grain and the grain boundary is an important finding of this research work. Because MoSi2 has a very high melting point (2030 deg C) and has a very high resistance to oxidation (due to the formation of a protective silica layer at the high temperature),the exploitation of this intermetallic compound for use as a ductile material is worthwhile. MoSi2 behaves in a ductile manner above its transition temperature (1000 deg C) and has a tetragonal crystal structure. The properties of this material can be improved at lower temperatures by a suitable crystallographic transformation into a hexagonal material through judicious processing techniques. Twinning and slip dislocations can occur which substantially increase its ductile properties [13,14]. Although sample A contained Mo, it had less ductility than did sample C which contained silicon and molybdenum.

CONCLUSION

1) Sample C containing Si and Mo had extensively broadened subgrain boundaries. This feature is not found in sample B material which contained only silicon.

2) Dark field microscopy confirmed the presence of MoSi2 precipitates within the grain boundaries and in the grain interior.

3) The subgrain size (1-3 microns) in the material containing Mo (sample A) was larger than was the subgrain size (0.4 micron) for the sample containing silicon (sample B).

4) Ledge dislocations are much more clearly seen in sample B which contained no Mo.

5) The ribbon-like polygonized structure, present in sample C which contained Mo and Si, is believed to be responsible for its higher ductility.

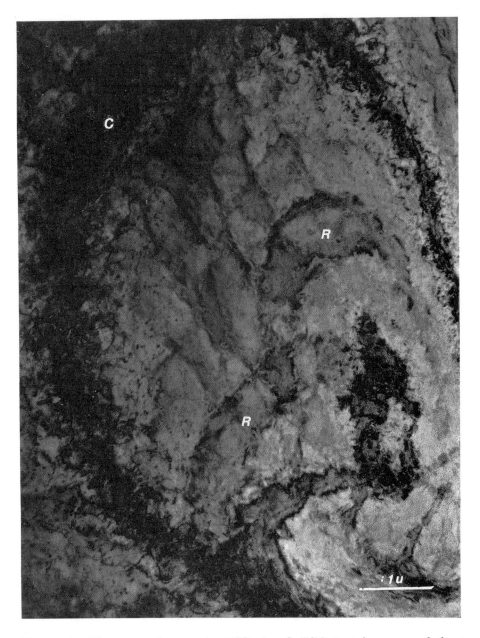

Figure 1: Electron micrograph (120kv) of P/M tantalum annealed at 1300°C shows a ribbon-like structure R and ben contour C. Magnification 210,000

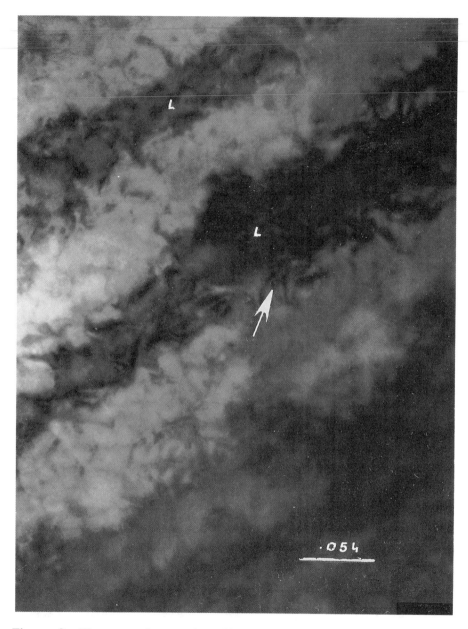

Figure 2: Electron micrograph (120kv) from the P/M tantalum as in figure 1 shows arrays of low angle grain boundaries containing numerous dislocation nodes as indicated by an arrow. Magnification 460,000

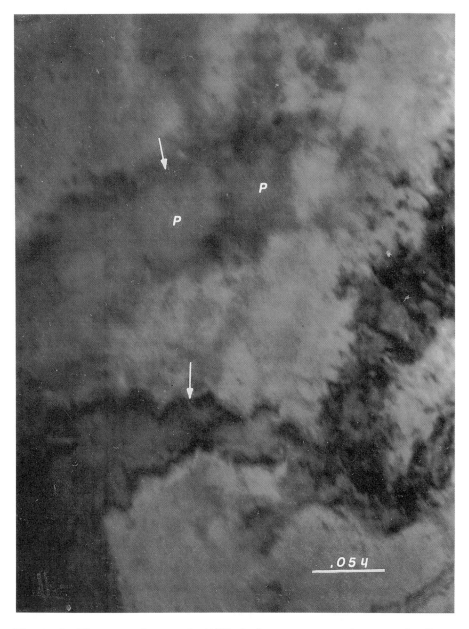

Figure 3: Electron micrograph (120kv) from same specimen as in figure 1 shows the formation of ribbon-like structure. Magnification 460,00

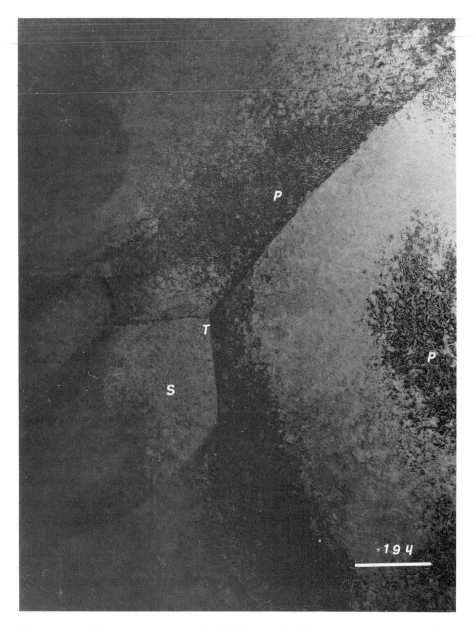

Figure 4: Electron micrograph (120kv) of P/M tantalum annealed at 1300°C (Si 400 ppm,), magnification 51,000 shows triple point grain boundary T, subgrain S and patches of precipitates of intermetallic compounds such as $TaSi_2$

Figure 5: Electron micrograph (120kv) of P/M tantalum sample as mentioned in figure 4, magnification 460,000 shows interfacial dislocation fringes D and strained grain boundary.

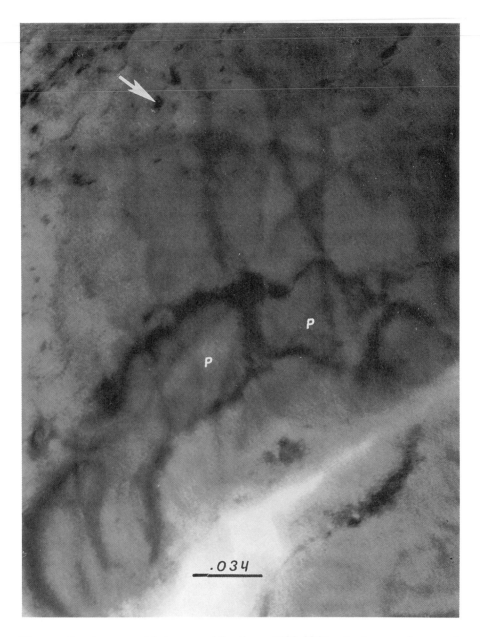

Figure 6: A very high magnification (600,000) electron micrograph from the sample mentioned in figure 4 showing dislocation nodes to produce a bent polygonized structure P.

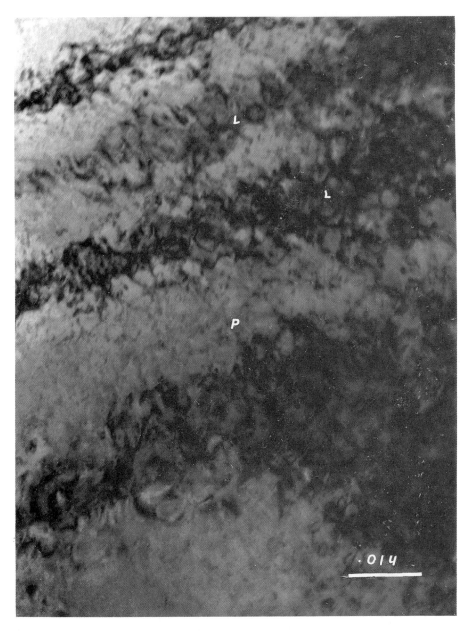

Figure 7: Another very high magnification (1,540,000) electron micrograph from the same sample in figure 4 shows an array of low angle grain boundaries L.

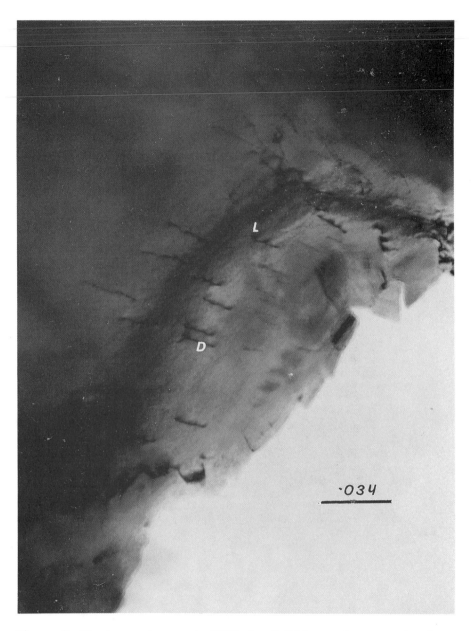

Figure 8: Electron micrograph (120kv) of P/M tantalum sample shows grain boundary ledges L and dislocation D, same sample in figure 4 but different area, magnification 125,000

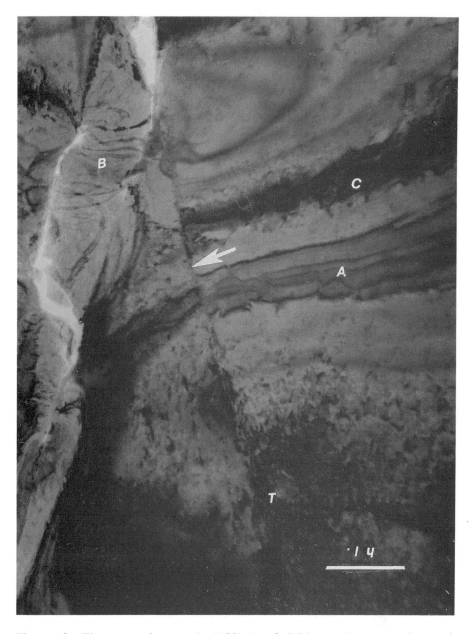

Figure 9: Electron micrograph (120kv) of P/M tantalum sample as in figure 4, magnification 172,000 shows large grain boundary indicated by an arrow meeting a subgrain at T, A, B and C, showing extinction and bend contours.

Figure 10: Electron micrograph (120kv) of P/M tantalum annealed at 1300°C (Mo 200 ppm, Si 400 ppm,), magnification 37,000 shows subgrain broadening effect and precipitate P.

Figure 11: Bright field electron micrograph (120kv) of P/M tantalum annealed at 1300°C as mentioned in figure 10, showing selected area SA and arrows to identify the precipitates, magnification 75,000.

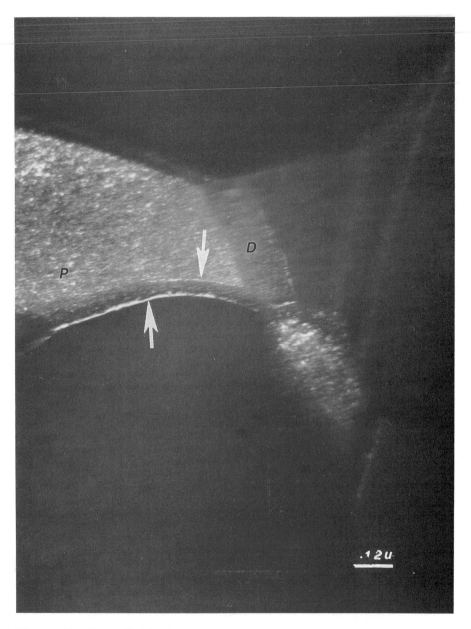

Figure 12: Dark field electron micrograph (120kv) showing $MoSi_2$ as precipitates on the grain boundary and grain interior.

ACKNOWLEDGMENTS

The authors acknowledge the assistance of Mr. Gil Barnes with the transmission electron microscopy work. Discussions with Dr. Charles Feng and Dr. Anthony J. Hickl were most helpful.

REFERENCES

1. H.G. Marsh and J.A. Pierret, " Embrittlement Resistant Tantalum Wire", U.S.Patent No. 4,062,679. December 13,1977.

2. H.G. Marsh and J.A. Pierret, "Method for Producing an Embrittlement - Resistant Tantalum Wire", U.S. Patent No. 4,235,629. November 25,1980.

3. C. Pokross, " Tantalum Microalloys",presented at the International Symposium on Tantalum and Niobium; sponsored by the Tantalum-Niobium International Study Center. November 1988.

4. B.A. Wilcox, " Basic Strengthening Mechanisms in Refractory Metals" pp. 1-39, " Refractory Metal Alloys". Editors: I. Machlin, R.T. Begley and E.D. Weisert. Plenum Press 1968.

5. D.L. Anton, D.B. Snow, L.H. Favrow and A.F. Giamei," Dispersion Strengthening of High Temperature Niobium Alloys", Report No. R89-917437- 3. United Technology Research Center, East Hartford, CT 06108, July 31,1989.

6. K.D. Moser, T.K. Chatterjee and P. Kumar," The Effects of Silicon on the Properties of Tantalum, J. of Metals 50-53, October 1989.

7. Lawrance E. Murr, Electron and Ion Microscopy and Microanalysis (New York: Marcel Dekker, Inc.) pp. 543-559

8. L.D. Kirkbride, J.A. Basmajian, D.R. Stoller, W.E. Ferguson, R.H. Perkins and D.N. Dunning," The Effect of Yttrium on the Recrystallization and Grain Growth of Tantalum"; J. of Less Common Metals, 393-408, Vol. 9, 1965.

9. P. B. Hirsch, R.B. Nicholson, A. Howie and D. W. Pashley, Electron Microscropy of Thin Crystals (Butterworths, 1965) p. 416

10. T. N. Lin and D. Mclean, "Changes Produced by Deformation in Grains and Grain Boundaries of Nickel", Metal Science Journal, Vol. 2, 1968, pp. 108-113.

11. C. T. Sims and W. C. Hagel, "The Superalloys", (John Wiley and Sons, 1972).

12. T. K. Chatterjee and C. Feng, "Defect Structure and Mechanical Twinning in Rolled and Annealed Tantalum", "Proc. of 45th Annual Meeting of the Electron Microscropy Society of America, ed G. W. Bailey" (San Francisco Press, Inc.)

13. A. E. Dwight, Trans. ASM 53, 479 (1961)

14. K. Inoue , T. Kuroda and K. Tachikawa, IEEE Trans. Mag. Mag-15, 635 (1979)

MICROSTRUCTURAL INVESTIGATION OF THE HEAT-AFFECTED ZONES IN NEWLY DEVELOPED ADVANCED AUSTENITIC STAINLESS STEELS

C. D. LUNDIN,[1] C. Y. P. QIAO[1] AND R. W. SWINDEMAN[2]

ABSTRACT

HAZ microstructures of a modified 316, standard 316 and a 17-14 CuMo weld were examined. A comparison study of HAZ liquation behavior was carried out by examining electric resistance spot welded specimens. Liquid distribution along liquated grain boundaries at elevated temperature is dependent on the wetting conditions. It was found that a liquid can easily cover the grain surface (liquid grain boundary region) at elevated temperature in modified 316. On the contrary, the liquid was preferentially located at the grain corners in 17-14 CuMo. A "Liquid pocket" phenomena was observed in the HAZ of standard 316 and 17-14 CuMo but not in the HAZ of modified 316. Different liquation behaviors are responsible for the various HAZ liquation cracking tendencies of the materials studied.

INTRODUCTION

Several advanced austenitic alloys were developed to meet the needs of modern power generating systems (650°C, 35 PMa) in the United States in recent years [1]. Oak Ridge National Laboratory (ORNL)

[1] Materials Science and Engineering Department, The University of Tennessee, Knoxville, TN 37996.

[2] Metals and Ceramics Division, Oak Ridge National Laboratory, Oak Ridge, TN 37831.

developed a modified 316 stainless steel as a candidate material for modern power generating systems. The improved elevated temperature creep and tensile strength of modified 316 were attained mainly due to compositional modification and application of a thermal-mechanical treatment technique. Carbide forming elements Ti, Nb and V were added and carbon content was increased in the modified 316 in order to form fine MC type carbides. The creep strength and high temperature tensile strength were significantly increased due to the interaction between the uniformly distributed MC type carbides and the dislocations in the matrix. It has been proven that the modified 316 possesses excellent high temperature creep and tensile properties [2]. However, the results of weldability evaluations indicate that a compositional adjustment is required in order to mitigate a HAZ liquation cracking tendency [3].

HAZ liquation behavior is a significant factor which affects HAZ liquation cracking susceptibility. The wetting action between liquated grain boundary and grain matrix at elevated temperature plays an important role in HAZ liquation formation. Quantitative investigations of precipitates dissolution and re-distribution in the HAZ of modified 316 was carried out by Lundin et al. [4]. It was determined that precipitate dissolution and re-distribution enhances HAZ liquation cracking tendency. As compared to standard 316 stainless steels, the modified 316 showed a higher HAZ liquation cracking tendency. Preliminary microstructural evaluation was conducted and it was found that migrated grain boundaries combined with grain boundary precipitation was responsible for the higher HAZ liquation cracking tendency of modified 316 [4-5].

Electric resistance spot welding was utilized to create a high HAZ cooling rate. Because electric resistance welding provided a relatively high cooling rate (compared to SMAW and GTAW), diffusional processes in the HAZ adjacent to the fusion zone are limited as compared to the GTA welding process. Thus, the microstructural and compositional changes which occur between elevated temperature and room temperature are minimized. A comparison of HAZ liquation behavior of the spot welded specimens of modified 316, conventional 316 and 17-14 CuMo was accomplished. Microstructural investigations of spot welded specimens provided a means to reveal HAZ liquation and thus HAZ liquation cracking behavior.

The base metal of modified 316 should possess a fully austenitic cold worked microstructure according to the basic alloy design tenets. Therefore, recrystallization and grain growth phenomena are expected to play a role in modified 316. Further, the segregation of S and P at the grain boundaries in the HAZ of modified 316 may be significant since the solubility of S and P in an austenitic microstructure is small. In addition, the interaction of grain growth and precipitates redistribution increases S and P segregation along grain boundaries in the HAZ.

HAZ cracks were found in GTA welds of modified 316 stainless steel. Microstructural investigations [4] have indicated that intergranular cracking in modified 316 occurred in a liquation related crack. Segregation of S and P along the grain boundaries and intergranular carbide precipitation was detected in the HAZ. It is a generally recognized that the S and P contents should be reduced in order to minimize the HAZ liquation propensity since S and P are more critical in regard to HAZ liquation cracking with fully austenitic solidification mode materials as compared to ferritic solidification mode austenitic stainless steels.

MATERIALS

The materials used in this study included one heat of modified 316, one heat of standard 316 and one heat of 17-14 CuMo. The composition of three materials is listed in Table 1. It is noted that the modified 316 heat studied contains relatively high P and S contrasted to conventional 316 stainless steel.

RESULTS AND DISCUSSIONS

The general HAZ microstructures of GTA welds in modified 316, standard 316 and 17-14 CuMo stainless steel are shown in Figures 1 to 3, respectively. It is evident that the coarse grained zone in the HAZ of the modified 316 and standard 316 is larger as contrasted to 17-14 CuMo. This is mainly due to the different cold worked conditions of the base material. Therefore, the migrated grain boundary phenomenon is more evident in modified 316 and standard 316. It is also noted that different HAZ liquation behaviors were observed among the three materials. Wetting between liquated grain boundaries and the grain matrix is considered an important factor affecting HAZ liquation cracking tendency. According to previous research [5], the intergranular rupture tendency is minimized when a liquid remains at the corners of the grains and on the contrary, HAZ liquation cracking tendency is the greatest if a grain boundaries are completely wetted by the liquid at elevated temperature. Wetting between an intergranular liquid and the grain matrix was observed in the HAZ of modified 316 as compared to 17-14 CuMo. Wetting of the grain boundary liquid with

753

Table 1. Chemical Composition of materials studied

Alloy Element	Modified 316 AX6/CE3890	Standard 316	17-14 CuMo
C	0.079	0.057	0.098
Si	0.27	0.58	0.95
Mn	1.77	1.86	0.83
P	0.040	0.024	0.014
S	0.010	0.019	0.005
Cr	14.29	17.25	16.50
Ni	16.87	13.48	13.80
Mo	2.27	2.34	1.69
Nb	0.10	<0.01	0.45
N	0.012	0.03	0.025
Cu	0.02	0.1	3.07
Al	0.006	0.023	-
Ti	0.21	0.02	0.21
Co	<0.01	0.02	-
V	0.52	-	0.07
B	0.006	0.005	0.005
Fe	Bal.	Bal.	Bal.

Figure 1. HAZ Microstructure of the GTA Weld in Modified 316.

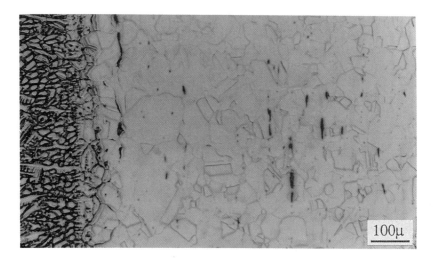

Figure 2.   HAZ Microstructure of the GTA Weld in Reference 316.

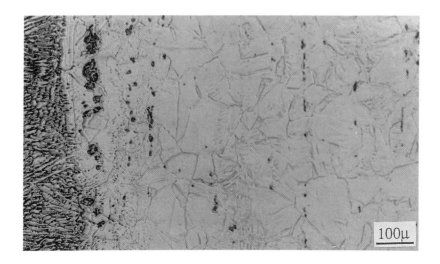

Figure 3.   HAZ Microstructure of the GTA Weld in 17-14 CuMo.

the grain matrix in standard 316 is intermediate among the three materials.

The morphologies of the HAZ microstructures of modified 316, standard 316 and 17-14 CuMo are shown in Figures 4 to 6. The so-called "liquid pocket" phenomenon is observed in the HAZ of 17-14 CuMo. The "liquid pocket" was observed to be located both intergranularly and intragranularly. An eutectic microstructural constituent was found inside the "liquid pockets". A beneficial influence of the "liquid pocket" phenomena on improving HAZ liquation cracking resistance is in evidence. Further discussion of the "liquid pocket" will be presented later.

The general OLM morphologies of the HAZ microstructural constituents in the resistance spot welded samples of modified 316, standard 316 and 17-14 CuMo are shown in Figures 7 to 9, respectively. As explained above, more significant liquation was found in the HAZ of resistance spot welds than in GTA welds for all three materials. The higher magnification constituent morphology in all three materials is shown in Figures 10 to 12, respectively. An appreciable "liquid pocket" (both intergranular and intragranular) in the HAZ of 17-14 CuMo was observed while no 'liquid pocket" was found in the HAZ of modified 316. The "liquid pocket" behavior was also observed in the HAZ of standard 316. However, the shape and amount of the "liquid pocket" constituent in standard 316 is different than that in 17-14 CuMo.

The SEM constituent morphologies of the HAZ of resistance spot welds in the three materials studied are shown in Figures 13 to 15. Both secondary electron imaging and backscattered electron imaging were employed in order to understand the segregation behavior in the liquated regions. Severe element segregation exists in the "liquid pocket" region in 17-14 CuMo according to secondary electron imaging micrograph (see Figure 15). Energy dispersive spectrum (EDS) analysis results indicated that the light phase in the "liquid pocket" is rich in Mo and Nb. The results from EDS analysis also revealed enrichment of S and P in the "liquid pocket" region. Based upon conventional solidification cracking theories, segregation of S and P along the grain boundary will significantly assist grain boundary (or cell boundary) rupture at elevated temperature. However, the S and P contents are concentrated in the "liquid pocket" instead of along the grain boundaries in 17-14 CuMo. This is apparently the reason that 17-14 CuMo showed a lower HAZ liquation cracking tendency. On the contrary, S and P are segregated along the grain boundaries in the HAZ of modified 316 stainless steel. Thus, modified 316 showed a higher HAZ liquation cracking tendency.

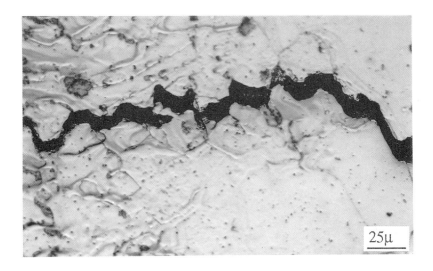

Figure 4. Liquated Region in the GTA Weld HAZ of Modified 316.

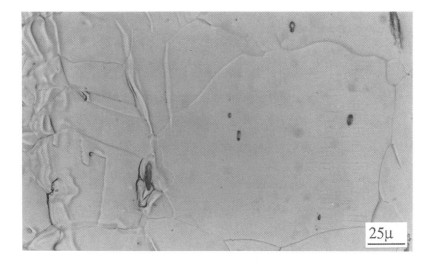

Figure 5. Liquated Region in the GTA Weld HAZ of Reference 316.

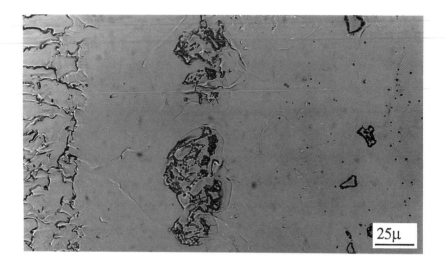

Figure 6.   Liquated Region in the GTA Weld HAZ of 17-14 CuMo.

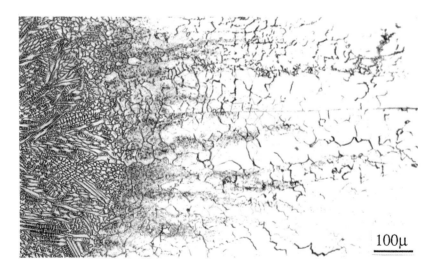

Figure 7.   OLM Morphology of HAZ Microstructure of Resistance Spot Weld in Modified 316.

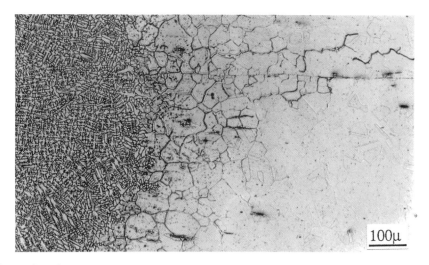

Figure 8. OLM Morphology of HAZ Microstructure of Resistance Spot Weld in Reference 316.

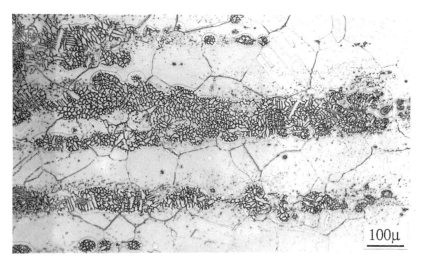

Figure 9. OLM Morphology of HAZ Microstructure of Resistance Spot Weld in 17-14 CuMo.

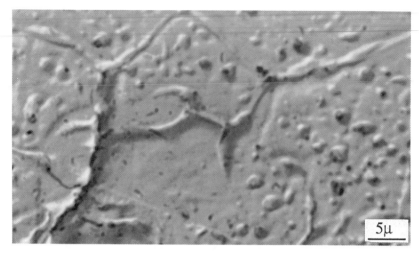

Figure 10.   OLM Morphology of Liquated Grain Boundary in HAZ of Modified 316.

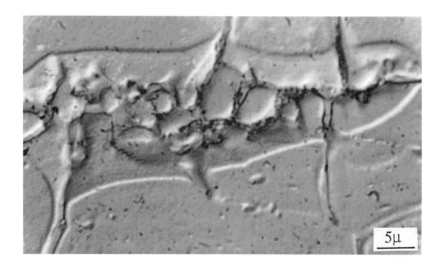

Figure 11.   OLM Morphology of Liquated Region in HAZ of Reference 316.

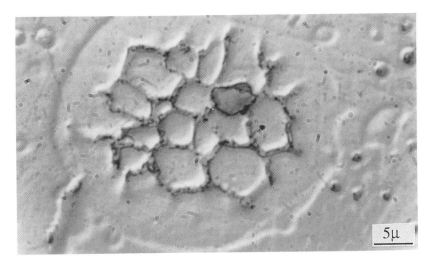

Figure 12. OLM Morphology of Liquated Region in HAZ of 17-14 CuMo.

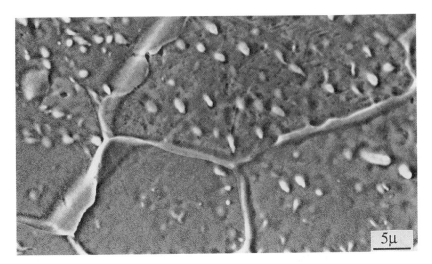

Figure 13. SEM Morphology of HAZ Microstructure of Resistance Spot Weld in Modified 316.

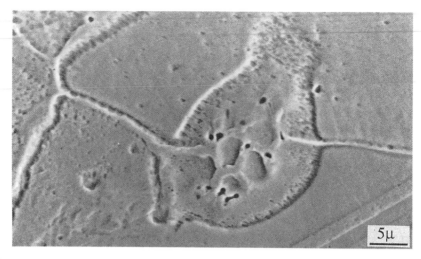

Figure 14. SEM Morphology of HAZ Microstructure of Resistance Spot Weld in Reference 316.

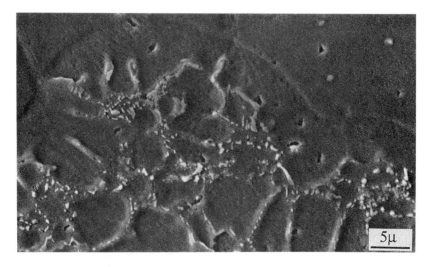

Figure 15. SEM Morphology of HAZ Microstructure of Resistance Spot Weld in 17-14 CuMo.

Constitutional liquation behavior in the weld HAZ of 18-Ni maraging steel weldments was addressed by Pepe and Savage in 1967 [7]. The fundamental theory proposed can be used to explain the behavior in the HAZ in the advanced austenitic stainless steels although the compositions of the materials are different. According to Pepe and Savage, the amount of constitutionally liquated materials in HAZ increased as the heating rate and cooling rate increased. This agrees with the results obtained in present work. A more extensive liquation region can be observed in resistance spot weld samples than in GTA weld samples. Analysis of liquation behavior is enhanced by the use of resistance spot welds since the liquated regions are "retained" to ambient temperature.

Figures 16 and 17 show the SEM constituent morphologies in the HAZ in the resistance spot welded specimen of modified 316 (secondary imaging and backscattered imaging, respectively). The center of the liquated grain boundary appears lighter than the matrix in the backscattered imaging. At higher magnification the morphology of the grain boundary liquation in the HAZ of modified 316 is shown in Figures 18 and 19, respectively. A dark region was clearly observed adjacent to the liquated grain boundary in this sample. The dark regions are the result of grain boundary migration. The grain boundary migration caused by grain growth has a so-called "sweeping up" effect which can increase segregation of the alloying elements (such as Mo and Nb) and trace elements (such as P and S) along the grain boundary. Thus, liquation more readily occurs along the segregated grain boundary since the segregated grain boundary usually has a lower melting point than the bulk melting temperature.

Figures 20 and 21 show the SEM constituent morphologies in the HAZ microstructure of a resistance spot weld sample of 17-14 CuMo stainless steel with secondary imaging and backscattered imaging, respectively. A clear morphology of the "liquid pocket" is shown in these figures. It is considered that the eutectic reaction takes place during "liquid pocket" solidification in the on-cooling portion of welding thermal cycle. A higher magnification morphology of location A in Figures 20 and 21 is indicated in Figures 22 and 23, respectively. As indicated in these figures, a cellular type structure is observed in the "liquid pocket". EDS analysis indicates that the particles presented are Nb-rich. It is also revealed that the trace elements P and S are concentrated in these locations.

Lundin et al. [8] summarized the effect of Nb on the HAZ liquation cracking tendency for nuclear grade and conventional stainless steels. They pointed out that Nb-bearing stainless steels generally show a high HAZ liquation tendency. The 17-14 CuMo heat used in this study has

763

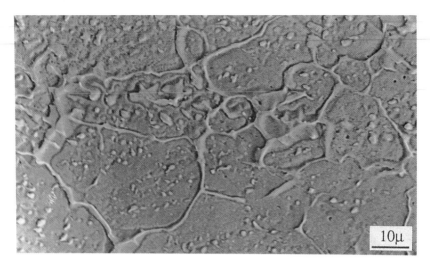

Figure 16. SEM Secondary Imaging of Liquated Grain Boundary Region in the HAZ of Resistance Spot Weld of Modified 316.

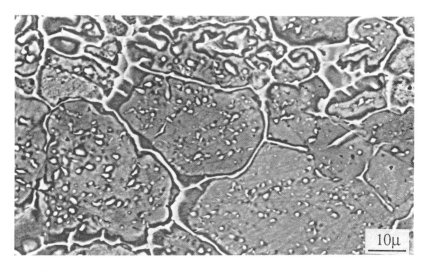

Figure 17. SEM Backscattered Imaging of Liquated Grain Boundary Region in the HAZ of Resistance Spot Weld of Modified 316.

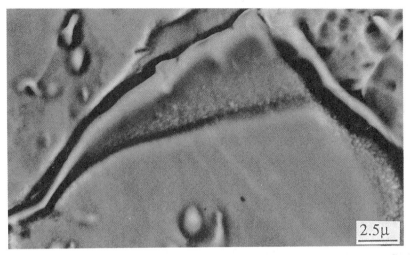

Figure 18. Higher Magnification of Secondary Imaging of Liquated Grain Boundary Liquation.

Figure 19. Higher Magnification of Backscattered Imaging of Liquated Grain Boundary Liquation.

765

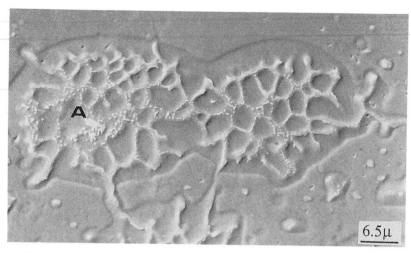

Figure 20. SEM Secondary Imaging of Liquated Grain Boundary Region in the HAZ of Resistance Spot Weld of 17-14 CuMo.

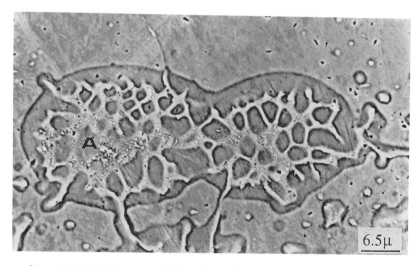

Figure 21. SEM Backscattered Imaging of Liquated Grain Boundary Region in the HAZ of Resistance Spot Weld of 17-14 CuMo.

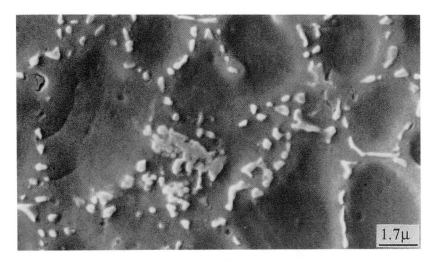

Figure 22. An Enlarged Morphology of Location A in Figure 20.

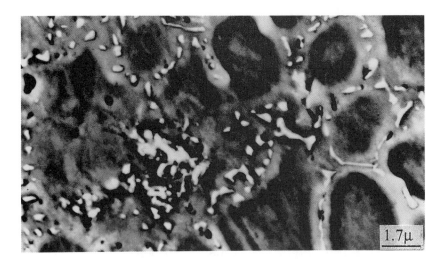

Figure 23. An Enlarged Morphology of Location A in Figure 21.

an appreciable higher Nb content than modified 316 and standard 316. However, the 17-14 CoMo has a slightly lower HAZ liquation cracking tendency than modified 316 [6]. The explanation is that modified 316 contains a higher P and S content than the 17-14 CuMo.

Different wetting behavior (between liquid and matrix) at elevated temperature is also considered to have an important influence on HAZ liquation cracking tendencies. It is evident that the liquid was preferentially located at the corners of grain boundaries at elevated temperature in 17-14 CuMo. The surface tension between the liquid and solid is relatively small at elevated temperature in the HAZ liquation region of 17-14 CuMo. The HAZ liquation cracking tendency is reduced due to limited liquid distribution at elevated temperature.

Figures 24 and 25 show the liquated grain boundary morphologies with secondary electron imaging and backscattered electron imaging in standard 316, respectively. The grain boundary wetting condition in standard 316 is intermediate among the three materials investigated. The "liquid pocket" phenomenon was observed in standard 316, however, it has a different morphology than that seen in 17-14 CuMo. Isolated voids were observed in the liquated grain boundary in standard 316 as indicated in this figure. The voids were created by volume contraction during liquid solidification. As mentioned above, standard 316 possesses a relative high HAZ liquation cracking resistance. Many factors contribute to the favorable characteristics in standard 316. The authors believe that the liquid distribution at elevated temperature is one of the most important factors that influence the behavior of standard 316.

SUMMARY

1. Different HAZ liquation behaviors were observed in modified 316, standard 316 and 17-14 CuMo stainless steel. The different HAZ liquation cracking tendencies of the three materials studied are directly related to the liquation behavior.

2. A so-called "liquid pocket" was observed in the HAZ of standard 316 and 17-14 CuMo, but not in the HAZ of modified 316. The existence of the "liquid pocket" improves of HAZ liquation cracking resistance.

3. Wetting conditions (liquid and grain matrix along the liquated grain boundary) control liquid distribution at elevated temperature. If the grain surface of a liquated HAZ boundary is completely wetted by liquid, HAZ liquation cracking is enhanced.

4. S and P are preferentially segregated in the "liquid pocket", and further reducing HAZ liquation cracking tendency.

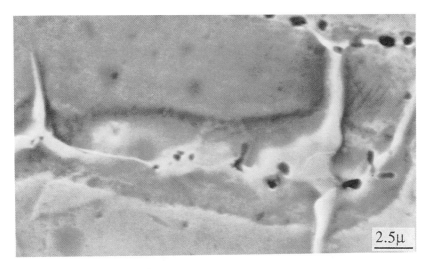

Figure 24. SEM Secondary Imaging of Liquated Grain Boundary Region in the HAZ of Resistance Spot Weld of Reference 316.

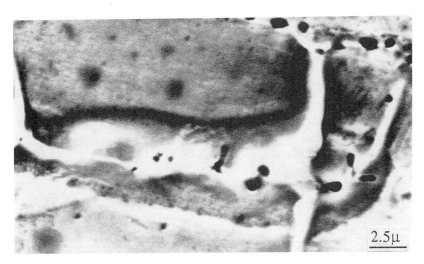

Figure 25. SEM Backscattered Imaging of Liquated Grain Boundary Region in the HAZ of Resistance Spot Weld of Reference 316.

769

ACKNOWLEDGEMENT

This work was part of an investigation of "The Joining and Welding Technology of Modified Austenitic Alloys" sponsored by the U. S. Department of Energy, AR&TD Fossil Energy Materials Program, DOE/FE AA 15 10 10 0, Work Breakdown Structure Element UTN-2 operated by Martin Marietta Energy Systems, Inc. at Oak Ridge National laboratory.

REFERENCES

1. R. W. Swindeman, P. J. Maziasz, E. Bolling and J. F. King, "Evaluation of Advanced Austenitic Alloys Relative to Alloy Design Criteria for Steam Service: Part 1 - Lean Stainless Steel," ORNL Technical Report, ORNL-6629/P1, May 1990.
2. P. J. Maziasz, "Developing an Austenitic Stainless Steel for Improved Performance in Advanced Fossil Power Facilities," JOM, pp. 14-20, July 1989.
3. C. D. Lundin, C. Y. P. Qiao, T. P. S. Gill and W. Ren, "The Hot Ductility and Hot Cracking Behavior of Modified 316 Stainless Steels Designed for High Temperature Service," Technical Report, ORNL/Sub/88-07685/01, July 1989.
4. C. D. Lundin, C. Y. P. Qiao, G. M. Goodwin and R. W. Swindeman, "HAZ Liquation Cracking Behavior in Newly Developed Lean 316 Stainless Steels," Proceedings of the Conference on New Alloys for Pressure Vessels and Piping, PVP-Vol. 201, MPC Vol. 31, pp. 155-163, edited by M. Prager and C. Cantzlar, June 1990.
5. C. D. Lundin and C. Y. P. Qiao, "HAZ Characterization in Modified 316 Stainless Steels," Proceedings of the Forth Anneal Conference on Fossil Energy Materials, May 1991.
6. C. D. Lundin, C. Y. P. Qiao and Y. Kikuchi, "Intergranular Liquation Cracking in An Advanced Austenitic Stainless Steel," will be published in Proceedings of Conference on Heat-Resistant Materials, August 1991.
7. J. J. Pepe and W. F. Savage, "Effect of Constitutional Liquation in 18-Ni Maraging Steel Weldments," Welding Journal, pp. s1-s12, September 1967.
8. C. D. Lundin, C. H. Lee and C. Y. P. Qiao, "Weldability and Hot Ductility Behavior of Nuclear Grade Austenitic Stainless Steels," Final Report of a Group Sponsored Study on weldability and hot ductility behavior of nuclear grade austenitic stainless steels, The University of Tennessee, December, 1988.

CREEP AND METALLOGRAPHY OF LOW COLUMBIUM

MODIFIED 9Cr-1Mo SHIELDED METAL ARC WELD METAL

F. V. Ellis*

ABSTRACT

Modified 9Cr-1Mo steel is a ferritic alloy possessing good elevated temperature strength. Currently, no "matching composition" specification has been adopted for either bare or coated electrodes. In order to achieve good impact toughness values, the Cb level of the as-deposited coated electrode welds are generally in the range of 0.018 to 0.030% Cb, which is well below the specified range of 0.06 to 0.10% Cb for the base material. Because Cb is a potent creep strengthener, there is some concern that the low Cb welds will have inadequate creep strength. Low Cb modified 9Cr-1Mo shielded metal arc welds were made in a pre-buttered 55 mm thick plate to evaluate this concern. Chemical analysis, microstructural characterization, tensile, Charpy and creep rupture testing was performed. The microstructure for the stress-relieved low Cb weld consisted of tempered martensite with an average hardness of approximately 98 HRB. The creep rupture strength for the low Cb welds was greater than mean for the wrought modified 9Cr-1Mo base material. The creep ductility was lower for the weld metal than for the wrought base material and an intergranular fracture mode was observed at longer test times.

*   Tordonato Energy Consultants, Inc.
     4156 S. Creek Road, Chattanooga, TN 37406

INTRODUCTION

Modified 9Cr-1Mo steel, or Grade 91, is a ferritic alloy possessing attractive mechanical, physical, and corrosion properties which have resulted in increasing applications in the petrochemical and utility industries. The elevated temperature strength for modified 9Cr-1Mo steel is comparable to that for TP304H stainless steel. This superior strength is achieved by controlled microalloying additions of Cb, V, and N, coupled with a normalizing and tempering heat treatment. Extensive testing has been performed on modified 9Cr-1Mo alloy, and its metallurgical and mechanical properties are fully documented [1].

Weldability studies have found no major concerns with joining Grade 91 steel, provided low hydrogen welding consumables and proper welding practices are used. However, modified 9Cr-1Mo steel is susceptible to HAZ cracking in the "soft" zone [2,3] which can lead to premature stress rupture failures under conditions of tensile loading normal to the weld fusion interface. Although initial laboratory studies indicated that hot-cracking was not a special problem with the Grade 91 alloy, subsequent weld qualification tests for thick-section welds have shown that for certain heats of filler metal hot-cracking may occur [4].

With regard to welding consumables, no "matching composition" specifications have yet been adopted by the American Welding Society (AWS) for either bare wire or coated electrodes. Most of the developmental effort for weld consumables has been dominated by a desire to achieve satisfactory impact properties with a goal of 68 Joules absorbed Charpy V-notch energy at room temperature. In this regard, it has been demonstrated that low oxygen content in the deposited weld and extended PWHT is beneficial [5]. Achieving weld metal tensile strength levels which fulfill the minimum requirements specified for wrought material has not been difficult. For coated electrodes, a major developmental effort demonstrated that it was necessary to go below the 0.06 - 0.10 percent Cb level to achieve the desired value of impact toughness; a 0.018 - 0.030 Cb range has been suggested for the deposited weld metal [6]. Because Cb is a potent creep strengthener, there is some concern that low Cb content welds will have inadequate creep strength. The purpose of the current study was to determine the creep strength and ductility of low Cb modified 9Cr-1Mo shielded metal arc weld metal.

MODIFIED 9Cr-1Mo WELD MATERIAL

In order to obtain sufficient undiluted weld metal for the Charpy, tensile, and creep testing programs, a shielded metal arc weld was made in the 1G weld position in a 55 mm thick Grade 91 plate (pre-buttered with a 10 mm thickness of low Cb weld metal) with a 40 mm wide groove. Figure 1 is the photomacrograph of the low Cb modified 9Cr-1Mo shielded metal arc weldment showing the Grade 91 base plate with the pre-buttered weld metal, the all-weld metal backing bar, and the weld nugget. The welding conditions were 26 volts, 120 amps, and a 3.2 mm diameter welding electrode. The preheat temperature was 204°C and the maximum interpass temperature was 343°C. The post weld heat treatment was extended considering the plate thickness and consisted of 4 hours at 749°C. The weld was made with the as-deposited target Cb content of less than .03 percent. Table 1 compares the measured chemical composition for the test weld and the specification for Grade 91 steel. The SMA weld metal is within specified ranges for all elements except the Cb content which was below the .03 percent level as desired for improved Charpy properties.

TABLE 1. Chemical Composition of Modified 9Cr-1Mo SMA Weld Metal.

	Weld Metal	ASME Gr. 91
C	0.095	.08 - .12
Mn	0.53	.30 - .60
P	0.007	.020*
S	0.011	.010*
Si	0.29	.20 - .50
Ni	0.08	.40*
Cr	8.38	8.0 - 9.5
Mo	0.92	.85 - 1.05
V	0.192	.18 - .25
Cb	0.029	.06 - .10
Cu	0.06	.04*
N	0.030	.030 - .070
O	0.040	

*Maximum

773

Figure 1. Photomacrograph of low Cb modified 9Cr-1Mo shielded metal arc weldment.

The microstructure of the low Cb modified 9Cr-1Mo SMA weld metal was fully tempered martensite (Figure 2) with an average hardness of 98 on the Rockwell B scale. The suggested etchants for Grade 91 include Vilella's reagent, super picral [1] and a 2.5% picric acid, and 2.5% HCl in alcohol mixture [7]. Good results can be achieved using any of these. The usual microstructure for modified 9Cr-1Mo steel is fully martensitic. However, the specified chemical composition range for the alloy is such that delta ferrite can form for some compositions. Kent [7] has studied the effect of weld metal composition on the potential for delta ferrite formation in 9Cr-1Mo and modified 9Cr-1Mo SMA welds using the Kaltenhauser ferrite factor. His results indicate that a fully martensitic microstructure would be expected for a ferrite factor of 7.5 or below. For the low Cb modified 9Cr-1Mo SMA weld metal, the calculated ferrite factor was 7.7 indicating that the delta ferrite content would be expected to be nil.

The yield and tensile strength of the low Cb weld metal was 612 MPa and 727 MPa, respectively. The room temperature tensile elongation was 20 percent (50.4 mm gage length). For the low Cb weld metal, the average Charpy absorbed energy at room temperature was 31 Joules and the upper shelf Charpy energy was 140 Joules at 121°C.

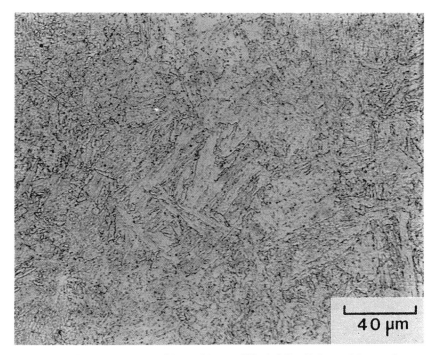

Figure 2.   Microstructure of low Cb modified 9Cr-1Mo weld metal.

CREEP PROPERTIES AND METALLOGRAPHY

The creep rupture test specimen had a 25.4 mm gage length and a 6.4 mm gage diameter.   The gage length was entirely in the undiluted low Cb weld metal and was oriented transverse to the welding direction.   The test specimens had machined ridges for extensometer attachment and threaded ends.   The test temperatures were 650°C, 675°C and 700°C with the test stresses from 65 MPa to 120 MPa selected to allow for time-temperature parameter correlation of the creep and rupture data.   Complete strain-time curves were measured using high resolution capacitance transducers and a computerized data acquisition system.

Figure 3 compares the observed rupture properties to a heat centered regression developed by ORNL [8] to describe the rupture properties of wrought modified 9Cr-1Mo base material given as,

$$\log t_r = C_h - 0.0231\sigma - 2.385 \log \sigma + 31,080/T \quad (1)$$

775

where t_r is the rupture time in hours, σ is the stress in MPa, T is the absolute temperature in °K, and C_h is the heat constant. The heat constant is a measure of the relative strength of the different heats of material based on the assumption of the regression model, that the stress and temperature dependence is the same for all heats. For the wrought base material, the heat constant for an average strength heat of material is -23.737. For the low Cb modified 9Cr-1Mo weld metal, the value for C_h was -23.535 and was used to determine the calculated life shown in Figure 3. The difference in these two heat constants is +.202 which is equal to a difference in log rupture time for the low Cb weld metal and the average wrought base material. In terms of rupture time, the weld metal rupture time is approximately 1.6 times that for the average strength wrought base material. In essence, this comparison shows that the low Cb weld metal has a higher creep rupture strength than that for an average heat of base material.

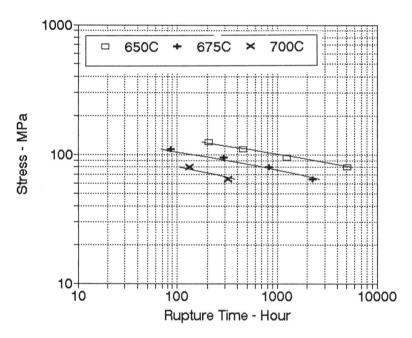

Figure 3. Stress rupture properties for low Cb modified 9Cr-1Mo weld metal.

The rupture elongations for the low Cb modified 9Cr-1Mo weld metal as a function of rupture time are shown in Figure 4. In general, these weld metal creep rupture elongations are less than those found for the Grade 91 wrought base material. At 650°C, the elongations are less than 10 percent for the longer term tests. Figure 5 shows the intergranular fracture mode found in the creep specimen tested at 80 MPa and 650°C.

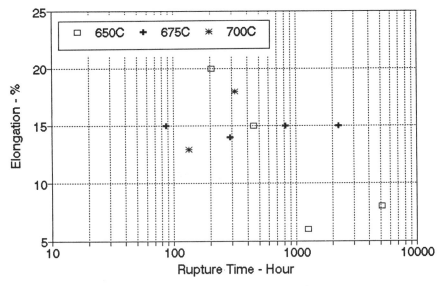

Figure 4. Creep rupture elongation versus rupture time.

The minimum creep rate was found by numerical differentiation of the strain versus time data. Figure 6 shows the observed minimum creep rate as a function of stress and temperature. The solid line shown in Figure 6 was calculated using the ORNL [8] heat centered regression for the wrought base material given as,

$$\log \dot{\epsilon}_m = C_h + 0.02549\ \sigma + 2.974 \log \sigma - 35{,}180/T \quad (2)$$

where $\dot{\epsilon}_m$ is the minimum creep rate in %/hour, σ is the stress in MPa, T is the absolute temperature in °K, and C_h is the heat constant. The ORNL relation was developed for temperatures from 600°C to 650°C and the average heat constant value was 27.391. For the low Cb weld metal, the temperature range was 650°C to 700°C and the average heat constant value was 27.801. Using these two heat constants, the minimum creep rate for the low Cb weld metal would be calculated to be approximately 1/4 of that for an average strength heat of wrought base material. As was previously shown for the rupture time data, the minimum creep rate data comparison shows that the low Cb weld creep strength is greater than that for an average strength heat of wrought base material.

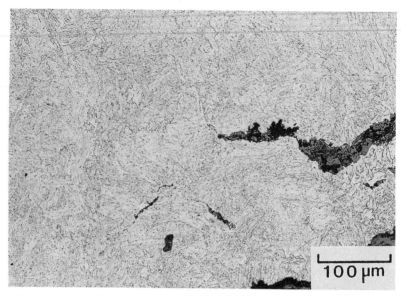

Figure 5. Intergranular cracking in low Cb modified 9Cr-1Mo weld metal creep rupture specimen tested at 650°C and 80 MPa.

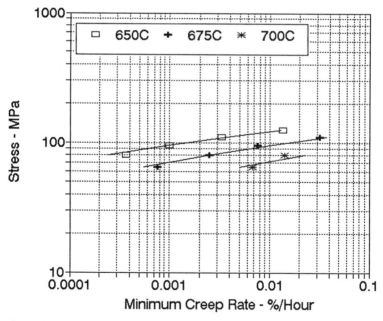

Figure 6. Minimum creep rate data for low Cb modified 9Cr-1Mo weld metal.

CONCLUSIONS

A low Cb (0.025% Cb) modified 9Cr-1Mo shielded metal arc weld was made in order to evaluate the effect of Cb content on microstructure and creep rupture properties. The microstructure for the low Cb weld was fully tempered martensite and had an average hardness of 98RB. The creep rupture time for the low Cb weld metal was approximately 1.6 times that for average strength wrought modified 9Cr-1Mo base material based on a heat-centered rupture data analysis. The minimum creep rate for the low Cb weld metal was approximately 1/4 of that for an average strength heat of wrought base material. The creep ductility for the low Cb weld metal was less than that for wrought base material. At 650°C, the rupture elongations for rupture times greater than about 1000 hours were less than 10 percent and an intergranular fracture mode was found at the longest test time.

ACKNOWLEDGEMENTS

This project was initiated by Mr. B. W. Roberts and Dr. D. A. Canonico and their guidance throughout the course of the work is gratefully acknowledged. The welding, tensile, Charpy, and creep rupture testing was performed as an internally funded research project at the ABB Combustion Engineering Metallurgical and Materials Laboratory.

REFERENCES

1. J. R. DiStefano, and V. K. Sikka, Summary of Modified 9Cr-1Mo Steel Development Program: 1975-1985, ORNL-6303, Oak Ridge National Laboratory, Oak Ridge, TN, 1986.
2. R. D. Townsend, "CEGB Experience and UK Developments in Materials for Advanced Plant," Advances in Material Technology for Fossil Power Plant, ASM International and EPRI, Materials Park, OH, 1987, p. 11.
3. B. W. Roberts and D. A. Canonico, "Candidate Uses for Modified 9Cr-1Mo Steel in an Improved Coal-Fired Power Plant," Conference Proceedings: First International Conference on Improved Coal-Fired Power Plants, EPRI Report CS-5581-SR, Electric Power Research Institute, Palo Alto, CA, 1988, p. 5.
4. F. V. Ellis, J. F. Henry, and B. W. Roberts, "Welding, Fabrication and Service Experience with Modified 9Cr-1Mo Steel," New Alloys for Pressure Vessels and Piping, ASME, NY, NY, 1990, p. 55.

5. H. Haneda, F. Masuyama, S. Kaneko, and T. Toyoda, "Fabrication and Characteristic Properties of Modified 9Cr-1Mo Steel for Header and Piping," Advances in Material Technology for Fossil Power Plant, ASM International and EPRI, 1987, p. 231.

6. J. F. King, V. K. Sikka, M. L. Santella, J. R. Turner, and E. W. Pickering, Weldability of Modified 9Cr-1Mo Steel, ORNL-6299, Oak Ridge National Laboratory, Oak Ridge, TN, 1986.

7. R. Panton-Kent, Unpublished Research, Welding Institute, Cambridge, United Kingdom.

8. C. R. Brinkman, V. K. Sikka, J. A. Horak, and M. L. Santella, "Long-Term Creep-Rupture Behavior of Modified 9Cr-1Mo Steel Base and Weldment Behavior," ORNL/TM-10504, Oak Ridge National Laboratory, Oak Ridge, TN, 1987.

APPLICATION OF OPTICAL METALLOGRAPHY TO CHARACTERIZE

UNIQUE FEATURES OF WELD MICROSTRUCTURES

S. A. David, M. J. Gardner, and J. M. Vitek

Oak Ridge National Laboratory
P.O. Box 2008
Oak Ridge, TN 37831-6095

In a weld, the microstructural characterization
of both the fusion zone (FZ) and the heat-affected
zone (HAZ) is critical in controlling the mechanical
behavior and hot-cracking tendencies of the weld.
Microstructures of both the FZ and the HAZ can be
found to be extremely complicated [1-5]. Significant gradations in composition and microstructure
are present in these regions. In the case of a
multipass weld, the multiple thermal cycles may
further add to the complexity of the microstructural
features and their characterization.

Optical metallography has been used extensively
and has played a critical role in characterizing and
understanding the development of microstructures in
weldments. This has been demonstrated using weldments of a wide variety of alloys. The materials
characterized include iridium alloys, austenitic
stainless steels, commercial and high-purity single
crystals, titanium alloys, and ferritic steels. The
experimental techniques used to characterize the FZ
and HAZ microstructures include special orthogonal

sectioning of weld specimens, weld simulation techniques, X-ray microradiography, and conventional metallographic techniques with special precautions in sectioning and etching to delineate the appropriate microstructural features.

In the FZ, epitaxial growth and the grain selection process for growth were demonstrated in an iridium alloy weld. Further, the affect of growth crystallography and the dendrite selection process on the development of the FZ microstructures was investigated by making welds on well-oriented single crystals of Fe-15Ni-15Cr alloy. Microstructural characterization of the single-crystal welds is critical to the understanding of the development of FZ microstructures. Also, it is essential that the dendritic substructure for various orientations be revealed with clarity so that the dendrite selection process can be understood. In austenitic stainless steel (type 308) welds, though the general duplex austenite (γ) + ferrite (δ) microstructures have been described and classified into various groups, it has not been possible to observe the clear and distinct ferrite forms. By orthogonal sectioning of volume elements from various locations in the weld and extensive characterization using optical metallographic techniques, four distinct types of ferrite morphologies have been identified. They are vermicular, lacy, acicular, and globular.

To understand and characterize the solidification behavior and solute segregation, it is essential to delineate the solidification substructure in welds. In α + β titanium alloy (Ti-6Al-1Nb-1Ta-0.8 Mo) welds, revealing the dendritic solidification substructure has been a challenge due to the solid-state transformation that masks the solidification substructure. Delineation of the solidification substructure in this alloy was made possible by X-ray microradiography. The chemical heterogeneity associated with the solute redistribution occurring during solidification was revealed by a variation in photographic blackening.

Characterization and understanding of the development of microstructures in the HAZ are also equally complicated. Due to the nature of the thermal excursions experienced by the HAZ, there is a significant gradation in composition and microstructural features. In combination with thermal simulations, optical microscopy has been used to identify the microstructural gradation in the HAZ of an Fe-3Cr-1.5Mo-0.1V ferritic steel. The microstructures ranged from banitic to bainite with large areas of ferrite to tempered bainite.

ACKNOWLEDGEMENT

The authors would like to thank C. P. Haltom, C. W. Houck, and G. C. Marsh for their help in metallographic analysis; J. F. King and M. L. Santella for reviewing the manuscript; and K. Gardner for preparing the manuscript. The research is sponsored by the Division of Materials Sciences, U. S. Department of Energy, under Contract DE-AC05-84OR21400 with Martin Marietta Energy Systems, Inc.

REFERENCES

1. S. A. David and J. M. Vitek, *Int. Mater. Rev.*, 34(5), 213 (1989).
2. S. Kou and Y. Le, *Metall. Trans.* **16A**, 1345 (1985).
3. T. Ganaha, B. P. Pearce, and H. W. Kerr, *Metall. Trans.* **11A**, 1351 (1980).
4. S. A. David, J. M. Vitek, M. Rappaz, and L. A. Boatner, *Metall. Trans. A*, **21A**, 1753 (1990).
5. J. M. Vitek and S. A. David, *Metall. Trans. A*, **21A**, 2021 (1990).

A MICROSTRUCTURAL EXAMINATION OF HOT CORROSION OF A Co-Cr-Fe ALLOY CAST BURNER NOZZLE FROM A COAL GASIFICATION PLANT

S. Cao[1], Charlie R. Brooks[1] and Gary Whittaker[2]

ABSTRACT

A microstructural analysis has been made of a burner nozzle removed from service in a coal gasification plant. The nozzle was a casting of a Co-29 wt. % Cr-19 wt.% Fe alloy. Extensive hot corrosion had occurred on the surface. There was penetration along grain boundaries, and corrosion products in these regions were particularly rich in S, and also contained Al, Si, O, and Cl. The grain boundaries contained Cr-rich particles which were probably $Cr_{23}C_6$ type carbides. In the matrix, corrosion occurred between the Widmanstatten e plates. Particles were found between these plates, most of which were rich in Cr and O, and probably were Cr_2O_3 oxides. Other matrix particles were found which were rich in Al, O and S. The corrosion was related to these grain boundary and matrix particles, which either produced a Cr-depleted zone around them or were themselves attacked.

---

1   Materials Science and Engineering Department and the Center for Materials Processing, The University of Tennessee, Knoxville, TN 37996 U.S.A.

2   Tennessee Eastman Corporation, Kingsport, TN 37662 U.S.A.

INTRODUCTION

In some coal gasification processes [1], the reactor burner nozzle is a crucial component. Through it coal slurry and oxygen are injected into the reaction chamber where they react exothermically producing high temperatures (i.e., up to 1500°C [2]). Thus this component is subjected to a corrosive environment involving oxidizing gases, and due to the impurities in the coal, reactive gases such as SO_2, H_2S and other species (e.g., Cl_2) [2-6]. In many cases, the corrosion products are liquid, and such corrosion is termed hot corrosion [7-8]. The materials used to withstand such an environment are usually alloys which have been successfully used in high temperature gas turbines, in which they are subjected to similar severe operating condition [9].

At Tennessee Eastman, some nozzles have been made of three alloys, as shown in Fig. 1. The region of the most severe attack is right at the tip, where a wrought plate of Co-Cr alloy (Stellite 188) was used; this tip can be replaced and the nozzle reused if the other components are not damaged. For the rest of the nozzle, a configuration was used which required two castings. In the top part, where temperatures are high, a cast Co-Cr-Fe alloy was used. In the lower portion, where the temperature is lower, Inconel 600 was used. These two castings were joined by welding, using Stellite 188 as a filler metal. Also these two cast parts had channels molded in them through which cooling water passed, so that only the outside was subjected to high temperatures.

The top casting was made of a Co-Cr-Fe alloy (typically (wt. %) 29 Cr, 19 Fe, 0.02 C, 0.7 Si, 0.5 Mn, 0.1 Ni and 0.006 S). Cr-Co alloys has been used in similar environments [10, 11]. However, the nozzles have shown a highly variable and unacceptable life, and a search was undertaken to explain the variation and to develop a material with acceptable operating life (e.g., 1000 hours). To assist in this procedure, a microstructural analysis was made of the cast components. This involved optical microscopy (OM), scanning electron microscopy (SEM), transmission electron microscopy (TEM) and scanning transmission electron microscopy (STEM). IN SEM and STEM, microchemical analyses were obtained of corrosion products and phases present using energy dispersive, fluorescent x-ray analysis spectrometry (EDS). In TEM and STEM, electron diffraction patterns were obtained from fine particles in the alloys. All of the microstructural information was used to attempt to identify the corrosion products and phases present.

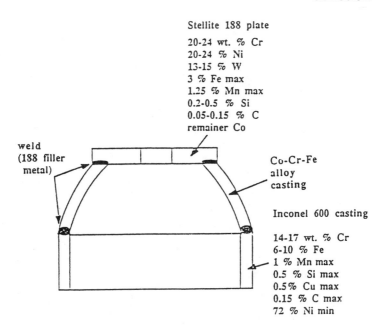

Stellite 188 plate

20-24 wt. % Cr
20-24 % Ni
13-15 % W
3 % Fe max
1.25 % Mn max
0.2-0.5 % Si
0.05-0.15 % C
remainer Co

weld
(188 filler
metal)

Co-Cr-Fe
alloy
casting

Inconel 600 casting

14-17 wt. % Cr
6-10 % Fe
1 % Mn max
0.5 % Si max
0.5% Cu max
0.15 % C max
72 % Ni min

Figure 1.   Schematic diagram of the burner nozzle configuration.

Table 1. Chemicals used in preparing metallographic samples.

Metallographic etchant used for optical and scanning electron
microscopy samples.

   100 ml   HCl + 5 ml 30% H_2O_2

   swab   1--5   sec,   25°C

Electrolytic etchant used to remove carbon films for extraction
replicas.

   10%   HCl in methanol

   use at 8--10 V,   0.2--0.5 A/cm^2,   25°C

Conditions used for preparation of thin foil transmission electron
microscopy samples.

   950 ml acetic acid + 50 ml $HClO_4$

   use at 70--80 V, 100-200 MA, 15°C (60F)

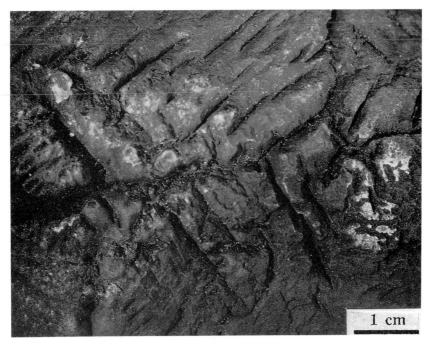

Figure 2. Photograph of the surface which was in contact with the corrosive gases, showing extensive attack.

EXPERIMENTAL PROCEDURE

Sections were cut from the casting, and mounted in bakelite and prepared by standard grinding and polishing procedures. They were etched with the chemical listed in Table 1.

For preparation of extraction replicas, the metallographic samples were etched, and then a layer of carbon (~200 nm thick) was deposited. The films were freed by electrochemical etching (Table 1), and then recovered on a Cu support grid.

For TEM and STEM observations of the structure of the bulk material, thin foils were made by cutting thin sections (e.g., 0.30 mm thick) from locations of interest, from which were punched discs about 3 mm in diameter. These cuts were made with a high speed, thin (e.g., 0.25 mm thick) abrasive wheel using copious water for cooling. The discs were then electrochemcially thinned (Table 1) in the central region until a small hole formed, the edge of which was transparent to the electrons.

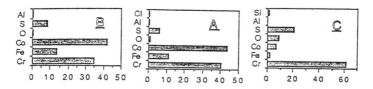

Figure 3.   Micrograph showing a corroded region at the surface.

Chemical analyses were obtained using EDS on SEM and STEM. Thin window or windowless detectors allowed detection of the light elements C, O and N. However, in carbon extraction replicas the carbon in the analysis region could not be separated from that from the carbon film, so carbon analyses are not reported here. The peak height or the integrated peak intensity of the major peak for each element was taken as a measure of element concentration. For each analysis, the values were normalized to 100, and the results are presented here as bar graphs.

On the micrographs, the method of imaging is identified by OM, SEM, TEM or STEM.

Microstructural Science, Volume 19

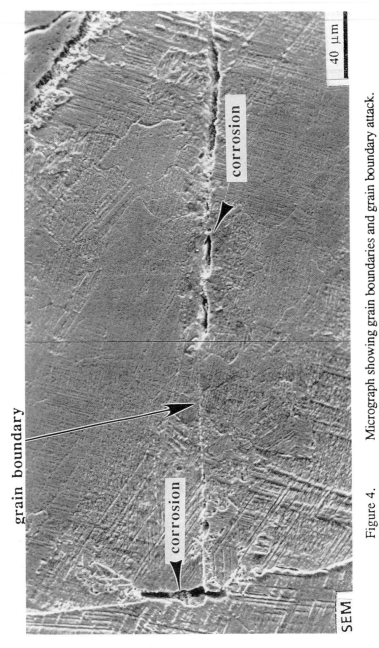

Figure 4.   Micrograph showing grain boundaries and grain boundary attack.

RESULTS

OM, SEM and EDS Results

The extensive corrosion on the surface is easily seen in cross-sectional views of the edge (Fig. 3). The areas A and B had similar analyses, being rich in Co, Cr and Fe, with some S present, and traces of Al, Si, O and Cl. However, the corrosion product at the surface (area C) was richest in Cr with lesser amounts of Co and Fe, but with prominent indications of S and O.

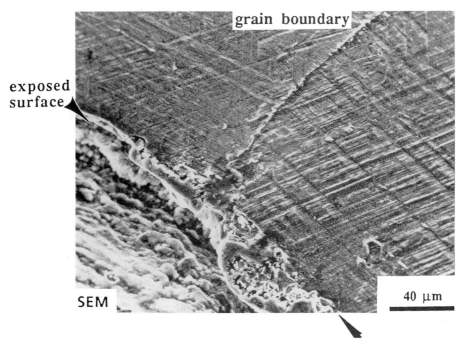

Figure 5.   Micrograph showing a grain boundary at the surface which was not attacked.

There was corrosive attack along grain boundaries (Fig. 4), but some boundaries did not show any attack at the surface (Fig. 5). Internal attack was found across the cross-sections, and such areas may be connected to the surface. They did not follow any obvious grain boundaries, but were straight, with straight branches (Fig. 6). Such regions are probably connected to the surface. The Widmanstatten

unetched

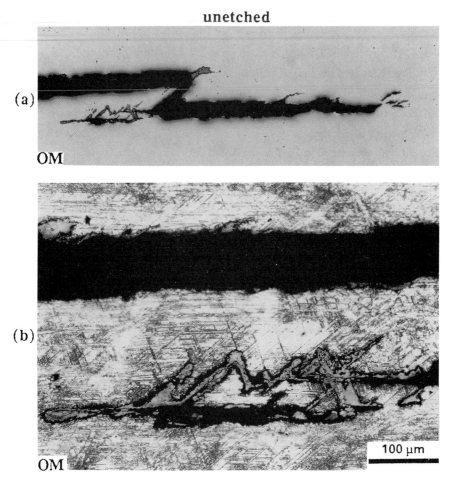

Figure 6.   Micrographs showing corrosion parallel to the Widmanstatten plates.

structure typical of cast Co-30% Cr alloys is revealed (also seen in Fig. 3), caused by the formation during cooling of the hexagonal-close packed phase ε in thin plates parallel to the {111} planes of the face-centered cubic γ phase, some of which may be retained [12]. The example in Fig. 7 clearly shows that these regions are parallel to the ε plates.

Fig. 8 shows the tip of an internally corroded region. The EDS analysis from the area removed from the region (A) showed strong Co and Cr, with some Fe, consistent with the chemical composition of the alloy. The analysis at the tip (B)

Widmanstatten pattern

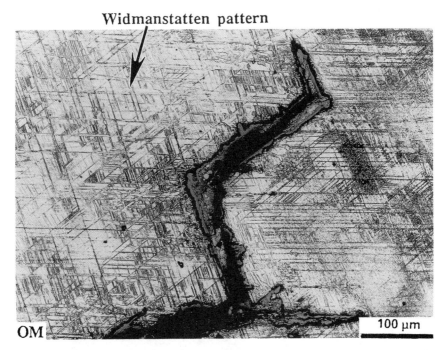

OM   100 µm

Figure 7.   Micrograph showing corrosion parallel to the Widmanstatten plates.

and the end of the side extension (C) were similar, rich in S and Cr. Region D, near
region C, had an analysis similar to that of the matrix (region A), but contained S,
Al and Si. Region E, further from the tip than B, was rich in Cr, O and S, with
traces of Al and Si, but with little Co and Fe. Region F, in the matrix just adjacent
to the tip, contained Co, Cr and Fe, but also gave strong S, O and Si peaks. Fig. 9
shows the tip of a corroded region, with EDS analysis of the corrosion products.
Note the high S and Cr levels, and the relatively low Co levels, and the high O level
in one region.

Fig. 10 shows the tip of a corroded region. The EDS analysis from the area
noted revealed a prominent S and O peak. Fig. 11 shows the tip of another corroded
region, and the EDS spectra revealed high S and Cr and low Co in the corrosion
product (area B). Some O, Al, Si and Cl were also present. The matrix analysis
(area A) was consistent with the composition of the alloy.

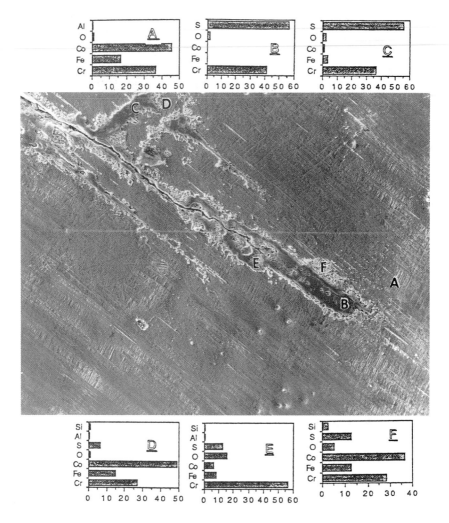

Figure 8.   Micrograph of the tip of a corroded region.

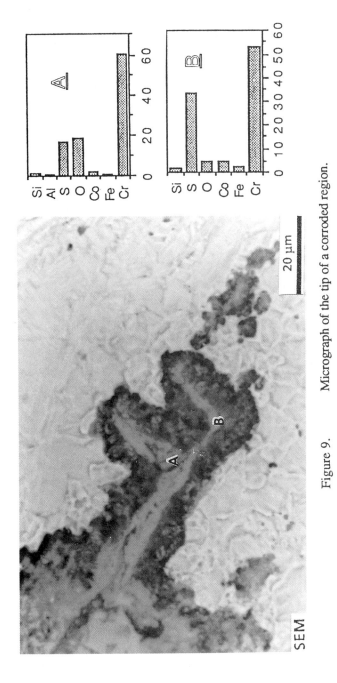

Figure 9.   Micrograph of the tip of a corroded region.

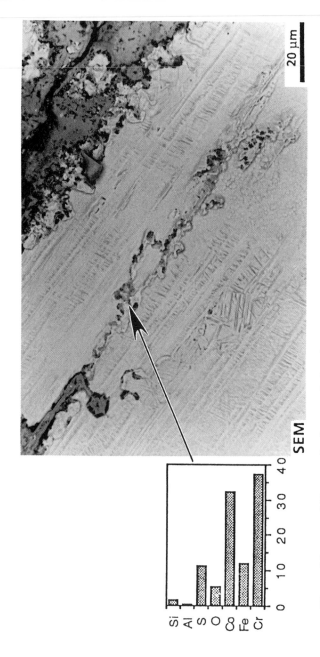

Figure 10.   Micrograph of a corroded region. The EDS spectrum of the areas noted revealed S, even though the microstructure showed little obvious attack here.

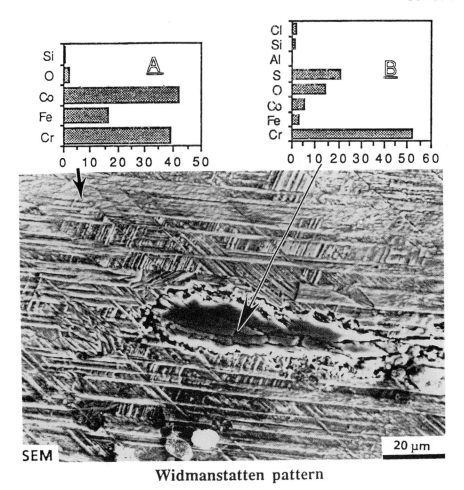

Widmanstatten pattern

Figure 11.   Micrograph of the tip of a corroded region.

TEM, STEM and EDS Results

Particles were extracted *in situ* from etched metallographic samples using carbon films. Fig. 12 shows clusters of particles, some of which were rich in S and Cr (in (b)) and others rich in Al, O, S and Cr (in (a)). Fig. 13 shows a cluster which contained mainly Al and O, probably Al_2O_3. Fig. 14 shows another cluster which gave similar analyses (area A). The large particle in Fig. 14 was rich in Al, Si and O, and probably was an aluminum silicate.

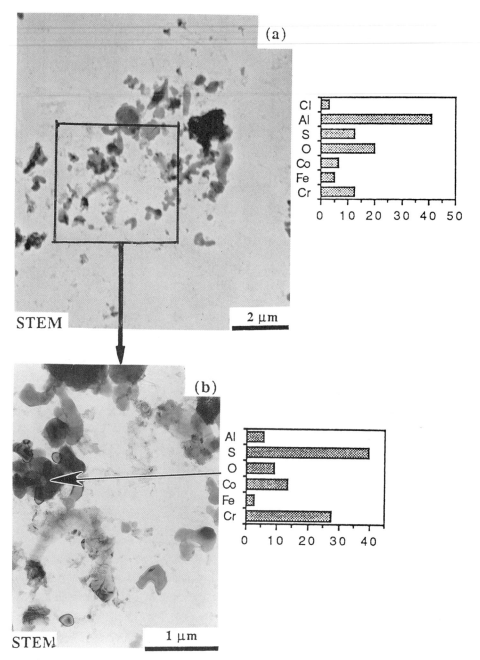

Figure 12.   TEM micrographs of carbon extraction replica, showing a cluster
                  of particles.

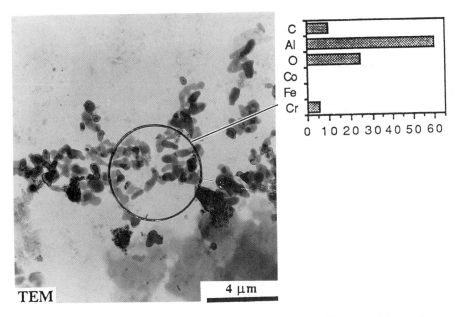

Figure 13.   TEM micrograph of carbon extraction replica containing a cluster
of particles.

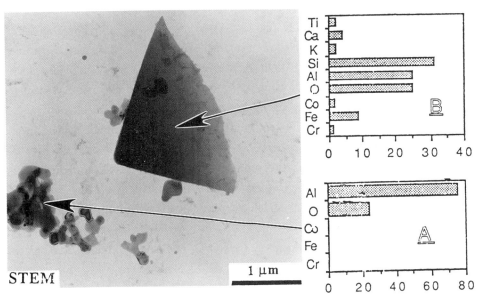

Figure 14.   TEM micrograph of carbon extraction replica containing particles.

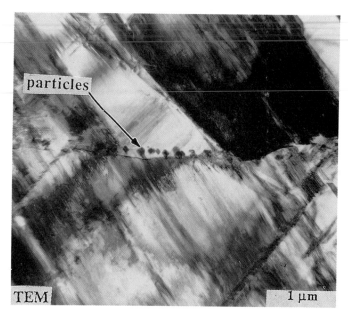

Figure 15.   TEM thin foil micrograph, showing fine particles along a grain boundary.

Fig. 15 shows a line of fine (~ 50 nm) particles along a grain boundary. Fig. 16 shows a line of particles in an extraction replica which is assumed to be along a grain boundary. The EDS analyses showed particles very rich in Cr, which may be Cr carbides. The grain boundary particles in Fig. 17 gave similar spectra. (These are carbon extraction replicas, so carbon analyses are not reported.)

Since corrosion was noted parallel to the Widmanstatten plates, it was suspected that this might be associated with particles on the ε interfaces. Fig. 18 shows particles in carbon extraction replica which are believed to lie on the e plate interfaces, since the elongated features of the Widmanstatten structure have been replicated from the etched metallographic surface. The two particles noted are rich in Cr and O, with lesser amounts of Fe, Co, S, and Cl, and are probably Cr_2O_3. Fig. 19 shows similar particles, which were also rich in Cr and O, but also contained some Co and Fe (with weaker indications of Al, Si, S and Cl).

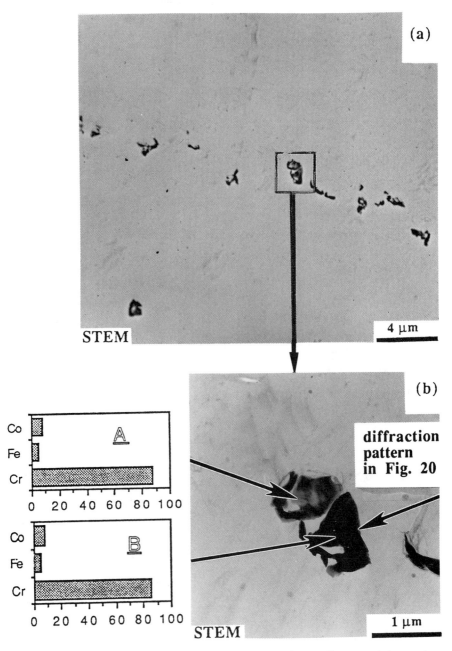

Figure 16.   TEM micrographs of carbon extraction replica containing grain
boundary particles.

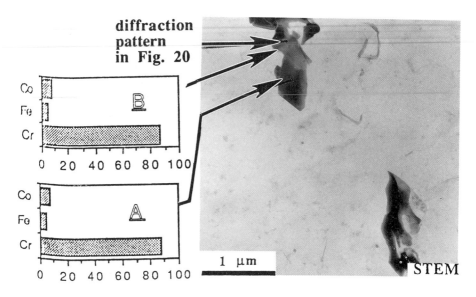

Figure 17.   TEM micrograph of carbon extraction replica containing particles.

Electron Diffraction Patterns

Electron diffraction patterns from one of the particles in Fig. 16 and one in Fig. 17 are shown in Fig. 20. The indexing of these two patterns (and another not shown) indicates that they are Cr-rich $Cr_{23}C_6$ type carbides (Fig. 21).

DISCUSSION

The gaseous environment generated in the combustion process of coal gasification is complicated [2-6], containing both sulfidizing and oxidizing gases; also the temperatures are high (e.g., up to 1500 °C) [2]. Thus the burner must withstand a very aggressive environment, and it is clear that the Co-Cr-Fe alloy casting did not have adequate corrosion resistance to the coal gasification combustion environment. Such alloys are relatively resistant to attack in some sulfur-containing gases [10,12], although under certain conditions the presence of Na and Cl may make them less resistance than Ni-based alloys [15]. The qualitative chemical analyses of the corrosion products in our study are consistent with those reported for Co-Cr alloys in similar environments [16-19]. The corrosion products on the surface

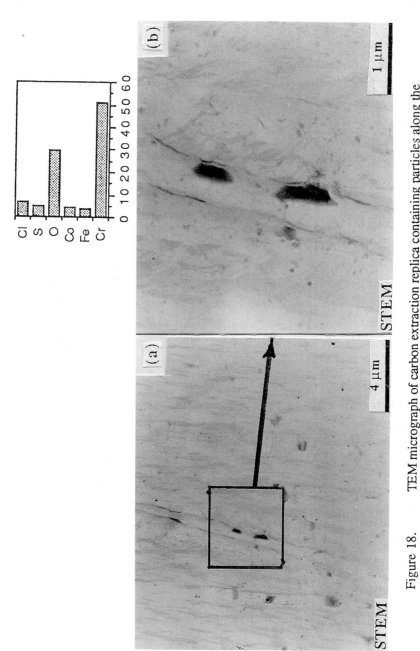

Figure 18.   TEM micrograph of carbon extraction replica containing particles along the
Widmanstatten plates.

803

are particularly rich in Cr and S, with little Co present; also present are smaller amounts of Cl, Si, O and Al (Fig. 3). The products in the intrusions into the surface contain more Co. The corrosion products in the interior, as revealed by the cross-sections, have a similar chemistry. Some of the products are rich in S and Cr, with little or no Co present (Figs. 8, 9 and 11). Some areas contain lesser amounts of O, Si, and Al. In these interior regions no Cl was found. Generally, the data are consistent with the removal of Co being due to the formation of Na_2SO_4 and $CoSO_4$, which form a low melting eutectic (670 °C) [15] and as a molten film would be removed into the gas by evaporation or by ablation. The removal of Co leaves behind the Cr- and S-rich corrosion products. However attractive this description is for the products at the surface, its application is not valid for the formation of products deep into the wall. Fig. 8 shows two areas (B and C) of a corroded region which are rich in Cr and S but contain no Co. Thus the Co must have been removed to the surface by a diffusion process.

The Co-Cr alloy with 19 wt. % Fe will transform from the face-centered cubic structure to the hexagonal-close packed ε structure upon cooling below about 800°C [20]. It is expected that the phase change will occur during cycling of the burner between periods of operation. Since corrosion occurred along planes parallel to the ε phase plates, the temperature at which it happened was below 800°C.

There is also concern about the formation of the σ phase in this alloy. The review by Rivlin [21] indicates that σ would form below 800°C, but no σ particles were found. Instead, extracted particles from the matrix were rich in Cr and S (Fig. 12), probably Cr sulfides, or Al, O and Si (Figs. 12-14), probably Al_2O_3 or silicates. The latter may have been present in the as-cast material, but the sulfides formed during service. The evidence points to the fine particles along the grain boundaries (Fig. 15) being Cr-rich carbides, based on the electron diffraction pattern analysis (Fig. 21), and there is a report that S-containing gases attack Cr-rich carbides in Co-Cr alloys [16, 22]. These carbides may have precipitated during service, but they may have been formed by carburization during service. The particles along the ε plates (Fig. 18) are rich in Cr and O (Fig. 18 and 19), and are probably Cr_2O_3; these also may have formed in service.

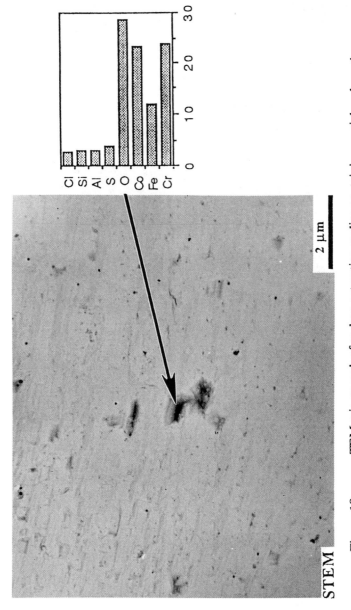

Figure 19.   TEM micrograph of carbon extraction replica containing particles along the
Widmanstatten plates.

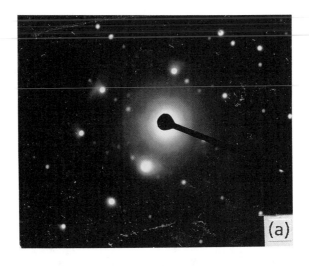

Figure 20.   Electron diffraction patterns from particles noted in (a) Fig. 17 and (b) Fig. 16

		measured	calculated
Fig.21b 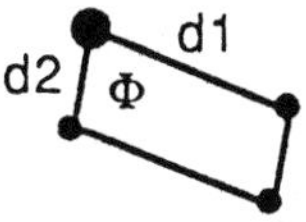	d1	5.28	d(200)=5.3
	d2	1.99	d(511)=2.04
	d1/d2	2.66	2.60
	Φ	78.2°	78.9°
Fig.21a 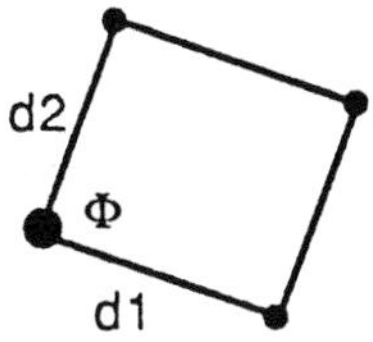	d1	1.99	d(511)=2.04
	d2	1.76	d(531)=1.79
	d1/d2	1.13	1.14
	Φ	84.4°	84.4°
	d1	2.18	d(422)=2.16
	d2	1.93	d(440)=1.87
	d1/d2	1.13	1.15
	Φ	74.5°	73.2°

Figure 21. Measured crystallographic data obtained from the electron diffraction patterns in Fig. 20, and from another not shown. The calculated values are for $Cr_{23}C_6$ type carbides [13].

CONCLUSIONS

The casting was subjected to severe hot corrosion. The analyses of the corrosion products indicates that the corrosion at the surface was due to the formation of $CoSO_4$ and Na_2SO_4, which formed a low melting eutectic. The attack also extended along boundaries into the interior of the grains and along the Widmanstatten e plate interfaces. The microstructural examination indicates that Cr-rich particles (oxides and carbides) which were present along grain boundaries and along the Widmanstatten plates of the matrix are related to the extensive corrosion in these regions. It is not known whether these formed during the high temperature service or were present in the as-cast condition.

It is interesting to note that one source [12] reports that the similar alloy UMCo50 (40 wt. % Co, 28 wt % Cr, 21 wt. % Fe, 0.12 % C max) is particularly resistant to hot corrosion by sulfur-containing oxidizing atmospheres, and another [10] states that this alloy is used extensively for furnace parts and fixtures (but not in gas turbine applications, although in such an application the lack of sufficient strength may preclude its use). The resistance to attack will partly rely on an oxidizing atmosphere allowing the Cr-rich oxides to form, and if the gas is too reducing this may not occur, which may be part of the difficulty in the specific coal gasification environment involved in the attack of the nozzle examined.

ACKNOWLEDGEMENT

Appreciation is expressed to Tennessee Eastman Company for partial support of this work, to Dr. John Dunlap for assistance in the STEM examinations, and to Bobby L. McGill for assistance in the SEM examinations.

REFERENCES

1. **Coal Gasification Processes**, P. Nowacki, ed., Noyes Data Corporation, Park Ridge, NJ (1981).

2. A. M. Beltran and D. A. Shores, "Hot Corrosion", **The Superalloys**, Wiley, New York, p. 317 (1972).

3. P. Kofstad, **High Temperature Corrosion**, Elsevier Applied Science, New York, Chapter 13 and 14 (1988).

4. C. T. Sims, "High-Temperature Alloys in High-Temperature Systems", **High Temperature Alloys for Gas Turbines**, Applied Science Publishers, London, p. 13 (1978).

5. S. Kanyluk, G. M. Dragel and D. Dubis, "Materials Problems Experience at the Synthan Coal-Gasification Pilot Plant", **J. of Engr. Mats. and Tech.**, Vol. 101, p. 105 (1979).

6. K. Natesan, "Corrosion Behavior of Materials in Low-and Medium-BTU Coal-Gasification Environments", **Proceeding Corrosion-Erosion-Wear of Materials in Emerging Fossil Energy Systems**, National Association of Corrosion Engineers, Houston, p. 100 (1982).

7. K. Natesan, "High-Temperature Corrosion in Coal Gasification Systems", **Corrosion-NACE**, Vol. 41, p. 646 (1985).

8. A. MacNab, "Design and Materials Requirements for Coal Gasification", **Chemical Engr. Progress**, Vol. 71, no. 11, p. 51 (1975).

9. **Superalloys - A Technical Guide**, E. F. Bradley, ed., ASM International, Materials Park, OH (1988).

10. **Metals Handbook, 9th edition, Vol. 3**, American Society for Metals, Metals Park, OH, p. 213 (1980).

11. K. Natesan, "High-Temperature Corrosion", **Prevention of Failures in Coal Conversion Systems**, U. S. Department of Commerce, Washington, p. 159 (1977).

12.   W. Betteridge, **Cobalt and its Alloys**, Wiley, New York (1982).

13.   K. W. Andrews, D. J. Dyson and S. R. Keown, *Interpretation of Electron Diffraction Patterns*, Plenum, New York, p. 172 (1967).

14.   C. Y. Lai, "Sulfidation Resistance of Various High-Temperature Alloys in Low Oxygen Potential Atmospheres", **High Temperature Corrosion in Energy Systems**, M. F. Rothman, ed., The Metallurgical Society, Warrendale, Penn., p. 227 (1985).

15.   K. L. Luthra and D. A. Shores, "Mechanism of Na_2SO_4 Induced Corrosion at 600-900 °C", **J. Electrochem. Soc.**, Vol. 127, p. 2202 (1980).

16.   J. Stringer and D. P. Whittle, "Hot Corrosion of Cobalt-Based Alloys", **Deposition and Corrosion in Gas Turbines**, Wiley, New York, p.197 (1973).

17.   G. H. Meier, N. Birks, F. S. Pettit and C. S. Giggins,"Thermodynamics Analyses of the High Temperature Corrosion of Alloys in Gases Containing More Than One Reactant", **High Temperature Corrosion**, National Association of Corrosion Engineers, Houston, p. 327 (1983).

18.   A. U. Seybolt and A. Beltran, "High Temperature Sulfur-Oxygen Corrosion of Nickel and Cobalt:, **Hot Corrosion Problems Associated with Gas Turbines**, American Society for Testing and Materials, Philadelphia, p. 21 (1967).

19.   A. K. Misra, "Hot corrosion of Nickel and Cobalt Base Alloys at Intermediate Temperatures", **10th International Congress on Metallic Corrosion, Vol. IV**, Trans Tech Publications, USA, p. 3533 (1987).

20.   H. J. Wagner and A. M. Hall, **The Physical Metallurgy of Colbalt-Base Superalloys**, DMIC Report 171, Defense Metals Information Center, Battelle Memorial Institute, Columbus (1962).

21.   V. G. Rivlin, "6: Critical Evaluation of Constitution of Cobalt-Chromium-Iron and Cobalt-Iron-Nickel Systems", **Int. Metals Rev.**, Vol. 26, p. 269 (1981).

22.   M. E. El Dahshan, D. P. Whittle and J. Stringer, "The Influence of Carbon on the Sulphidation and Hot Corrosion of Co-Cr Alloys", **Deposition and Corrosion in Gas Turbines**, Wiley, New York, p. 210 (1973).